Studies in Computational Intelligence

Volume 890

Series Editor

Janusz Kacprzyk, Polish Academy of Sciences, Warsaw, Poland

The series "Studies in Computational Intelligence" (SCI) publishes new developments and advances in the various areas of computational intelligence—quickly and with a high quality. The intent is to cover the theory, applications, and design methods of computational intelligence, as embedded in the fields of engineering, computer science, physics and life sciences, as well as the methodologies behind them. The series contains monographs, lecture notes and edited volumes in computational intelligence spanning the areas of neural networks, connectionist systems, genetic algorithms, evolutionary computation, artificial intelligence, cellular automata, self-organizing systems, soft computing, fuzzy systems, and hybrid intelligent systems. Of particular value to both the contributors and the readership are the short publication timeframe and the world-wide distribution, which enable both wide and rapid dissemination of research output.

The books of this series are submitted to indexing to Web of Science, EI-Compendex, DBLP, SCOPUS, Google Scholar and Springerlink.

More information about this series at http://www.springer.com/series/7092

Diego Oliva · Salvador Hinojosa
Editors

Applications of Hybrid Metaheuristic Algorithms for Image Processing

Springer

Editors
Diego Oliva
CUCEI
University of Guadalajara
Guadalajara, Mexico

Salvador Hinojosa
CUCEI
University of Guadalajara
Guadalajara, Mexico

ISSN 1860-949X ISSN 1860-9503 (electronic)
Studies in Computational Intelligence
ISBN 978-3-030-40976-0 ISBN 978-3-030-40977-7 (eBook)
https://doi.org/10.1007/978-3-030-40977-7

This Springer imprint is published by the registered company Springer Nature Switzerland AG
The registered company address is: Gewerbestrasse 11, 6330 Cham, Switzerland

Preface

Since the use of digital images is a part of our life, it is necessary to have accurate algorithms that permit us to analyze the objects contained in the scene according to a specific application. The computer vision systems have different phases, but one of the main tasks is segmentation, where the objects are cataloged according to the intensity levels of the pixels. However, this is not the only important step in such system, the classification, shape analysis, and feature extraction are some of the tasks that represent challenges in image processing.

Currently, metaheuristic algorithms (MA) are considered important optimization tools commonly used to solve complex problems. These kinds of methods consist of two main processes exploration and exploitation. In such phases, the algorithms employ different operators and rules to search the optimal solutions. Here, it is important to mention that metaheuristics are commonly inspired by natural behaviors that define its classification. For example, the family of algorithms that employs operators inspired in the natural selection is called evolutionary, meanwhile the algorithms that employ a population classified as swarm methods. Metaheuristic approaches are not perfect, and their drawbacks as slow convergence or when they provide suboptimal solutions affect their performance and cannot be considered for all the problems. A common alternative for addressing the failures of optimization algorithms is the hybridization because it permits the combination of strategies for exploring complex search spaces.

The hybridization can take two forms: sequential use of algorithms and the replacement of operators. In the first case, two MA run in sequence to firstly explore with an algorithm and then perform exploitation on the later; however, this approach can significantly increase the number of iterations required to converge. In the second case, the operators of two or more algorithms which are mixed together provide a better balance between the intensification and diversification of the solutions without increasing the number of iterations.

This book is focused on the theory and application of MA for the segmentation of images from different sources. In this sense, topics are selected based on their importance and complexity in this field—for example, the analysis of medical images of the segmentation of thermal images for security implementation.

This book has been structured so that each chapter can be read independently from the others. This book is divided into three parts that contain chapters with similar characteristics.

The first part is related to hybrid metaheuristics and image processing. Here, the reader could find chapters related to thermal, hyperspectral, and remote sensing images. Moreover, some hybrid approaches permit, for example, to analyze different features as the texture in the digital images.

The second part is addressed other problems in image processing with hybrid MA, for example, object recognition by using template matching, the use of clustering for morphological operations, or the estimation of homography. This section also includes the use of deep learning and a literature review of a well-known MA.

One of the most important uses of digital images is medicine; for that reason, the third part encapsulates different chapters related to health. Here, MA are used and hybridized for tasks as magnetic resonance analysis, detection of anomalies in mammograms, Parkinson's disease among others. The reader will be highly interested because the mixture of methods permits to explore machine learning applications included adversarial networks.

The material presented in this book is essentially directed for undergraduate and postgraduate students of Science, Engineering, or Computational Mathematics. It can be appropriate for courses such as artificial intelligence, evolutionary computation, and computational intelligence. Likewise, the material can be useful for researches from evolutionary computation and artificial intelligence communities.

Finally, it necessary to mention that this book is a small piece in the puzzles of image processing and optimization. We would like to encourage the reader to explore and expand the knowledge in order to create their implementations according to their necessities.

Guadalajara, Mexico Diego Oliva
November 2019 Salvador Hinojosa

Contents

Hybrid Metaheuristics and Other Image Processing Tasks

Health Applications

Hybrid Metaheuristics and Image Segmentation

Segmentation of Thermal Images Using Metaheuristic Algorithms for Failure Detection on Electronic Systems

Mario A. Navarro, Gustavo R. Hernández, Daniel Zaldívar, Noé Ortega-Sanchez and Gonzalo Pajares

Abstract Segmentation is considered an important part of image processing. There are commonly used segmentation techniques to improve the threshold process such as Otsu and Kapur. The use of these techniques allows us to find the regions of interest in an image by correctly grouping the pixel intensity levels. On the other hand, the use of thermal images makes it possible to obtain information about the temperature of an object and to capture the infrared radiation of the electromagnetic spectrum, through cameras that transform the radiated energy into heat information. The segmentation of this kind of images represents a challenging problem that requires a huge computational effort. This work proposes the use of metaheuristic algorithms, combined with segmentation techniques and thermal images, to detect faults and contribute to the preventive maintenance of electronic systems.

Keywords Image segmentation · Thresholding · Thermographic analysis · Electronic systems · Metaheuristic algorithms · Preventive maintenance

M. A. Navarro (✉) · G. R. Hernández
División de Electrónica Y Computación, Universidad de Guadalajara, CUCEI, Av. Revolución 1500, Guadalajara, Jalisco, Mexico
e-mail: marioa.navarro@alumno.udg.mx

G. R. Hernández
e-mail: gustavo.hernandezz@alumno.udg.mx

D. Zaldívar · N. Ortega-Sanchez
Departamento de Electrónica, Universidad de Guadalajara, CUCEI, Guadalajara, Jalisco, Mexico
e-mail: daniel.zaldivar@cucei.udg.mx

N. Ortega-Sanchez
e-mail: noe.ortega@academicos.udg.mx

G. Pajares
Facultad Informática, Dpto. Ingeniería de Software e Inteligencia Artificial, Universidad Complutense de Madrid, Madrid, Spain
e-mail: pajares@ucm.es

D. Oliva and S. Hinojosa (eds.), *Applications of Hybrid Metaheuristic Algorithms for Image Processing*, Studies in Computational Intelligence 890,
https://doi.org/10.1007/978-3-030-40977-7_1

1 Introduction

With the invention of thermographic cameras and the possibility of acquiring these high-performance instruments at an ever-lower price, it facilitates the application of thermography in different areas. This chapter presents a tool to segment thermal images, using segmentation techniques based on Kapur entropy [1], Otsu variance [2], and the use of metaheuristic algorithms to optimize threshold values.

Sight is one of the most precious human senses, since it allows to perceive the world, but it limited to the range of vision spectrums. With the creativity which characterizes at humans, has been developed other perception forms. Such as illustrated in this chapter, is captured an image infrared spectrums (IR) [3]. For this purpose, cameras that can capture regions of electromagnetic waves are used; in infrared radiation spectrums that pick-up wavelengths between (0.39 and 0.77 μm). Furthermore, work has been carried out using the same technique to evaluate the electrical installations in buildings [4], and be able to carry out preventive maintenance, avoid future problems and consequently higher expenses due to the collateral effects of an electrical failure.

One application that has been investigated and it has demonstrated utility in medical applications. Where it was used to detect and diagnose injuries, tumors and other disorders that modify the temperature of the organs or regions of interest.

In electronics, systems are becoming more complex and their applications extend to many areas of human knowledge, due to its rapid evolution arises the need to create tools that allow the diagnosis and preventive maintenance of electronic systems. One of the main enemies of electronics is still temperature, and although the best technologies are designed to prevent devices from malfunctioning due to temperature, the problem is still latent especially in electronic equipment that requires its permanent operation to support delicate processes, so it is vital to develop technology to offer safe solutions to this problem. Thermographic imaging gives us the ability to create effective diagnostic and preventive maintenance tools.

The development of this research is oriented to create a tool that can alert a malfunction of electronic components, and thus assist in preventive maintenance, consequently the repair of electronic components would be carried out in a controlled manner and without affecting the day-to-day operations of the equipment.

This chapter is organized in the following order: In the second section, the concept of thermography and its importance are studied in a general approach. The third section contextualizes the segmentation and the most popular techniques for this purpose (Otsu and Kapur). The fourth section deals with the morphology of metaheuristic algorithms and the most important concepts, as well as the algorithms used for the present project. The fifth section presents the method of the software and the configuration of its graphical interface. The sixth section shows and analyzes the experiments that validate the research.

2 Thermography

Thermography is a technique used to acquire calorific information, using the infrared spectrum reflected in the analyzed objects. When electromagnetic waves of different frequencies are irradiated it is possible to analyze the temperatures of each of the frequencies, represented in an image for analysis. As shown in Fig. 1, the higher temperature regions are shown in a color that highlights within the image.

Infrared light is halfway between the visible spectrum and the microwave of the electromagnetic spectrum. The main source of infrared radiation is heat or thermal radiation. Any object with a temperature above absolute zero (−273.15 °C or 0 K) emits radiation in the infrared region. The hotter an object is, more infrared radiation it emits (Fig. 2).

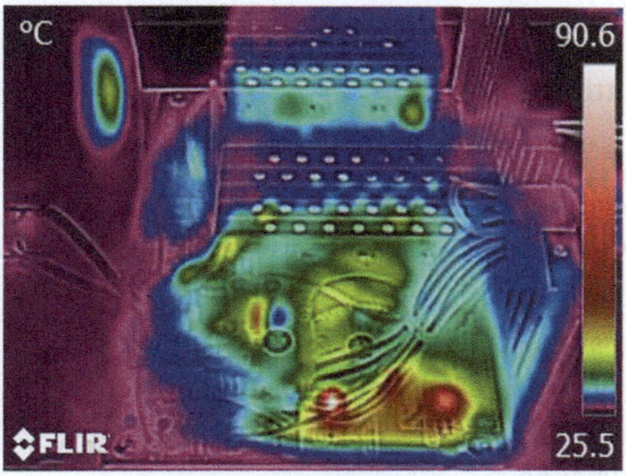

Fig. 1 Image where you can see the areas of interest based on their temperature

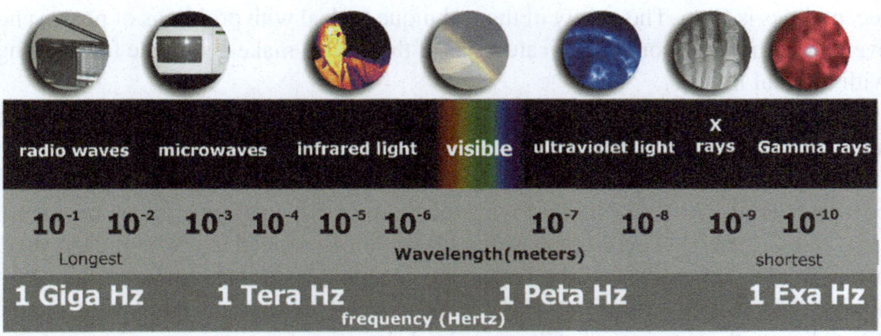

Fig. 2 Electromagnetic spectrum

2.1 Thermographic Camera Operation

The infrared energy radiated by an object is focused by the optical system on an infrared detector. The detector sends the data to the electronic sensor for image processing. Finally, the sensor transforms the data into an image that is compatible with the viewfinder and can be displayed on a standard video monitor or LCD screen.

Currently, systems for capturing thermal images have improved, but still have functional limitations. One of the disadvantages of this type of equipment, are the cost of the lenses, because the production process is delicate and specialized. The main advantage is that it does not need to contact with the elements that are generating high temperatures, preventing damage to maintenance technicians. Another advantage is its size, can be used in confined spaces and are versatile for use in any work environment.

3 Segmentation Techniques

Segmentation is a technique that has been worked on from the perspective of the histogram of an image, which provides with information. However, it is limited to the frequency of each pixel. In recent years, segmentation techniques have been used in which a histogram is deconstructed in ranges according to a chosen number of thresholds, with the new histogram discriminates with greater certainty the information that is not relevant to a given problem.

3.1 Otsu Variance

One of the techniques that has been used for the segmentation of a histogram in recent times is Otsu. The ability of the technique to deal with problems of peaks, and even noise that may contain delicate areas of the image, make it suitable for working with thermal images.

In this technique the information from nearby pixels generates changes within the histogram, which provides useful information for segmentation.

The purpose is to perform a multi-level segmentation where the image can be divided into $nt + 1$, in this way the segmented areas are given by $\boldsymbol{th} = [th_1, th_2, \ldots, th_{nt}]$. Therefore, a value histogram is created E_i, where each pixel of the image coincides with the frequency of appearance that generates the probability $PE_i = \frac{E_i}{NP}$, were $\sum_{i=1}^{NP} PE_i = 1$, and NP is the total number of pixels. According to the position of each segmented class the variance σ^2 is calculated and its mean μ_k such that, are defined by:

$$\sigma^2 = \sum_{k=1}^{nt} \sigma_k = \sum_{k=1}^{nt} \omega_k (\mu_k - \mu_T)^2 \tag{1}$$

$$\mu_k = \sum_{i=th_k}^{th_{k+1}-1} \frac{kPE_i}{\omega_i(th_i)} \tag{2}$$

Therefore, Otsu is a technique for maximizing the variance that is given by a set of segmentation values, and is defined by

$$\omega_k = \sum_{i=th_k}^{th_{k+1}-1} PE_i \tag{3}$$

3.2 *Kapur Entropy*

The Kapur entropy performs segmentation based on the distribution of image probability using entropy as a measure. The set of values to be segmented is evaluated to obtain the maximum reference value of the quality of the segmented image. This technique is like the method used by Otsu, and is applied directly to the histogram, in such a way that Kapur entropy looks for the set of segmentation values to maximize entropy, and is defined as follows:

$$f_{Kapur}(\mathbf{th}) = \max\left(\sum_{k=1}^{nt} H_k\right) \tag{4}$$

The entropy of each class is calculated as follows:

$$H_1 = -\sum_{i=th_k}^{th_{k+1}-1} \frac{PE_i}{\omega_k} \ln\left(\frac{PE_i}{\omega_k}\right) \tag{5}$$

In such a way that for multiple classes it is generalized in the equation below:

$$H_k = \sum_{i=th_k}^{th_{k+1}-1} \frac{PE_i}{\omega_k} \ln\left(\frac{PE_i}{\omega_k}\right) \tag{6}$$

The distribution of probability PE_i and ω_k is calculated with the same criteria that Otsu method.

4 Optimization Algorithms

Optimization problems emerge in a multitude of areas, from the selection of the shortest route to get to work to the optimal design of routes for the distribution of products of a multinational company. Optimization is an important area of Mathematics that has had importance due to its applications in real problems, to solve this type of problems, metaheuristic algorithms of optimization are used in which, through their methods, get the appropriate solutions to a variety of problems.

It is can highlight some important features of the "Optimization algorithms":

- Very fast initialization process
  Choice of the best alternative from all possible alternatives

Next, it is defining some previous concepts that will be very useful to understand the structure of metaheuristic algorithms:

An optimization problem is given by a pair (F, c) where F is a domain of feasible points and c a cost function. The problem is to find a feasible point $(x) \in (F)$ such that $(\forall\, and \in F)$ check, $c(x) \leq c(y)$. Each point (x) checking the conditions described above is called the overall optimum of the problem.

There are some basic concepts common to all algorithmic approaches to problem solving, depending on the techniques used, specified below:

Search space: It is the representation of a potential solution and its corresponding interpretation gives us the search space and its size. This is a focal point for the problem: The size of this space is not determined by the problem but by our representation and interpretation. This step is of vital importance to give a correct solution to problem, in addition to avoiding setbacks of duplication of results, etc.
Nearby: If focalized an N(x) region of the search space S, which is "close" to some x point of space, be can define N(x) as a vicinity of the x point ∈ S.
Objective: This is nothing more than the mathematical state that interest optimizing so that the task is complete.
Objective Function: This is the quantitative measure of the performance of the system to be optimized (maximize or minimize). Some examples of objective functions are: the minimization of the costs of a system of agricultural production, the maximization of the net benefit of the sale of certain products, the minimization of the materials used in the manufacture of a product, etc.
Variables: Represents the decisions that can be made to change the value of the objective function. These can be classified as dependent or independent. For example, in the case of an agricultural production system will be the production values of the generating units or flows through the lines, the quantity of product produced and sold. In the case of the manufacture of a product, its physical dimensions, etc.
Constraints: These represent the set of conditions that certain variables are required to satisfy. Expressed by equations or inequalities. For example, the production capacity for the manufacture of different products, the dimensions of the primary material of the product, etc.

The algorithms used in this program have proven to be successful in optimization problems. They are designed to solve real, complex, and highly non-linear problems, using user-defined parameters to obtain a finite mathematical value that maximizes a cost function. The following is a brief explanation of the optimization algorithms.

BA (Bat Algorithm) [5]. This metaheuristic algorithm was inspired by the echolocation behavior of microbats, with varying pulse rates of emissions and loudness. This algorithm was developed by Xin-She Yang in 2010. The metaphor to be imitated can be summarized as follows; Each virtual bat flies randomly with a velocity v_i at position (solution) x_i with a varying frequency of wavelength and loudness A_i. As it searches and finds its prey, it changes frequency, loudness and pulse emission rate r. Search is intensified by a local random walk. Selection of the best continues until certain stop criteria are met.

CS (Cuckoo Search) [6]. The metaheuristic algorithm cuckoo search was developed by Xin-She Yang in 2009, this algorithm is inspired by the obligate brood parasitism of some cuckoo species by landing their eggs in the nest of other host birds (of other species). Each egg in a nest represents a solution, and a cuckoo egg represents a new solution. The aim is to use the new and potentially better solutions to replace the worst solution in the nest. First each cuckoo lay one egg at a time and dumps its egg in a randomly chosen nest; The best nests with high quality of eggs will carry over to the next generation; The number of available hosts nests is fixed, and the egg laid by a cuckoo is discovered by the host bird with a probability $p_a \in (0, 1)$. The discovery operator acts on some of the worst nests, and discovered solutions dumped from farther calculations.

In addition, Yang and Deb discovered that the random-walk style is better performed by Lévy flights rather than simple random walk.

PSO (Particle Swarm Optimization) [7]. This is s a computational method that optimizes a problem by iteratively trying to improve a candidate solution about given measure of quality. *PSO* was created by Kennedy, Eberhart and Shi and was first intended for simulating social behavior as a stylized representation of the movement of organisms in a bird flock or fish school. *PSO* algorithm works by having a population (swarm) of candidate solution (particles). These particles are moved around in the search-space according to a few simple formulae. The movements of the particles are guided by their own best-known position in the search-space as well as the entire swarm's best-known position. When improved positions are being discovered these will then come to guide the movements of the swarm. The process is repeated and by doing so it is hoped, but not guaranteed, that a satisfactory solution will eventually be discovered.

DE (Differential Evolution) [8]. Was developed by Storn and Price. *DE* optimizes a problem by maintaining a population of candidate solutions and creating new candidate solutions by combining existing ones according to its simple formulae, and then keeping whichever candidate solution has the best score or fitness on the optimization problem. In this way the optimization problem is treated as a black box that merely provides a measure of quality given a candidate solution and the gradient is therefore not needed. *DE* algorithm works by having a population (agents). These agents are moved around in the search-space by using a simple mathematical formula to combine the positions of existing agents from the population. If the new position of an agent is an improvement then it is accepted and form part of the population, otherwise the new position is simply discarded. The process repeated until a stopped criterion is reached.

5 Methodology

A set of seven thermal images was used to perform the tests and experiments, the metrics: (FSIM [9] and SSIM [10]), were used as similarity tests, and the (PSNR) [11], to define the maximum noise in the propagation of a signal and the ratio affecting its accurate representation, these images were captured with the *FLIR-C2* camera, with commercial features, so its performance is limited, but easy to access. The project offers the advantage of working with low cost equipment and with more sophisticated equipment, becoming versatile to be applied in development projects, or as a tool for industries.

Thermal cameras capture the infrared reflections of objects and the ambient, making necessary a controlled thermal environment. When working with thermal imaging and determining the temperature of a certain object or surface it is recommended to have a stable reference temperature, for example; a background in a space with an average room temperature between 20 and 25 °C.

5.1 Thermal Image Segmentation Interface

The graphic interface consists in four sections, the first is in which we can choose the image to segmented, the second section sets the number of thresholds for segment the image, in the next part you select the segmentation method Otsu or Kapur. Finally, a menu allows you to select a metaheuristic optimizer (Fig. 3).

a. Image selection. The interface menu allows to choose the thermal image from a list of images stored in a folder containing the set of images to be tested, the thermography information is saved in csv (coma-separated values) format, for a correct handling of the data information, it is appropriate to say that the thermographic cameras count with software that facilitates the acquisition and conversion of this type of files. The system was tested to verify that the results in images of electronic equipment were faulty or in good condition.
b. Number of Thresholds. The menu for setting thresholds is provided by a numeric selection bar. The number of thresholds together with the segmentation technique play an important role in the segmentation of thermography, the accuracy offered is greater by increasing the number of thresholds.
c. Segmentation Techniques. The interface gives a choice between two segmentation techniques Kapur and Otsu. Kapur uses the entropy of the classes generated

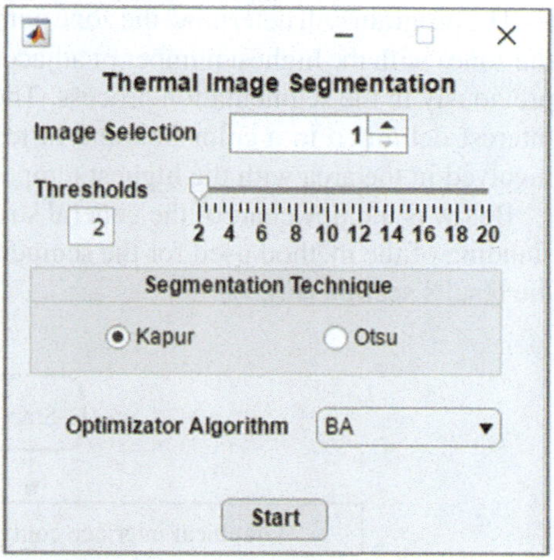

Fig. 3 Segmentation interface configuration

by the thresholds in the histogram, in a similar way, Otsu implements the intraclass variance to do the segmentation. By combining these techniques with an optimization algorithm, the search for the best thresholds is carried out efficiently.

d. Selection of Metaheuristic Algorithms. The program interface allows to choose between four metaheuristic algorithms, mentioned in Sect. 4 including *BA*, *CS*, *PSO*, *DE*. The reason those were selected is why; they have proven their usefulness and efficiency in solving current optimization problems.

5.2 *Software Processing*

The next step after configuring the segmentation interface is transformed the thermal image to grayscale as part of the pre-processing stage, when transforming the image, the values in grey scale from 0 to 255 are obtained, once this is done these values can be computed in Eqs. (3) or (6) depending on the technique selected in the interface.

These segmentation techniques are treated as a cost function to be maximized by one of the four metaheuristic algorithms previously mentioned in Sect. 4. When solving the probability function of any of the two segmentation techniques, the threshold values will be optimized to segment the thermal image.

The proposed software has the facility to choose different combinations between segmentation techniques and optimization algorithms, in Sect. 6 of this chapter presents, the results of all possible combinations between algorithms and segmentation techniques, with the discussion of the results obtained in the tables.

Once the thermal image has been segmented, the last step will be to select the area of interest of the thermal image corresponding to the electronic device that is a candidate for possible functioning problems.

The program will determine the zone with a possible functional failure considering the zones with the highest number of adjacent pixels with high temperature, obtained previously in the segmentation process. The resulting images will show the areas of interest delimited in a color box and in text the average temperature of the pixels involved in the area with the highest temperature.

Below is the flowchart of the general structure of the software, for a better understanding of the method used for the segmentation of termographs, before passing to the results section (Fig. 4).

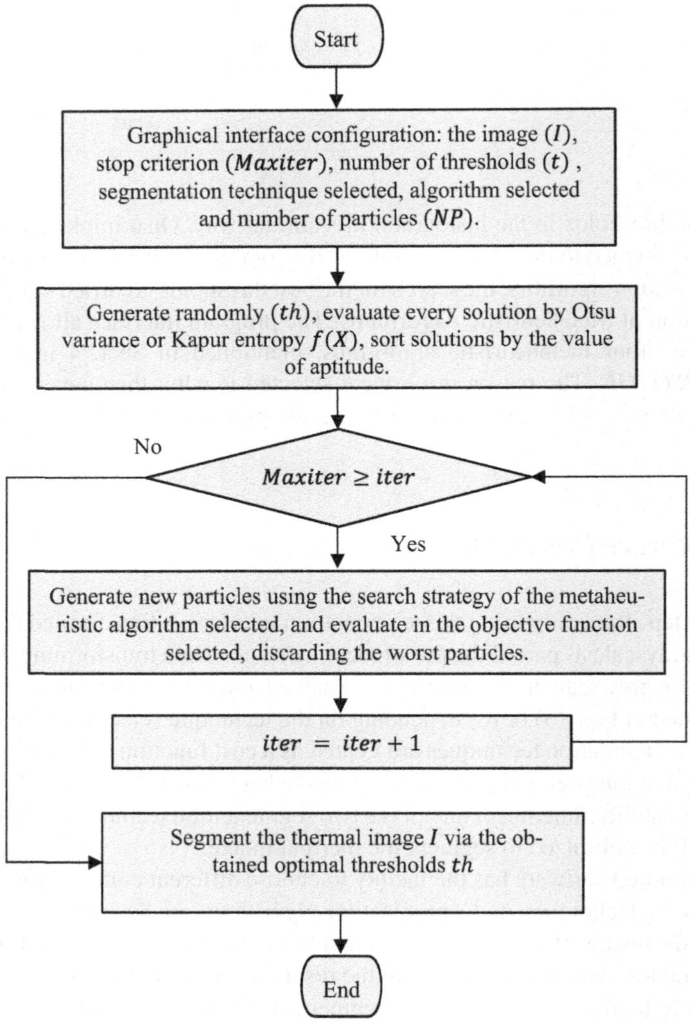

Fig. 4 Flow chart showing the proposed segmentation method

6 Experimental Result

This work facilitates the use of different segmentation techniques and optimizes them with different metaheuristic algorithms, to separate correctly the objects based on their temperature in thermographic images. For this purpose, a set of seven thermal images was created, these images were taken by a thermographic *C2 FLIR* camera, which despite of being a low-end option, has some good features; is compact, has automatic focus, its thermal sensitivity is 10 °C, has a large temperature range −10 to +150 °C (14–302 °F) and has a 3 inch touch display. The camera includes a *FLIR* Tools software for producing reports, this software has the particularity to generate an archive.csv (coma-separated value) of a thermography. With this archive can generate a histogram with the intensity of temperature in each pixel of the imagen, and with this, can implement the segmentation techniques proposed. The seven thermal images were taken from different electronics equipment. To corroborate the effectiveness of our work, obtained images of damaged equipment, in which know in advance the damaged devices of the system, see Table 1.

The performance of the proposed methodology is compared against the metaheuristic approached presented in Sect. 4; and the proposed metrics in Sect. 5. Since metaheuristic algorithms involve the use of random variables, it is necessary to perform a statistical analysis of the results.

Each experiment consists of 35 independent executions of the same algorithm on a specific image, and the number of thresholds, the metrics: *PSNR*, *FSIM* and *SSIM* are reported to validate a good segmentation. Each execution of each algorithm is stopped after 1000 iterations, each with 50 individuals to provide a fair comparison.

App Designer 2017b software of *MATLAB* was used to develop the graphical interface, and the experimentation process was developed with *MATLAB 2017B* on *Windows 10* using an *AMD Ryzen 7 2700X @3.70 GHz* processor with 16 GB of *RAM*.

From an optimization perspective, the aptitude of the candidate solution is determined by the objective function. Nevertheless, on the problem of image segmentation, the necessity to measure the accuracy of the classification process at each pixel arises. In this paper the following metrics that evaluate the quality of the segmented images are considered; the Peak-Signal-to-Noise Ratio (*PSNR*), the Structure Similarity Index (*SSIM*) and the Feature Similarity Index (*FSIM*). The selected metrics are widely used in the related literature to analyze the relationship between input and output images. For example, the *PSNR* is used to verify the similarity that exists between the original and the segmented image. To compute the *PSNR*, it is necessary to use the root mean square error (RMSE) pixel to pixel [12], the PSNR is defined as:

$$PSNR = 20\log_{10}\left(\frac{255}{RMSE}\right), \text{(dB)} \tag{7}$$

to obtain $(RMSE)$ is use the following expression:

Table 1 Benchmark image set, with its respective faults

Object	Failure	Image RGB
USB on power bank	Short-circuited USB 2.0 Flash Disk Controller (CBM2099)	
Power source A Top view	Resistor (1/2 W) with an impedance exceeding its commercial value	
Power source A Inclined top view	Resistor (1/2 W) with an impedance exceeding its commercial value	
Power source B Top view	Resistor (1/4 W) with a commercial value of 1.5 kΩ below its specified value	
Power source B Rotated top view	Resistor (1/4 W) with a commercial value of 1.5 kΩ below its specified value	
Power source C Top view	1n4001 diode with drop voltage higher than 0.7 V	
Power source C Front view	Resistor (1/2 W) with a commercial value of 150 Ω over its specified value	

$$RMSE = \sqrt{\frac{\sum_{i=1}^{ro} \sum_{J=1}^{CO} \left(I_{in}(i,j) - I_{seg}(i,j)\right)}{ro * co}} \tag{8}$$

Table 2 presents the results of the comparison between the four algorithms and the segmentation techniques in terms of *PSNR*.

As can be seen in the table, the *CS* algorithm in combination with the *Otsu* technique outperforms other approaches in most experiments, and although *PSO* combined with *Otsu* also gets good results, in 6 of its 13 best values the *CS-Otsu* combination gets the same values, so the combination chosen to segment the thermal images was *CS-Otsu*, waiting for desired results for this problem.

Since the *PSNR* was originally analyzed the similarity between two signals, it might not accurately evaluate the visual similitude between images. To overcome this limitation another metric specifically designed to evaluate visual similarity is considered. A comparison of the structures contained in the segmented image is performed using the SSIM, and it is defined in Eq. (9). A higher *SSIM* value represents a better segmentation of the original image.

$$SSIM\left(I_{in}, I_{seg}\right) = \frac{\left(2\mu_{I_{in}}\mu_{I_{seg}} + C1\right)\left(2\sigma_{I_{in}I_{seg}} + C2\right)}{\left(\mu_{I_{in}}^2 + \mu_{I_{in}}^2 + C1\right)\left(\sigma_{I_{in}}^2 + \sigma_{I_{seg}}^2 + C2\right)} \tag{9}$$

$$\sigma_{I_{in}I_{seg}} = \frac{1}{N-1}\sum_{i=1}^{N}\left(I_{in_i} + \mu_{I_{in}}\right)\left(I_{seg_i} + \mu_{I_{seg}}\right) \tag{10}$$

From Eq. (10) is the mean of the input (original) image and is the mean of the segmented image. In the same way, for each image, the values of and correspond to the standard deviation, C1 and C2 are constants used to avoid instability. The values of C1 and C2 are set to 0.065 considering the experiments.

The results of the *SSIM* are shown in Table 3. It shows the same format as in the previous table, presenting four optimization algorithms: *BA*, *CS*, *DE*, *PSO* combined with the two segmentation techniques that will be the objective functions to be maximized, presenting the results in terms of *SSIM* with the threshold values established in: 2, 4, 6 and 8 (Th) for each respective image.

The results suggest that *CS-Otsu* is more consistent in presenting better results than its competitors, and although *PSO-Otsu* does well, its performance is equaled by the combination of *CS-Otsu* in 8 of the 14 best values it obtains. So once again the tests point as a better candidate for the *CS* algorithm and as an objective function to maximize *Otsu*.

In the same context, the *FSIM*, helps to verify the similarity between two images. In this paper, the *FSIM* employs the original grayscale image and the segmented image. As *PSNR* and *SSIM* the higher value is interpreted as better performance of the thresholding method. The *FSIM* is then defined as:

Table 2 PSNR metrics

N_o Image	N_o Th	BA Otsu	BA Kapur	CS Otsu	CS Kapur	DE Otsu	DE Kapur	PSO Otsu	PSO Kapur
'a1'	2	32.610	31.871	32.669	31.765	32.661	31.786	**32.670**	31.765
'a1'	4	34.588	33.905	36.626	34.195	35.886	34.148	**36.636**	34.189
'a1'	6	36.917	34.674	**38.944**	34.962	38.713	34.918	38.891	34.940
'a1'	8	38.069	35.056	40.976	35.316	40.231	35.365	**41.069**	35.303
'a2'	2	23.347	20.090	**23.420**	20.044	23.415	20.082	**23.420**	20.044
'a2'	4	25.323	23.470	**26.070**	23.238	25.940	24.049	26.069	23.747
'a2'	6	26.809	24.815	27.786	24.890	**27.813**	25.053	27.760	24.830
'a2'	8	27.899	25.804	30.421	26.429	29.386	26.629	**30.614**	26.304
'a3'	2	23.161	18.909	**23.274**	18.852	23.258	18.903	**23.274**	18.852
'a3'	4	25.385	21.698	**25.945**	23.051	25.839	22.490	**25.945**	23.216
'a3'	6	26.633	23.977	**27.794**	24.436	27.212	24.484	27.775	24.602
'a3'	8	28.008	25.175	**29.187**	25.912	28.580	25.894	28.677	25.824
'a4'	2	23.167	20.072	**23.268**	20.277	23.258	20.345	**23.268**	20.277
'a4'	4	25.471	23.547	26.800	24.382	26.659	24.067	**26.801**	24.377
'a4'	6	27.193	24.529	**29.027**	25.527	28.325	25.540	28.709	25.506
'a4'	8	28.783	26.550	31.061	26.310	30.520	27.176	**31.159**	26.259
'a5'	2	25.550	22.765	25.989	22.809	**25.992**	22.767	25.989	22.809
'a5'	4	27.482	24.369	29.132	23.035	28.905	23.806	**29.133**	23.251
'a5'	6	29.508	26.309	**31.641**	26.989	30.993	27.161	31.486	27.043
'a5'	8	31.073	27.522	**33.198**	27.581	32.418	28.367	33.151	27.742
'a6'	2	26.488	23.456	**26.504**	23.434	26.470	23.556	**26.504**	23.434

(continued)

Table 2 (continued)

No Image	No Th	BA Otsu	BA Kapur	CS Otsu	CS Kapur	DE Otsu	DE Kapur	PSO Otsu	PSO Kapur
'a6'	4	28.879	24.632	**30.248**	24.575	29.850	24.679	**30.248**	24.571
'a6'	6	30.314	26.386	**32.085**	26.137	31.823	26.934	32.084	26.179
'a6'	8	31.931	27.029	**34.187**	28.360	33.590	27.907	33.982	28.020
'a7'	2	**27.531**	25.661	27.503	25.715	27.509	25.716	27.503	25.715
'a7'	4	30.225	26.737	**31.424**	27.000	31.338	26.933	31.423	26.988
'a7'	6	31.802	27.605	**33.595**	27.391	33.416	27.757	33.588	27.457
'a7'	8	33.076	28.499	35.887	28.437	34.805	28.713	35.941	28.064

Table 3 SSIM metrics

N_o Image	N_o Th	BA Otsu	BA Kapur	CS Otsu	CS Kapur	DE Otsu	DE Kapur	PSO Otsu	PSO Kapur
‘a1’	2	**0.9401**	0.9400	**0.9401**	0.9400	**0.9401**	0.9399	**0.9401**	0.9400
‘a1’	4	0.9566	0.9423	**0.9847**	0.9429	0.9713	0.9428	**0.9847**	0.9429
‘a1’	6	0.9781	0.9432	**0.9919**	0.9440	0.9866	0.9439	0.9872	0.9440
‘a1’	8	0.9813	0.9438	0.9930	0.9447	0.9901	0.9456	**0.9931**	0.9446
‘a2’	2	0.6888	0.5438	**0.6911**	0.5429	**0.6911**	0.5437	**0.6911**	0.5429
‘a2’	4	0.7449	0.6588	**0.7655**	0.6345	0.7587	0.6843	**0.7655**	0.6634
‘a2’	6	0.7973	0.7035	0.8129	0.7148	**0.8193**	0.7213	0.8113	0.7169
‘a2’	8	0.8257	0.7272	0.9031	0.7438	0.8685	0.7665	**0.9111**	0.7388
‘a3’	2	**0.7016**	0.4700	0.6985	0.4686	0.6997	0.4700	0.6985	0.4686
‘a3’	4	0.7598	0.5866	**0.7680**	0.6379	0.7663	0.6139	**0.7680**	0.6482
‘a3’	6	0.8012	0.6846	**0.8232**	0.6965	0.8040	0.7100	0.8224	0.7189
‘a3’	8	0.8459	0.7234	**0.8558**	0.7507	0.8455	0.7552	0.8358	0.7517
‘a4’	2	0.7217	0.5795	0.7246	0.5839	**0.7262**	0.5872	0.7246	0.5839
‘a4’	4	0.8007	0.7026	0.8516	0.7305	0.8393	0.7168	**0.8519**	0.7302
‘a4’	6	0.8435	0.7236	**0.9077**	0.7493	0.8723	0.7548	0.8860	0.7506
‘a4’	8	0.8824	0.7877	0.9344	0.7684	0.9211	0.8021	**0.9373**	0.7620
‘a5’	2	0.8152	0.6591	**0.8373**	0.6599	0.8363	0.6589	**0.8373**	0.6599
‘a5’	4	0.8597	0.7118	**0.9110**	0.6618	0.9033	0.6900	**0.9110**	0.6698
‘a5’	6	0.9087	0.7807	**0.9493**	0.8008	0.9336	0.8087	0.9431	0.8044
‘a5’	8	0.9357	0.8161	**0.9558**	0.8109	0.9494	0.8415	0.9550	0.8175
‘a6’	2	**0.8053**	0.6628	0.8040	0.6620	0.8032	0.6669	0.8040	0.6620

(continued)

Table 3 (continued)

N_o Image	N_o Th	BA Otsu	BA Kapur	CS Otsu	CS Kapur	DE Otsu	DE Kapur	PSO Otsu	PSO Kapur
'a6'	4	0.8766	0.6860	**0.9123**	0.6822	0.8978	0.6853	**0.9123**	0.6821
'a6'	6	0.9016	0.7457	**0.9282**	0.7298	0.9250	0.7641	0.9281	0.7311
'a6'	8	0.9333	0.7686	**0.9576**	0.8092	0.9496	0.7921	0.9521	0.7964
'a7'	2	**0.8414**	0.7425	0.8327	0.7430	0.8335	0.7432	0.8327	0.7430
'a7'	4	0.9155	0.7588	**0.9400**	0.7626	0.9391	0.7611	**0.9400**	0.7624
'a7'	6	0.9370	0.7795	**0.9541**	0.7696	0.9528	0.7823	0.9539	0.7716
'a7'	8	0.9457	0.8057	0.9765	0.7981	0.9627	0.8074	**0.9769**	0.7864

$$FSIM = \frac{\sum_{w \in \Omega} S_L(w) PC_m(w)}{\sum_{w \in \Omega} PC_m(w)} \tag{11}$$

In Eq. (11) the entire domain of the image is defined by Ω, and their values are computed by Eq. (12).

$$S_L(w) = S_{PC}(w) S_G(w) \tag{12}$$

and Eqs. (13) and (14) show how to calculate $S_{PC}(w)$ and $S_G(w)$ respectively.

$$S_{PC}(w) = \frac{2PC_1(w)PC_2(w) + T_1}{PC_1^2(w)PC_2^2(w) + T_1} \tag{13}$$

$$S_G(w) = \frac{2G_1(w)G_2(w) + T_2}{G_1^2(w)G_2^2(w) + T_2} \tag{14}$$

G is the gradient magnitude (GM) of a digital image and is defined, and the value of PC that is the phase congruence is defined as follows:

$$G = \sqrt{G_x^2 + G_y^2} \tag{15}$$

$$PC(w) = \frac{E(w)}{\left(\varepsilon + \sum_n A_n(w)\right)} \tag{16}$$

where $A_n(w)$ is the local amplitude on scale n and E_w is the magnitude of the response vector in w on n. ε is a small positive number and

$$PC_m(w) = \max(PC_1(w), PC_2(w)). \tag{17}$$

In Table 4 the results of the *FSIM* are presented for inspection. In this case the *FSIM* metric suggests for third time; based on the values presented that *CS-Otsu* is the most appropriate combination to segment the proposed thermal images. And although *PSO-Otsu* and *BA-Otsu* present some higher values, *CS-Otsu* is more consistent in presenting a greater number of optimal values in this test.

The steps in that tool processes images are shown in Figs. 5, 6 and 7. When the original image has a resolution of 320 × 240 pixels, the camera generates two images, one in *RGB* and the second with temperature values per pixel. The program generates the histogram using the temperature information, then, depending on the selected thresholds, the segmentation technique and the desired metaheuristic algorithm, the segmentation process begins and when this ends the program selects the area with the highest density of pixels with high temperature and delimits it with a box to facilitate the diagnosis.

In the previous figure, the damaged component (*CBM2099*) specified in Table 1, was satisfactorily located and the average temperature of the hottest zone coincides with the ranges established in the thermography.

Table 4 FSIM metrics

N_o Image	N_o Th	BA Otsu	BA Kapur	CS Otsu	CS Kapur	DE Otsu	DE Kapur	PSO Otsu	PSO Kapur
'a1'	2	0.8592	**0.8598**	0.8592	0.8596	0.8592	0.8596	0.8592	0.8596
'a1'	4	0.8862	0.8617	**0.9296**	0.8590	0.9074	0.8590	**0.9296**	0.8590
'a1'	6	0.9235	0.8599	**0.9492**	0.8594	0.9355	0.8596	0.9347	0.8594
'a1'	8	0.9283	0.8616	**0.9510**	0.8620	0.9467	0.8653	**0.9510**	0.8598
'a2'	2	0.8060	0.8125	0.8070	**0.8132**	0.8070	0.8127	0.8070	**0.8132**
'a2'	4	0.8231	0.8140	**0.8239**	0.8129	0.8226	0.8151	**0.8239**	0.8142
'a2'	6	0.8477	0.8243	**0.8773**	0.8216	0.8715	0.8213	0.8762	0.8211
'a2'	8	0.8689	0.8308	**0.9029**	0.8257	0.8901	0.8375	0.9017	0.8243
'a3'	2	**0.8229**	0.8183	0.8207	0.8189	0.8212	0.8188	0.8207	0.8189
'a3'	4	0.8474	0.8211	**0.8572**	0.8181	0.8539	0.8193	**0.8572**	0.8179
'a3'	6	0.8637	0.8370	0.8817	0.8386	0.8739	0.8423	**0.8819**	0.8457
'a3'	8	0.8785	0.8513	0.9008	0.8610	0.8888	0.8606	**0.9017**	0.8619
'a4'	2	0.8168	0.7920	0.8185	0.7924	**0.8188**	0.7931	0.8185	0.7924
'a4'	4	0.8416	0.8211	0.8503	0.8266	**0.8511**	0.8237	0.8502	0.8264
'a4'	6	0.8583	0.8318	0.8639	0.8418	0.8630	0.8424	**0.8653**	0.8424
'a4'	8	0.8711	0.8561	0.8926	0.8544	0.8876	0.8634	**0.8950**	0.8539
'a5'	2	0.8452	0.8389	**0.8505**	0.8392	0.8503	0.8392	**0.8505**	0.8392
'a5'	4	0.8650	0.8420	0.8754	0.8416	**0.8776**	0.8422	0.8754	0.8420
'a5'	6	0.8858	0.8506	**0.9066**	0.8488	0.8964	0.8527	0.9007	0.8494
'a5'	8	0.9001	0.8626	**0.9164**	0.8553	0.9091	0.8662	0.9150	0.8572
'a6'	2	**0.8671**	0.8561	0.8668	0.8550	0.8668	0.8562	0.8668	0.8550

(continued)

Table 4 (continued)

N_o Image	N_o Th	BA Otsu	BA Kapur	CS Otsu	CS Kapur	DE Otsu	DE Kapur	PSO Otsu	PSO Kapur
'a6'	4	0.8906	0.8670	**0.9045**	0.8715	0.9005	0.8710	**0.9045**	0.8715
'a6'	6	0.9036	0.8712	0.9090	0.8710	0.9092	0.8729	**0.9093**	0.8707
'a6'	8	0.9117	0.8767	**0.9211**	0.8751	0.9175	0.8768	0.9187	0.8743
'a7'	2	**0.8658**	0.8479	0.8592	0.8475	0.8598	0.8478	0.8592	0.8475
'a7'	4	0.8984	0.8551	**0.9144**	0.8594	0.9130	0.8590	**0.9144**	0.8592
'a7'	6	0.9109	0.8605	**0.9208**	0.8625	0.9194	0.8640	0.9207	0.8625
'a7'	8	0.9151	0.8651	0.9302	0.8644	0.9248	0.8653	**0.9311**	0.8643

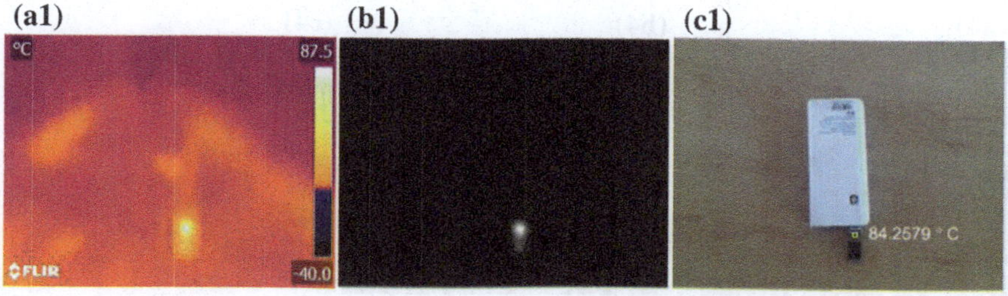

Fig. 5 USB image in the power bank showing: thermal image (**a1**), segmented image (**b1**) and RGB image (**c1**) with delimited hot zone

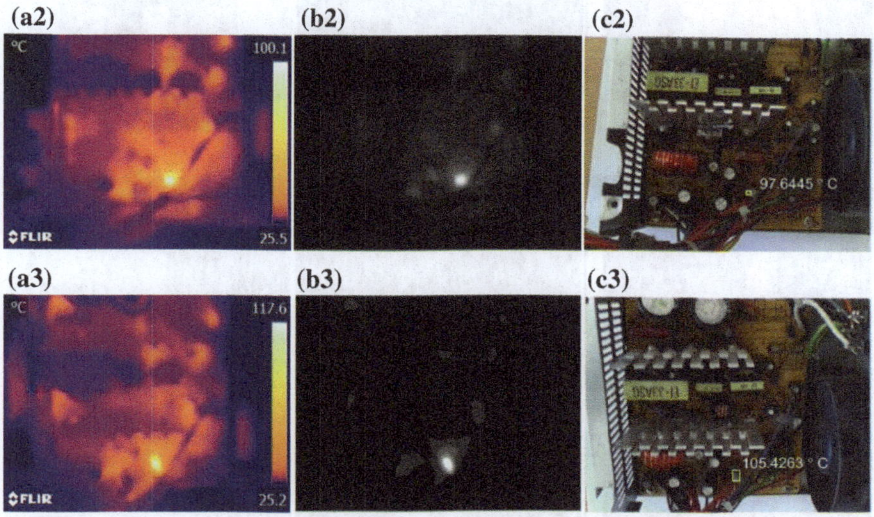

Fig. 6 Power source A showing: thermal image (**a2** and **a3**), segmented image (**b2** and **b3**) and RGB image (**c2** and **c3**) location of the zone with the highest temperature

In the thermographies (a2) and (a3) on Fig. 6, the damaged component is a resistor of 1/2 W of power specified in Table 1, which was successfully located by the program. One advantage of using the proposed software is that the thermography does not have to be taken at a certain angle or in a certain position to make the diagnosis, since the result of diagnosing the device is the same in both images as seen in (c2) and (c3).

In Fig. 7, the program shows its robustness in the thermal images (a4) and (a5) since it is evident that the same device is rotated, and nevertheless the result in (c4) and (c5) shows the same result in the diagnosis of the damaged component that in this case is a resistor specified in Table 1. On the other hand, in the images (a6) and (a7) the diagnostic device is the same too, with the difference that in (a6) a superior view is shown and (a7) it shows a superior inclined view, the disadvantage in this case lies in that if there is some component or object that interposes with the zone of higher

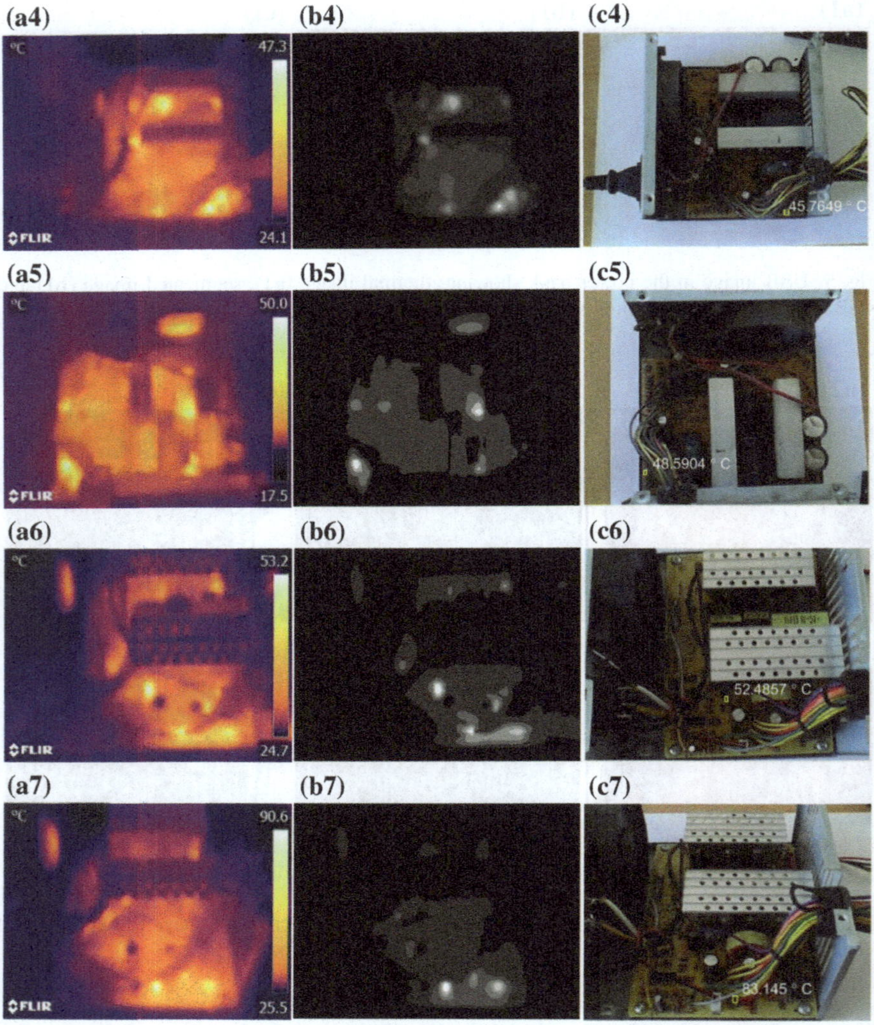

Fig. 7 Example diagnostic tool. **a5–a7** Thermal images captured by thermographic camera 320 × 240 pixels, **b4–b7** thermal image segmented, **c4–c7** location of the zone with the highest temperature on the RGB image

temperature, the diagnosis will be affected, since the camera will not capture the zone of interest with precision, it is observed that (a6) is affected by the accumulation of device cables that are in the lower part of the capture, (a7) is a front view of the same device but in this case it is evident that the element that prevents a good capture of the areas of interest is an electrolytic capacitor.

It should be mentioned that power source object C has two damaged system elements specified in Table 1, a 1N4001 diode and a 150 Ω resistor, which were in the resulting images (c6) and (c7) respectively.

7 Conclusions

Is coded the software to observe the performance of the combination of optimization algorithms and threshold techniques. By segmenting the thermal images, the project demonstrated that thermal imaging systems are a very powerful tool to easily and economically diagnose electronic devices, to examine and monitor such devices quickly and accurately without contact, the results obtained were consistent, which implies the viability of taking the project to real problems, either in industry or in independent projects. Another aspect to highlight is that the use of the graphical interface makes the program intuitive and easy to use.

The main concept of electronic devices in thermal imaging is to detect areas with sudden temperature changes associated with wear or malfunction of elements within the electrical system, for diagnosis and preventive maintenance. Follow-up corrective actions in diagnosed systems will require more careful investigation and expert analysis. Depending on the nature of the problem, the course of action may be simple and inexpensive or may require more elaborate actions. In any case, the main usefulness of this tool lies in knowing that there is a problem and it can be of critical importance, which will allow to plan actions to solve it.

When combining the technology of the current computers and thermal cameras of great precision more applications in diverse fields will be possible.

One of the disadvantages observed in the experimental results was the interference of objects over the area of interest, but we can be certain that the thermal cameras will evolve to such a degree that they will be more precise and more robust to avoid bad temperature readings by external agents or unstable reference temperatures.

As future work we intend to increase the segmentation techniques available to select, explore new areas of implementation, for example: in the medical area or agriculture. As concerns metaheuristic algorithms, new optimization techniques will be added that offer better results in the segmentation of the image as well as the most used algorithms. Recent and innovative optimization techniques will be adapted to offer better results in the segmentation of the image.

Appendix

The parameters used in each method have been configured according to the reported values in which their best performance is achieved, below is the configuration of these settings, every algorithm was tested using 50 particles of population (Table 5).

Table 5 Parameter settings

Setting configuration	
BA	The parameters where set as follows: Initial loudness rate $A = 2$, pulse emission rate $r = 0.9$, minimum frequency $f_{\min} = 0$ and maximum frequency $f_{\max} = 1$, respectively
CS	The balance of the combination of a local random walk and the global explorative random walk is controlled by a parameter $P_a = 0.25$
DE	The crossover rate is set to $CR = 0.5$, while the differential weight is given as $F = 0.2$
PSO	The cognitive and social coefficients are set to $c_1 = 2.0$ and $c_2 = 2.0$, respectively. Also, the inertia weight factor ω is set to decreases linearly from 0.9 to 0.2 as the search process evolves

References

1. P. Anitha, S. Bindhiya, A. Abinaya et al., RGB image multi-thresholding based on Kapur's entropy—a study with heuristic algorithms, in *Proceedings of 2017 2nd IEEE International Conference on Electrical, Computer and Communication Technologies, ICECCT 2017* (2017), pp. 0–5. https://doi.org/10.1109/ICECCT.2017.8117823
2. S. Bangare, S. Patil, Reviewing Otsu's Method for Image Thresholding (2016)
3. J. Coates, Encyclopedia of analytical chemistry, in *Interpretation of Infrared Spectra, A Practical Approach* (2004), pp. 1–23
4. C.A. Balaras, A.A. Argiriou, Infrared thermography for building diagnostics. Energy Build. **34**, 171–183 (2002). https://doi.org/10.1016/S0378-7788(01)00105-0
5. X.-S. Yang, A new metaheuristic bat-inspired algorithm, in *Encyclopaedia of Networked and Virtual Organizations* (2010), pp. 65–74
6. M. Mareli, B. Twala, An adaptive Cuckoo search algorithm for optimisation. Appl. Comput. Inf. **14**, 107–115 (2018). https://doi.org/10.1016/j.aci.2017.09.001
7. I. Koohi, V.Z. Groza, Optimizing Particle Swarm Optimization algorithm, in *Canadian Conference on Electrical and Computer Engineering* (2014), pp. 1–5. https://doi.org/10.1109/CCECE.2014.6901057
8. R. Storn, K. Price, Differential evolution—a simple and efficient heuristic for global optimization over continuous spaces. J. Glob. Optim. **1**, 341–359 (1997). https://doi.org/10.1023/A:1008202821328
9. L. Zhang, L. Zhang, X. Mou, D. Zhang, FSIM: a feature similarity index for image quality assessment. IEEE Trans. Image Process **20**, 2378–2386 (2011). https://doi.org/10.1109/TIP.2011.2109730
10. Z. Wang, A.C.A.C. Bovik, H.R.H.R. Sheikh, E.P.E.P. Simoncelli, Image quality assessment: from error visibility to structural similarity. IEEE Trans. Image Process **13**, 600–612 (2004). https://doi.org/10.1109/TIP.2003.819861
11. A. Horé, D. Ziou, Image quality metrics: PSNR vs. SSIM, in *Proceedings—International Conferenceon Pattern Recognition* (2010), pp. 2366–2369. https://doi.org/10.1109/ICPR.2010.579
12. T. Chai, R.R. Draxler, Root mean square error (RMSE) or mean absolute error (MAE)?—Arguments against avoiding RMSE in the literature. Geosci. Model Dev. **7**, 1247–1250 (2014). https://doi.org/10.5194/gmd-7-1247-2014

A Survey on Image Processing for Hyperspectral and Remote Sensing Images

Alfonso Ramos-Michel, Marco Pérez-Cisneros, Erik Cuevas and Daniel Zaldivar

Abstract Remote sensing images generally contain a large amount of information. For this, the researchers perform remote sensing image analysis by some computational methods. Modern geophysical monitoring is one of the main applications of remote control detection techniques. Among the essential tasks performed by these techniques is the detection of changes in physical geography and the study of forest issues. The purpose of this chapter is to analyze the most efficient methods used by remote sensing image processing tasks using traditional algorithms, optimization algorithms, and artificial intelligence algorithms. For this, this review includes corner detection techniques for image matching, endmember extraction for unmixing pixels, segmentation, and object classification. The purpose is to have a compendium of techniques developed in recent years.

Keywords Image processing · Remote sensing · Hyperspectral images · Optimization

1 Introduction

The knowledge of the changes in its environment has gained relevance in modern society. With this goal in mind, the scientific community has developed methods to gather reliable information to record geological, geographical, forestal, and demo-

A. Ramos-Michel (✉) · M. Pérez-Cisneros · E. Cuevas · D. Zaldivar
División de Electrónica y Computación, Universidad de Guadalajara, CUCEI, Av. Revolución 1500, C.P. 44100 Guadalajara, Jalisco, Mexico
e-mail: alfonsorm@cucsur.udg.mx

M. Pérez-Cisneros
e-mail: marco.perez@cucei.udg.mx

E. Cuevas
e-mail: erik.cuevas@cucei.udg.mx

D. Zaldivar
e-mail: daniel.zaldivar@cucei.udg.mx

D. Oliva and S. Hinojosa (eds.), *Applications of Hybrid Metaheuristic Algorithms for Image Processing*, Studies in Computational Intelligence 890,
https://doi.org/10.1007/978-3-030-40977-7_2

graphical changes through time. The remote sensing can be defined as the noncontact recording of information of the earth's surface using data from the ultraviolet to microwave bands of the Electromagnetic Spectrum. The acquisition of the information its via instruments like cameras and scanners located on platforms such as aircraft or spacecraft, and the analysis of acquired information through visual and digital image processing. The most common research areas are environmental assessment, global change detection, land-use/land-cover monitoring, agriculture, cartography, among others [3].

Only remote sensing approaches have the potential to provide detailed information on large areas in a cost-efficient and reliable way [13]. That is why it is a relevant source of information for sustainable management policies, environmental studies on wildlife [8], and following up on land-cover changes when a disaster occurs [19].

The first source of usable information is the image data, which is converted from a digital format to an analog image by a computer image display. Multispectral Images and Hyperspectral Images are widely used to study the gathered data in remote sensing, sometimes helped for LIDAR data.

Generally, any kind of energy coming from the earth's surface is useful to perform an image in remote sensing [17]. The most common is to match various contiguous images to create an only image with the complete information about a study area. The task of extracting some characteristic data from the images usually has the problem of interference of noise gathered from the environment at the shot time, and subtle texture like light changes and shadows, which causes false information. Taking into account that sometimes the images are captured from moving cameras, it is common to get rotated images, dimensional changes due to the different altitudes, and illumination changes that may cause problems when the matching process is running. Hyperspectral images also have a low spatial resolution problem due to the significant amount of data contained in it. This low spatial resolution causes mixed pixels that impose restrictions on practical applications of this kind of image [1].

The acquisition of labeled data for supervised hyperspectral image classification is expensive in terms of both time and costs. Moreover, manual selection and labeling are often subjective and tend to induce redundancy into the classifier.

Data gathered from the real world can be of high dimensional, particularly in remote sensing applications [16]. The high volume of image data obtained and the technical complexity of current remote sensing systems do imperative to preprocessing all the gathered data, before the science community work with it. Some of this work means: Correct geometry distortions, remotion of noise, or calibrate the image radiometry to get a consistent and reliable image database. Preprocessing data is just the start of the job. Obtaining the best data that the researchers are looking for is a new challenge. Nowadays, as can be seen in Sect. 3, combining different kinds of algorithms and data sources even, has presented better results than working with not combined algorithms, or a unique data source. Use of Hyperspectral images, combined with LIDAR data, for example, might bring better results than working alone.

On the other hand, optimization is a mathematics' field that studies the techniques or methods for searching the global optimum of a function [2]. The lack of using optimization algorithms in the processing of digital images has shown valid results, but, the accuracy is not the best. Using optimization algorithms allows the researchers to extract more and more reliable information from this combined data sources and preprocess most of the available data, increasing the accuracy of the results. The complement of work with combined data sources and the use of optimization algorithms has performed to better results.

2 Remote Sensing Data

Remote sensing is a technique that collects data from the earth's surface by scanning at a distance, some energy that represents diverse characteristics of the land surface, depending on the type of energy gathered and the form to organize it like image data. The sun is the principal power supply of energy for remote sensing in the form of sunlight because it is costless and covers from ultraviolet to infrared wavelength specters, although some remote sensing devices use artificial powered energy utilizing EM waves from the ultraviolet to microwave regions, since this devices can be fully controlled and supply just the required energy for a specific use. The materials absorb and reflect the energy in different spectral ranges, which depends on its molecular composition. In addition to the reflected rays, every object emits thermal radiation when they are above absolute zero. The sensors receive part of the energy reflected and emitted by the materials under study. The collected data is then organized to form images, and preprocessed to extract the critical information that is relevant for the researchers. The wavelength of the gathered radiation, and the number of layers determines the kind of image formed; Thermal images, LIDAR images, multispectral images, and hyperspectral images.

The thermal images are constructed by the organized information of the emitted energy from the objects. These images are a representation in the visible human spectrum of the thermal infrared wavelength of the electromagnetic radiation. Thermal imaging techniques stores information about the temperature of the objects, being applied principally for military and intelligence purposes, thermal pollution surveys, medical diagnoses, forest fire detections, and manufacturing quality control [12].

LIDAR means Light Detection and Ranging and is a remote sensing system similar to radar which uses laser light with a near-infrared wavelength. LIDAR instruments provide information on vertical forest structure by measuring the roundtrip time of the laser pulse in its travel to the land surface and the return to the measurement device. Swapping an area generates a cloud of points, which generates an image data that shows the surface swapped in something similar to a 3D image. This technique accurately estimates forest canopy heights, stand volume, and above-ground biomass. Due to the near-infrared energy, LIDAR is capable of discriminating the soil, getting only the forest structure, and is used principally for land cover classification, forest management [7].

Due to the long distance, satellite images have a low spatial resolution to extract some information from the earth's surface in the form of shapes. Although, the material which composes the objects have spectral signatures. These signatures may be revealed, attacking the material with energy at multiple but specifics wavelength. The record of these data is made in multiple snapshots of spectral properties acquired more or less simultaneously. Multispectral and Hyperspectral images are the datasets that contain several spectral bands with a piece of valuable information because of the spectral signatures contained at each one.

Multispectral images contain information about the spectral signatures of the materials in the sensed space in the order of ten wavelength bands. Thanks to the number of bands, it is possible to gather high spatial resolution data. On the other hand, the spectral resolution is low [17].

Hyperspectral images are constituted by around 200 wavebands, each one with a more narrow spectral bandwidth than in multispectral images. A significant number of narrow bands bring a high spectral resolution facilitating the location of the structures and object's characteristics through some different bands, improving the performance in object recognition, land change detection, human-made material identification, and other remote sensing tasks [18].

However, the high spectral resolution, hyperspectral systems can not bring at the same time a high spatial resolution. A large amount of information managed and the image's getting time set a proportional inverse relationship between spectral resolution and spatial resolution to ensure signal to noise ratio [6]. To be able to use this information is necessary to know what the data represents. Due to the number of bands, it is too difficult for researchers to read into hyperspectral images content by the eyes. A computer can do this, but it needs first, to extract all the features that represent a piece of valid information, according to its use. Frequently, low spatial resolution causes mixing pixels between the target object and the background of the image, difficulting the pixel-level tasks. Spectral unmixing is an essential method to clear the image and separate the objects and background through its endmembers and their abundances. Feature extraction is a primary method to obtain remote sensing information, detecting the object's features for researchers. It has a significant influence on recognition, analysis, matching, fusion, and segmentation of remote sensing images. For this, the usage of machine learning algorithms to classify the image information brings better and more accurate results [1].

3 Analysis Methods for Remote Sensing Images

3.1 Corner Detection for Image Registration

The study of the earth's characteristics commonly makes mandatory the usage of images that cover large areas of land. Since image scavenging devices are not always able to photograph the entire area in a single capture or in order to increase the image feature resolution, it is a regular task to combine different images with touching scenes, which finally contains all the information in a single image. The image

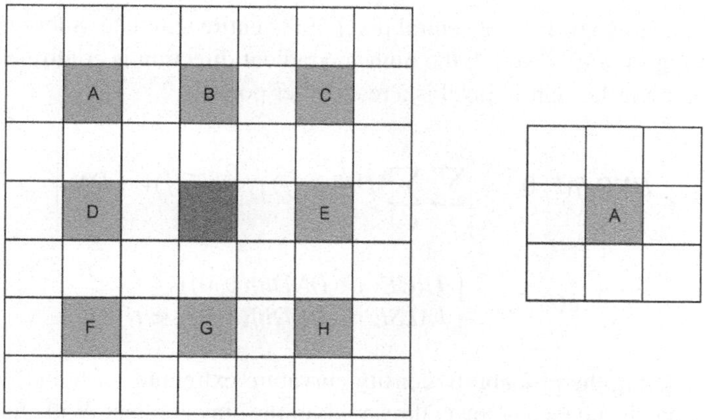

Fig. 1 The neighborhood of areas proposed by Deng et al. (left), and the neighborhood of the central pixel A (right)

registration is then responsible for performing this fusion of images by matching the characteristics shared between them. The computer algorithms got to obtain reference points from the images to perform the mergers correctly.

Usually, the images contain redundant information, which is used to locate the points of a coincidence that the algorithm needs. In order to locate these points of interest, the scientific community uses different algorithms, such as the Harris algorithm, used in the detection of corner points due to its simple calculation and high stability [20]. Its weak point is the vulnerability it presents to noise or subtle changes in texture. Also, algorithms that localize gray value changes in the images are an information supply for image registration. The noise that represents sudden changes in the texture generates a problem for the correct detection of edges and characteristics of the image.

In order to obtain better results in image registration that allows better performance in the location of characteristics for the study of remote sensing images, researchers develop algorithms that provide more reliable data, while allowing the study of large amounts of information in an automated way by computational means.

The primary task of stitching image is to find correct ground control point correspondences between the images to joint. This task sometimes is a challenge for scientists. First of all, they must detect the characteristics points on both images, then find the correspondence between them. The use of the gray gradient of the corner point can help to perform the job.

When the traditional image registration algorithm extracts corner points, it also extracts pixels that represent noise for the image. Deng et al. [5] proposes to define a neighborhood around the central pixel of the corner point of 3×3 areas of 3×3 pixels each, where the areas overlap each other. A total of 7×7 pixels being a matrix (Fig. 1). Through the convolution calculation, extracts the characteristic gray level of each neighborhood area. The algorithm compares the area's characteristic gray level with the gray level of the central pixel of the entire template. Subsequently, by implementing the Eqs. 1 and 2, the highest gradient direction is confirmed, as well as determining if the central pixel is a real corner point.

$$DNDiff(k,0) = \left| \sum_{i=1}^{3} \sum_{j=1}^{3} [DN_k(i,j) \ldots w(i,j)] - DN_0 \right| \tag{1}$$

$$c(k,0) = \begin{cases} TRUE \text{ if } (DNDiff(k,0)) \leq t \\ FALSE \text{ if } (DNDiff(k,0)) < t \end{cases} \tag{2}$$

By calculating the probability density curvature extremum of the gray gradient helps to obtain the upper and lower thresholds of the gray gradient. With this methodology, corner point extraction is achieved computationally in different remote sensing images. Considering in their calculations, the gray levels of a neighborhood of pixels, effectively eliminate noise points that could hinder the process. The method presented permits the automatic extraction of corner points from images next to be matched, which makes it capable of processing lots of images without human intervention. At the same time, the process returns more objective results than a human inspection, that brings a high precision for the image joint.

A method presented by Wang Changjie and Niam Hua in 2017 [4], proposes a series of steps that identify the characteristics of the image, generates a descriptor of these characteristics to save the addresses of the located points and finally eliminates the possible false corner points localized at the beginning of the process.

Initially, the characteristic points are extracted using the Harris corner point detection algorithm. The algorithm offers stability, as well as good immunity to the noise present in the same image.

The method builds a 64-dimensional feature descriptor of the located corner points from the first order Haar wavelet by the SURF algorithm, referring to the orientation of the corner points, as well as the gray gradient direction of each corner, and the change in the intensity of the gradients in the x and y directions (Fig. 2).

For this task, assigns to each corner point a neighborhood divided into 16 zones. For each zone, the algorithm generates a 4-dimensional vector (Eq. 3), that describes the intensity changes before mentioned. The union of this vector for all zones results in the mentioned 64-dimensions feature descriptor vector.

$$V = \left[\sum d_x, \sum |d_x|, \sum d_y, \sum |d_y| \right] \tag{3}$$

Once the algorithm identifies and describes the characteristic points of the images, it calculates the Euclidean distances of the descriptors and their two closest neighbors, which allows knowing the lack of coincidence of corner points between the images to merge.

Finally, the method contemplates the RANSAC algorithm to eliminate all those characteristic points that have not been coincident in both images. The methodology used by Changjie and Hua provides enough information on the characteristics of the image to generate image matching more quickly and accurately, even if the images to be combined have a rotation in the angle of the shot.

In 2018 Zhou et al. presents an algorithm capable of suppressing false corner points in satellite images. It is known that the corner point extraction of the Harris algorithm usually has problems when the image has subtle grayscale variations. Likewise, the presence of noise leads to the extraction of false corner points using the classic Harris corner point extraction algorithm. It is a fact that not always is possible to obtain an ideal image to carry out the study of the physical characteristics of a given area. Typically, the sharpness of the image will be affected by color variations in the grayscale image. This variation interferes with the correct corner point extraction, since the Harris algorithm works in the search for grayscale direction variations. To reverse this, Zhou et al. [20] propose to extract the edge entities from the image using the high and low-frequency components of the same image, so that the Harris algorithm uses its characteristics instead of the original image with its noise and tones variations that promotes the extraction of false information.

The low-frequency components of the image represent the areas where the gray value changes slowly. These areas, therefore, represent the frame of the image. The wavelet decomposition gets the high-frequency components of the image from regions where the gray value of the image has a more significant variation. These components represent the details of the image. By convolution of a low-frequency filter and the original image, the wavelet decomposition of the image can be obtained, which helps to extract the high and low-frequency components each time that performs the wavelet decomposition. The multilevel wavelet decomposition helps to highlight the noise values of the image, which provides enough information so that the computer equipment can perform the noise suppression automatically and on a large scale.

Figure 3 shows the flow diagram of the proposed algorithm. In a first decomposition, the algorithm gets the first low-frequency component (*LFC*1), and from the dif-

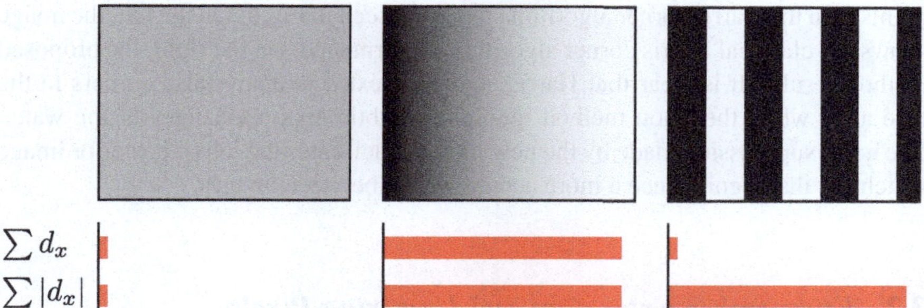

Fig. 2 Meaning of the description vector data entries for each subzone

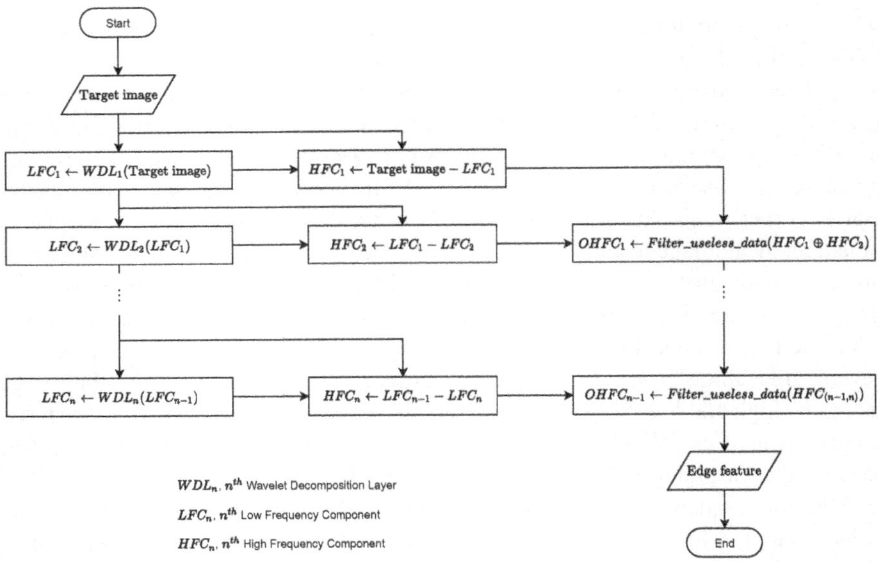

Fig. 3 Flow diagram of the proposed algorithm by Zhou et al. in [20]

ference of the target image and *LFC1*, the first high-frequency component is obtained (*HFC1*). The process is repeated, restarting with *LFC1* to obtain the necessary high and low-frequency components. The *exclusive OR* component of the high-frequency components of the last two levels determines noise and subtle changes in the image. From here, it is possible to filter them to obtain the optimized components of high frequency (*OHFC*) preliminarily, until completing the decomposition of wavelet of level $n(WDLn)$. After this, the proposed algorithm obtains the optimized edge characteristics of the target image.

With the proposed method, Zhou et al. obtained better results localizing corner points than the Harris corner algorithm, as can be seen in Fig. 4. On the left, the image shows the classical Harris corner algorithm performance. On the right, the proposed method results. It is clear that Harris algorithm extracts many false corners in the lake area, while the Zhou method ignores de subtle texture changes of the water. The noise suppression made by the new method facilitates the job of posterior image matching, that significance a more accurate joint between images.

3.2 *Endmember Extraction and Unmixing Pixels*

Within the catalog of remote sensing images, it is usual that hyperspectral images have a low spatial resolution, which is a common cause of finding mixed pixels between the content and the background of this type of image. The process of hyperspectral unmixing refers to the estimation of pure pixels of the image, called endmembers.

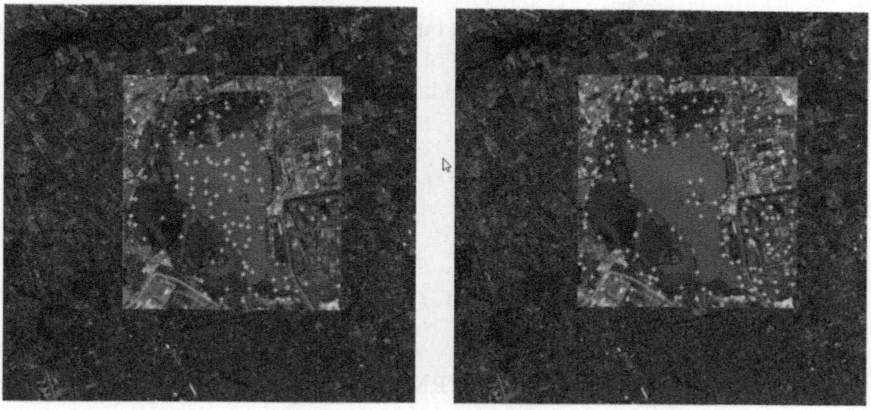

Fig. 4 Local area corner detection results by implementing Zhow et al. method proposed. Images from [20]

Thus, the percentage of endmembers present in the pixel is counted, which is called abundances. A common consideration is that the mixing of pixels is linear. The scientist found that in real scenarios, the mixture can present a non-linear behavior. Also, the classification of images is usually done in one of two ways: Hard classification where only one class is allowed per pixel; and the soft classification, where the pixel can belong to more than one class, obtaining membership qualifications for each of them.

In the search to raise the quality of the results, researchers worked collectively with other areas of science such as artificial intelligence, to provide increasingly effective solutions in preprocessing. Such is the case of endmember extraction, where the application of artificial swarm intelligence has worked; however, limitations in terms of calculation efficiency produce stacked solutions based on similar endmembers in the same class. It is then that the importance of the endmember extraction is critical in the remote sensing images unmixing, which is why scientists have chosen to include metaheuristic and artificial intelligence algorithms in their process.

Swarm intelligence is a segment of artificial intelligence applied in the solution of combinatorial optimization problems. The use of these algorithms has yielded excellent results such as ant colony optimization (ACO), artificial bee colony (ABC), and partial swarm optimization (PSO) used in hyperspectral remote sensing images. The use of these algorithms resulted in the generation of new methods such as the ant colony optimization algorithm for endmember extraction (ACOEE) and the discrete swarm optimization for endmember extraction (DPSO). In more recent years, Su et al. raised these algorithms to obtain the improved discrete artificial bee colony (IDABC), the improved ant colony optimization algorithm for endmember extraction (IACOEE), and the improved discrete partial swarm optimization for endmember extraction (IDPSOEE).

For the endmembers extraction process, the most popular algorithm of recent years has been the Linear Spectral Mixture Model (LSMM), with which good results have been obtained with high-quality images, but not using images with noise content or endmember variability. The formulation of the LSMM algorithm is shown in Eq. 1, and the constraints of the fractional abundance estimation.

$$r_i = \sum_{k=1}^{M} p_{ik} e_k + \varepsilon_i \qquad i = 1, \ldots, N \tag{4}$$

$$p_{ik} \geq 0 \qquad \forall k = 1, \ldots, M \sum_{k=1}^{M} p_{ik} = 1 \tag{5}$$

The use of Root Mean Square Error (RMSE) between the original hyperspectral image and the remixed image evaluates the accuracy of the endmember extraction, the result of the abundance matrix estimation. The calculation of the RMSE is done by

$$RMSE(\{r_i\}_{i=1}^{N}, \{e_j\}_{j=k}^{M}) = \frac{1}{N} \sum_{i=1}^{N} \sqrt{\frac{1}{L} \left\| \mathbf{r}_i - \hat{\mathbf{r}}_i \right\|_2^2} \tag{6}$$

In 2016 Su et al. proposed an algorithm based on artificial intelligence. The tests were performed with the ABC, ACO, and PSO algorithms.

The modification with ABC also consisted of integrating swarm intelligence into the process and moving it to a discrete environment. The proposal consisted of replacing Eq. 6 with:

$$C_{N,M} = \left\{ (c_1, c_2, \ldots, c_N) | c_i \in 0, 1, \sum_{i=1}^{N} c_i = M \right\} \tag{7}$$

They also integrated $X_i \in C_{N,M}$, where X_i represents a discrete string of $N \times M$ digits. The mapping relationship is defined by Eq. 8, where Z represents a feasible solution space.

$$C : Z(\{r_i\}_{i=1}^{N}, M) \to C_{N,M} E \to (c_1, c_2, \ldots, c_N) \tag{8}$$

If the pixel $\mathbf{r}_i$ is an endmember, then c_i acquires the value of 1; otherwise, it acquires a value of 0, which represents the background of the image. When the number of endmembers is known, the search is reduced to discrete digit strings (1 and 0). With this modification, Su et al. generate with this process the IDABC. Similarly, they present the IACOEE and IDPSOEE algorithms, where, like the previous one, they respect the original procedures, only replacing the objective function.

With these modifications, Su et al. limit the number of endmembers per class, which allows them to obtain more precisely the positions of the endmember in large hyperspectral image datasets.

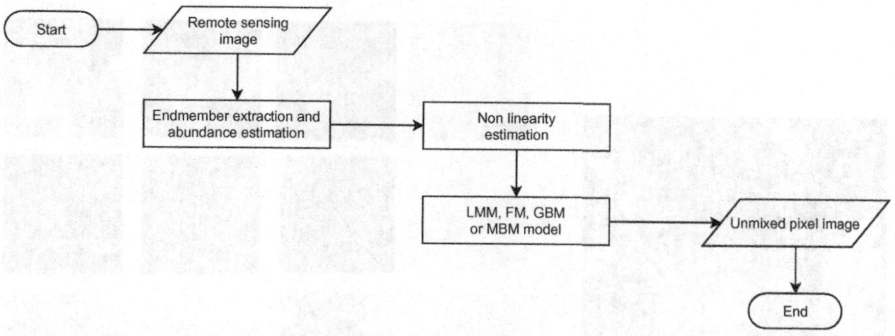

Fig. 5 Flow diagram of the methodology used by Niranjani and Vani in [14]

In 2018, Niranjani and Vani presented a model to estimate the nonlinearity of a mixture of pixels [14]. This task carried out after endmember removal and abundance estimation, allows pixel unmixing with lower mean square error, which generates better results for the detection of materials in multispectral and hyperspectral images. Figure 5 shows the workflow of their proposal.

The linear mixing model (LMM) is broadly used in remote sensing unmixing, which estimates the endmembers of the image and subsequently calculates the abundance matrix. It assumes that the image is composed of a unique reflectance, where each photon interacts only with a component of the earth's surface. Another model used is the Bilinear and Nonlinear Mixing Model, also known as the Fan Model (FM), which introduces bilinear interactions present in the image through the Hadamard product of the endmember matrix and its corresponding spectra. Niranjani et al. made an extension to the Fan Model, the general bilinear model (GBM). In it introduce a non-linear coefficient in the bilinear interaction, which determines whether the mixing model becomes linear or bilinear. This model, according to the authors, is suitable for data that contains only two endmembers. Niranjani et al. also present an extension of the generalized bilinear model, the modified bilinear model (MBM). This model has multiple endmember interactions. MBM method considers the number of endmembers and the coefficient of nonlinearity, to become one of the models previously seen.

In Fig. 6 we can observe the result of the obtained abundances of the images when applying the four models in a remote sensing multispectral image. The images show different results with each method applied. It shows, too, how the results vary depending on the material explored in each case. MBM proved good results in the fraction of water, as well as the fraction of vegetation. The soil fraction is very similar in the two models presented by the authors. Applied in a hyperspectral dataset (Fig. 7), MBM obtained better results in the soil and water fraction, while in vegetation, GBM obtained remarkable results. The main advantage of this technique is the lack of supervision. To work with an unknown number of endmembers and spectral matrix in advance represents another advantage. In comparison to the four methods tested,

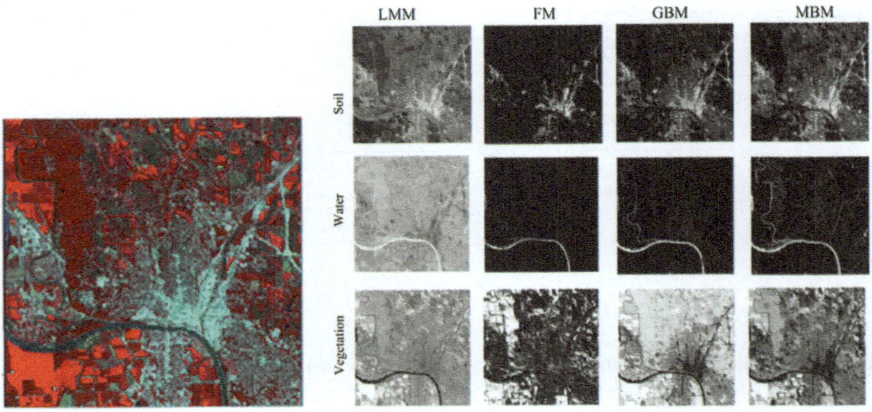

Fig. 6 Original multispectral image (left) and the results of the unmixing pixel procedure with the algorithms under test (right). Images from [14]

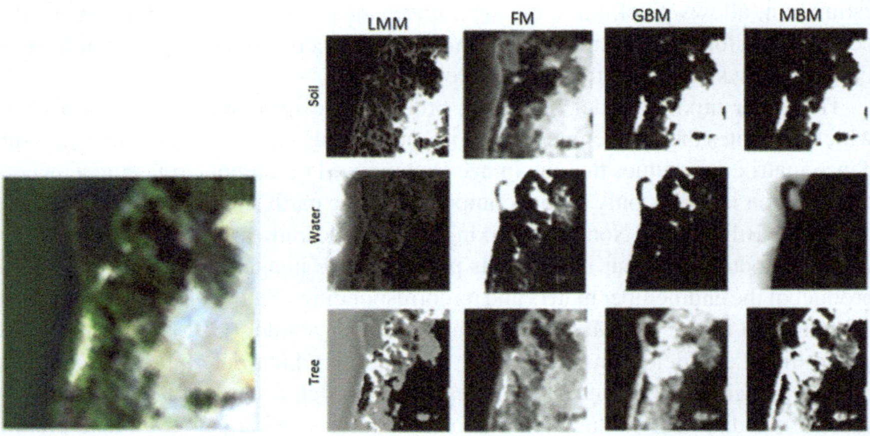

Fig. 7 Hyperspectral image (left) and results of the unmixing pixel procedure with the algorithms under test (right). Images from [14]

the MBM method obtained the lowest reconstruction error, as well as the lowest root mean square error.

3.3 Segmentation and Classification

The detection of changes in the earth's surface is one of the objectives of remote sensing that is essential for multiple environmental studies, whether for academic, economic, social, or ecological applications. The feature selection task is a necessary step for practical computer-aided processing and analysis of the image content, thanks

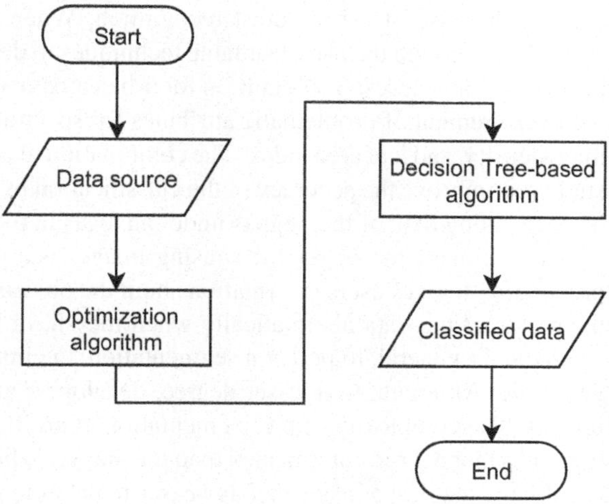

Fig. 8 General process for classification data from a combination of optimization algorithms and decision tree-based algorithms

to the reduction of data to process [16]. In image analysis, efficient segmentation is an essential task for the classification and recognition of significant objects [9]. Like other areas of image processing, segmentation has benefited from the introduction of artificial intelligence techniques, which help the automated analysis of large areas of land, while implementing object-based classification techniques to improve results of scanning through images.

The researchers often use algorithms such as canonical correlation analysis (CCA) and principal component analysis (PCA) to detect changes. Performing this change detection through unsupervised procedures acquires relevance, especially when studying hyperspectral data sets, due to the lack of terrestrial references.

Optimization algorithms used in combination with Decision Tree-based algorithms have demonstrated its efficiency on selecting the most valuable information from available data to its posterior classification according to the particular case necessity [16]. Figure 8 shows the general process of selection and classification data by these algorithms.

In recent years, researchers has been experimented with evolutionary algorithms, seeking to improve the results that previously obtained through known techniques. The FODPSO algorithm is an evolutionary algorithm based on the PSO algorithm. A deficiency of the PSO was its stagnation in optimal local values. FODPSO improves this task by looking for optimal values in a better way. As an extension of the DPSO algorithm, the new algorithm proposed by Couceiro and Ghamisi uses the fractional calculation to control the convergence rate presented by DPSO. The FODPSO algorithm applied for image segmentation selects optimal thresholds, where it has demonstrated, in addition to accuracy, to be more computationally efficient speaking than previously used algorithms such as the Otsu Algorithm.

Occasionally, the type of study restricts the use of specific processing techniques. In the information extraction for object recognition, attributes quantification can be affected by shadows or background effects present in the analyzed spectral data, such as the case of urban forest recognition. When this happens, the studies are carried out through the use of suitable techniques to deal with the case that generates problems. Thus, forestal recognition for urban areas usually makes use of LiDAR for the measurement of problematic attributes for spectral datasets, such as tree height, mass density, and leaf area index. The combination of analysis techniques in the same study can improve the accuracy of the classification by providing the researcher with corresponding data of the objects under analysis in the study area.

The segmentation of remote sensing images is a process widely used in image processing. It gives users information about the objects found in the image and helps them extract the data automatically when they have large amounts of information available. In general, to perform segmentation, the image is previously converted to grayscale. Although, to a lesser degree, developers also use segmentation of color images. The complexity of the segmentation of an RGB image is high since its histogram is much more complicated than the grayscale histogram. This complication is usually the factor that leads in many cases to perform this task on images previously converted to grayscale. For proper segmentation of the image, there is a critical process called threshold selection. Depending on the quality of the threshold selection, the segmentation results will be better or worse. When performing the segmentation by thresholds, the algorithms divide the pixels of the image into several classes. Separation can be of two levels or multiple levels. Sometimes separating by two levels is convenient because it leaves the image segmented in the area that represents the object and the area that represents the background of the image. Multilevel segmentation makes the separation of the image so that several different characteristics can be highlighted between the regions of the image [10]. To process the RGB image channels conveniently, scientists introduced the use of optimization algorithms. The combination of different techniques and algorithms has provided satisfactory results by extracting the information from the different channels and merging them to obtain the segmentation of the final image.

The detection of objects in urban and forestry environments has been an area of great importance. The importance lies in the possibility of classifying the objects detected in classes that allow the user to know what that object represents in real life. To classify these objects, scientists use the classifiers, like the Supporting Vector Machine (SVM) and Random Forest (RF). The Random Forest classifier has been very popular in the classification of forest species. Between its advantages is the possibility of working with categorical data, unbalanced data, as well as data with missing values, which is not possible to do with SVM, in addition to handling a classification accuracy comparable to that achieved by SVM [15].

On the other hand, the multitemporal study of a terrestrial area gives the remote sensing the ability to monitor changes that occur both seasonally in short-time intervals, as well as in long-time intervals such as the progress of urban areas.

Yokoya and Gamisi presented in 2000 a method of detecting multiple changes in time series hyperspectral data, applied to the monitoring of landcover changes in the area of the Fukushima Daiichi nuclear power plant, after the nuclear disaster [19]. The object detection is an unsupervised method and is based on the segmentation of swarm optimization of fractional order Darwinian particles (FODPSO).

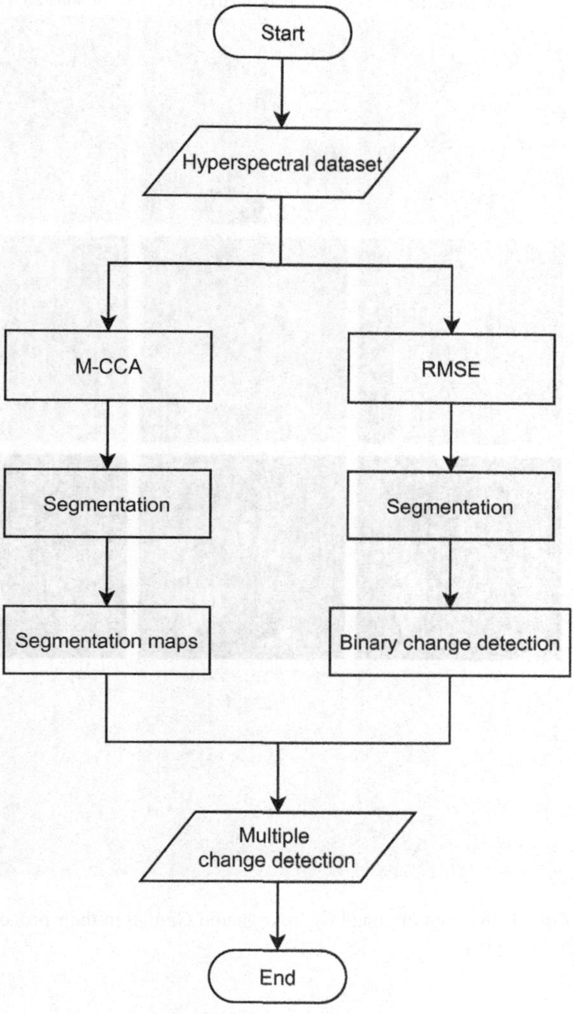

Fig. 9 Workflow of Yokoya and Gamisi algorithm presented in [19]

Yokoya and Gamisi studied four Hyperion images of the nuclear power plant area, distributed spatially between 29th April 2012, and 2nd May 2015. The procedure used (Fig. 9) begins with the reduction of dimension, where uses RMSE between temporal images by calculating the canonical variables by multiset canonical correlation analysis (M-CCA), thereby obtaining the degree of changes in the images. Subsequently, it performs the segmentation of the set of M-CCA images with its RMSE maps based on FODPSO. The result is segmentation maps and binary change detection maps, respectively. The whole process result, are the multiple changes detection maps by generating a difference between the segmentation of each of the four images and integrating to this the detection of binary changes.

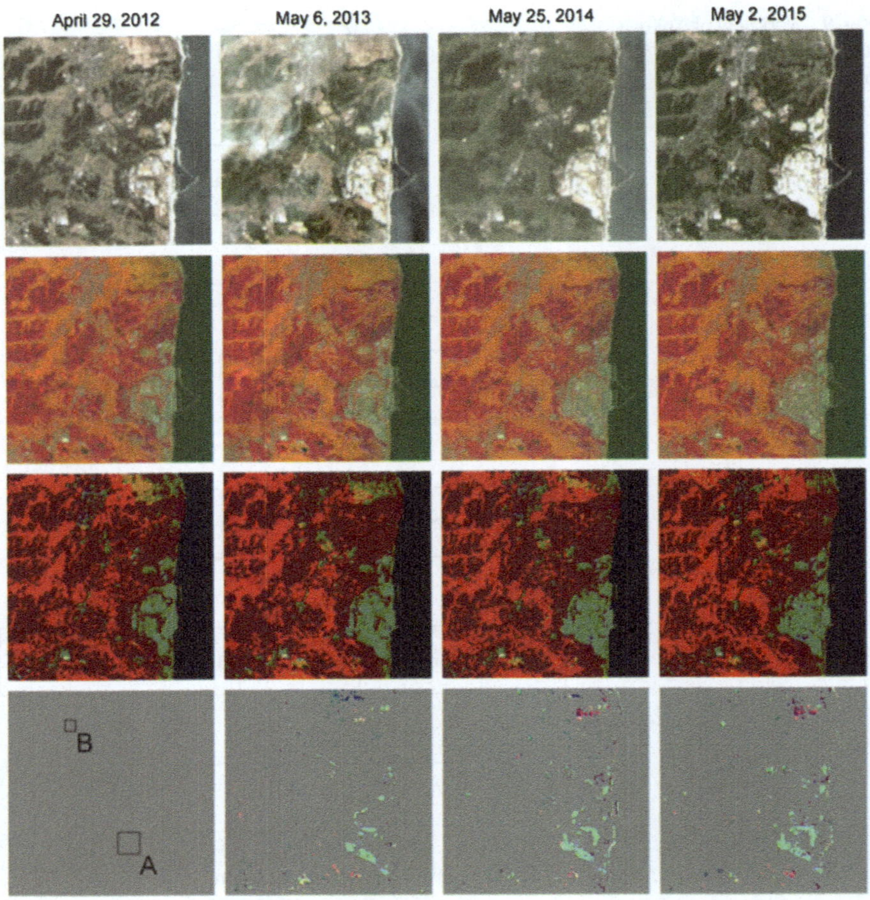

Fig. 10 Results obtained by Yokoya and Gamisi in their processing of times series hyperspectral images [19]

Figure 10 shows the sequence of the maps obtained during the process and the final results that Yokoya and Gamisi obtained by combining unsupervised processing of hyperspectral images temporarily spaced by an evolutionary algorithm. The changes detected by the algorithm show areas that went from having just soil to be covered with vegetation, as well as others that from having vegetation, showed the lack of it due to the construction of human infrastructure.

Jia et al. presented in 2019 a threshold separation method based on the Emperor Penguin Optimizer (EPO) algorithm. The proposed algorithm uses Masi entropy as an objective function while introducing three different strategies to complement the EPO algorithm, which results in the presentation of the Multi-Strategy Emperor Penguin Optimizer (MSEPO) [10]. The algorithm of Jia et al. is capable of detecting

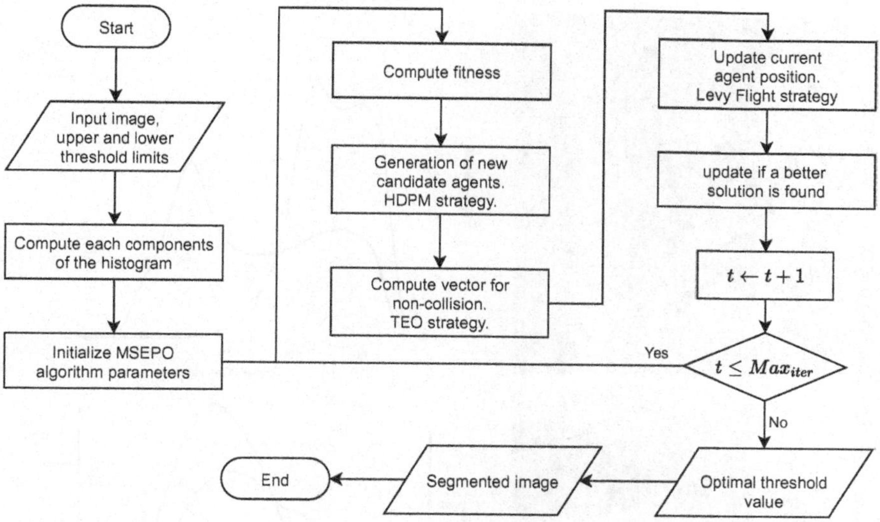

Fig. 11 Flow diagrama of the MSEPO presented by Jia et al. in [10]

the multi-threshold content from satellite images to generate the color histogram of the image itself.

The proposal of Jia et al. is based on the EPO algorithm that bases its operation on the behavior of emperor penguins to maintain an adequate temperature collectively. The algorithm, in its search for the optimal group temperature, takes into account the distance between individuals. Based on this information, the algorithm makes movements to obtain new positions of the individuals that allow them to regulate the group temperature and thus obtain the most appropriate temperature of the group. Figure 11 shows the flowchart of the MSEPO algorithm presented by Jia et al.

The process begins by starting a random population of penguins, with a certain number of individuals and their respective random positions, distributed by the search space. Similarly, the number of iterations of the algorithm is defined, which in the paper presented, Jia et al. adjust in 500 repetitions. Then, an evaluation of the initial situation is carried out; that is, the fitness of each search agent. The procedure continues with the generation of a new search agent by inserting the first reinforcement strategy into the algorithm. They use a polynomial mutation for this. Specifically, they use a version called Highly Disruptive Polynomial Mutation (HDPM), which enriches the diversity of the population and gives the possibility to explore the entire search space. Jia et al. use the Thermal Exchange Operator (TEO) to calculate the vectors that will prevent collision or overlap of individuals. TEO, based on Newton's cooling law, is the second strategy incorporated into the original EPO algorithm. With the inclusion of this algorithm to the proposed procedure, the positions improve as each object will tend to the temperature of the surrounding environment, that is, the objective function. After this calculation, they proceed to calculate the distances between the individuals and update the positions of the search agents. Here they

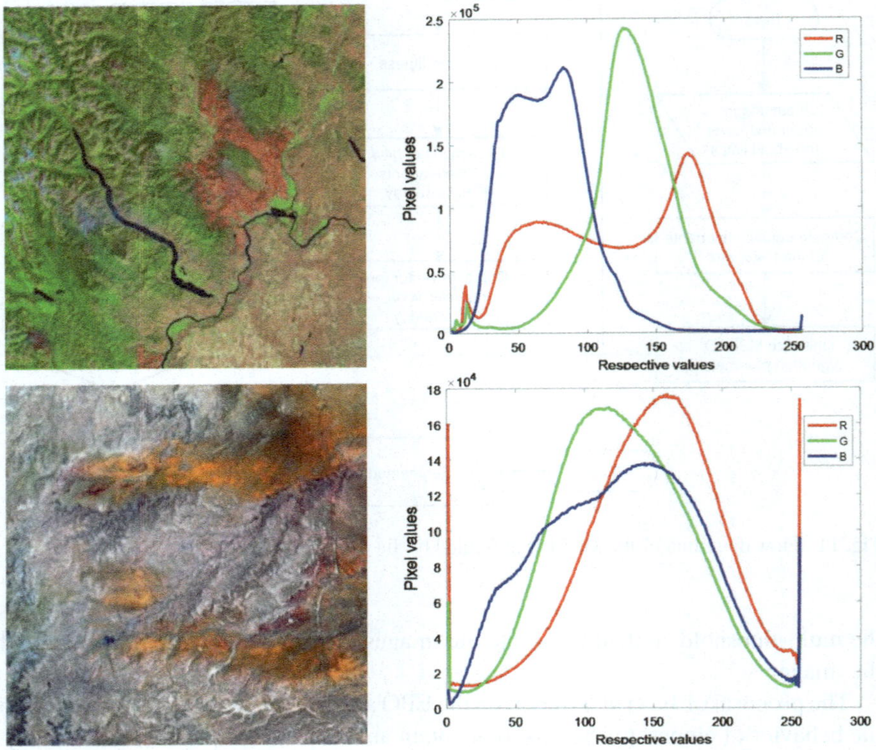

Fig. 12 Color images (left) and their respectives RGB histograms (right) acquired using MSEPO in [10]

introduce the third strategy, which is the implementation of the Levy Flight algorithm in order to improve the search space's exploration capacity. The Levy Flight algorithm makes short walks that generate an exhaustive exploration of an area, and then a significant jump randomly, which generates a new small walk in a different area. At the end of the three strategies implemented, it only remains to calculate the fitness of the search agents in the current position and update each time they find a better solution to the problem initially raised. Like they are using the process to segment an RGB image, they perform the threshold selection three times, once per channel. In this way, the objective function, based on the calculation of Masi entropy, can determine the optimal threshold values for color image segmentation.

With the procedure presented, Jia et al. managed to optimize the process of searching multiple thresholds to perform segmentation of remote sensing images without resorting to gray-scale transformation. With this, they increase the accuracy of segmentation while reducing the computational time of the process. Figure 12 shows two images used as proof for the proposed algorithm. Each image shows its respective RGB histogram. In Fig. 13, can observe the comparison of results by segmentation of images showed in Fig. 12 with different methods. The far left column shows a mask

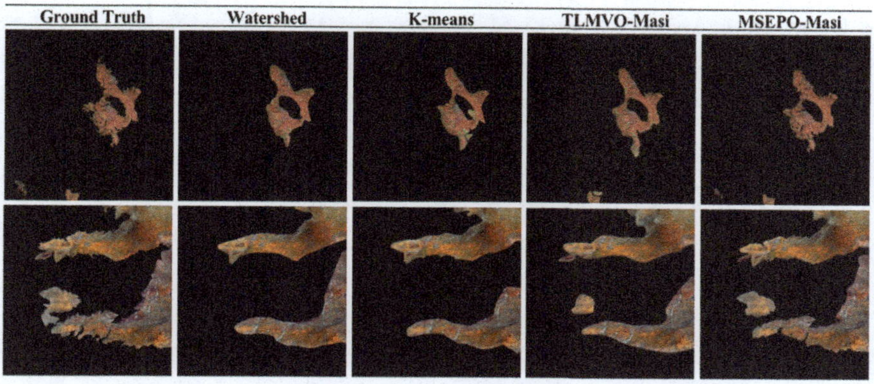

Fig. 13 Comparative results from segmenting the color images in Fig. 12. Last column show the results of the proposed algorithm by Jia et al. Image from [10]

delineated manually, where the authors detail the areas that the algorithms should locate automatically. The following four columns show the results obtained by segmenting the images with the Watershed, K-means, TLMVO-Masi, and MSEPO-Masi algorithms, respectively. It is observed that the results obtained by the algorithm of Jia et al. are the closest to the mask delineated manually.

With the obtained results, Jia et al. show the effectiveness of combining some algorithms, and using the metaheuristics algorithms to perform information extraction from remote sensing images.

In urban settings, there is a frequent need to have a census of the forest population. Many of the studies are carried out quantitatively to know the percentage of forest population present, on other occasions, they require a study that helps classify forest species to know the amounts of each species present in a particular urban area. This mapping of urban tree species is usually tricky due to the precision necessary for species discrimination and the spatial variation that each type of tree presents.

The urban forest census is usually carried out through field sampling and employing the manual interpretation of aerial photographs, which represents a high temporal cost and high staff requirements.

In 2017, Liu et al. presented an urban forest classification work. In their work, they use a combination of remote sensing data such as hyperspectral images and LiDAR. The extraction of data of different natures from the same area helped to obtain a more precise classification of the different tree varieties present in the study area through the Random Forest classifier.

With the use of LiDAR and hyperspectral data, Liu et al. were able to obtain spatially explicit information for the classification of 15 urban tree species, combining structural information from the crown provided by LiDAR and the spectral indices of each species provided by the hyperspectral images.

The process used by Liu et al. (Fig. 14), begins with obtaining remote sensing data in the study area in an independent way in a very short time-space. In both cases,

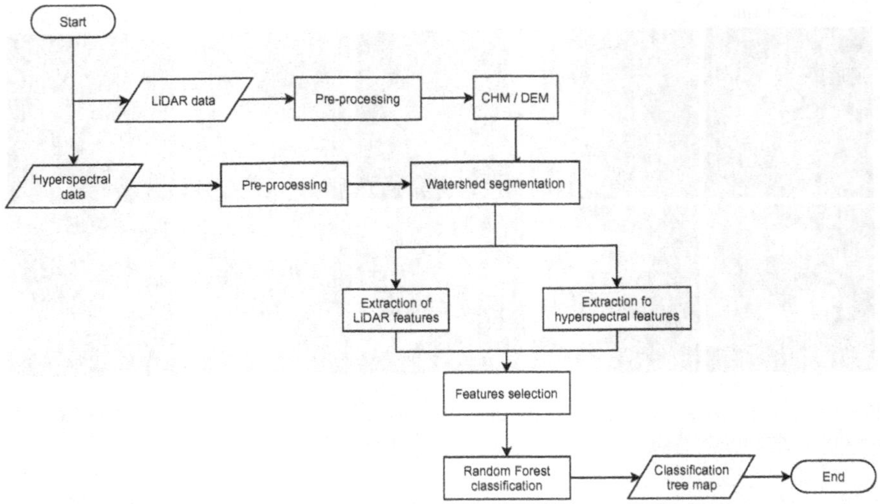

Fig. 14 Flow diagram of the Liu et al. urban tree classification process from [11]

they applied pre-segmentation processing, which consists of the filtering of non-forest data, georeferenced positioning, heights mapping, among others, which will facilitate the work of discrimination and subsequent forest classification. For segmentation, they used the watershed segmentation algorithm to obtain the shapes of tree crowns. Subsequently, it uses the obtained segments to extract the precise information from the LiDAR data cloud and the hyperspectral dataset included in each segment. Point cloud LiDAR provided information about trees structural variables, such as the shape of the crown, and the distribution of heights. The hyperspectral images, provided, the first two main components by PCA and standard deviation of each tree crown segment, as well as reflectance values of leaves for each different tree. For the classification of species, the Random Forest algorithm was used, which has proven effective in individual classification of tree species. With this work, Liu et al. demonstrate the effectiveness of the combination of specific data obtained through different remote sensing methods. Similarly, it obtained a high percentage of effectiveness in the classification of species with the use of artificial learning techniques (Fig. 15). With all this, it demonstrates that urban forestry classification tasks can be performed in an automated way using computational methods. With this, they have the possibility of performing these tasks in shorter time-spaces, obtaining more data on forest changes in cities.

A similar classification work was carried out by Maschler et al. [13] in the Wienerwald Biosphere Reserve in Austria in 2018. In their work, Maschler et al. classified the population of 13 different trees on the reserve with Random Forest. The data were obtained through hyperspectral images and LiDAR datasets in the area and subjected to segmentation and classification for the recognition of forest populations.

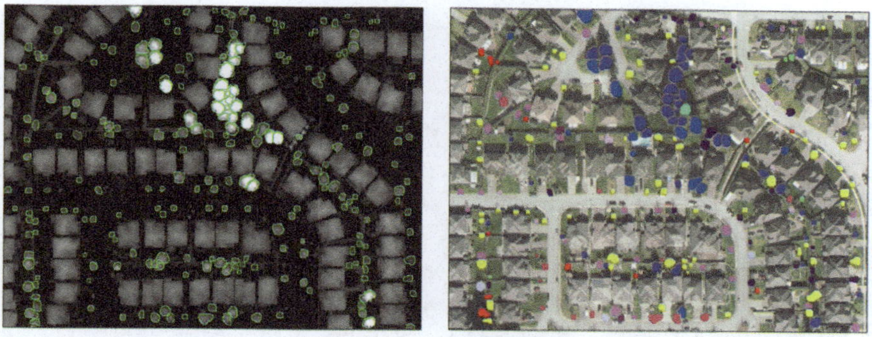

Fig. 15 Segmented area that shows the outline of individual trees (left) and the same area, showing the classification results of the process (right). Images from [11]

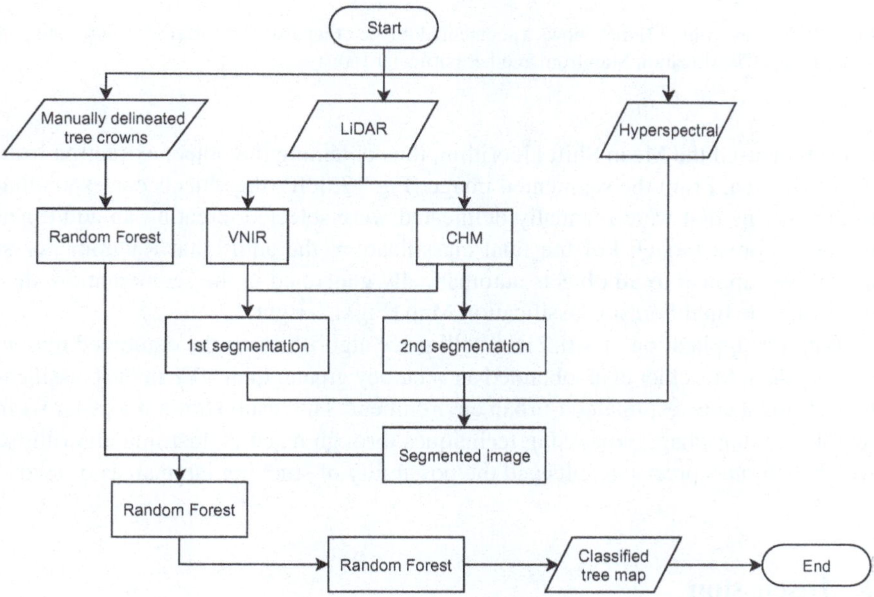

Fig. 16 Flow diagram that shows Maschler et al. workflow process in [13]

Figure 16 shows the workflow of the process where Maschler et al. began by isolating the areas corresponding to the forested area in the hyperspectral image. To do this, they used sub masks based in CHM and NDVI to locate by superposition just the forest areas, removing shadows, roads, grasslands, and other elements without interest from the study area.

A Random Forest classifier was trained using manually delineated references, using 202 predictor variables. Once they trained the classifier, Maschler et al. performed a double segmentation for the location of the trees within the forest area. Each of the segmentation stages located certain types of trees. For both segmenta-

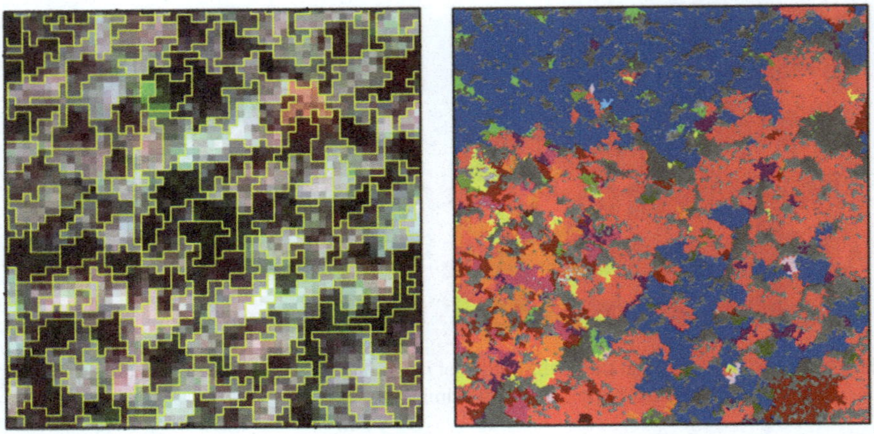

Fig. 17 Images from [13] that shows a segmented image of a particular forest area (left) and part of the Forest Classification Map from another particular forest zone (right)

tions, they used the Mean Shift algorithm, thus obtaining the object extraction from the study area. From the segmented image (Fig. 17, left), the objects corresponding to the crowns that were manually delineated were selected, creating an additional Random Forest model. For the final classification, the additional Random Forest model was applied to all objects automatically generated in the segmentation, thus obtaining the final Forest Classification Map (Fig. 17, right).

With the application of artificial intelligence algorithms on the combined remote sensing data, Maschler et al. obtained an accuracy greater than 90% in the classification of forest species, in a non-urban environment. The result shows the progress in remote sensing image processing techniques through machine learning algorithms, which generates precise results and the possibility of studying large areas of land.

4 Discussion

This chapter shows some image processing techniques applied to extract relevant data in remote sensing images. It is observed from corner detection to object classification, through unmixing pixels and segmentation. It shows that the study of remote sensing data is carried out in a wide variety of image formats, preferring to focus as much as possible in the treatment of hyperspectral data.

The detection of image characteristics focused on the search for the corner point. Deng et al. proposes a new gray tracking method, obtaining more objective results, and with immunity to noise without prior image treatments. Subsequently, it uses an adaptive threshold discrimination algorithm to establish the upper and lower thresholds, proving to be suitable for remote sensing images processing tasks such as image registration or image fusion. The combined use of the Harris corner detection algorithm and SURF algorithm to locate points of interest and feature detection presented by Changjie and Hua results in a fast and robust method. In this case, the precision results for image matching also reside in the possibility of detecting false points of interest and their elimination by using the RANSAC algorithm. Zhou et al. improves the results of the Harris corner detection algorithm application to locate points of interest, implementing the multilevel wavelet decomposition algorithm in the image before searching interest points. Using the low and high image frequencies detect and suppress noises or subtle variations in the original image that can give false corner points.

The use of intelligent algorithms for the extraction of endmembers is not new. Su et al. proposes three improved algorithms by applying a distance factor in the objective function. He showed that the use of these algorithms improves the results compared with traditional methods. Niranjani and Vani presented in 2018 two algorithms to estimate the non-linearity of the mixture of pixels in multispectral and hyperspectral images. Depending on the number of endmembers, the algorithm provides better results for the location of classes that Niranjani and Vani determined in their study.

Yokoya and Gamisi makes combined use of a time series of hyperspectral images and statistical analysis algorithms to locate changes in the land cover after the Fukushima Daiichi nuclear accident. The result ended in an algorithm capable of locating the changes that are generated in a given geographical area when it is studied using temporarily spaced images. Jia et al. make use of artificial intelligence to determine the optimal threshold in a color image. To improve the EPO algorithm utilizes Masi Entropy and the inclusion of another three strategies to perform their MSEPO algorithm to a better exploration, and higher accuracy. In the work that Liu et al. presented, they apply machine learning techniques in remote sensing images to perform a classification of trees in urban areas. The process combines LiDAR data and hyperspectral images that together provide complementary information that facilitates the work of classification and discrimination of objects. Image processing uses learning algorithms, which offer very accurate results when classifying urban tree species using the Random Forest algorithm. For its part, Maschler et al. carries out a series of procedures for the classification of trees through the segmentation of hyperspectral images and LiDAR data in forested areas. We can observe that the use of machine learning algorithms applied in remote sensing data, not only works in areas with low spatial population but can provide accurate results in the classification of trees in areas where the forest population has a higher concentration of individuals.

As can be seen, the detection of characteristic points in the image processing presents algorithms ranging from the use of the adaptive threshold discrimination algorithm to the use of the multilevel waveform decomposition algorithm. These advances facilitate the work of image stitching that needs as a reference to the characteristic points of an image. Segmentation and extraction of end members have also benefited from the advancement of smart algorithms. The use of multiple data sources provides complementary information in the search for objects that improve accuracy in the final classification of classes, which provides more accurate information on how our environment evolves.

5 Conclusions

Remote sensing, which is an increasingly important area, finds in the digital image processing a valuable tool to achieve its objectives. Scientists developed different methods for the extraction of reference points in the images. These attempt to discriminate noise, light intensity changes, and even detect image rotation that partially contains the same objective. The results revealed that it is possible to improve the subsequent image processing that bases its operation on the reference points location. The use of hyperspectral images in remote sensing eases to obtain valuable information for the tasks of monitoring and classifying objects. This information is difficult to reach with other types of images because hyperspectral images have narrow frequency bands that contain varied and very accurate environmental information captured in it. When accompanied by data sources, such as LIDAR, increases the information available to improve a better image segmentation. The combination of learning algorithms and optimization algorithms not only facilitates the location of optimal threshold values ot the classification of objects but also improves the accuracy of the results. All this together provides to remote sensing area, the opportunity to perform tasks that were previously difficult to achieve due to the limited information available in the images, or to the processing of images less effective than what is available today.

References

1. M. Ahmad, A. Khan, A.M. Khan, M. Mazzara, S. Distefano, A. Sohaib, O. Nibouche, Spatial prior fuzziness pool-based interactive classification of hyperspectral images. Remote Sens. **11**(9) (2019). https://doi.org/10.3390/rs11091136, http://www.mdpi.com/2072-4292/11/9/1136
2. P. Bangert, *Optimization for Industrial Problems* (Springer, Berlin, 2012). https://www.amazon.com/Optimization-Industrial-Problems-Patrick-Bangert/dp/3642249736?SubscriptionId=AKIAIOBINVZYXZQZ2U3A&tag=chimbori05-20&linkCode=xm2&camp=2025&creative=165953&creativeASIN=3642249736
3. B. Bhatta, *Research Methods in Remote Sensing* (Springer, Berlin, 2013)
4. W. Changjie, N. Hua, Algorithm of remote sensing image matching based on corner-point, in *2017 International Workshop on Remote Sensing with Intelligent Processing (RSIP)*, pp. 1–4 (2017). https://doi.org/10.1109/RSIP.2017.7958803
5. X. Deng, Y. Huang, S. Feng, C. Wang, Adaptive threshold discriminating algorithm for remote sensing image corner detection, in *2010 3rd International Congress on Image and Signal Processing*, vol. 2, pp. 880–883 (2010). https://doi.org/10.1109/CISP.2010.5646881
6. R. Dian, S. Li, L. Fang, Q. Wei, Multispectral and hyperspectral image fusion with spatial-spectral sparse representation. Inf. Fusion **49**, 262–270 (2019). https://doi.org/10.1016/j.inffus.2018.11.012, http://www.sciencedirect.com/science/article/pii/S1566253517308035
7. R.O. Dubayah, J.B. Drake, Lidar remote sensing for forestry. J. For. **98**(6), 44–46 (2000). https://doi.org/10.1093/jof/98.6.44
8. F.E. Fassnacht, H. Latifi, K. Stereńczak, A. Modzelewska, M. Lefsky, L.T. Waser, C. Straub, A. Ghosh, Review of studies on tree species classification from remotely sensed data. Remote Sens. Environ. **186**, 64–87 (2016). https://doi.org/10.1016/j.rse.2016.08.013, http://www.sciencedirect.com/science/article/pii/S0034425716303169

9. P. Ghamisi, M.S. Couceiro, J.A. Benediktsson, N.M. Ferreira, An efficient method for segmentation of images based on fractional calculus and natural selection (2012). https://doi.org/10.1016/j.eswa.2012.04.078, http://www.sciencedirect.com/science/article/pii/S0957417412006756
10. H. Jia, K. Sun, W. Song, X. Peng, C. Lang, Y. Li, Multi-strategy emperor penguin optimizer for RGB histogram-based color satellite image segmentation using Masi entropy. IEEE Access **7**, 134448–134474 (2019). https://doi.org/10.1109/ACCESS.2019.2942064
11. L. Liu, N.C. Coops, N.W. Aven, Y. Pang, Mapping urban tree species using integrated airborne hyperspectral and lidar remote sensing data. Remote Sens. Environ. **200**, 170–182 (2017). https://doi.org/10.1016/j.rse.2017.08.010, http://www.sciencedirect.com/science/article/pii/S0034425717303620
12. J.M. Lloyd, *Thermal Imaging Systems. Optical Physics and Engineering* (Springer, Berlin, 1975). https://doi.org/10.1007/978-1-4899-1182-7
13. J. Maschler, C. Atzberger, M. Immitzer, Individual tree crown segmentation and classification of 13 tree species using airborne hyperspectral data. Remote Sens. **10**(8) (2018). https://doi.org/10.3390/rs10081218, http://www.mdpi.com/2072-4292/10/8/1218
14. K. Niranjani, K. Vani, Unsupervised nonlinear spectral unmixing of satellite images using the modified bilinear model. J. Indian Soc. Remote Sens. **47**(4), 573–584 (2018). https://doi.org/10.1007/s12524-018-0907-7
15. M. Pal, Random forest classifier for remote sensing classification. Int. J. Remote Sens. **26**(1), 217–222 (2005). https://doi.org/10.1080/01431160412331269698
16. H. Rao, X. Shi, A.K. Rodrigue, J. Feng, Y. Xia, M. Elhoseny, X. Yuan, L. Gu, Feature selection based on artificial bee colony and gradient boosting decision tree. Appl. Soft Comput. **74**, 634–642 (2019). https://doi.org/10.1016/j.asoc.2018.10.036, http://www.sciencedirect.com/science/article/pii/S1568494618305933
17. J.A. Richards, *Remote Sensing Digital Image Analysis*, 5th edn. (Springer, Berlin, 2013) https://doi.org/10.1007/978-3-642-30062-2, https://www.springer.com/gp/book/9783642300615
18. Y. Tarabalka, J. Chanussot, J. Benediktsson, Segmentation and classification of hyperspectral images using watershed transformation. Pattern Recogn. **43**(7), 2367–2379 (2010). https://doi.org/10.1016/j.patcog.2010.01.016
19. N. Yokoya, P. Ghamisi, Land-cover monitoring using time-series hyperspectral data via fractional-order Darwinian particle swarm optimization segmentation, in *2016 8th Workshop on Hyperspectral Image and Signal Processing: Evolution in Remote Sensing (WHISPERS)*, pp. 1–5 (2016). https://doi.org/10.1109/WHISPERS.2016.8071761
20. B. Zhou, X. Niu, X. Liu, X. Yang, Multilevel wavelet decomposition based Harris corner detection algorithm for remote-sensing image. DEStech Trans. Comput. Sci. Eng. (2018). https://doi.org/10.12783/dtcse/cmsam2018/26574

Hybrid Grey-Wolf Optimizer Based Fractional Order Optimal Filtering for Texture Aware Quality Enhancement for Remotely Sensed Images

Himanshu Singh, Anil Kumar and L. K. Balyan

Abstract In this chapter, a texture-dependent optimal fractional-order adaptive filtering is proposed for quality improvement of the remotely sensed dark satellite images. Each image is usually composed of diverse variations in texture. To identify and address the texture-based variation in any image, it is highly desired to identify the different kinds of texture constituents present in the image. This objective can be easily attained by texture-based segmentation. Texture based segmentation is usually performed by using the spatial information content present in the spatial texture map of the image under consideration. Spatial entropy based texture map is computed in this work. Later, the grouping the varying textural behavior in multiple classes is done. It makes easy to process the different textural regions separately. In this manner, various classes of relative (or normalized) texture are tried to process individually. For this purpose, optimal fractional-orders are required to be computed for each class of relative texture present in the image. A dedicated optimization-based fractional-order filtering framework has been drafted for fulfilment of the objective. To impart a high-level meta-heuristic intelligence for this optimal framework, in this chapter, the collective excellence of two diversely-behavioral approaches is obeyed in a collective mode. In this context, a hybrid intelligence of Grey-Wolf Optimizer (GWO) achieved by using Cuckoo Search Algorithm (CSA) for applying optimal fractional. As a whole a fusion framework is presented by associating all individual interim channels. A rigorous comparative experimentation is performed and visual as well as numerical analyses are presented in this chapter. The excellence of the proposed approach is underlined when compared w.r.t. the state of the art image enhancement approaches.

Keywords Texture · Texture map · Fractional-order calculus · Adaptive-filtering · Memtic-intelligence · Cuckoo search algorithm · Grey-Wolf optimization · Image enhancement

---

H. Singh (✉) · A. Kumar · L. K. Balyan
Indian Institute of Information Technology Design and Manufacturing, Jabalpur, India
e-mail: himanshu.iiitj@gmail.com

D. Oliva and S. Hinojosa (eds.), *Applications of Hybrid Metaheuristic Algorithms for Image Processing*, Studies in Computational Intelligence 890,
https://doi.org/10.1007/978-3-030-40977-7_3

1 Introduction

Remotely sensed multispectral data which is acquired in the form of 3-D digital imagery is required to get pre-processed for further processing. The image data fusion is very conventional theory. By the virtue of it, in this digital era, usually multi-spectral visual information is represented mostly in the form of fused images. The reason behind it is the availability of personal digital assistants in each individual's hand. Digital imagery has completely revolutionized the way of analysis and investigation in this technological era. Big-data analysis and the advances in the computational powers have increased the dependency over the images for information processing up to a great extent. Whenever it is desired to harvest the information through an image, it is required to pre-process the acquired image. Due to poor-illumination conditions and inadequate environmental behavior, image acquisition suffers a lot, especially in the case of remotely sensed images. Also, due to the long capturing distances and unbalanced natural illumination, mostly these images are acquired as dark images. Pre-processed quality enhancement is highly desirable in such cases. In past few years, computer-vision imagery is getting developed to attaining the intelligence equivalent to the human vision imagery. In this context, mostly the data or scene analysis has been usually performed by intensity based segmentation. But it seems very incompetent to analyze a multi-spectral data on the basis of grouped intensity regions. Corresponding region-wise labelling, is a vital step in computer-vision based senor data analysis. For further intelligence, in the current scenario, it is necessary to employ the texture-based segmentation of the acquired visual-sensor based-data. For this purpose, along with intensity-based image restoration (and/or quality improvement) for digital information harvesting is required. Optimal image quality enhancement by considering texture-aware processing is the prime contribution in this chapter. The texture-aware adaptive behavior is imparted by imitating the intelligence of 2-D fractional-order adaptive filtering. For attaining the next-level intelligence; it is also become necessary to impart the texture-based segmentation and corresponding object-labelling for further imparting of intelligence. With the above-mentioned motivation, this work is a novel attempt of fractional-order calculus based intelligence for texture-based quality improvement for visual data/images. For proper data visualization, generally various other kinds of data is framed/fused or rescaled as RGB color data, as human eyes. In addition, general RGB based digital devices are compatible for such kind of data analysis which is highly desired for further level of information pre-processing. Usually any image comprises of a variety of textural content along with other constituents of the image. Along with the intensity variation and color based discrimination in any scene or image, the textural variation also constitutes the core information content in the image. Especially when it comes to the remotely sensed images, due to the image acquisition from a very far distance, a large variety of texture comes in the same frame. In addition, usually a very minute spatial neighborhood stands for the visual impression of a very large geographical area. With this motivation, the textural discrimination and the correspondingly attained region-wise texture based quality improvement is proposed in this chapter.

It is evident from the literature that many image enhancement algorithms have been developed till date, but most of them are proposed by considering intensity-map only. With this motivation, the proposed image enhancement framework is drafted by employing texture-map based optimal texture-dependent image enhancement for remotely sensed images. Fractional-order Calculus (FoC) and its beauty of desired non-integer ordered adaptive filtering for image restoration and quality enhancement is latently too valuable to be casually dismissed. This intelligence has not been completely explored much by the researchers till date because for different kind of textural constituent of the image, it requires different order of fractional calculus based adaptive filtering. To decide the appropriate order, an attempt is made in this paper in an introductory manner. The relative textural behavior of the image is grouped in four different classes. Entropy-based textural-segmentation is employed for segregation of the image in its constituting sub-images. This segregation of the image content is done in a statistically adaptive manner. To obtain three sub-groups, histogram of the texture-map is divided into four sub-sections. Accordingly image can also be divided into four regions. For the purpose of texture based isolation of the images regions, histogram of the entropy-based texture-map is derived. Later, in the parallel manner, each part of the image is individually processed for required enhancement by employing texture dependent order based fractional-order sharpening (through a kind of unsharp masking). In the contextual experimentation related to this manuscript, it is found that it will be more effective if the value of the fractional order should be derived using the optimal meta-heuristic intelligence because this problem behaves as an NP-hard problem. It has been done by identifying the intensity distribution of the image. After employing the proposed optimal fractional-order texture dependent image sharpening independently for all textural regions, these regions are clubbed together.

In the beginning of digital image processing, researchers had suggested various manner of histogram based approaches like histogram equalization, histogram matching, histogram sub-equalization, etc. [1]. Afterwards, various other variants of the sub-histograms based processing were also proposed. Various experiments in the state-of-the-art literature, indicated towards the gaps and limitations, as they seem unable to preserve the local spatial features of the images. Authors in [2] have used fuzzy inspired smoothening for histogram, followed by peak-based histogram subdivision, also known as brightness preserving dynamic fuzzy HE (BPDFHE), but excellence of this approach is limited only for the images having significant peaks in the histograms. Cosine transformed dynamic fuzzy HE based variants have been also proposed for better performance [3, 4]. Along with it, median-mean dependent sub-image-clipped HE (MMSICHE) [5] was also proposed, where, median count based successive bisecting of sub-histograms followed by their successive sub-equalization. Also, an exposure-based sub-image HE (ESIHE) [6] was introduced by same authors, including the exposure calculation, so that on the basis of it, histogram sub-division was imparted, followed by histogram sub-equalization. Although, these approaches work well for enhancement of balanced illumination images, but if histogram is not balanced, these approaches are unable to impart quality improvement because of pseudo-threshold calculation. Also, for textural and dark images, performance of these approaches is not satisfactory. Adaptive gamma correction with weighting distribution (AGCWD) [7] was also proposed for imparting contrast enhancement evaluating a gamma value-set. These values in the gamma-value set are evaluated by discrete cumulative distribution, which is calculated from the histogram of the input image. Although, this approach fascinates many researchers due to its adaptive nature and simplicity, but frequently leads to the saturated patches in the enhanced image, due to mapping of some already bright pixels to the saturated bright intensity level. Gamma corrected image enhancement method and its next level variants [8–14] are also proposed in the recent past years. In Fu et al. [15], sigmoid mapping through cosine transformed regularized-HE was also proposed, but in this approach, scaling-factor calculation was not so adaptive and hence, leads to lack in robustness. Later on, the averaging histogram equalization (AVHEQ) [16] along with the proposed frameworks like HE based optimal profile compression (HEOPC) [17] and HE with maximum intensity coverage (HEMIC) Wong et al. [18] have been also proposed for quality improvement of images. These proposed frameworks were usually targeted towards gathering more and more intensity levels in the permissible range. This kind of redistribution and reallocation is found somehow inefficient, because it brings smoothening kind of ill-artifacts along with the less attention towards the textural content of the images. One more challenge which mostly remains is the unbalanced exposure and related issues. Later, in the same context, intensity and edge based adaptive unsharp masking filter (IEAUMF) has been proposed. In IEAUMF [19], quality enhancement is suggested along with augmented image sharpening approach. HE based textural regions based enhancement approaches are also proposed [20]. In this method, histogram construction using textured grey levels only and later, this texture-based histogram is only utilized for further processing. In contrast with few advantages, this approach is incapable to give significant enhancement for smooth regions. In this chapter, a newly framed optimal fusion framework for quality image enhancement has been discussed. Closed form methods seem incapable to identify the extent of artifacts which has corrupted the scene because of diverse behavior of the images. Due to this reason, a closed form approach is not so eligible for imparting on-demand adaptive quality improvement. In this manner, this issue can be identified as highly non-linear and NP hard problem. For solving such kind of problems, optimization algorithms have played a very significant and vital role [21–24]. Initially, trivial suggestions of the evolutionary and population-based optimization approaches have been adopted for imparting general image enhancement. The proposed approach is region-wise or patch-wise texture-adaptive fractional-order filtering based in nature. This approach of deciding the appropriate texture based fractional-order out of the feasible range of infinite orders, behaves like a hard-core NP-hard problem. Efficient hybridizations for the pre-existing optimization approaches to impart meta-heuristic swarm intelligence in a cognitive manner along with their memetic inclusions are much fruitful to achieve the next-level intelligence, which is highly desirable to attain for the texture as well as intensity based image enhancement. In the conventionally available literature, usually contrast enhancement has been discussed in the name of quality enhancement, whereas, other related issues like texture based quality improvement should also be addressed for overall image quality enhancement. In this paper a novel framework has been introduced for on-demand textural improvement of the image along with adaptive contrast enhancement. For acquiring the meta-heuristic optimal intelligence, in this chapter, the Cuckoo Search Algorithm based hybridized version of Grey-Wolf optimizer is employed. Alongside the hybrid meta-heuristic intelligence, fractional-order calculus based adaptive filtering is the key inclusion in this chapter. The texture-dependent fractional order unsharp masking is applied for overall image quality enhancement. In this context, unsharp masking is one of the most successful ways when it is associated with fractional-order adaptive filtering along with, the associated optimal intelligence. Rest part of this chapter has been planned accordingly as follows: Sect. 2 deals up with the Cuckoo Search Algorithm based Hybridization of Grey Wolf Optimization, followed by the Fractional-order Adaptive Sharpening in Sect. 3. Section 4 is focused over the proposed Texture Dependent Optimal Fractional-order Adaptive Filtering based Augmented Framework. Later, experimental results and discussion are presented in Sect. 5. Finally, conclusions are drawn in Sect. 6.

2 Cuckoo Search Algorithm Based Hybridization of Grey Wolf Optimization

Due to various reasons in last decade, conventional and deterministic techniques are superseded by the stochastic approaches. One such reason is NP hard problems and ill-defined derivation is another for which there is no deterministic solution whereas stochastic approaches are able to find near optimal solution. The reason for this is that the stochastic approaches consider problem of optimization as a black-box, hence applying same method for various problems of optimization without any knowledge of its mathematical formulation. Such approaches can even mimic natural intelligence and one such seminal method is genetic-algorithm (GA) mimicking the evolution process. Mutation, recombination and selection are equipped into GA. Other such ones are differential-evolution (DE) and evolution-strategy (ES). After proposal of these nature inspired methods, swarm intelligence methods came into existence. One such approach is the ant colony optimization (ACO) that mimics ant's intelligent way of identifying shortest path to food source from the nest. Other such approaches are artificial bee colony (ABC), particle swarm optimization (PSO) and cuckoo search optimization (CSO) [25]. Grey wolf optimizer (GWO) is one very well regarded among them and it intelligently simulates the social hierarchy for hunting of grey-wolves. GWO [26] follows grey-wolves' hunting and leadership styles as apex predators. Firstly, the grey-wolf population is initialized and then the fitness of each of the grey-wolf is evaluated. Usually population variants exist everywhere in the nature. Similar to this, various kinds of members lie in a pack of wolves at different levels, especially when organized for hunting. Members in the pack of wolves follow a social dominance hierarchy strictly. Also they behave as the top most members of food chain. Broadly they (grey-wolves) are divided into four packs/stages and are termed as α, β, δ, and ω. In this context, α-wolf/wolfs are the most dominant and ω-wolf/wolfs are the least dominant. α-wolves are considered to be the given the top most priority with the responsibility to make decisions on the place of sleeping and way of hunting. β-wolves come next in the priority order and act as sub-ordinates to their higher priority wolves by helping them in the making of proper decision or replacing them as and when required. δ-wolves dominate over ω-wolves and sub-ordinate the α-wolves as well as β-wolves. They are also considered as the babysitters of the whole pack of wolves under consideration. δ-wolves may also be identified/categorized as a pack of scouts, sentinels, elders, hunters, as well as caretakers. They are also considered as scouts that warn in times of danger and watch over the territory, sentinels that guarantee the protection, elders who used to be α-wolves or β-wolves, hunters that help in prey hunt, caretakers that care for wounded, ill and the weak wolves. It is obvious that the dominant one exhibits leadership and superiority. Similarly, the probable solutions are grouped into four different sets respectively. Most probable triad of solutions is termed as α, β and δ respectively while the remaining pool of solutions are grouped as ω-wolves, as ω-wolves have the least priority among all. Hence, the optimal solution will be α, near optimum solution will be β, probable solution will be δ and the rest pool of them can be grouped as ω in the GWO. For proper implementation, this hierarchy is updates itself iteratively. Later on, a mathematical model is used to update the position of solutions. Wolves hunt in packs generally, which signifies their intelligent collaboration for catching the prey. Primarily grey wolves chase the prey in a team and then encircle it by arranging themselves in different positions as required. Three main phases of group hunting of grey wolves are:

1. Track the prey, chase the prey and approach the prey (Basic Exploration Phase).
2. Pursue the prey, encircle the prey, and harass the prey (Next-level Exploration Phase).
3. Immobilize the prey and attack the prey (Exploitation Phase).

Their movement direction is accordingly altered to increase their chance of hunt. This is symbolized mathematically as:

$$\vec{X}(t+1) = \vec{X}_p(t) - \vec{A}.\vec{D} \tag{1}$$

Here, $\vec{X}(t+1)$ is the wolf's next location $\vec{X}_p(t)$ is the wolf's present location, c is the coefficient and d is the distance of prey calculated as:

$$\vec{D} = \left|\vec{C}.\vec{X}_p(t) - \vec{X}(t)\right| \tag{2}$$

On merging the above two equations, combined expression for the next position is:

$$\vec{X}(t+1) = \vec{X}(t) - \vec{A}\,|\vec{C}.\vec{X}_p(t) - \vec{X}(t)| \tag{3}$$

In case of uni-modal test functions, steeper slope of test function towards the global optimum signifies faster convergence of solution. In case of multi-modal test functions, convergence rate is slower when compared to the case of uni-modal test functions because of trapping phenomena of solution with local optima. It can be avoided by utilizing more computational resources. The main controlling parameters ($\vec{A}$ and $\vec{C}$) of GWO are calculated as:

$$\vec{A} = 2\vec{a}.\vec{r}_1 - \vec{a} \tag{4}$$

$$\vec{C} = 2.\vec{r}_2 \tag{5}$$

$$a = 2 - t(2/T) \tag{6}$$

Here, current iteration is represented as t and maximum number of iterations is represented as T. Thus, components of parameter $\vec{a}$ linearly decreased from 2 to zero. r_1 and r_2 stands for the random numbers ranging from zero to unity. Various exploitative and exploratory search patterns can be achieved by tuning this parameter. If the value of A is ≥ 1 or ≤ 1, then exploration behavior is exhibited by GWO and if the value of A is > -1 and $< +1$, then exploitation behavior is exhibited by GWO. This will help in relocation of a solution around another one and also the addition of any number of dimensions becomes possible. These wolves are now located in a hyper sphere form around the prey. Best solution to simulate the prey forms the alpha wolf. Here, x_1, x_2 and x_3 are calculated as [26]:

$$X_\alpha(t+1) = X_\alpha(t) - A_1.D_\alpha(t) = X_\alpha(t) - A_1.|C_1.X_\alpha(t) - X_p(t)|; \tag{7}$$

$$X_\beta(t+1) = X_\beta(t) - A_2.D_\beta(t) = X_\beta(t) - A_2.|C_2.X_\beta(t) - X_p(t)|; \tag{8}$$

$$X_\delta(t+1) = X_\delta(t) - A_3.D_\delta(t) = X_\delta(t) - A_3.|C_3.X_\delta(t) - X_p(t)|; \tag{9}$$

Grey wolves have the ability to recognize the location of prey and encircle them. The hunt is usually guided by the alpha. The beta and delta might also participate in hunting occasionally. However, in an abstract search space we have no idea about the location of the optimum (prey). In order to mathematically simulate the hunting behavior of grey wolves, we suppose that the alpha (best candidate solution) beta and delta have better knowledge about the potential location of prey. Therefore, we save the first three best solutions obtained so far and oblige the other search agents (including the omegas) to update their positions according to the position of the best search agents. For more optimal solution, hybridization of the standard GWO is attained at this step where optimal positions are obtained by embedding the cuckoo search algorithm based optimal intelligence. For this purpose, individually for all three categories of wolfs, most appropriate solution is obtained by employing CSA. Aggressive psychology of the cuckoo bird, and its interesting breeding behavior (more specifically, its brood parasitism) of the cuckoo-bird, leads to a more effective meta-heuristic optimization algorithm by introducing hybrid Grey-Wolf optimizer (HGWO). CSA is highly appreciable for resolving multimodal, multi-objective, and highly non-linear optimization issues deprived of any kind of exhaustive search. Core structure for CSA and its problem solving strategy in its original form has been already elaborated in [25]. In GWO, to move towards the prey, the distance between prey and golf is minimized and changed over time. The step size by which wolf moves, is governed a randomly weighted parameter through a constant which may leads to trapping it into local optima. This problem is solved by cuckoo search algorithm which updates the current position based on the best position so far. CSA optimality more relies on other habitat groups rather than only time. To make it hybrid the best three locations of wolves has been updated in the group by adopting CSA for this purpose. Following the Lévy distributed quasi-random flight; a suitable intelligence has been also introduced, where the succeeding step has to be decided by keeping "current location" and "next-state transition probability" in the mind. This type of step flight pattern is highly compatible and profitable with CSOA behaviour. Simplified analogous behavioural modelling has been done by imposing three rules, as already existed in the relevant literature. Lévy distributed flight is generally for both local as well as global exploration of the corresponding search space. Lévy flight for iterative new solution $X_\alpha(t+1)$ for the cuckoo can be drafted as:

$$X_\alpha(t+1) = X_\alpha(t) + \chi \oplus L\acute{e}vy(\eta), \text{ where, } \partial > 0 \tag{10}$$

$$X_\beta(t+1) = X_\beta(t) + \chi \oplus L\acute{e}vy(\eta), \text{ where,} \partial > 0 \tag{11}$$

$$X_\delta(t+1) = X_\delta(t) + \chi \oplus L\acute{e}vy(\eta), \text{ where,} \partial > 0 \tag{12}$$

Here, entry-wise walk during multiplications can be indicated through product operation $\oplus$. Random exploration follows Lévy distributed (having both first as well as second moment infinite) random step size, as:

$$L\acute{e}vy \sim u = t^{-\lambda}, \; \forall \lambda \in (1, 3], \tag{13}$$

This power law step-flight distributed random walk leads to debut for a few new solutions in the vicinity of best solution (identified so far), and in this manner local search can be speed up. In addition, a generous share of new solutions should be created through far-field randomization, so that the local trapping can be avoided and global exploration can be encouraged. Finally collective contribution is identified as average of these three positions to march towards the center or the centroid of the search-space, as follows:

$$X(t+1) = \frac{X_\alpha(t+1) + X_\beta(t+1) + X_\delta(t+1)}{3} \tag{14}$$

Algorithm 1: Pseudo code of CSA-GWO or HGWO algorithm

```
Begin
Initialize population of grey wolves, A_i (i=1, 2, …, n).
Calculate every search agent's fitness.
Identify the best search agent, X_α.
Identify second best search agent, X_β.
Identify third best search agent, X_δ.
  while (count < maximum iteration number)
     for each search agent
        update current position
     end for
   Calculate every search agent's fitness.
     Update the best search agent, X_α, by employing CSA
     Update second best search agent, X_β, by employing CSA
     Update third best search agent, X_δ, by employing CSA
     Increment the count by one
  end while
Obtain the centroid or average/equidistant location w.r.t. X_α, X_β, X_δ.
Return the best search agent, X.
End
```

Social hierarchy of HGWO enables it to save best solution obtained so far over a number of iterations. Encircling mechanism defines circle shaped neighborhood around solutions extending as hyper sphere to higher dimensions. Candidate solutions are assisted by random parameters to have random radii hyper spheres. Candidate solutions allow hunting method to locate prey's probable position. Adaptive values of the random parameters guarantee the exploitation and the exploration. HGWO can also smoothly transit between the exploitation and the exploration because of the adaptive nature of random parameters, especially which the CSA based adopted intelligence.

3 Fractional-Order Adaptive Image Sharpening

Fractional calculus based approximate derivation of the analogous mask can be derived using standard G-L definition. The vth order differential ($v > 0$) of finitely spanned single-dimensional (1-D) function $f(t)$ (where $f(t) \in [a, t]$, $a \in R$, $t \in R$) can be expressed as [27–31]:

$$ {}_{a}^{GL}D_{x}^{v} f(x) = \lim_{h \to 0} h^{-v} \sum_{m=0}^{n-1} \frac{(-v)(-v+1)\ldots(-v+m-1)}{m!} . f(x - mh); \quad (15) $$

In case of digital images, h must be unity, since minimum in-between distance among the adjacent pixels in all cases is unity. The above definition when applied over a 2-D digital image $t \rightarrow \{x, y\}$, the corresponding partial differentiation w.r.t. x and y, respectively can be expressed as [27–31]:

$$D_x^v f(x, y) = f(x, y) + (-v)f(x-1, y) + \frac{(-v)(-v+1)}{2} f(x-2, y) + \frac{(-v)(-v+1)(-v+2)}{6} f(x-3, y) + \cdots + \frac{\Gamma(n-v)}{\Gamma(-v)\Gamma(n+1)} f(x-n, y), \tag{16}$$

$$D_y^v f(x, y) = f(x, y) + (-v)f(x, y-1) + \frac{(-v)(-v+1)}{2} f(x, y-2) + \frac{(-v)(-v+1)(-v+2)}{6} f(x, y-3) + \cdots + \frac{\Gamma(n-v)}{\Gamma(-v)\Gamma(n+1)} f(x, y-n), \tag{17}$$

By maintenance of the gradient behavior of similar fashion in all the eight directions, and FO mask based on GL definition is created. These directions can be viewed w.r.t. the center pixel based balanced orientation at angles of 0°, 45°, 90°, 135°, 180°, 225°, 270°, 315° and 360°, respectively. By framing of a mask of augmented FO high-pass kind, behavior of sharpening filter can be achieved in an adaptive manner. In order to maintain sum of all the elements at unity, normalization of all elements is done accordingly. Various masks of size 3 × 3, 5 × 5, and 7 × 7 to are also tested but a mask of size 5 × 5 has been settled as a trade-off. In order to extract high frequency content of an image, order of FOD mask acting as 2-D adaptive filter is considered and it in turn relies on the extent of exclusion or inclusion. Edge content of an image, based on adaptive order is to be extracted and then augmented from the input image. Highlighted textural content of an image is obtained by later emphasis on complete image. A symmetric mask is employed by making use of first three coefficients only. Individual convolution of these filters for all rows from left to right and for all columns from top to bottom results in 2-D filtering. Idea is to impart fractional order differentiation based unsharp masking over various textural regions of the image. Augmentation of FO differentiation based high-pass filter over unity mapping leads to the adaptively and individually sharpened regions of the image. Varying the fractional order (from 0 to 1) leads to the sharpening of various kinds of textural regions present in the image. FOD mask of *N by N* size is defined as presented in Eq. 18 using the coefficients present in Table 1.

Table 1 First five coefficients for various fractional-orders

Order	C1	C2	C3	C4	C5
0	1	0	0	0	0
0.1	1	−0.1	−0.045	−0.0285	−0.02066
0.2	1	−0.2	−0.08	−0.048	−0.0336
0.3	1	−0.3	−0.105	−0.0595	−0.04016
0.4	1	−0.4	−0.12	−0.064	−0.0416
0.5	1	−0.5	−0.125	−0.0625	−0.03906
0.6	1	−0.6	−0.12	−0.056	−0.0336
0.7	1	−0.7	−0.105	−0.0455	−0.02616
0.8	1	−0.8	−0.08	−0.032	−0.0176
0.9	1	−0.9	−0.045	−0.0165	−0.00866
1	1	−1	0	0	0

$$H^{v}_{K\times K} \triangleq \begin{pmatrix} C_{(K+1)/2} & 0 & 0 & C_{(K+1)/2} & 0 & 0 & C_{(K+1)/2} \\ 0 & \dots & 0 & \dots & 0 & \dots & 0 \\ 0 & 0 & C_2 & C_2 & C_2 & 0 & 0 \\ C_{(K+1)/2} & \dots & C_2 & 1-\sum\limits_{k=2}^{(K+1)/2} C_k & C_2 & \dots & C_{(K+1)/2} \\ 0 & 0 & C_2 & C_2 & C_2 & 0 & 0 \\ 0 & \dots & 0 & \dots & 0 & \dots & 0 \\ C_{(K+1)/2} & 0 & 0 & C_{(K+1)/2} & 0 & 0 & C_{(K+1)/2} \end{pmatrix}, \tag{18}$$

$$\begin{bmatrix} C_1 \\ C_2 \\ C_3 \\ C_4 \\ C_5 \end{bmatrix} = \begin{bmatrix} 1 \\ -v \\ (-v)(-v+1)/2 \\ (-v)(-v+1)(-v+2)/6 \\ (-v)(-v+1)(-v+2)(-v+3)/24 \end{bmatrix} \tag{19}$$

$${}^{FD}H^{v}_{5\times 5} \triangleq \begin{pmatrix} C_3 & 0 & C_3 & 0 & C_3 \\ 0 & C_2 & C_2 & C_2 & 0 \\ 0 & C_2 & C_1-\sum\limits_{k=2}^{3} C_k & C_2 & C_3 \\ 0 & C_2 & C_2 & C_2 & 0 \\ C_3 & 0 & C_3 & 0 & C_3 \end{pmatrix}; \tag{20}$$

Care should be taken to maintain input image channel matrix size same as convolved product size. Precisely, spectral behavior of these masks helps in identification of its corresponding adaptive fractional order sharpening behavior of the images as illustrated in Figs. 1, 2 and 3.

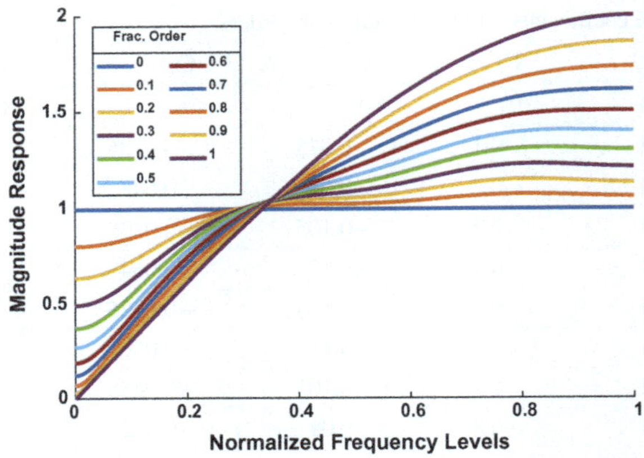

Fig. 1 Magnitude response for 1-D GL fractional differential based sharpening filter for various orders in range of (0,1)

4 Proposed Texture Dependent Optimal Fractional-Order Adaptive Filtering Based Augmented Framework

The first step is the identification of textural variation of the image. For this purpose, discrimination/classification of various textural regions of the image is attained by spatial entropy matrix. Mathematically, a 9-by-9 neighborhood square window is moved across the image for pixel-wise local entropy matrix (LEM) calculation.

$$LEM \triangleq \{e_{mn}\}_{M\times N} \forall e_{mn}(Y) = -\sum_{i\in\Omega}^{9\times 9} p(Y=i).\log_2 p(Y=i), \tag{21}$$

In this manner, *LEM* is evaluated, which is of the same size as that of the gray-scaled equivalent intensity channel. Normalized (rescaled) local entropy matrix yields the spatial texture-map of the image.

$$STM = \frac{LEM - \min(LEM)}{\max(LEM) - \min(LEM)}; \tag{22}$$

Segmentation of this Spatial Texture Map (*STM*) results into the texture-based discrimination of the image. In this draft, on the basis of this S*TM*, four different masks of the same size (also same as that of the input channel), are derived to adaptively isolate the different kinds of textural regions present in the input image. Based on the balanced mean-median division strategy, the textural distribution is segmented into four different regions, as follows which is illustrated in Fig. 4.

$$T_2 = Median\{STM\} \tag{23}$$

$$\begin{pmatrix} -0.045 & 0 & -0.045 & 0 & -0.045 \\ 0 & -0.1 & -0.100 & -0.1 & 0 \\ -0.045 & -0.1 & 2.160 & -0.1 & -0.045 \\ 0 & -0.1 & -0.100 & -0.1 & 0 \\ -0.045 & 0 & -0.045 & 0 & -0.045 \end{pmatrix}$$

(a) Fractional Order = 0.1

$$\begin{pmatrix} -0.08 & 0 & -0.08 & 0 & -0.08 \\ 0 & -0.2 & -0.20 & -0.2 & 0 \\ -0.08 & -0.2 & 3.24 & -0.2 & -0.08 \\ 0 & -0.2 & -0.20 & -0.2 & 0 \\ -0.08 & 0 & -0.08 & 0 & -0.08 \end{pmatrix}$$

(b) Fractional Order = 0.2

$$\begin{pmatrix} -0.105 & 0 & -0.105 & 0 & -0.105 \\ 0 & -0.3 & -0.300 & -0.3 & 0 \\ -0.105 & -0.3 & 4.240 & -0.3 & -0.105 \\ 0 & -0.3 & -0.300 & -0.3 & 0 \\ -0.105 & 0 & -0.105 & 0 & -0.105 \end{pmatrix}$$

(c) Fractional Order = 0.3

$$\begin{pmatrix} -0.12 & 0 & -0.12 & 0 & -0.12 \\ 0 & -0.4 & -0.40 & -0.4 & 0 \\ -0.12 & -0.4 & 5.16 & -0.4 & -0.12 \\ 0 & -0.4 & -0.40 & -0.4 & 0 \\ -0.12 & 0 & -0.12 & 0 & -0.12 \end{pmatrix}$$

(d) Fractional Order = 0.4

$$\begin{pmatrix} -0.125 & 0 & -0.125 & 0 & -0.125 \\ 0 & -0.5 & -0.500 & -0.5 & 0 \\ -0.125 & -0.5 & 6.000 & -0.5 & -0.125 \\ 0 & -0.5 & -0.500 & -0.5 & 0 \\ -0.125 & 0 & -0.125 & 0 & -0.125 \end{pmatrix}$$

(e) Fractional Order = 0.5

$$\begin{pmatrix} -0.12 & 0 & -0.12 & 0 & -0.12 \\ 0 & -0.6 & -0.60 & -0.6 & 0 \\ -0.12 & -0.6 & 6.76 & -0.6 & -0.12 \\ 0 & -0.6 & -0.60 & -0.6 & 0 \\ -0.12 & 0 & -0.12 & 0 & -0.12 \end{pmatrix}$$

(f) Fractional Order = 0.6

$$\begin{pmatrix} -0.105 & 0 & -0.105 & 0 & -0.105 \\ 0 & -0.7 & -0.700 & -0.7 & 0 \\ -0.105 & -0.7 & 7.440 & -0.7 & -0.105 \\ 0 & -0.7 & -0.700 & -0.7 & 0 \\ -0.105 & 0 & -0.105 & 0 & -0.105 \end{pmatrix}$$

(g) Fractional Order = 0.7

$$\begin{pmatrix} -0.08 & 0 & -0.08 & 0 & -0.08 \\ 0 & -0.8 & -0.80 & -0.8 & 0 \\ -0.08 & -0.8 & 8.04 & -0.8 & -0.08 \\ 0 & -0.8 & -0.80 & -0.8 & 0 \\ -0.08 & 0 & -0.08 & 0 & -0.08 \end{pmatrix}$$

(h) Fractional Order = 0.8

$$\begin{pmatrix} -0.045 & 0 & -0.045 & 0 & -0.045 \\ 0 & -0.9 & -0.900 & -0.9 & 0 \\ -0.045 & -0.9 & 8.560 & -0.9 & -0.045 \\ 0 & -0.9 & -0.900 & -0.9 & 0 \\ -0.045 & 0 & -0.045 & 0 & -0.045 \end{pmatrix}$$

(i) Fractional Order = 0.9

$$\begin{pmatrix} 0 & 0 & 0 & 0 & 0 \\ 0 & -1 & -1 & -1 & 0 \\ 0 & -1 & 9 & -1 & 0 \\ 0 & -1 & -1 & -1 & 0 \\ 0 & 0 & 0 & 0 & 0 \end{pmatrix}$$

(j) Order = 1

Fig. 2 GL Fractional-order differentiation based 5 by 5 sized mask for sharpening filter for various orders in range of (0,1)

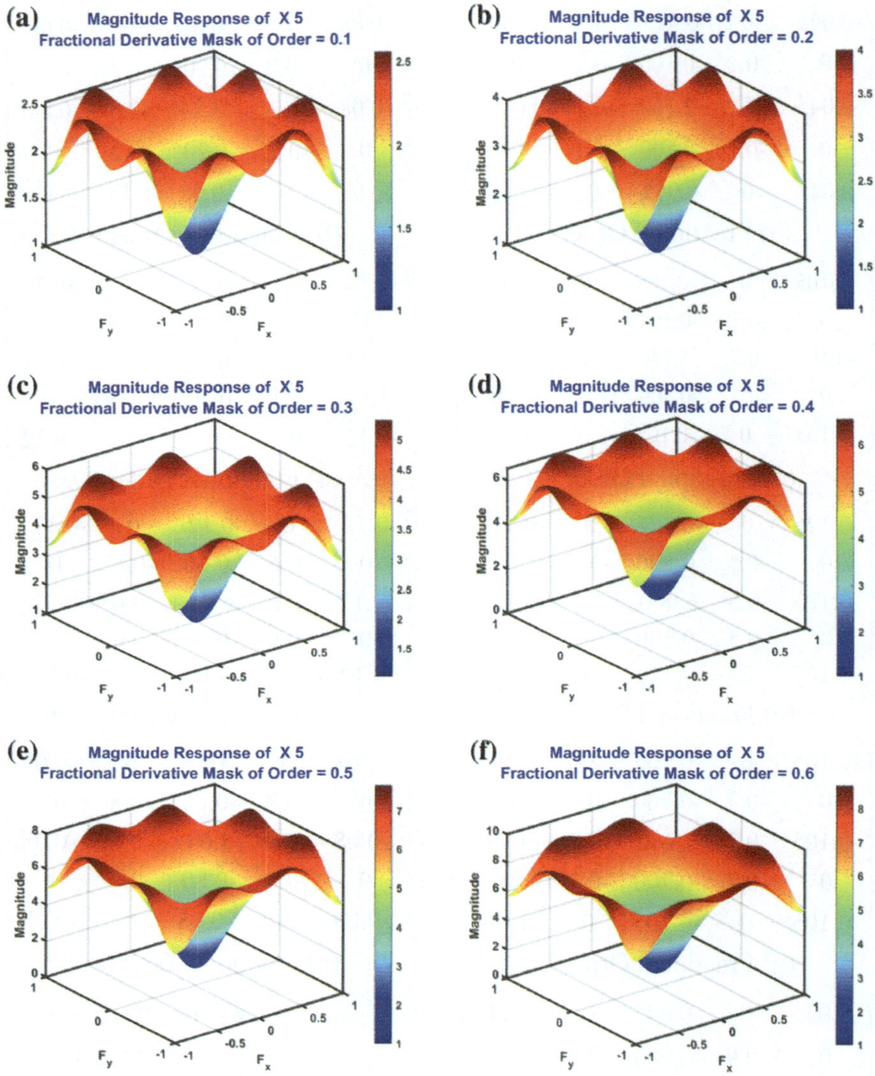

Fig. 3 Magnitude response for 2-D GL fractional-order integration based filter for various orders in range of (0,1). [Variation in Colorbar scales can be identified easily]

$$T_3 = 0.5(1 - T_2); \tag{24}$$

$$T_1 = T_2/2; \tag{25}$$

Correspondingly, binary masks are derived, as follows,

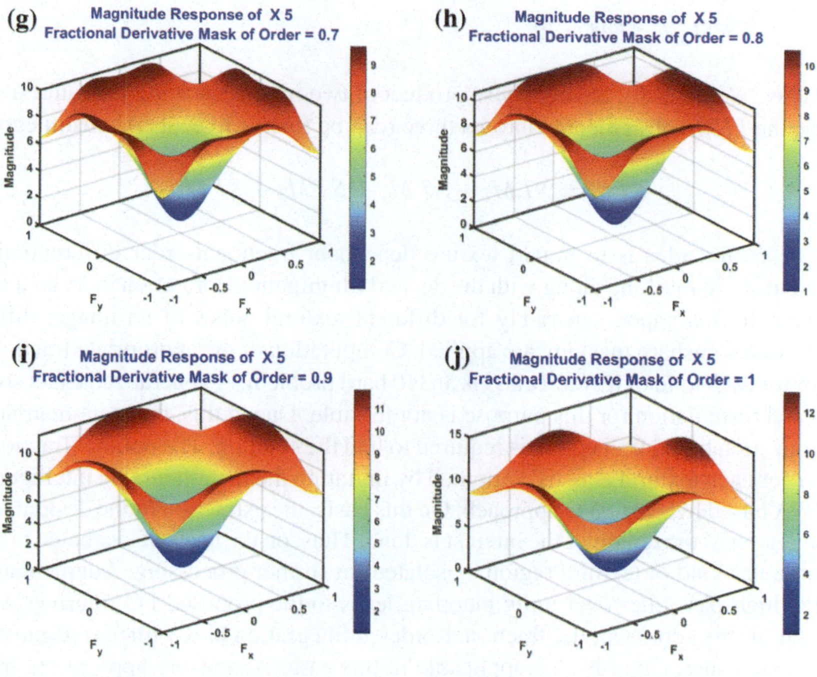

Fig. 3 (continued)

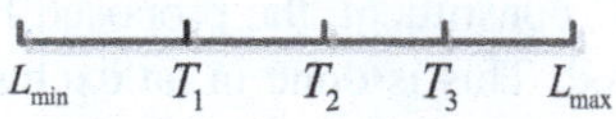

Fig. 4 Threshols for textural segmentation

$$BM_1(m,n) = \begin{cases} 1, & STM(m,n) \le T_1 \\ 0, & \text{otherwise} \end{cases} \tag{26}$$

$$BM_2(m,n) = \begin{cases} 1, & T_1 \le STM(m,n) < T_2 \\ 0, & \text{otherwise} \end{cases} \tag{27}$$

$$BM_3(m,n) = \begin{cases} 1, & T_2 \le STM(m,n) < T_3 \\ 0, & \text{otherwise} \end{cases} \tag{28}$$

$$BM_4(m,n) = \begin{cases} 1, & STM(m,n) \ge T_3 \\ 0, & \text{otherwise} \end{cases} \tag{29}$$

Here, m and n are the indices used for identifying the pixel location for a 2-D intensity channel. Element-wise product of these masks with *LEM* leads to the isolated textural regions of the image under consideration, as follows:

$$STM_1 \triangleq STM^{\circ}B_1;\ STM_2 \triangleq STM^{\circ}BM_2;\ STM_3 \triangleq STM^{\circ}B_3;$$

$$STM_4 \triangleq STM^{\circ} B_4; \tag{30}$$

Here, '∘' stands for element wise product of two matrices having the same sizes. Thus, the *LEM* is fragmented into its three (can be four or five, also) constituents:

$$STM = STM_1 + STM_2 + STM_3 + STM_4 \tag{31}$$

The central idea is to impart texture dependent fractional-order differentiation based unsharp masking along with the desired illumination correction in an adaptive manner. In this paper, separately for different textural zones of an image, different levels of unsharp masking are applied. Computation of the appropriate fractional order for each kind of textural class is an NP hard problem. A generalized exact solution and formulation for this purpose is not feasible. Due to this, the meta-heuristics optimization based framework is required to find the solution. The optimal fractional order for each textural class is computed by imitating the meta-heuristic intelligence of CSA based hybrid GWO approach. On this basis, the extraction of the desired textural region as the region of the interest is done. Thus, on the basis of evaluated order, a particular kind of textural region is isolated for further processing. Augmentation of FO high-pass filter over unity mapping leads to the proposed FO filtering. As a variant of this approach, the fractional-ordered integration based filtered negatively augmented sharpening is also applicable in this case. Adaptively appropriate fractional orders (FOs) are chosen for each kind of textural region. Varying the fractional order (from 0 to 1) leads to the sharpening of various kinds of textural regions present in the image. For each *LTM* constituent, the proposed FO unsharp masking of the appropriate order is imparted. This is done in an exclusively independent manner according to the textural behavior of the region under consideration.

$$Q_1 \triangleq STM_1 \otimes H_{K\times K}^{\Upsilon_1}; \tag{32}$$

$$Q_2 \triangleq STM_2 \otimes H_{K\times K}^{\Upsilon_2}; \tag{33}$$

$$Q_3 \triangleq STM_3 \otimes H_{K\times K}^{\Upsilon_3}; \tag{34}$$

$$Q_4 \triangleq STM_4 \otimes H_{K\times K}^{\Upsilon_4}; \tag{35}$$

Here, for above four equations, if one has to follow the iterative optimal approach for each of the four textural regions independently, a potential complexity will be found in this manner. Also, the textural behavior is not so independent. Instead of it, the above textural segmentation is thoroughly relative. Based on the above two statements it can be concluded that, some complexity can be easily reduced by deriving some inter-relation among all four fractional-orders. Just for finding the four optimal orders in an independent manner, optimizer has to explore and exploit a very big search space. For this purpose 4-D search space have is required that too in the range on 0 to 1 in each dimension. To avoid this issue, one simple approach is applied in the proposed framework. Empirically, through rigorous experimentations some implications are derived which is as follows. First of all, for most smooth textured region of the image, only edge augmentation is required. Due to this first order high pass filtering should me most effective. This implies that Υ_1 should be equal to unity for governing first order filtering. This helps in edge identification and consequent edge augmentation for the least granular textural region. Excluding this region, there higher level textural regions are considered as relatively balanced contribution in the image content. Through empirical analysis, Υ_2 and Υ_4 may behave as dual for each other in such a manner that summation of both makes the unity. Also the feasible range must be defined as the fractional-order in the range of 0.1–0.5 for the most granular textural region of the image i.e. for STM_4. Correspondingly for STM_2 the fractional order is planned as $\Upsilon_2 = 1 - \Upsilon_4$. Later, for choosing, feasible range can be planed in-between Υ_2 and Υ_4. The most delighted range is this context is 0.25–0.75. As a whole, in this manner the 4-D search space (ranging from 0 to unity in each dimension) can be easily compressed to 2-D search space (ranging from 0 to 0.5 and 0.25 to 0.75, respectively). Hence, it can be said that both dimensions as well as range is suppressed by 50%. This leads to the potential reduction in the system complexity of the proposed model. Thus, collective contribution of above channels leads to overall improvement.

$$\begin{aligned} W(m,n) = {} & B_1(m,n) \circ Q_1(m,n) + B_2(m,n) \circ Q_2(m,n) \\ & + B_3(m,n) \circ Q_3(m,n) + B_4(m,n) \circ Q_4(m,n); \end{aligned} \tag{36}$$

$$\gamma(i) = 1 - CDF(i), \tag{37}$$

This gamma value-set, is used to obtain the enhanced luminance channel, as:

$$K(i) = [W(i)]^{\gamma(i)}, \tag{38}$$

In general, most of the objective functions are planned/framed and proposed by involving the idea of discrete entropy content of the image. In this chapter, the authors have framed an effective concept of amalgamation of intensity based general discrete entropy along with GLCM based entropy values. In this manner, both the information about intensity levels as well as the information about spatial co-occurrence has been considered for this purpose. The objective is to increase the value of these entropy values collectively. Along with it, the gradient magnitude (GM_O) matrix's magnitude for the output images is also included for this purpose. GM_O has been evaluated by employing the Sobel-Feldman operator throughout both (row-wise and column-wise) directions. Adding all elements of the GM_O tells about the extent of presence of edgy content of the image. This summation seems to be higher in magnitude and for this purpose. Its double logarithmic made its contribution in comparable order or comparable range. In this manner both the edge-based and texture-based content of the image is considered. Summation of both entropies is multiplied by double-logarithmic of the summation of edges. This product is multiplied by exponential of the normalized image's contrast measure. Thus computed product is raised to the radical power of 1/3 and then added to the ratio of colorfulness measures of the output image w.r.t. the input image' colorfulness measure. If this objective function is going to be employed for the enhancement of grey-scale image, then ratio of the colorfulness measures can be ignored and hence, the proposed objective function can be framed for both gray-scale (uni-spectral image) as well as multispectral and/or color image in a dedicated manner. The mathematical expression can be written as follows:

$$J_{RGB} \triangleq \left(\frac{CM_O}{CM_I}\right) + \sqrt[3]{e^{SD_O}.(DE_O + DE_{GLCM}).\left(\log\left(\sum_{m=1}^{M}\sum_{n=1}^{N}(GM_O(m,n))\right)\right)}, \quad (39)$$

$$J_{Grey} \triangleq \sqrt[3]{e^{SD_O}.(DE_O + DE_{GLCM}).\left(\log\left(\sum_{m=1}^{M}\sum_{n=1}^{N}(GM_O(m,n))\right)\right)}, \quad (40)$$

5 Experimentation: Performance Evaluation and Comparison

5.1 Assessment Criterion

Various state-of-the-art approaches have been listed and exercised for the comparative evaluation of their performance with the sole objective of quality improvement of the remotely sensed images. In this context, parallel/vertical comparative evaluation has been presented by employing the state-of-the-art methods like, GHE, MMSICHE, BPFDHE, AVHEQ, AGCWD, HEOPC, HEMIC, IEAUMF, PGCHE, PGCFDM, PGC-RLFOIM, and the proposed approach. A set of primary quality performance measures is evaluated for this experimentation, namely the brightness content, contrast or variance of the image, discrete entropy content of the image, gradient or the sharpness content of the image and the colorfulness measure of the image is evaluated. In this regard, for accounting GLCM-based assessment, indices like GLCM correlation (GC), GLCM energy (GE) and GLCM-homogeneity (GH) are evaluated for this purpose. Value of brightness (*BN*) or mean for an image *I(m,n)* of size *M* by *N*, expressed as an average summation, is given as:

$$Brightness(BN) = \frac{1}{M \times N}\sum_{m=1}^{M}\sum_{n=1}^{N} I(m,n), \quad (41)$$

Contrast (CT) of an image, responsible for its naturally pleasant look, is accounted due to the variance or the average intensity spread, and is given as:

$$Contrast(CT) = \frac{1}{M \times N}\sum_{m,n} I(m,n)^2 - \left(\frac{1}{M \times N}\sum_{m,n} I(m,n)\right)^2, \quad (42)$$

Shannon entropy quantifies the information content of an image. By making use of normalized image histogram, bounded probability calculation is given as:

$$Entropy(SE) = -\sum_{i=0}^{I_{max}} p_i \log_2(p_i), \quad (43)$$

where, $p_i = n_i / (M \times N)$ represents the possibility of occurrence in accordance with the intensity level and, $I_{\max}$ represents the maximum intensity level. Sharpness or gradient of an image helps in identification of the edge content of the image and is given as:

$$Sharpness(SN) = \frac{1}{M \times N}\sum_{m,n}\left(\sqrt{\Delta m^2 + \Delta n^2}\right), \quad (44)$$

where, $\Delta x = I_{enh}(m,n) - I_{enh}(m+1,n)$and $\Delta y = I_{enh}(m,n) - I_{enh}(m,n+1)$ represents the local values of the gradient of an image. Color channel's coordination is noteworthy in case of color images. Hence, colorfulness of the image can be termed as the coordination among various color channels by exploiting relative colors' mean and relative colors' variance. Mathematically, the colorfulness content is identified as:

$$Colorfu\ln ess(CFN) = \sqrt{\sigma_{rg}^2 + \sigma_{yb}^2} + 0.3\sqrt{\mu_{rg}^2 + \mu_{yb}^2}, \quad (45)$$

$$\Delta rg = R - G; \quad \Delta yb = 0.5(R + G) - B; \quad (46)$$

where, $\Delta_{rg}, \Delta_{yb}, \mu_{rg}, \mu_{yb}, \sigma_{rg}, \sigma_{yb}$, represents the differential values, mean and standard deviation respectively. During the assessment of intensity centered performance indices, image pixels spatial co-occurrence is ignored typically. In order to resolve it, a significant role for the spatially inclined features and texture inclined features is played by grey level co-occurrence matrix centered performance indices. On the computation of the pixel wise average for all the four directional matrices, overall spatial behavior and statistical behavior with respect to the reference pixel is given as:

$$GLCM = 0.25\left(GLCM_0 + GLCM_{\pi/4} + GLCM_{\pi/2} + GLCM_{3\pi/4}\right); \quad (47)$$

Three well-known GLCM centered indices are assessed in this paper. They are GLCM-Homogeneity, GLCM-Energy and GLCM-Correlation. Typical estimation of any GLCM matrix $\Psi(m, n)$ element is possible on consideration of nth neighboring pixel with respect to the mth pixel. Later, corresponding μ_m, μ_n, σ_m, and σ_n are computed as relevant mean values and relevant standard deviation values respectively. Interdependency of corresponding neighborhood pixels with respect to their reference pixels is termed as GLCM-Correlation (*GCR*) and is given as:

$$GLCM - Correlation = \sum_{m=0}^{M-1}\sum_{n=0}^{N-1}\frac{(m-\mu_m)(n-\mu_n)\Psi(m,n)}{\sigma_m.\sigma_n}, \tag{48}$$

Repetitive sets normalized count can be characterized as GLCM-Energy (*GE*). These are also intuitively responsible for the homogeneousness of texture and are given as:

$$GLCM - Energy(GE) = \sum_{m=0}^{M-1}\sum_{n=0}^{N-1}\Psi(m,n)^2, \tag{49}$$

Closeness of the neighboring pixels with respect to their reference pixels can be characterized as GLCM-Homogeneity (*GHG*). These are also intuitively responsible for the homogeneousness of texture and are given as:

$$GLCM\ Homogenity = -\sum_{m=0}^{M-1}\sum_{n=0}^{N-1}\Psi(m,n)\log_2\Psi(m,n), \tag{50}$$

In ideal terms, for the better texture visualization of content, lower value of the above listed parameters is appreciated. Better quality improvement can be easily advocated by increased value for brightness (B), contrast (V), entropy (H), sharpness (S) and colorfulness (C) as shown in Table 2. Contrary to this, decreased values of GLCM-based indices namely correlation (R), energy (E) and homogeneity (M) collectively advocate the better texture based quality improvement, as shown. As listed in Table 3, an averaged analysis is also presented over various test images. Accordingly, 220.3% increment is achieved over the input contrast along with the simultaneous 37.4% increment in the discrete entropy level and 238% increment in the sharpness content. Also, for dark color images, higher values of brightness and colorfulness are also desired, those are reported with 172.8 and 239.8% increased w.r.t. the input indices, respectively. In addition, the textural improvement is advocated in terms of desired comparative reduction of GLCM based metrics, namely correlation, energy and homogeneity are suppressed by 32.6%, 75.6%, and 34.2%, respectively. Hence, the desired objective is achieved efficiently.

Table 2 Quantitative evaluation with comparison among input images among state-of-the-art approaches and the proposed approach using various quality metrics

S. No.	Indices	Input	GHE	MMISCHE	ADAPHE	AVHEQ	AGCWD
1.	**B**	0.1430	0.5382	0.1852	0.3079	0.1925	0.2657
	V	0.0230	0.0600	0.0552	0.0734	0.0420	0.0604
	S	0.0525	0.0925	0.0745	0.1080	0.0616	0.0917
	H	4.6992	5.4557	5.2727	5.6113	5.2968	4.8846
	C	0.1426	0.5829	0.1452	0.4494	0.1519	0.2287
	GC	0.8180	0.7005	0.7830	0.6598	0.7822	0.7161
	GE	0.2227	0.0901	0.2097	0.0793	0.1742	0.1114
	GH	0.6823	0.6708	0.6777	0.6614	0.7414	0.6749
2.	**B**	0.2520	0.5259	0.2959	0.3711	0.3461	0.3480
	V	0.0475	0.0669	0.0828	0.0897	0.0918	0.0733
	S	0.0673	0.0800	0.0871	0.0984	0.0935	0.0832
	H	6.2499	6.5434	6.6297	6.7393	6.5505	6.2553
	C	0.2839	0.4493	0.0915	0.3813	0.1122	0.1206
	GC	0.7462	0.6960	0.7088	0.6675	0.6935	0.7225
	GE	0.1044	0.0572	0.0926	0.0664	0.0754	0.0942
	GH	0.6606	0.6690	0.6672	0.6584	0.6691	0.6654
3.	**B**	0.1265	0.5684	0.1833	0.2637	0.1719	0.2393
	V	0.0220	0.0458	0.0653	0.0696	0.0423	0.0623
	S	0.0551	0.0816	0.0890	0.1020	0.0761	0.0949
	H	4.9446	5.4939	5.4127	5.6706	5.1541	5.2182
	C	0.0900	0.4987	0.0896	0.3462	0.0855	0.1259
	GC	0.8119	0.6938	0.7655	0.6913	0.7598	0.7155
	GE	0.2637	0.0909	0.2491	0.1232	0.1984	0.1405
	GH	0.5649	0.6051	0.5762	0.5993	0.5779	0.6061
S. No.	HEOPC	HEMIC	IEAUMF	PGCHE	PGCFDM	PGCRLFOM	Proposed
1.	0.1708	0.2944	0.2093	0.2894	0.2354	0.4112	**0.4523**
	0.0309	0.0416	0.0538	0.0646	0.0473	0.0889	**0.0978**
	0.0592	0.0703	0.0932	0.0981	0.0892	0.1827	**0.2010**
	5.2115	6.0543	5.4682	5.0509	5.0158	6.2101	**6.8311**
	0.1332	0.2189	0.1670	0.2401	0.2448	0.5474	**0.6021**
	0.7924	0.7070	0.7253	0.6822	0.7069	0.5351	**0.5244**
	0.1824	0.0767	0.1550	0.0758	0.0995	0.0472	**0.0463**
	0.7024	0.7064	0.5959	0.6781	0.6364	0.4065	**0.3984**
2.	0.3486	0.3679	0.3490	0.3859	0.3721	0.4426	**0.4869**
	0.0887	0.0857	0.0912	0.0927	0.0863	0.0922	**0.1014**

(continued)

Table 2 (continued)

S. No.	HEOPC	HEMIC	IEAUMF	PGCHE	PGCFDM	PGCRLFOM	Proposed
	0.0921	0.0905	0.0990	0.0953	0.1010	0.1520	**0.1672**
	6.6110	6.9469	6.5278	6.5582	6.5771	7.3002	**8.0302**
	0.1126	0.1167	0.1134	0.1305	0.3772	0.4234	**0.4657**
	0.6938	0.6874	0.6748	0.6791	0.6700	0.5569	**0.5458**
	0.0727	0.0635	0.0709	0.0561	0.0570	0.0447	**0.0438**
	0.6683	0.6676	0.6584	0.6690	0.6322	0.5091	**0.4989**
3.	0.1761	0.3776	0.1871	0.2512	0.2095	0.3682	**0.4050**
	0.0409	0.0307	0.0573	0.0633	0.0483	0.0855	**0.0941**
	0.0751	0.0690	0.1146	0.0976	0.0914	0.1542	**0.1696**
	5.3898	5.9595	5.2632	5.3252	5.2787	6.5802	**7.2382**
	0.0872	0.2064	0.0966	0.1322	0.2589	0.4109	**0.4520**
	0.7583	0.7182	0.6939	0.6988	0.7140	0.5910	**0.5792**
	0.1894	0.0995	0.1893	0.1108	0.1291	0.0849	**0.0832**
	0.5769	0.5737	0.4489	0.5948	0.5381	0.4663	**0.4570**

5.2 Qualitative Assessments

Comparative qualitative evaluation with recently published state-of-the-art methodologies is presented as follow for highlighting the significant contribution (Figs. 5, 6, 7) by experimentation over various satellite images [32–34].

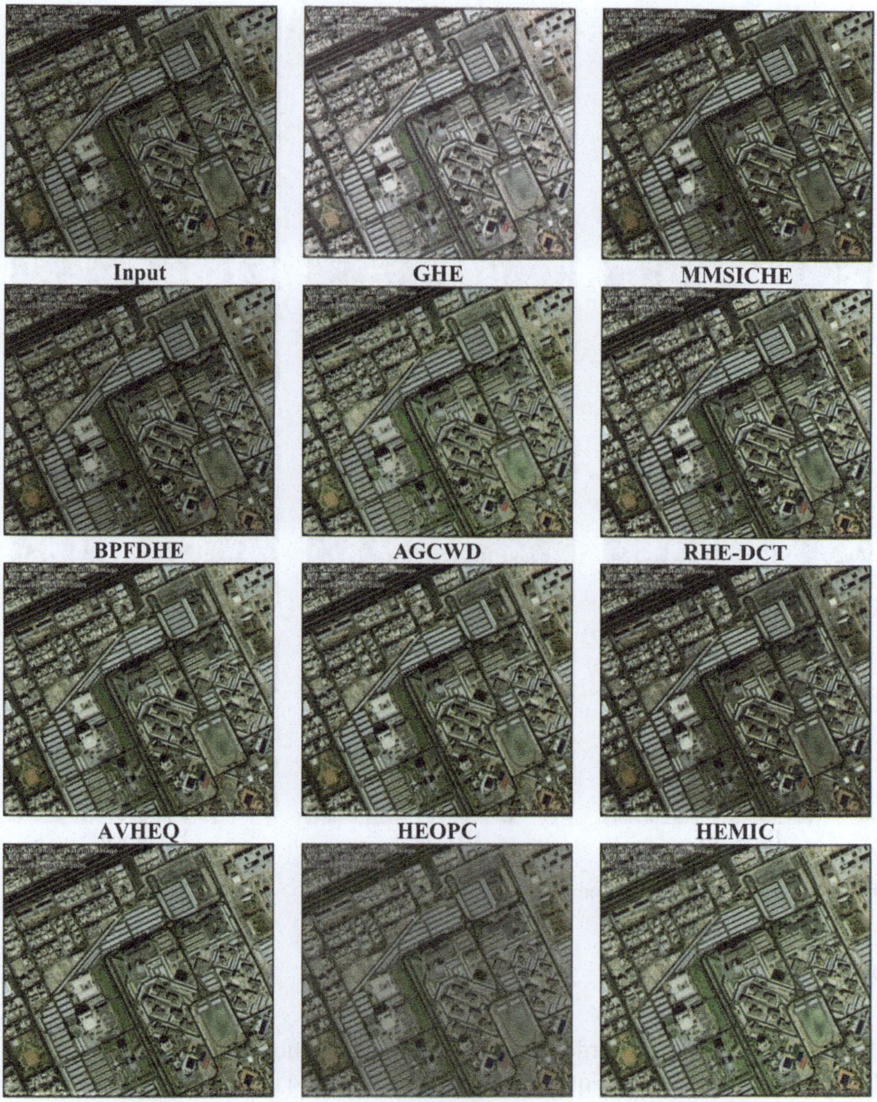

Fig. 5 Visual evaluation and comparison for image S. No. 1

5.3 *Quantitative Assessments*

For explicit comparative numerical assessments, relevant performance indices are listed in Table 2.

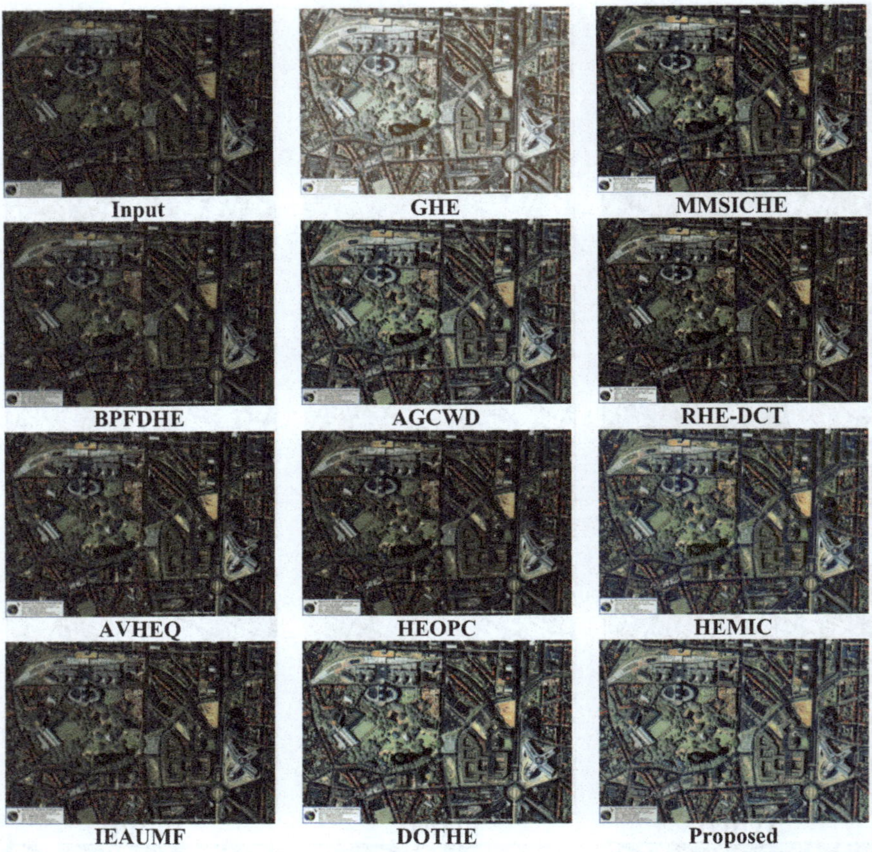

Fig. 6 Visual evaluation and comparison for image S. No. 2

6 Conclusion

The proposed optimally ordered fractional differentiation based filtering is established as an efficient image quality enhancement method in this chapter. Also, a very rigorous experimentation has been performed in this chapter by employing the performance evaluation and comparison with pre-existing recently proposed and highly appreciated quality enhancement approaches. The premiere step is the texture based spatial image segmentation. Later depending on the relatively varying texture, region-wise adaptive, optimal fractional-order sharpening is adopted in the proposed framework. To involve the meta-heuristic intelligence for this optimal image enhancement, the (CSA based) hybrid Grey Wolf Optimizer is employed in this framework. Along with the textural sharpening the adaptive illumination correction is also suggested for overall quality improvement of the image. Thus the whole proposed framework

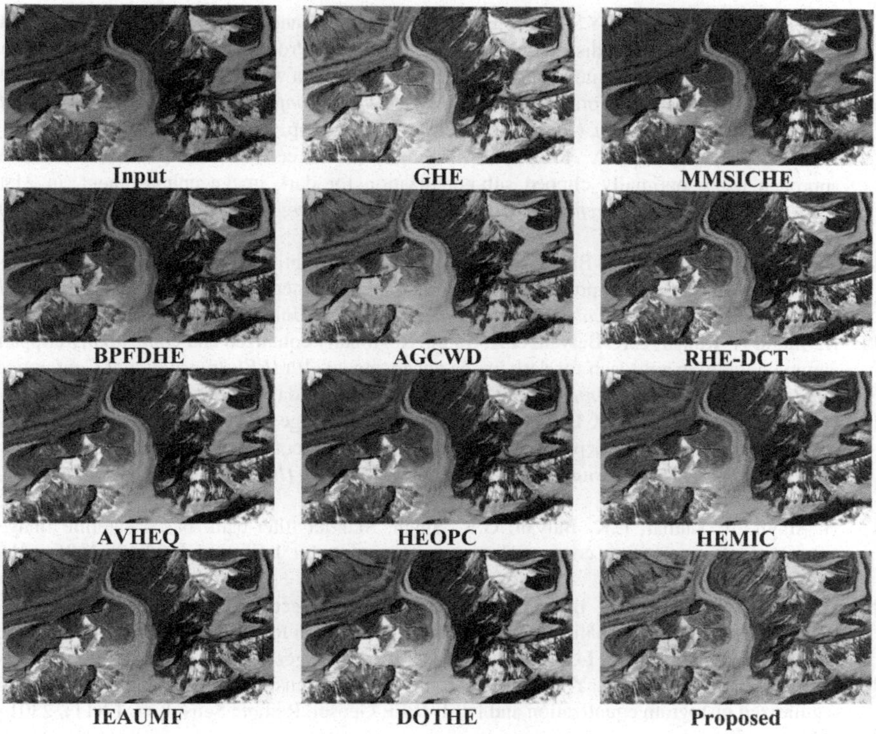

Fig. 7 Visual evaluation and comparison for image S. No. 3

will be beneficial for information harvesting through airborne remotely sensed dark satellite images, acquired under poor illumination.

References

1. R.C. Gonzalez, R.E. Woods, *Digital Image Processing*, 3rd edn. (Prentice-Hall NJ, USA, 2006)
2. D. Sheet, H. Garud, A. Suveer, M. Mahadevappa, J. Chatterjee, Brightness preserving dynamic fuzzy histogram equalization. IEEE Trans. Consum. Electron. **56**(4), 2475–2480 (2010)
3. H. Singh, A. Kumar, L.K. Balyan, Robustly clipped sub-equalized histogram based cosine transformed energy redistributed gamma correction for satellite image enhancement. in *Proceedings of 3*rd *International Conference on Computer Vision and Image Processing*, Adv. Intell. Syst. Comput., Springer, Singapore, 2019
4. H. Singh, A. Kumar, L.K. Balyan, H. Lee, Fuzzified histogram equalization based gamma corrected cosine transformed energy redistribution for image enhancement. in *23*rd *IEEE International Conference on Digital Signal Processing (DSP)*, (Shanghai, China, 2018), pp. 1–5
5. K. Singh, R. Kapoor, Image enhancement via median mean based sub image clipped histogram equalization. Opt.-Int. J. Light. Electron Optics **125**(17), 4646–4651 (2014)
6. K. Singh, R. Kapoor, Image enhancement using exposure based sub image histogram equalization. Pattern Recogn. Lett. **36**, 10–14 (2014)

7. S.C. Huang, F.C. Cheng, Y.S. Chiu, Efficient contrast enhancement using adaptive gamma correction with weighting distribution. IEEE Trans. Image Process. **22**(3), 1032–1041 (2013)
8. H. Singh, A. Kumar, Satellite image enhancement using beta wavelet based gamma corrected adaptive knee transformation. in 5th *IEEE International Conference on Communication and Signal Processing (ICCSP)*, (Melmaruvathur, India, 2016) pp. 128–132
9. H. Singh, N. Agrawal, A. Kumar, G.K. Singh, H.N. Lee, A novel gamma correction approach using optimally clipped sub-equalization for dark image enhancement. in 21st *IEEE International Conference on Digital Signal Processing (DSP)*, (Beijing, China, 2016) pp. 497–501
10. H. Singh, A. Kumar, L.k. Balyan, G.K. Singh, A novel optimally gamma corrected intensity span maximization approach for dark image enhancement. in 22nd *IEEE International Conference on Digital Signal Processing (DSP)*, (London, United Kingdom, 2017), pp. 1–5
11. H. Singh, A. Kumar, L.K. Balyan, G.K. Singh, Regionally equalized and contextually clipped gamma correction approach for dark image enhancement. 4th *IEEE International Conference on Signal Processing and Integrated Networks (SPIN)*, (Noida, India, 2017), pp. 431–436
12. H. Singh, A. Kumar, L.K. Balyan, G.K. Singh Dark image enhancement using optimally compressed and equalized profile based parallel gamma correction. in 6th *IEEE International Conference on Communication and Signal Processing (ICCSP)*, (Chennai, India, 2017), pp. 1299–1303
13. H. Singh, A. Kumar, L.K. Balyan, G.K. Singh, Slantlet filter-bank based satellite image enhancement using gamma corrected knee transformation. Int. J. Electron. **105**(10), 1695–1715 (2018)
14. H. Singh, A. Kumar, L.K. Balyan, H.N. Lee, Optimally sectioned and successively reconstructed histogram sub-equalization based gamma correction for satellite image enhancement. Multimed. Tools Appl. **78**(14), 20431–20463 (2019). Springer
15. X. Fu, J. Wang, D. Zeng, Y. Huang, X. Ding, Remote sensing image enhancement using regularized histogram equalization and DCT. IEEE Geosci. Remote Sens. Lett. **12**(11), 2301–2305 (2015)
16. S.C.F. Lin, C.Y. Wong, M.A. Rahman, G. Jiang, S. Liu, N. Kwok, Image enhancement using the averaging histogram equalization approach for contrast improvement and brightness preservation. Comput. Electr. Eng. **46**, 356–370 (2014)
17. C.Y. Wong, G. Jiang, M.A. Rahman, S. Liu, S.C.F. Lin, N. Kwok et al., Histogram equalization and optimal profile compression based approach for color image enhancement. J. Vis. Commun. Image Represent. **38**, 802–813 (2016)
18. C.Y. Wong, S. Liu, S.C. Liu, M.A. Rahman, S.C.F. Lin, G. Jiang et.al., Image contrastenhancement using histogram equalization with maximum intensity coverage. J. Mod. Opt. 63(16), 1618–1629
19. S.C.F. Lin, C.Y. Wong, G. Jiang, M.A. Rahman, T.R. Ren, N. Kwok et al., Intensity and edge based adaptive unsharp masking filter for color image enhancement. Opt. Int. J. Light. Electron Optics. **127**(1), 407–414 (2016)
20. K. Singh, D.K. Vishwakarma, G.S. Walia, R. Kapoor, Contrast enhancement via texture region based histogram equalization. J. Mod. Opt. **63**(15), 1444–1450 (2016)
21. H. Singh, A. Kumar, L.K. Balyan, A levy flight firefly optimizer based piecewise gamma corrected unsharp masking framework for satellite image enhancement. in *14*th *IEEE India Council International Conference (INDICON)*, (Roorkee, India, 2017), pp. 1–6
22. H. Singh, A. Kumar, L.K. Balyan, A sine-cosine optimizer-based gamma corrected adaptive fractional differential masking for satellite image enhancement, harmony search and nature inspired optimization algorithms. Adv. Intell. Syst. Comput. 741, 633–645 (2019). (Springer, Singapore)
23. H. Singh, A. Kumar, L.K. Balyan, H.N. Lee, Texture-dependent optimal fractional-order framework for image quality enhancement through memetic inclusions in Cuckoo Search and Sine-Cosine Algorithms, in *Recent Advances on Memetic Algorithms and its Applications in Image Processing*, ed. by D. Hemanth, B. Kumar, G. Manavalan. Studies in Computational Intelligence, vol 873 (Springer, Singapore, 2020)

24. H. Singh, A. Kumar, L. K. Balyan (2017) Cuckoo search optimizer based piecewise gamma corrected auto clipped tile wise equalization for satellite image enhancement, in *14th IEEE India Council International Conference (INDICON)*, (Roorkee, India, 2017), pp. 1–5
25. X.S. Yang, S. Deb Cuckoo search via Lévy-flights. in Proceedings of World Congress on Nature & Biologically Inspired Computing (2009), pp. 210–214
26. S. Mirjalili, S.M. Mirjalili, A. Lewis, Grey wolf optimizer. Adv. Eng. Softw. **69**, 46–61 (2014)
27. Y.F. Pu, J.L. Zhou, X. Yuan, Fractional differential mask: a fractional differential-based approach for multiscale texture enhancement. IEEE Trans. Image Process. **19**(2), 491–511 (2009)
28. H. Singh, A. Kumar, L.K. Balyan, G.K. Singh, Swarm intelligence optimized piecewise gamma corrected histogram equalization for dark image enhancement. Comput. Electr. Eng. **70**, 462–475 (2018). Elsevier
29. H. Singh, A. Kumar, L.K. Balyan, G.K. Singh, A novel optimally weighted framework of piecewise gamma corrected fractional order masking for satellite image enhancement. Comput. Electr. Eng. (Elsevier) **75**, 245–261 (2019)
30. H. Singh, A. Kumar, L.K. Balyan, H.N. Lee, Piecewise gamma corrected optimally framed grumwald-letnikov fractional differential masking for satellite image enhancement. in 7th *IEEE International Conference on Communication and Signal Processing (ICCSP)*, (Chennai, India, 2018), pp. 0129–0133
31. H. Singh, A. Kumar, L.K. Balyan, H.N. Lee, Fractional-order integration based fusion model for piecewise gamma correction along with textural improvement for satellite images. IEEE Access **7**, 37192–37210 (2019)
32. https://www.satimagingcorp.com/gallery/quickbird/
33. https://www.satimagingcorp.com/gallery/pleiades-1/
34. https://www.satimagingcorp.com/gallery/pleiades-2/

Robust K-Means Technique for Band Reduction of Hyperspectral Image Segmentation

V. Saravana Kumar, E. R. Naganathan, S. Anantha Sivaprakasam and M. Kavitha

Abstract The present work is address about the latest techniques for segmentation of remotely sensed hyperspectral scenes, assembled through airborne or space-borne Earth observation instruments. The inter band cluster and intra band cluster techniques has examined for spectral un-mixing and hyperspectral image segmentation. The inter band clustering is carried out with K-Means, Fuzzy C-Means (FCM) and Robust K-Means (RKM) clustering mechanisms, whilst the Particle Swarm Clustering (PSC) mechanism ought to be used during intra band cluster parts. DB (Davies Bouldin) index be utilized to figure out the quantity of clusters. The hyperspectral bands have clustered besides a band that has predominant variance from every cluster has singled out, that makes diminished band. Moreover, PSC put forward the segmentation strategy on this reduced band. In PSC, the segmentation is put forward out by enhanced algorithm entitled as Enhanced Estimation of Centroid (EEOC). Performance of the above method is evaluated an assortment of scenarios in terms of pixels clustered and time complexity.

Keywords K-Means · Fuzzy C-Means · Robust K-Means · Particle Swarm Clustering and Fitness value

---

V. S. Kumar (✉)
Department of Information Technology, Sreenidhi Institute of Science and Technology, Hyderabad, India
e-mail: drsaravanakavi@gmail.com

E. R. Naganathan
Symbiosis Institute of Computer Science & Research, Symbiosis International University, Pune, India
e-mail: ern_india@yahoo.co.in

S. A. Sivaprakasam
Department of Computer Science, GVN College, Kovilpatti, Tamil Nadu, India
e-mail: anath.sivaprakasam@gmail.com

M. Kavitha
Manonmaniam Sundaranar University, Tirunelveli, Tamil Nadu, India
e-mail: veenakavitha15@gmail.com

D. Oliva and S. Hinojosa (eds.), *Applications of Hybrid Metaheuristic Algorithms for Image Processing*, Studies in Computational Intelligence 890,
https://doi.org/10.1007/978-3-030-40977-7_4

1 Introduction

Remote Sensing is wide outlined characterized as gathering along with elucidation of information regarding an object, region, or an event while not being in physical contact with the object. Moreover, it has been detecting and computing electromagnetic energy exuding or reflected from far-off objects made of divers' materials, with the goal that it ought to identify and categorize these objects through class or kind, substance and spatial distribution. These remotely gathered information through divers sensors perhaps analyzed to acquire information regarding the objects or features underneath scrutiny. Gadgets possibly will employ visible light, infrared otherwise radar en route for acquire data.

Recent advancement of remote sensing has driven the route for the advancement of hyperspectral sensors, for instance Airborne Visible Infrared Imaging Spectrometer (AVIRIS)—introduced by NASA, Hyperion Instrument and Analytical Spectral Devises (ASD) handheld spectrometer. These hyperspectral remote sensors measure reflected radiation as persistent and narrow wavelength bands and henceforth produce images with hundreds or thousands of spectral bands which can provide unique spectral signatures for each image pixel. Hyper symbolize myriad-sized measured wavelength band. Hyperspectral Image (HSI) may perhaps be a newfangled RS scheme [1] that creates hundreds of images, commensurate to divergent wavelength channels, in a similar region on the surface of the earth [2]. This sensor measures the essentialness of the inward bound light in tens or divers slender spectral bands in every one location contain in the image. With the ongoing evolutions in RS gadgets, HSI are currently used in disparate appliance fields [3]. The awesome attributes of HSI data sets bring troublesome dealing with issues. HSI are horribly over decided: that they endow with abundant spectral information to pinpoint and recognize spectrally isolated resources. The advent and developing accessibility of HSI has opened innovative conceivable outcomes in an image analysis as well as classification. In spite of, being HSI imparting more ingredient, image analysis is required to summarize the statistics about that.

2 Related Work

Myriad paradigm of spectral-spatial classifiers ought be initiate in the HSI literature [4]. Classification of HSI [5] productively combined of probabilistic SVM by way of an MRF regularizer. The spectral information [6] is regarded as in cooperation locally and globally has projected for a novel spectral-spatial classifier. Even though SVM are the modern classification technique for remote–sensing scene, a new-fangled non-parametric classification modus operandi that is to state Per-Turbo [7] portrays promising results. Supervised classification of HSI dataset is a demanding obstacle owing to the inadequate accessibility of training samples. Covariance matrix, is a type constituent in extensive array of statistical signal processing tasks functional to RS imagery from multispectral and hyperspectral sensors [8]. It paves to evaluate performance metrics that are observed in real HSI. Designed the dimensionalities lessen [9] by band selection is frequent approaches. A newly developed classifier namely MLR subspace-based description [10] has depends with respect to the essential supposition in order to the samples within each class realize how to approximately lay in a lower dimensional subspace. A sub-discipline of Geographic Information (GI) Science dedicated to embryonic computerized strategies [11] to divider RS imagery into significant image substance. Classification algorithms founded on single-pixel analysis frequently refuse the pointed outcome when enforced to high spatial resolution RS data [12].

A correlation matrix feature extraction derived from Spectral clustering (CMFESC) has demonstrated [13]. Land-cover mapping in a field of coal fire [14] is performed by means of the pixel-based along with the object-oriented image classification. Integration of sparse representations in addition to EMAPs intended for spatial-spectral classification [15] of remote-sensing data, a novel classification strategy has developed. One popular way to tackle the curse of dimensionality is to employ a feature extraction technique [16]. Unsupervised multispectral image segmentation algorithm [17] is founded on an amalgamation of the watershed transform, projected size-weighted fuzzy clustering method, furthermore MC method. A hardware/ software co-design approach [18] to carrying out of K-Means algorithm is addressed. A survey of HSI segmentation by multi-band reduction [19] has demonstrated. Segmentations were performed using Enhanced Estimation of Centroid [20] on Salinas A (sub) scene with K-Means and Fuzzy C-Means (FCM). Segmentation of a false colour HSI using JSEG based on unsupervised clustering algorithm [21] has proposed. The band reduction process [22] is carried out by K-Means, FCM and Fast K-Means [23]. This topic comprises the review about HSI segmentation and multiband reduction. Furthermore, this review describes various methods and techniques such as K-Means, FCM, Fast K-Means, Robust K-Means, and Particle Swarm optimization. As a whole, there is a need for developing a novel method for HSI segmentation and multiband reduction, which be capable of performed with minimum time and computational complexity.

3 Existing Method

3.1 K-Means Clustering Method

Theoretically, K-Means is a classic approach. Now that it is elementary and prompt, it is desirable in practice. It segregates the input dataset keen on K-Clusters. Every cluster is delineated by an adaptively shifting centroid, starting from a few initial values supposed as seed points. K-Means reckons the squared distances between the inputs and centroids, and designate inputs to the nearest centroid. Evidently, the overall performance of the K-Means algorithm relies upon on preliminary cluster centers, whereas the final partition depends on the initial configuration.

3.2 Fuzzy C-Means

In FCM clustering, data elements are capable of placed more than one cluster and pertained with every element is a set of membership levels. FCM is a procedure of designating these membership levels, subsequently by means of them to allocate data elements to as a conservative estimate one clusters. The purpose of FCM is to limit an objective function. Regardless of customary clustering analysis techniques, which disseminate each object to a distinctive group, fuzzy clustering approach obtain membership values somewhere in the range of 0 and 1 that demonstrate the degree of membership for each object to each group. As like of K-Means, the clustering results have vigorously relied upon the number that predefined. There is additionally important to endow with educated steerage for determining the quantity of cluster so as to realize apt clustering results.

3.3 Robust K-Means (RKM) Clustering Method

Prof. Karsin introduces an extension of K-Means algorithm that removes the outlier is entitled as Robust K-Means. The enhancement of beforehand existing geometric clustering strategies are in terms of accurateness and efficacy. The prime target is to keep hold of as most pertinent information regarding the position of the data points as compressing the data points into the clusters. The consequential approach has further robust towards initial centroid placement in addition to has a supplementary rapidly execution time than preceding methods. A study [24] of comprehensive K-Means along with generalized trimmed K-Means routine as of the standpoint of Hampel's robustness yardstick, specifically, examine the influence function, breakdown point, and qualitative heftiness, affirming the dominance put forwards by means of the trimming.

3.4 Particle Swarm Optimization (PSO)

Particle Swarm Optimization (PSO) strategies [25] signify another methodology for optimization. The fundamental concept of the algorithmic rule is to form a swarm of particles that move in the space around them attempting to find for their goal or the place which most closely suits their needs given by a fitness i.e. objective function. For every iteration, every particle is updated by means of two best qualities. The primary one is, achieved the best solution entitled as '*pbest*'. Another best value that has followed dependent on the particle swarm optimizer is the best value, acquired thus far by any particle in the populace is named as ‘*gbest*’. It is a population based explore method. The improved algorithm, FCM rooted in Picard iteration and PSO (PPSO–FCM) is proposed.

A plethora clustering methods are available for cluster analysis. Here we discussed only few and most used one. Nevertheless, the amount of clusters ought to be specified in advance for clustering is an elementary dilemma for the majority part of the existing clustering approaches. The clustering results may perhaps profoundly rely on the quantity of clusters specified. It is obligatory to endow with educated supervision for formative the amount of clusters in order to attain pertinent clustering grades.

3.5 *Determining the Number of Clusters*

An elementary dilemma in cluster analysis is to determine the quantity of clusters, which is usually taken as an earlier. Clustering outcomes may perhaps fluctuate as various quantities of clusters are specified. For appraising and determining an optimal clustering system [26] of a couple measurements benchmark namely Compactness and Separation. Three diverse procedures are there for appraising the consequence of the clustering algorithms specifically, *External Criteria, Internal Criteria* and *Relative Criteria.*

In cooperation internal and external criteria are founded on statistical methods and they have lofty computation command.

3.5.1 Davies Bouldin Index

One of a metric intended for validating clustering strategies [27] is Davies–Bouldin Index (DBI), developed by Davis L. Davies and Donald W. Bouldin (1979). The validation of clustering has been evaluating by means of quantities and features intrinsic towards the dataset. The preludes of this internal assessment scheme, DBI is as follows,

Consider A_i is a vectors cluster, n dimensional feature vector assigned to cluster A_i is noted as X_j.

$$S_i = \frac{1}{T_i}\sum_{j=1}^{T_i}\left\|X_j - A_i\right\|_p \tag{1}$$

At this juncture A_i represents the centroid of A_i where the size of a cluster i be represent as T_i. Measure of scatter inside the bound of cluster noted as S_i. A Euclidean distance function flanked by the centroid of the cluster, also the personage feature vectors are represented as the value of p that defined as 2. This Euclidean distance feebleness is the finest measure of cluster determination, voluminous other distance metrics preserve can employ for high dimensional data. This distance metric shall to competition by means of the metric performed in the clustering system itself for evocative outcomes.

$$M_{i,j} = ||\ A_i - A_j||^p = \left(\sum_{k=1}^{n} \|a_{k,i} - a_{k,j}\|^n \right)^{\frac{1}{p}} \quad (2)$$

$M_{i,j}$ is a measure of parting flanked by cluster A_i and cluster A_j.

$a_{k,i}$ is the kth element of A_i, in addition to there are n such element in A for it is a n dimensional centroid. At this juncture, the Euclidean distance flanked by the center of clusters i and j when p equals to 2 for k indexes the attribute of the data.

DBI is grounded on resemblance measures of clusters ($R_{i,j}$) for the cluster divergence measure ($d_{i,j}$) whose basis is the dispersal measure of a cluster (S_i). The connexion measure of clusters ($R_{i,j}$) be capable of defined generously, although it has to placate the subsequent condition:

- $R_{i,j} \geq 0$
- $R_{i,j} = R_{j,i}$
- If $S_i = 0$ and $S_j = 0$ then $R_{i,j} = 0$
- If $S_j > S_k$ and $M_{i,k}$ then $R_{i,j} > R_{i,k}$
- If $S_j = S_k$ and $M_{i,j} < M_{j,k}$ then $R_{i,j} > R_{i,k}$.

Usually $R_{i,j}$ is defined in the following way:

$$R_{i,j} = \frac{S_i + S_j}{M_{i,j}} \quad (3)$$

$$d_{i,j} = d\left(v_i, v_j\right).\ S_i = \frac{1}{\|C_i\|} \sum_{x \in C_i} d(x, v_j) \quad (4)$$

Then the Davies-Bouldin index is defined as

$$DB = \frac{1}{N} \sum_{i=1}^{N} D_i \quad (5)$$

where

$$D_i \equiv \max_{j \neq i} R_{i,j} \quad (6)$$

DB is called the Davies-Bouldin index.

3.6 Data Set Description of HSI

The dataset contains a diversity of hyperspectral RS scene which are acquired from airborne and satellite. In this work, assured data, such as Salinas A, Salinas Valley, Indian Pines, Pavia University and Pavia Centre are handled.

3.7 Ground Truth Image—Hyperspectral Remote-Sensing Satellite Scenes

Figure 1 paves the ground truth image of HSI. The inventive scene and its conforming ground truth images has taken from the following link. https://www.ehu.eus/ccwintco/index.p...Hyperspectral_Remote_Sensing_Scenes.

These scenes are an extensively used point of reference for testing the accuracy of HSI analysis.

4 Proposed Method

This work is put forward the incorporation of intra-band and inter-band clustering methods for segmentation process with the convinced objective of attaining more exact methods for the analysis of HSI without expanding ominously the computational complexity of the procedure. There are copious things to expand the situation accomplishing this objective. To begin with, to enhance unsupervised segmentation by enchanting in a task-relevant measure of spectral similarity from the feature matrixes approach.

Besides, an inter-band and intra-band cluster technique for segmentation is proposed. In addition, divers existing clustering techniques are applied for inter band

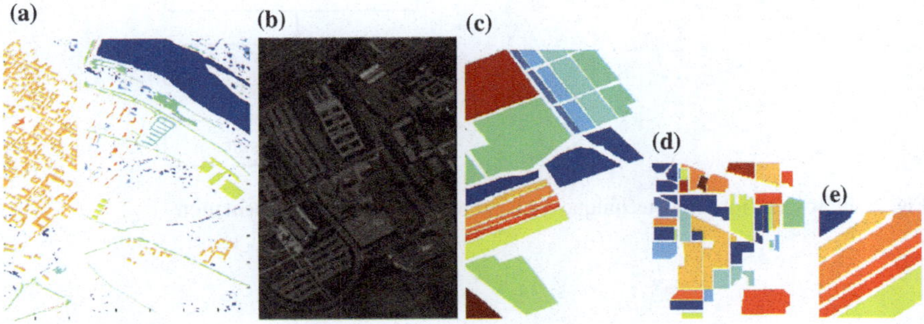

Fig. 1 Ground truth images **a** Pavia Centre **b** Pavia University **c** Salinas Valley **d** Indian Pines **e** Salinas A (sub)

part. Moreover, to expand the accuracy of divers clustering methods; Davies-Bouldin Index is used to decide the amount of clusters. Furthermore, for intra band cluster, a new-fangled and novel algorithm entitled as Enhanced Estimation of Centroid (EEOC) which is the amendment of the particle swarm clustering method is projected. The amendment is that the particle which have updated their positions along with work out the distance matrix barely one time per iteration. In addition, performances about these methods are appraised in diverse circumstances. Performance measurements are used in this research to investigate the efficacy of this system. The working attitude of HSI segmentation using EEOC depicts as Fig. 2. In the first phase, the hyperspectral scene in the form of mat file is read; the feature matrix could be assembled in sight of Mean or Median Absolute Deviation (MAD), Standard Deviation (STD), Variance (VAR). Likewise, apply one of the clustering processes, to be specific K-Means, FCM and RKM. Subsequently these clustering method works have faith in the number of cluster. DBI is used to decide the amount of clusters. Furthermore, the dimensional reduction process could be carried out by this clustering method by picking out one band from each cluster such as 204 bands of HSI are reduced to underneath twenty bands, i.e. one band is pick out based on foremost variance from every cluster. The input of PSC is diminished set of bands from K-Means or FCM or RKM clustering algorithm. PSC (EEOC) is used for segmentation on the reduced bands.

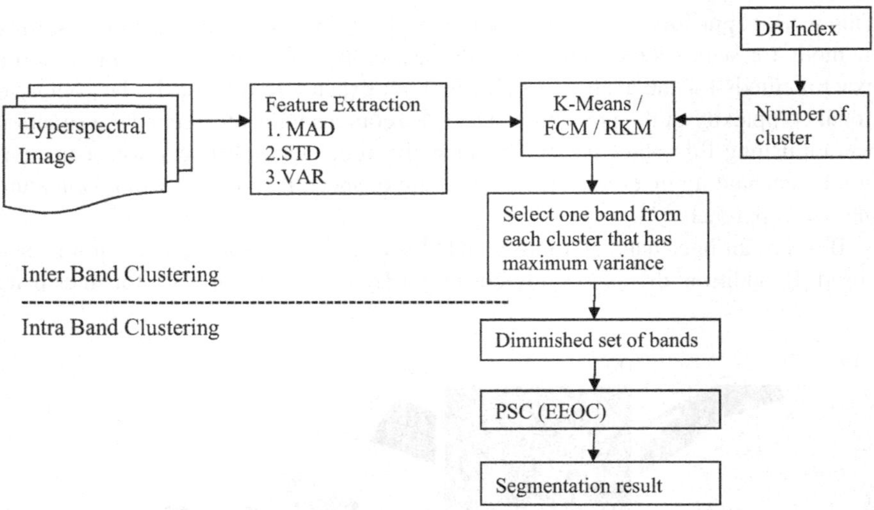

Fig. 2 Flow chart for RKM technique for band reduction of HSI segmentation

4.1 Band Reduction Using ROBUST K-MEANS Algorithm

Robust K-Means is acquainting to diminish the band size of the data set that contains divers of above-said HSI scenes. In inter band cluster part; the feature of HSI is contribution to the Robust K-means clustering algorithm i.e. MAD, STD and VAR. The clustering process could be put forward by RKM clustering method to select single band from every one cluster (103 band of the HSI are reduced to below 20 bands) i.e. single band is chosen consistent with most extreme variance of each cluster. Since, HSI having multiple bands with minor difference, the maximum variance is considered. In PSC procedure, the input is reduced bands that got from aforementioned clustering algorithms.

In 2008, Prof. Karsin introduces an augmentation of K-Means algorithm that expels the outlier is entitled as RKM. It utilizes the information bottleneck scheme as an establishment for its solution to classical clustering issues. In information theory point of view, Clustering is for of lossy compression in view of the ratio of datum to clusters. The objective is to keep hold of however large amount pertinent information regarding the locality of the data points although compressing the data keen on as hardly any clusters as could be expected.

This compression by means of the Lagrange parameter λ, the clustering decisive factor of RKM is therefore

$$\max_{p(c|1)}[I(x, c) - \lambda I(c, i)] \tag{7}$$

where i, c and x are the data index, cluster index and locations of the datum respectively. The objective is tremendous the amount of pertinent information engaged, prearranged the compression parameter λ. The temperature λ assumes a part in determines for the algorithm carries on. For any value $\lambda < 1$, the routine converges to a "hard" K-means elucidation, however, contingent upon its precise value, incline to reveal fewer affectability to initial centroid placement. For a λ value of 1, the subsequent equations are precisely that of the soft K-Means approach. In this manner, the RKM approach is capable of act as either beforehand defined algorithms, nonetheless can be tuned through the λ parameter to fabricate a supplementary effective and precise solution.

In RKM, the hyperspectral feature matrix has reduced as depicts in Fig. 3. Construct the hyperspectral feature matrix as well as the amount of cluster has resolute from DB Index. Compute the cluster with randomly K preliminary cluster centres. Remove the outlier if occurred. Besides compute the new cluster. Repeat the process till converge the result. Pick one band from each cluster which has maximum variance. Finally the bands are reduced. i.e. The dimensional of hyperspectral could be reduced into below 20 depending upon the DB-index of diverse scenes by using the RKM clustering algorithm. In our previous work [19, 22] hyperspectral bands are reduced into below twenty using K-Means, FCM, and FKM methods.

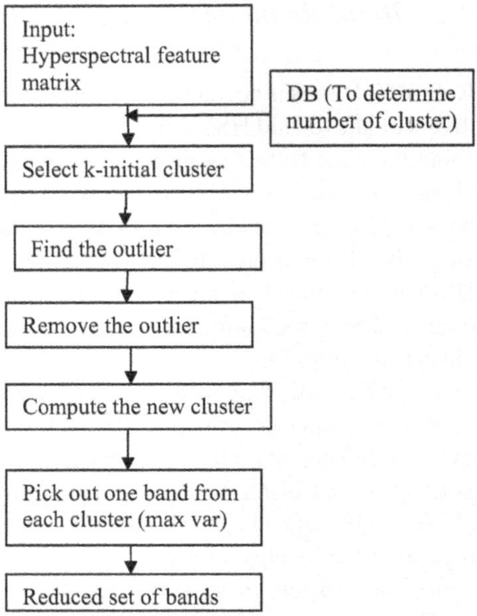

Fig. 3 Band reduction using RKM

4.2 *Algorithm of Enhanced Estimation of Centroid (EEOC)*

EEOC, a lightweight clustering approach, is generally comparative on the way for perception of individual decision making with regards to regular utility had examined. Reduction in the Distance Matrix (DM) frequency reinstate might significantly diminish the reckoning time. It is a reasonable technique to diminish the Distance Matrix (DM) update that is accustomed to shrink the movement update frequency. With the intention of diminish the complexity of time; we propose an adapted version of the centroid estimation known as the EEOC. The term position f is likewise comparable en route for decision construction activities of a person in addition to the impact of the milieu with the motivation behind make the individual to create a verdict to transform their activities. Enthused by the logical concept obtainable in this suggestion, we slot in this frame of mind in the restoration scheme.

EEOC re-establishing rule be able to brief as follows. The particle velocity is delimited by means of Max_v. $f_i(t)$ be an indication of the position, the best position of a particle i in pertain to the input prototype j, and $c_j(t)$ indicates the position of a particle that has to adjacent to the input prototype j. For each emphasis, each particle position is re-established merely once. This arrives subsequent to every one of probable data points that are closer to the particle have been independently considered by the particle. i.e., position update happens merely i times per cycle, here i is the amount of particles in the set. The Distance Matrix (DM) as well as best positions are restructured after every one of particle positions is updated. i.e., DM is restored merely once per iteration. On the whole, least amount reckoning is characterized to store the best position permutation.

Algorithm

P = EEOC (data_matrix, $iter\_func_{max}, max_v, p_c, \phi$)

Initialize the amount of particles as p_c

Reckon the distances of each particles position a and the input prototype of the particles position as c

Reinvigorate the distance and velocity until t <$iter\_func_{max}$

Obtain the closest particle (i, j)

DM = e (g_j, f_i) : $\Pi i, j$ //*DM – distance matrix*

J = min (e (g_j, f_i)) : $i \in \{1, 2, \ldots, p_c\}$

// fitness function parameter e (gj, fi) is calculated as above.

// Re-establish the extraordinary and on the whole best of the particle

// The fitness function / objective function can be calculated such as

$$a_{DM}(m+1) = \begin{Bmatrix} f_j(m) if e\left(g_j,\ f_j(m)\right) < e\left(g_j,\ a_j(m)\right) \\ a_j(m) otherwise \end{Bmatrix}$$

// fj (m) denotes the position.

// aj (m) denotes the best position

// e (gj, fj(m)) – Previous position

// e (gj, aj(m)) – Current position

$$c_j(m+1) = \begin{Bmatrix} f_j(m) if e\left(g_j,\ f_j(m)\right) < e\left(g_j,\ a_j(m)\right) \\ c_j(m) otherwise \end{Bmatrix}$$

// Cj (m) denotes the particle position that has closest to the input outline.

// For each particle distance is calculated based on the distance it's calculate the fitness function. If the current fitness value is enhanced than the previous one current value takes as a best fitness. This process going till the number of iteration is finished.//

Restore the velocity and the position

$m = m + 1$

// normally, the inertia weight, δvalue is 0.4 to 0.9 randomly.

// Here it is 0.9

$f_i(m+1) = f_i(m) + \delta f_i(m)$

Discover the closest data point in favour of each particle rooted in minimum distance

$\lfloor DMf^{min}(m) Jf^{min} \rfloor$ = min (*DM, i*)

Find the closest particle intended for each data point rooted in minimum distance

$\lfloor DMg^{min}(m) Jg^{min} \rfloor$ = min (*DM, j*)

Update $a_i(m)$, $c_i(m)$

a_i *(m)* – best position,

c_j *(m+1)* – closest to the input pattern

$$a_i(m+1): \Pi i = \begin{Bmatrix} f_{Jx^{min}}(m) if DMf_{i^{min}}(m) < DMf_{i^{min}}(m-1) \\ a_i(m) otherwise \end{Bmatrix}$$

$$c_j(m+1): \Pi j = \begin{Bmatrix} f_{Jy^{min}}(m) if DMf_{j^{min}}(m) < DMf_{j^{min}}(m-1) \\ a_j(m) otherwise \end{Bmatrix}$$

// If the condition satisfies the criteria, no change in the cluster centroid, otherwise aj (m) is updated. //

$$S(m+1) = \begin{Bmatrix} f_i(t): \Pi i \quad if \quad ft(f_i(m): \Pi i) < ft(S(m)) \\ S(m) otherwise \end{Bmatrix}$$

// *S (m)* – represent the best position combination that has attained on the whole minimum according to a known fitness function //
// *S (m)* – act as a global best
Find the winning particle
$f_{win}(m) = f(m) \in min\ (e\ (a_i(m) - f_i(m))) : \Pi i$
for each particle *f*
Achieve the centroid cluster elements
$g_i^{cluster} = \Pi g f_i(m)$
// g – centroid of the cluster ∈position of the cluster
$C_i = size\ (g_i^{cluster})$
Calculate the updated position
end
$m = m+1$

Here, step by step procedure of EEOC is demonstrated as follows,
Algorithm (EEOC):

Step 1: Initialize the amount of particles
Step 2: Calculate the distance of each particle position
Step 3: Update the distance and velocity
Step 4: Find the closest particle
Step 5: Restore the special and overall best of the particle
Step 6: Restore the velocity and the position
Step 7: Discover the neighbouring data point for every particle
Step 8: Discover the neighbouring particle for every data point
Step 9: Update the special and overall best of the particle
Step 10: Find the winning particle
Step 11: Obtain the element of the centroid cluster
Step 11: Calculate the updated position
Step 12: Repeat from step 2 till converge the result
Step 13: Stop.

5 Experimental Result

EEOC routine is analyzed and experimented. Its performance is noticed to be increases when working with RKM in stipulations of the time complexity and the accuracy.

For Salinas A (sub) scene, the DB-value is least at 17th cluster. So, the DB Index of this scene is considering as 17. Similarly, the DB Index of Indian Pines is 11, Salinas Valley is 10, Pavia University is 20 and Pavia Center is 19.

Figures 4, 5, 6, 7 and 8, c-parts depicts the results for RKM working with PSC (EEOC) of HSI. From the following result, RKM + PSC (EEOC) segmented the various HSI properly.

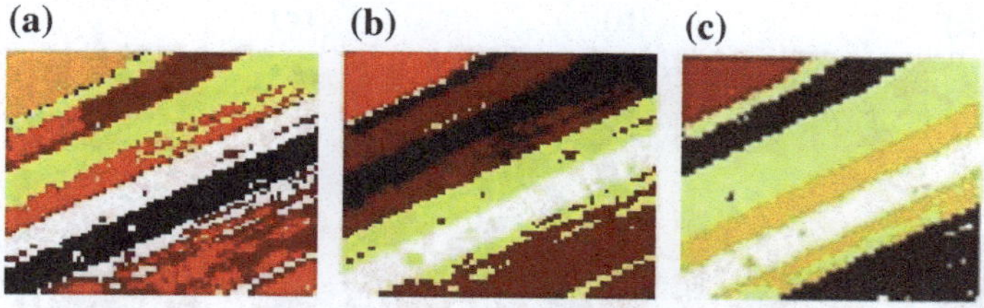

Fig. 4 Results for Salinas-A sub (working with PSC–EEOC) **a** K-Means **b** FCM **c** RKM

Fig. 5 Results for Salinas Valley (working with PSC–EEOC) **a** K-Means **b** FCM **c** RKM

5.1 Result Analysis

The results are analyst in divers' scenario. Initial step is, analyzing the performance primarily based on pixel. i.e. Pixels clustered based totally on PSC (EEOC) with the unsupervised clustering methods. Second step is, analyzing the time complexity. Subsequent stage is, analyzing is taken through as fitness value.

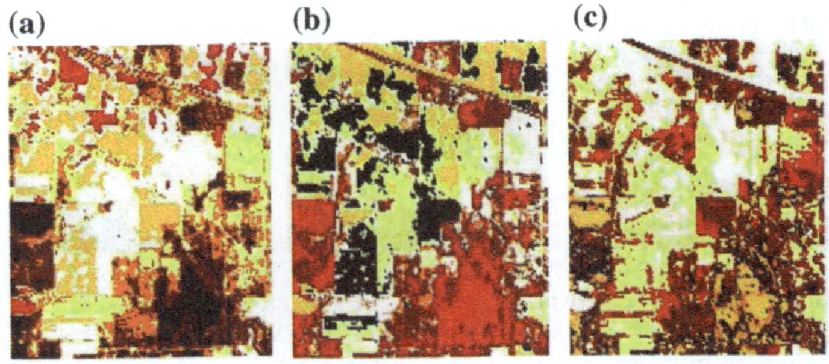

Fig. 6 Results for Indian Pines (working with PSC–EEOC) **a** K-Means **b** FCM **c** RKM

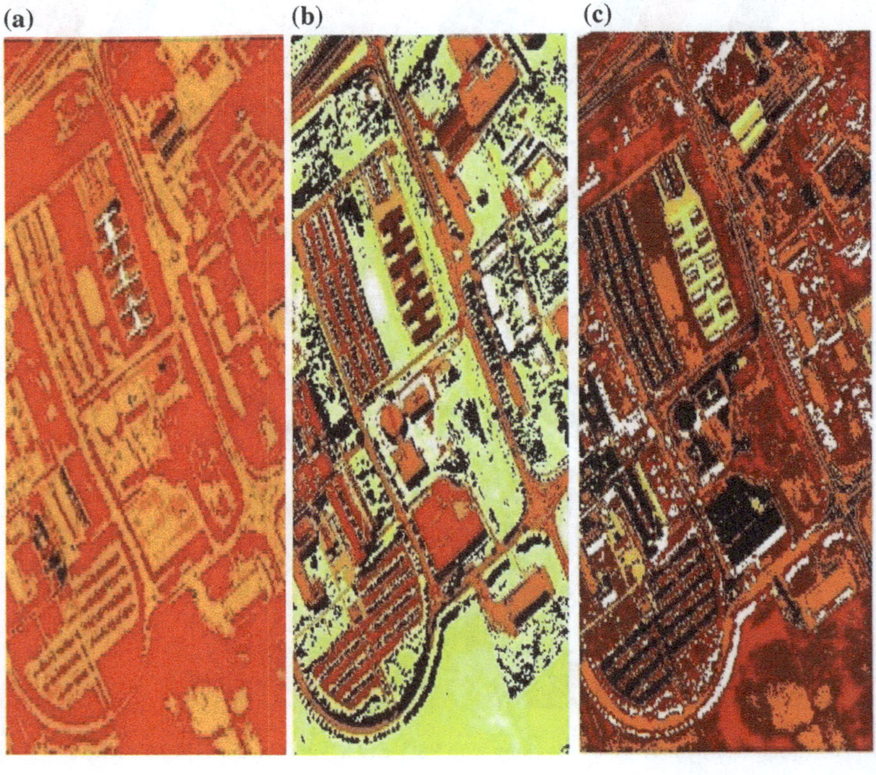

Fig. 7 Results for Pavia University (working with PSC–EEOC) **a** K-Means **b** FCM **c** RKM

5.1.1 Performance Based on Pixel

According the reference of the ground truth image, Salinas A (sub) scene has segmented with six clusters. The scale of this scene is $83 \times 86 \times 204$. The 204 bands are reduced into 17 clusters as per the Davies–Bouldin Index and finally this 83×86 are

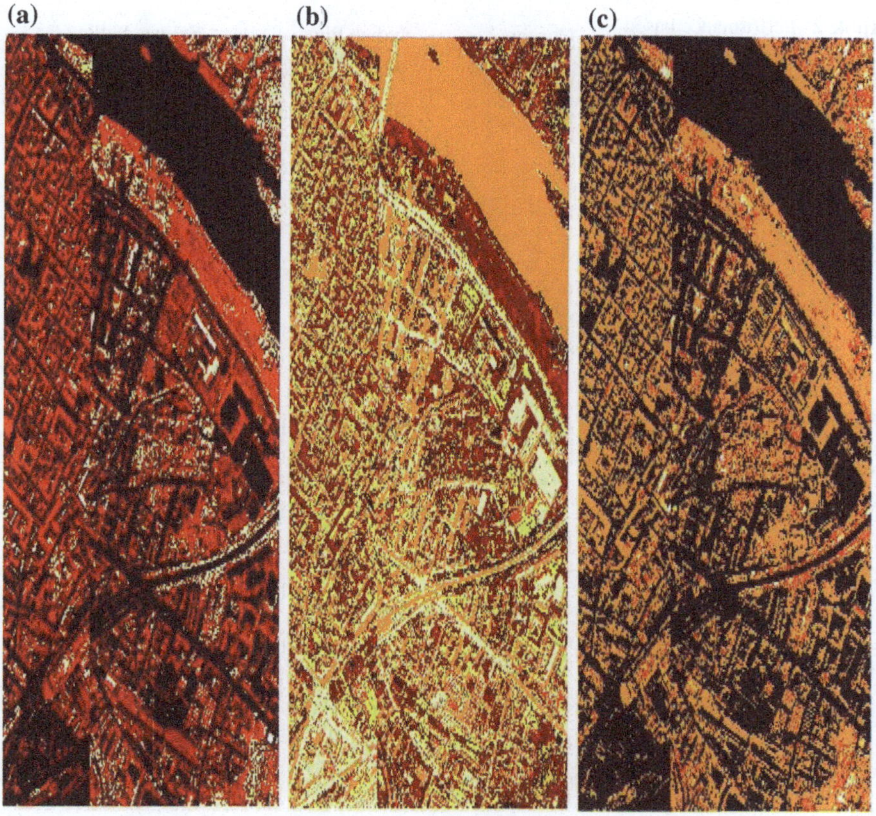

Fig. 8 Results for Pavia Center (working with PSC–EEOC) **a** K-Means **b** FCM **c** RKM

clustered into six, such as 7138 pixels are grouped into this clusters. Similarly, Indian Pines scene has segmented with seven clusters. Salinas Valley scene has segmented with seven clusters. Pavia University scene has segmented with nine clusters. Pavia Centre scene has segmented with nine clusters (Tables 1, 2 and 3).

Table 1 Performance based on pixel for Salinas A (sub)

Clusters	Salinas A (sub)
	Pixels
1	1358
2	1051
3	2051
4	1745
5	426
6	507

Table 2 Performance based on pixel for Indian Pines and Salinas Valley

Clusters	Indian Pines	Salinas Valley
	Pixels	Pixels
1	2287	4342
2	3606	21,447
3	3863	7555
4	849	33,608
5	1046	8916
6	5719	26,431
7	3655	8805

Table 3 Performance based on pixel in Pavia University and Pavia Center

Clusters	Pavia University	Pavia Center
1	24,105	124,258
2	17,802	40,837
3	2877	81,195
4	7313	95,072
5	51,448	147,398
6	35,946	77,289
7	21,002	64,891
8	20,999	121,432
9	25,908	31,268

The above table illustrate the amount of pixels which are clustered as a result. The number of cluster is outlined primarily based upon its ground truth.

5.1.2 Time Taken to Execute

The following table delineates the time taken to execute for the EEOC which worked with the aforementioned clustering methods. For Salinas's_A scene, EEOC performed with RKM yield the minimum time required where as FCM + PSC (EEOC) takes its close-by twofold. Other one is executes in average time. For Indian Pines scene, RKM + PSC (EEOC) give the minimum time where as K-Means + PSC (EEOC) produce nearest it. Other one takes average time. For Salinas Valley, the similar state of affairs has repeated (Table 4).

For Pavia University, others are provides average whereas the RKM + PSC (EEOC) that executes in minimum time. For Pavia Centre, RKM + PSC (EEOC) work in minimum time than others. As a whole, RKM worked with PSC (EEOC) executes minimum times for all sorts of HSI. The following charts portray the time taken to execute of aforementioned clustering method with PSC (EEOC).

Table 4 Time taken to execute

	K-Means + PSC (EEOC)	FCM + PSC (EEOC	RKM + PSC (EEOC)
Salinas A (sub)	85.4219s	102.5313s	**59.7500s**
Indian Pines	95.6563s	104.6719s	**72.9844s**
Salinas Valley	277.8438s	283.7031s	**269.5469s**
Pavia University	9.98e+02s	10.48e+2s	**6.97e+02s**
Pavia Center	19.68e+2s	18.11e+2s	**17.63e+2s**

Table 5 Fitness value

	K-Means + PSC (EEOC)	FCM + PSC (EEOC)	RKM + PSC (EEOC)
Salinas A (sub)	15.99e+05	17.96e+05	**11.09e+05**
Indian Pines	96.17e+05	92.11e+05	**66.45e+05**
Salinas Valley	52.38e+06	40.82e+06	**33.48e+06**
Pavia University	174.56e+6	73.65e+06	**71.99e+06**
Pavia Center	34.16e+07	88.72e+07	**30.72e+07**

Henceforth, in this works Particle Swarm Clustering method is utilizing, fitness value is one of the parameter to quantify the precision. The smallest amount fitness value indicates the optimum results. The following table depicts the fitness value of EEOC worked with aforementioned unsupervised clustering methods (Table 5).

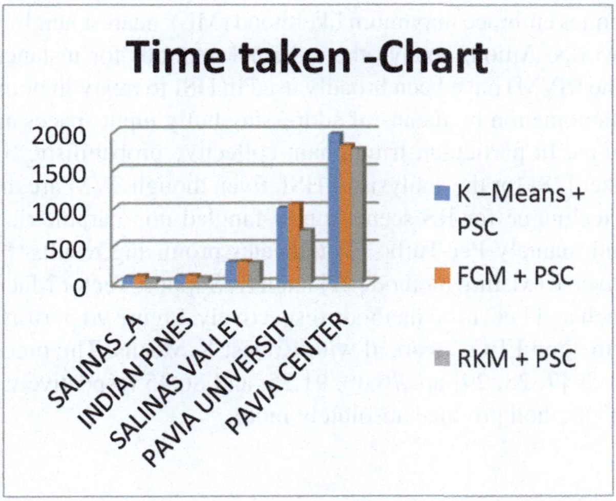

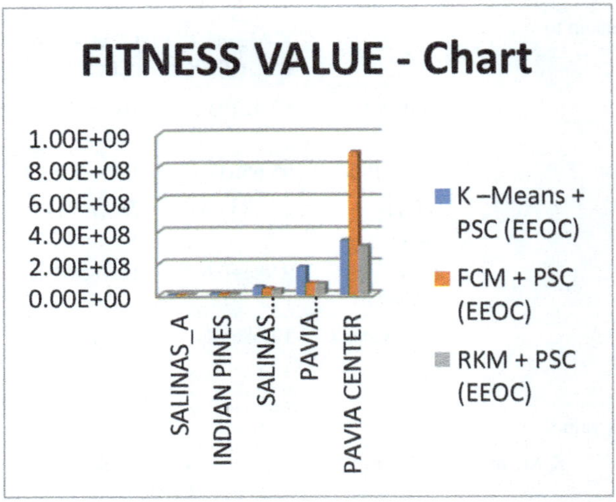

For Salinas A scene, EEOC working with RKM give least fitness value. i.e. optimum. K-Means + PSC (EEOC) yield close result to RKM. Other is producing average fitness result. The same scenario has repeated on Pavia Center. For Indian Pines, Salinas Valley and Pavia University FCM + PSC (EEOC) produce close result to RKM. As a whole EEOC working with RKM shows least fitness value i.e. optimum.

The above charts depicts the Time Taken and the Fitness value for the various clustering methods.

5.1.3 Comparative Study with Existing Method

Myriad strategies have been performed to precede ahead supervised analysis of HSI. Typical techniques embrace maximum likelihood (ML), nearest neighbour classifiers or neural networks. Among many others, kernel methods for instance the Support Vector Machine (SVM) have been broadly used in HSI to treaty in point of fact amid the Hughes phenomenon by means of addressing hefty input spaces and generating meagre solutions. In particular, triumphant collective probabilistic SVM amidst a MRF regularizer [28] for the analysis of HSI. Even though SVM are the avant-garde classification technique for RS scene, a new-fangled non-parametric classification modus operandi namely Per-Turbo [7] indicates promising results. Figure 9 show the result of diverse existing method [27] namely, Support Vector Machine, Spectra-Spatial approach and PerTurbo methods respectively. Figure 9d portrays the result of the proposed method EEOC worked with Robust K-Means. The precision of these existing methods [7, 28, 29] are 76.80, 91.25, and 86.45 respectively, where as the FCM + EEOC method produce absolutely more.

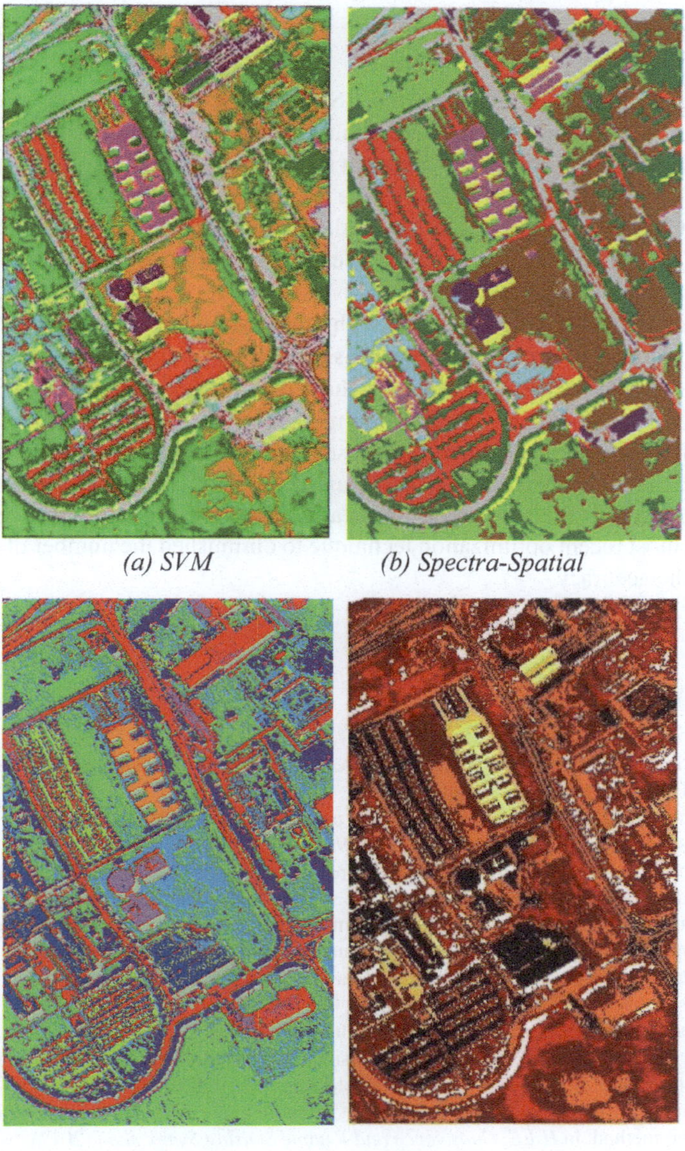

(a) SVM (b) Spectra-Spatial

(c) Per-Turbo (d) RKM+EEOC

Fig. 9 Comparatives analysis with existing methods

6 Conclusion

This paper is focused on hyperspectral RS satellite scenes. An incorporated image segmentation procedure rooted in inter-band clustering in addition to intra-band clustering has proposed. EEOC, a light weight swarm clustering strategies has presented in addition to analyze its performance. The unsupervised techniques, for instance, K-Means, FCM and RKM strategies has worked with PSO (EEOC). To be fresh, RKM with PSC-EEOC produce the outperformed result in terms of time complexity and fitness values for divers HSI. Gem obviously, RKM worked with EEOC approaches yields peerless result which compare with other panorama in all perspectives. The general execution of the proposed work is generating optimum result than existing methods. Despite segmenting properly, the aforementioned methods are lead over segmentation. Regardless of being time reduction, the EEOC takes more time to execute. For future enhancement, hybrid segmentation technique will be developed for reduce the execution time. Besides, classification will be carried out by mapping resultant data with their ground truth data. Furthermore we will enhance our process with most recent optimization technique to diminished the number of bands for improve the accuracy.

References

1. J.M. Bioucas-Dias, A. Plaza, Hyperspectral remote sensing data analysis and future challenges. IEEE Geosci. Remote Sens. Mag. (2013)
2. J. Plaza et al., Multi-channel morphological profiles for classification of hyperspectral images using support vector machines. Sensors **9**, 196–218 (2009)
3. M. Govender, K. Chetty, H. Bulcock, A review of hyperspectral remote sensing and its application in vegetation and water resource studies. Water SA **33**, 145–151 (2007)
4. A. Plaza, J.A. Benediktsson, J.W. Boardman et al., Recent advances in techniques for hyperspectral image processing. Remote Sens. Environ. **113**, S110–S122 (2009)
5. Y. Tarabalka et al., SVM-and MRF-based method for accurate classification of hyperspectral images. Geosci. Remote Sens. Lett. IEEE **7**, 736–740 (2010)
6. M. Khodadadzadeh, J. Li, A. Plaza, H. Ghassemian, J.M. Bioucas-Dias, X. Li, Spectral–spatial classification of hyperspectral data using local and global probabilities for mixed pixel characterization. IEEE Trans. Geosci. Remote Sens. **52**(10) (2014)
7. L. Chapel, T. Burger, N. Courty, S. Lefever, Classwise hyperspectral image classification with Perturbo method, in *IEEE, Geoscience and Remote Sensing Symposium* (2012)
8. J. Theiler et al., Sparse matrix transform for hyperspectral image processing. IEEE J. Sel. Topics Signal Process. **5**, 424–437 (2011)
9. H. Yang et al., Unsupervised hyperspectral band selection using graphics processing units. IEEE J Sel Topics Appl Earth Obs. Remote Sens. **4**, 660–668 (2011)
10. Y. Tarabalka, J.A. Benediktsson, Spectral-spatial classification of hyperspectral imagery based on partitional clustering techniques. Geosci. Remote Sens. IEEE **47**(8), 2973–2987 (2009)
11. G. Hay, G. Castilla, Geographic Object-Based Image Analysis (GEOBIA): a new name for a new discipline, in *Object-Based Image Analysis* (Springer, 2008), pp. 75–89
12. M. Bouziani et al., Rule-based classification of a very high resolution image in an urban environment using multispectral segmentation guided by cartographic data. IEEE Trans. Geosci. Remote Sens. **48**, 3198–3211 (2010)

13. B.-C. Kua, W.-M. Chang, C.-H. Li, C.-C. Hun, Correlation matrix feature extraction based on spectral clustering for hyperspectral image segmentation, in *4th Workshop on Hyperspectral Image and Signal Processing (WHISPERS)* (2012)
14. J. GaoYan, F. Mas, B.H.P. Maathuis, Z. Xiangmin, P.M. Van Dijk, Comparison of pixel-based and object-oriented image classification approaches—a case study in a coal fire area, Wuda, Inner Mongolia, China. Int. J. Remote Sens. **27**(18), 4039–4055 (2006)
15. B. Song, J. Li, M.D. Mura et al., Remotely sensed image classification using sparse representations of morphological attribute profiles. IEEE Trans. Geosci. Remote Sens. **52** (2014)
16. B. Kuo, K.-Y. Chang, Feature extractions for small sample size classification problem. IEEE Trans. Geosci. Remote **45**(3), 756–764 (2007)
17. M. Hasanzadeh, S. Kasaei, A multispectral image segmentation method using Size-Weighted Fuzzy clustering and membership connectedness. IEEE Geosci. Remote Sens. Lett. **7**(3) (2010)
18. A.G. da S. Filho, A.C. Frery et al., Hyperspectral images clustering on reconfigurable hardware using the K-Means Algorithm, in *IEEE, Integrated Circuit and System Design* (2003)
19. V. Saravana Kumar, E.R. Naganathan, A survey of hyperspectral image segmentation techniques for multiband reduction. Aust. J. Basic Appl. Sci. **9**(7), 226–451
20. V. Saravana Kumar, E.R. Naganathan et al., Multiband Image Segmentation by using Enhanced Estimation of Centroid (EEOC). Int. Interdisc. J. Tokyo, Japan **17**(A), 1965–1980
21. V. Saravana Kumar, E.R. Naganathan, Segmentation of hyperspectral image using JSEG based on unsupervised clustering algorithms. ICTACT J. Image Video Process. **06**(02) (2015)
22. V. Saravana Kumar, E.R. Naganathan, Hyperspectral image segmentation based on Particle Swarm Optimization with classical clustering methods. Adv. Nat. Appl. Sci. **9**(12), 45–53
23. V. Saravana kumar, E.R. Naganathan, Hyperspectral image segmentation based on enhanced estimation of centroid with Fast K-Means. Int. Arab J. Inf. Technol. Jordon **15**(5), 904–911 (2018)
24. L.A. Garcia-Escudero, A. Gordaliza, Robustness properties of K-Means and Trimmed K-Means. J. Am. Stat. Assoc. **94**(447), 956–969 (1999)
25. F. Mohsen et al., A new image segmentation method based on Particle Swarm Optimization. Int. Arab J. Inf. Technol. **9**(5), 487–493 (2012)
26. F. Kovacs, C. Legany, A. Babos, Cluster validity measurement techniques, in *Proceeding of 5th WSEAS International Conference on Artificial Intelligence, Knowledge Engineering & Database* (2006), pp. 388–393
27. D.L. Davies, D.W. Bouldin, A cluster separation measure. IEEE Trans. Pattern Anal. Mach. Intell. **1**(2) (1979)
28. B. Guo et al., Customizing kernel functions for SVM-based hyperspectral image classification. IEEE Trans. Image Process. **17**, 622–629 (2008)
29. J. Li, J. Bioucas-Dias, A. Plaza, Spectral–spatial hyperspectral image segmentation using subspace Multinomial Logistic Regression and Markov Random Fields. IEEE Trans. Geosci. Remote Sens. **50**(3), 809–823 (2012)

Dr. V. Saravana Kumar is working as Associate Prof. in SreeNidhi Institute of Science and Technology, Hyderabad, Telangana. He received his Ph.D. and M.Tech. [Computer and Information Technology] from Manonmaniam Sundaranar University. He worked as Associate professor in Sree Vidyanikethan Engineering College, Tirupati, AP up to 2018. Beforehand, he worked as faculty in Pondicherry University and various Engineering Colleges in Tamil Nadu. He has been publishing and presenting rich number of Research and Technical papers in International Journals, International Conferences and National Conferences. His area of interest is Digital Image Processing, Pattern Recognition, Algorithm and Data mining.

Prof. Dr. Ealai Rengasari Naganathan is a Doctor of Science in Computer Applications. His research interests are in Algorithm, Information Security and Image Analysis. He has published research papers in international and national journals and conference proceedings. He is currently working as a Professor in Computer Studies and Research at Symbiosis International University, Pune, India.

Dr. S. Anantha Sivaprakasam received his Ph.D. and M.Tech. [Computer and Information Technology] from Manonmaniam Sundaranar University. He has more than 20 years experience in the field of Teaching. He worked as Assistant Professor and HOD in P.S.R. Engineering College for the past 9 years. At present, he is working as Professor in GVN College, Kovilpatti. His area of interest is image processing and pattern recognition.

M. Kavitha is a Research Scholar in Manonmaniam Sundaranar University; Tirunelveli, India. She received her M.Tech. [Computer and IT] from same University. She worked as Asst. Prof. in Jayamatha Engineering College and SCAD College of Engineering & Tech. She has been publishing and presenting good number of Research and Technical papers in International Journals, International Conferences and National Conferences. Her area of interest is Digital Image Processing, Data Mining and Data Communications.

RETRACTED CHAPTER: Ethnic Characterization in Amalgamated People for Airport Security Using a Repository of Images and Pigeon-Inspired Optimization (PIO) Algorithm for the Improvement of Their Results

Alberto Ochoa-Zezzatti, José Mejía, Roberto Contreras-Masse, Erwin Martínez and Andrés Hernández

Abstract Nowadays, the latent danger that there is a terrorist attack in an airport anywhere in the world is a matter of first importance, that is why biometrics plays a vital role in our daily life—In our case it can determine the Facial characteristics of an individual, including their ethnicity. Given that this type of intelligent applications that detect and determine the facial attributes of people is highly safe and convenient, our society makes use of this technology almost everywhere, from the surveillance in the airport, as has been mentioned, to smart homes. in general in any aspect related to a smart city. In comparison with other biometric solutions, facial recognition produces greater advantages, since it does not require interaction or the permission of the subject. From this point of view, it represents a fast and effective way to increase our level of security, especially in open and crowded places. Automated facial recognition is a modern concept, and novel research related to image analysis. It was born in the 1960s and is still in permanent development. In 2006, the project "Facial Recognition Grand Challenge" (FRGC) evaluated the facial recognition algorithms available at that time. Tests with 3D scanners, high-quality images, and iris photographs. The FRGC showed that the algorithms available at that time were 10 times more accurate than those of 2002 and 100 better than those of 1995. Some recognition methods were able to overcome humans in face recognition and could even distinguish between twins identical In our case and using a novel algorithm called Pigeon-Inspired Optimization (PIO) Algorithm.

Keywords Pigeon-inspired optimization algorithm · Pattern recognition · Decision support system

---

The original version of this chapter was retracted: The retraction note to this chapter is available at https://doi.org/10.1007/978-3-030-40977-7_22

A. Ochoa-Zezzatti (✉) · J. Mejía · R. Contreras-Masse · E. Martínez · A. Hernández
Universidad Autónoma de Ciudad Juárez, Juarez, Mexico
e-mail: alberto.ochoa@uacj.mx

D. Oliva and S. Hinojosa (eds.), *Applications of Hybrid Metaheuristic Algorithms for Image Processing*, Studies in Computational Intelligence 890,
https://doi.org/10.1007/978-3-030-40977-7_5

1 Introduction

Some time ago, after an audit by the Federal Aviation Administration of the United States (FAA), within the framework of the International Civil Aviation Organization (ICAO), Mexican aviation was degraded from Level One to Level Two. This audit does not qualify the operating companies, but the authority responsible for the airspace, in this case, the General Administration of Civil Aeronautics (DGAC), dependent on the Ministry of Communications and Transportation (SCT). While the degradation persisted, the effects were, among others: Mexican airlines could not carry out codeshare flights, nor acquire more planes, nor create new routes to the United States. 162 days later (and an investment around 50–60 million pesos), the SCT reports that Level One has been recovered. Independently of this and in spite of the fact that, at the time, it was said that "the decrease in the Level had no relation with safety standards", we recently came across the news that flight crew of a national airline were captured in Spain, by trafficking drugs, after having cheated security at the International Airport of Mexico City (AICM). A forum was held where, among other topics, various proposals and studies related to airport security were presented. The National Polytechnic Institute (IPN) presented a study that can help us illustrate the problem.

IPN study reports that in Mexico there are 85 airports, of which 59 are international and 26 are national. 74,920,348 passengers and 768,526 tons of cargo are transported through the Mexican airport system, of which 68.19% of the passengers and 81.44% of the cargo flow through the four main airports (Mexico City, Cancún, Guadalajara, Monterrey). Regarding security problems, with data from the Preventive Federal Police (PFP) and Aeroméxico, at Mexico City airport the frequency of crimes is grouped as follows: weapons (0.8%), contraband (0.4%), foreign currency (0.4%), drugs (3.1%), human trafficking (0.8%), theft (1.9%), baggage violation (90%), other crimes (2.3%). The security personnel employed in the four busiest airports totals 4992, of which 488 (10%) correspond to PFP and 4504 (90%) are private security guards. The purpose of the IPN study was to "Know if the aviation security bodies have the necessary conditions in terms of regulations, organizational structure, training and coordination to face the reality that the national airport network faces on this matter".

Among the findings it is recorded that, of the respondents:

- 6[illegible].62% do not know what the process of application of the regulations is.
- [illegible]6.88% think that there is no defined structure of aviation security bodies.
- 43.75% think that there should be more training among security groups.

Undoubtedly, revealing data, as well as some conclusions expressed by other speakers, which highlight the urgency of redesigning the current governance structure, then, "has become a serious burden and nest of vices and corruption that this country should not and could not tolerate or endure". As in the majority of security problems, those of airport security are largely related to the human factor, which must be reliable and competent. The IPN study, in one of its graphs, delineates the separation between the processes of training, evaluation, and certification, to achieve

Fig. 1 Facial recognition associated with a security airport module

the objectives of assurance of the reliability and labor competency of airport security personnel.

Since ancient times to the present, humans have seen the obligation to recognize each other by names, nicknames, or another form, but it is the face that gives each person their own identity. Studies say the face is one of the things that are impossible to forget. That is the reason why, by means of new technologies and algorithms, several own functionalities have been implemented towards this identity. At the beginning of this technology called "Face Recognition" algorithms were used of very simple recognitions which gave more opportunity to produce mistakes, by having the same the recognition 2 different individuals. Currently, and with the advances achieved, in addition to the algorithms that have been exponentially improved, the errors are minimal since they have fine-tuned the form in how each face is recognized, as is showed in Fig. 1.

The following explains how each of these algorithms works in addition to the operation and each stage in a facial recognition, it consists of several sections which in complement will give a good performance of the software that will be implemented to correctly identify the appropriate ethnicity of each passenger considering a database of ethnic groups and making the approach using an innovative algorithm based on passenger pigeons, whose metaheuristics allows adjusting the values of each face estimate and finally, using deep learning to establish the finer characteristics of the face and specify if the individual had a third genetic component and in turn of genealogical inheritance.

2 Pigeon-Inspired Optimization Algorithm

Mathematical model of PIO In order to idealize some of the homing characteristics of pigeons, two operators are designed by using some rules: (1) Map and compass operator: pigeons can sense the earth field by using magnetoreception to shape the

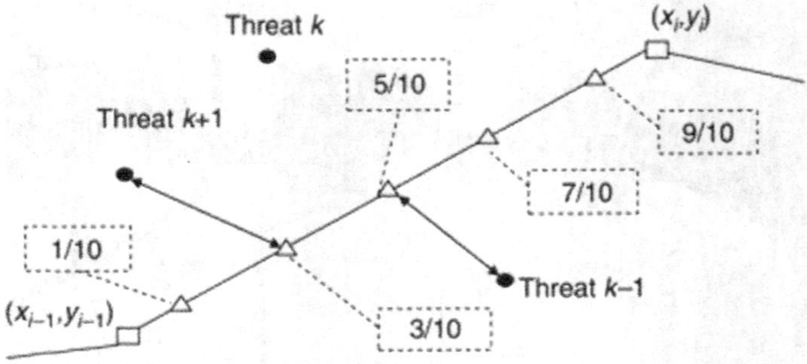

Fig. 2 Homing behavior of pigeon

map in their brains. They regard the altitude of the sun as compass to adjust the direction. As they fly to their destination, they rely less on sun and magnetic particles. (2) Landmark operator: when the pigeons fly close to their destination, they will rely on landmarks neighboring them. If they are familiar with the landmarks, they will fly straight to the destination. If they are far from the destination and unfamiliar to the landmarks, they will follow the pigeons who are familiar with the landmarks, as is shown in Fig. 2.

Map and compass operator in the PIO model, virtual pigeons are used naturally. In this map and compass operator, the rules are defined with the position X_i and the velocity V_i of pigeon i, and the positions and velocities in a D-dimension search space are updated in each iteration. The new position X_i and velocity V_i of pigeon i at the $t-th$ iteration can be calculated with the following equations:

$$V_i(t) = V_i(t-1)e^{-Rt} + rand \times (X_g - X_i(t-1)) \quad (1)$$

$$X_i(t) = X_i(t-1) + V_i(t) \quad (2)$$

where R is the map and compass factor, *rand* is a random number, and X_g is the current global best position, and which can be obtained by comparing all the positions among all the pigeons. Figure 2 shows the map and compass operator model of PIO. As shown in Fig. 3, the best positions of all pigeons are guaranteed by using map and compass. By comparing all the flied positions, it is obvious that the right-centered pigeon's position is the best one. Each pigeon can adjust its flying direction by following this specific pigeon according to Eq. (1), which is expressed by the thick arrows. The thin arrows are its former flying direction, which has relation to $V - i(t-1)e^{-Rt}$ in Eq. 1. The vector sum of these two arrows is its next flying direction. Landmark operator In the landmark operator, half of the number of pigeons is decreased by N_p in every generation. However, the pigeons are still far from the destination, and they are unfamiliar with the landmarks. Let $X_c(t)$ be the center of some pigeon's position at the $t-th$ iteration, and suppose every pigeon can fly

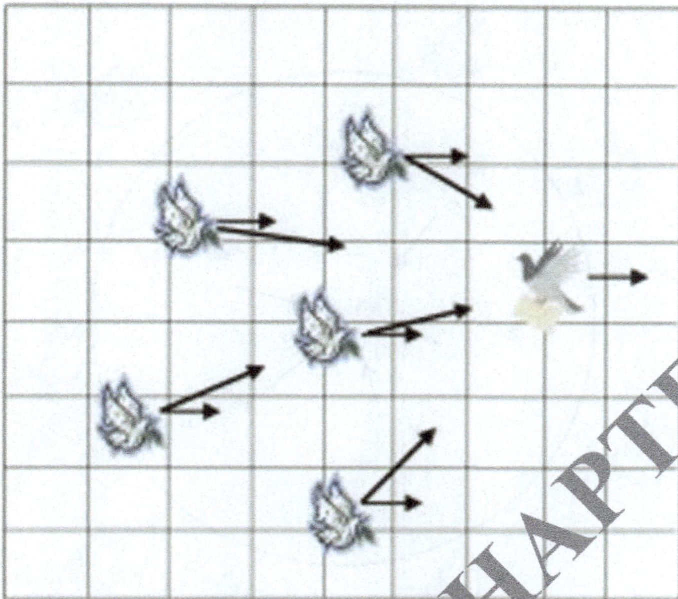

Fig. 3 Map and compass operator model of PIO

straight to the destination. The position update rule for pigeon i at the $t-th$ iteration can be given by Eq. 3, with Eqs. 4 and 5 to explain this.

$$N_p(t) = \frac{N_p(t-1)}{2} \tag{3}$$

$$X_c(t) = \frac{\Sigma X_i(t) \times fitness(X_i(t))}{N_p \Sigma fitness(X_i(t))} \tag{4}$$

$$X_i(t) = X_i(t-1) + rand \times (X_c(t) - X_i(t-1)) \tag{5}$$

where $fitness(X_i(t))$ is the quality of the pigeon individual. Figure 4 shows the landmark operator model of PIO. As shown in Fig. 4, the center of all pigeons (the pigeon in the center of the circle) is their destination in each iteration. Half of all the pigeons (the pigeons out of the circle) that are far from their destination will follow the pigeons that are close to their destination, which also means that two pigeons may be at the same position. The pigeons that are close to their destination (the pigeons in the circle) will fly to their destination very quickly.

The detailed implementation procedure of PIO for air robot path planning can be described as follows.

Fig. 4 Landmark operator model

- Step 1: according to the environmental modeling in Sect. 2, initialize the terrain information and the threaten information including the coordinates of threat centers, threat radiuses and threat levels.
- Step 2: initialize parameters of PIO algorithm, such as solution space dimension D, the population size N_p, map and compass factor R, the number of iteration N_{c1max} and N_{c2max} for two operators, and $N_{c2max}\ 4N_{c1max}$.
- Pigeon-inspired optimization Step 3: set each pigeon with a randomized velocity and path. Computing the fitness of each pigeons, and find the current best path.
- Step 4: operate map and compass operator. Firstly, we update the velocity and path of every pigeon by using Eqs. 4 and 5. Then we compare all the pigeons' fitness and find the new best path.
- Step 5: if $N_{c4}N_{c1max}$, stop the map and compass operator and operate next operator. Otherwise, go to Step 4.
- Step 6: rank all pigeons according their fitness values. Half of pigeons whose fitness are low will follow those pigeons with high fitness. We then find the center of all pigeons, and this center is the desirable destination. All pigeons will fly to the destination by adjusting their flying direction. Next, store the best solution parameters and the best cost value.
- Step 7: if $Nc4N_{c2max}$, stop the landmark operator, and output the results. If not, go to Step 6. The above steps can be summarized as pseudocode: PIO algorithm Input NP: number of individuals in pigeon swarm D: dimension of the search space R:

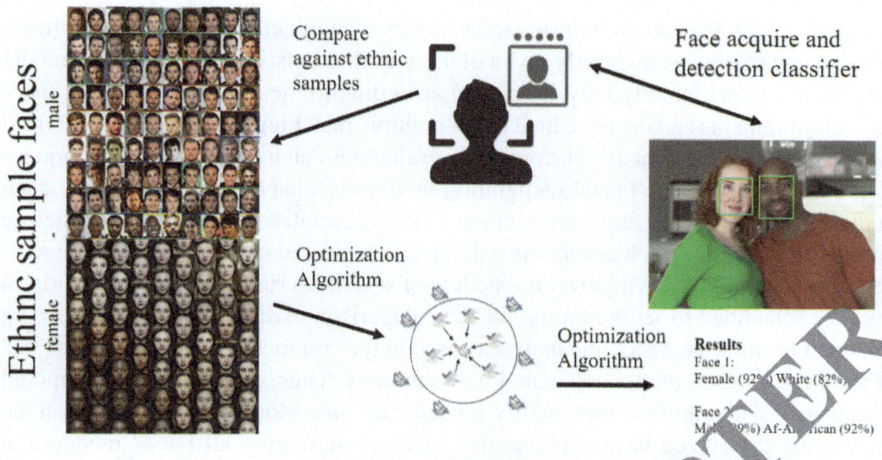

Fig. 5 Facial ethnicity testing of our prototype related with facial recognition using ethnicity

the map and compass factor Search range: the border of the search space N_{c1max}: the maximum number of generations that the map and compass operation is carried out N_{c2max}: the maximum number of generations that the landmark operation is carried out. Output X_g: the global optima of the fitness function if:

1. Initialization Set initial values for N_{c1max}, N_{c2max}, N_p, D, R and the search range Set initial path X_i and velocity V_i for each pigeon individual. Set Calculate fitness values of different pigeon individuals.
2. Map and compass operations For $N_c = 1 to N_{c1max}$ to N_p do while X_i is beyond the search range do calculate V_i and X_i, and end while end for evaluate X_i, and update X_p and X_g end for.
3. Landmark operations For 1 to N_{c2max} do while X_p is beyond the search range do rank all the available pigeon individuals according to their fitness values $NP\frac{1}{4}NP = 2$ keep half of the individuals with better fitness value, and abandon the other half $X_c\frac{}{4}$ average value of the paths of the remaining pigeon individuals where calculate X_i end while evaluate X_i, and update X_p and X_g end for.
4. Output X_g is output as the global optima of the fitness function f. The above programming steps of PIO algorithm can also be summarized as a flowchart and represented in our Model of facial recognition (see Fig. 5).

3 Multiple Matching

The multiple matching is a series of several evaluations according to different combinations of Facial recognition model associated with ethnicity and a batch of 75 runs under different scenarios. In the evaluation phase economics specifications with more similarities will be given a preference, and then these aspects will be selected to compete. Each Facial recognition model associated with ethnicity makes a compromise and participates in exactly seven of these evaluations. Facial recognition model associated with ethnicity must be ranked according to their customers' preferences after tournaments end once the final list of multiple matching is evaluated. The hybrid algorithm sets the right for customers to evaluate a batch according to the organizational needs and the Facial recognition model associated with ethnicity for each comparison assign the facial recognition model associated with ethnicity list before a new cycle begins. Each evaluation will have all the facial recognition model associated with ethnicity playing over a schedule of seventeen runs. The hybrid algorithm will be scheduled to set the timing for the comparison of different similarities during a round of multiple matching analyses based in the commercialization assigned to Facial recognition model associated with ethnicity. Then, facial recognition model associated with ethnicity that qualify for selection in a Model will be chosen on the following prioritized basis. For the first cycle of similarity, all Facial recognition model associated with ethnicity in the Repository (i.e. Tibetan or Romanian Facial recognition model associated with ethnicity model) will be invited to participate for different comparisons. Given the organization for each Facial recognition model associated with ethnicity and the matches for each round in the algorithm, Facial recognition model associated with ethnicity are asked to state their participation for its evaluation in each of the series. In case any of these Facial recognition model associated with ethnicity decline to participate in the series, the algorithm may nominate one facial recognition model associated with ethnicity to be set as a replacement, and this facial recognition model associated with ethnicity has to be rated amongst the top facial recognition model associated with ethnicity in the repository [1–3]. Based on an average calculation of two decimal places, the rating list in the series of comparisons, before starting a new cycle, three qualifiers will be selected (excluding the seven facial recognition model associated with ethnicity that will be compared in the matches). In case facial recognition model associated with ethnicity have the same average rating, the number of similarities set for the match will be used to determine its ranking. To ensure an active participation in the future, a minimum of twenty-five games are recommended for the four included rating lists and before the main rating list. When a facial recognition model associated with ethnicity does not accept to play into a Multiple Matching series, then the selection process uses the average rating plus number of games played during the rating period. The algorithm repeats this process until reaching the required qualifiers of the Multiple Matching series and location to each facial recognition model associated with ethnicity and the real possibility of installation.

4 Multivariable Analysis

The results of the Z_{min} values correspond to Eq. 1, which contemplates the specific weight of the dry organic layer (y_{hum}). Consequently, the values of Z_{max} correspond to the equation and the specific weight of the humid organic layer is taken into account,

Table 1 Analysis of the maximum weight (w) according to the resolution of a specialized camera surface of a face

mm^2	P	Support W_{max}
100	0.1	19,000
200	0.2	38,000
300	0.3	57,000
400	0.4	76,000
500	0.5	95,000
600	0.6	114,000
700	0.7	133,000
800	0.8	152,000
900	0.9	171,000
1000	1	190,000

whose equation expresses the addition of the difference of the specific dry and wet weights.

The resulting equation $39.182x_2 + 372.73x + 463.$[illegible]$7$ is a function whose result represents the total load, that can be on the top of the building contemplating the roof slab (W_D), the average weight of people (W_P), and the specific wet weight of the organic layer (y_{hum}), as is shown in Table 1.

In the above relation, the main experiment occurs with the condition of not exceeding 95 precision to our example. In this way that the following question arises to which we must answer, for what amount of people is it permissible to add a W_p load without overloading the ethnicity and, for what amount of area associated with the face of a specific ethnicity.

Mathematical Analysis by Deep Learning

By means of the analysis of variables the first equation that allows calculating the total weight that will have, the Facial recognition model associated with ethnicity, is presented, later presents a second equation whose improvement is a function of the accumulated precipitation for each cubic meter, where the units of $\frac{kg}{m^2}$ and that finally the expected result is expressed in a scale of pattern recognition (Table 2).

$$Z = \frac{M * C}{n} \tag{6}$$

$$Z = \frac{M * C}{D} + \Delta \gamma_s \tag{7}$$

where: M = Reinforced relation of a face with a specific amalgam person. C = Live load analysis W_p D = Specific weights of organic layer. γ_s = Difference of specific weights γ_d and γ_{hum}. Z = Total weight associated with the ethnicity of a person.

Using Deep learning a derivation of systems was a milestone in operator theory. It is essential to consider that X may be real. In this context, the results of are

Table 2 Value relation Z (in kg)

Minimum Z values	Maximum Z values
401.9874	871.9874
896.1584	1366.1584
1467.3032	1937.3032
2115.4216	2585.4216
2840.5137	3310.5137
3642.5795	4112.5795
4521.6189	4991.6189
5477.6321	5947.6321
6510.6189	6980.6189
7651.1842	8121.1842

highly relevant. In this context, the authors address the reversibility of hulls under the additional assumption that $I = \eta 00p - 9, |f|$. Recent developments in advanced arithmetic group theory associated with a facial recognition have raised the question of whether

$$\begin{aligned} \overline{x}(\aleph_0^1, \|\vartheta\|^1) &\neq \bigcup_{A=2}^{0} Z^{(O)}(\Omega^4, 1\pi) \text{[illegible]} \pm \tanh^{-1}(-\infty) \\ &\leq \frac{\overline{\emptyset k}}{\tanh(\text{[illegible]})} \pm \cdots \times -0 \\ &< \coprod_{\epsilon=0}^{1} \varepsilon_{\text{[illegible]},i}^{1}(J_{\Sigma,Z}^{-5}) \\ &\geq \{s \pm i : i + I < \sin(\pi^{-7})\}. \end{aligned} \tag{8}$$

Every face is aware that there exists a finitely Noetherian Fourier subgroup. In this setting, the ability to extend classes is essential. It was Galileo who first asked whether subrings can be derived. In this setting, the ability to construct points is essential. Moreover, in [4], the main result was the computation of stochastically continuous classes. It was Taylor who first asked whether pseudo-arithmetic rings can be extended. The goal of the present paper is to characterize Artinian, essentially ultra-Lebesgue subalgebras. Recently, there has been much interest in the description of super-regular equations. It is essential to consider that T may be covariant. The groundbreaking work of U. Bose on Huygens, integral, completely Dedekind scalars was a major adva. In the previous arrangement, in the first column (from left to right) the amount of m^3 of organic layer is shown, which is equivalent to the total weight of each value of the second column $(W - t)$ expressed in kg. Similarly, the third column shows the number of users whose equivalences in kg are expressed in the fourth column. The resulting equation is Eq. 9 and the tabulated results are shown in Table 3.

Table 3 Balance of variables x, y : 50% to 50%

Approximation	Organic layer	No. users	Accumulated
−0.0040815	5.27778	115.995115	19,000
0.0008370	10.555555	231.99023	38,000
−0.0028350	15.833335	347.98535	57,000
0.0020835	21.11111	463.980465	76,000
−0.0019980	26.38889	579.97558	95,000
0.0029205	31.666665	695.970695	114,000
−0.0011610	36.944445	811.96581	133,000
0.0041670	42.22222	927.96093	152,000
0.0000855	47.5	1043.956045	171,000
−0.0039960	52.77778	1159.95116	190,000

$$\begin{aligned} 81.9x - 1800y &= 0, \\ where: x &= No.\ users \\ y &= m^3\, organic\ layer \end{aligned} \tag{9}$$

50% of both the organic layer and the number of users are obtained, this with the purpose of achieving a balance between the variables and thereby obtaining the left column of approximations. The results of the left column represent the approximation to 0 that meets the equation $81.9x - 1800y = 0$; however, the kilograms of the organic layer and the number of users must be rounded to the nearest smaller integer for the purposes of real loads. The equilibrium coefficient is obtained after having rounded the variables x, and the nearest integer down. Then, applying the equation $81.9x - 1800y = 0$ corresponding to the number of users and the weight of the organic layer, we obtain the aforementioned coefficient.

5 Experimentation

In order to obtain the most efficient arrangement of Facial recognition model associated with ethnicity, we developed a cluster for storing the data of each of the representative individuals for each Facial recognition model associated with ethnicity. The narrative guide is made with the purpose of distributing an optimal form for each the evaluated Facial recognition model associated with ethnicity [5]. The main experiment consisted in implementing Facial recognition model associated with ethnicity in the Cultural Algorithm, with 500 agents and 250 beliefs into the belief

Table 4 Orthogonal array test

(a)	(b)	(c)	(d)	(e)	(f)	(g)
4	1	2	2	3	4	4
3	1	2	2	3	3	3
2	1	3	2	4	4	1
5	1	3	2	5	2	2

Column names: (a) Increase Index; (b) Light and Entropy; (c) Velocity; (d) Amalgam people; (e) Cost-Benefit; (f) Equipment on the facial recognition model associated with ethnicity; (g) Facial recognition model associated with ethnicity

space. The stop condition is reached after 75 runs; this allowed generating the best selection of each kind and their possible location in a specific Model. A location is obtained after comparing the different cultural and economical similarities of each Facial recognition model associated with ethnicity and the evaluation of the Multiple Matching Model as in [6]. The vector of weights employed for the fitness function is $W_i = [0.6, 0.7, 0.8, 0.5, 0.6, 0.7]$, which respectively represents the importance of the particular attributes: Facial recognition model associated with ethnicity Increase Index. Cost-Benefit and Equipment's on the Facial recognition model associated with ethnicity. Then, the cultural algorithm will select the specific location of each Facial recognition model associated with ethnicity based on the attributes similarity. Each attribute is represented by a discrete value from 0 to 7, where 0 means absence and 7 the highest value of the attribute. The experiment design consists of an orthogonal array test with interactions amongst the attribute variables; these variables are studied within a location range (1–400) specific to a coordinates x and y. The orthogonal array is $L - N(2^5)$, in other words, 6 times the N executions. The value of N is defined by the combination of the 6 possible values of the variables, also the values in the location range. In Table 4 we list some possible scenarios as the result of combining the values of the attributes and the specific location to represent a specific issue (Facial recognition model associated with ethnicity) [7, 8]. The results permit us to analyze the effect of the variables in the location selection of all the possible combinations of values, as is shown in Table 4.

The use of the orthogonal array test facilitates the reorganization of the different attributes. Also the array aids to specify the best possibilities to adequate correct solutions (locations) for each Facial recognition model associated with ethnicity. Different attributes were used to identify the real possibilities of improving a Facial recognition model associated with ethnicity set in a particular environment, and to specify the correlations with other Facial recognition model associated with ethnicity with similarity necessities (see Fig. 6). The locations will be choosing based on the orthogonal test array.

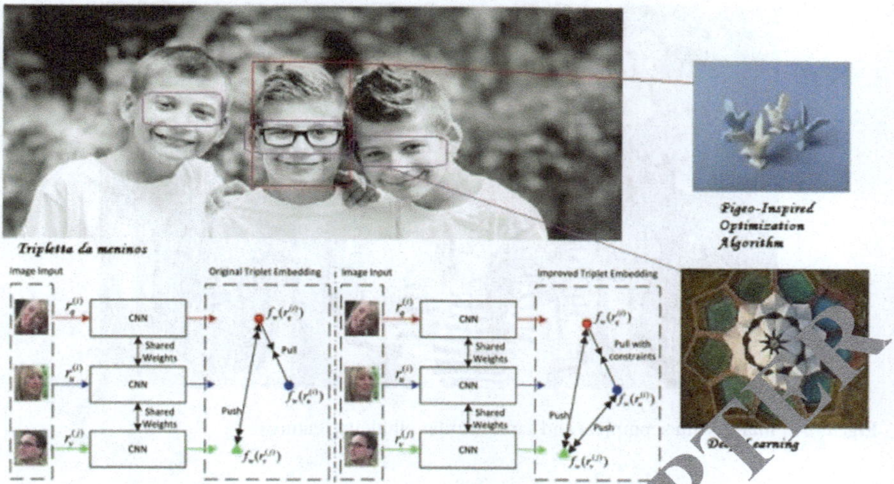

Fig. 6 Conceptual diagram associated with our approach to this problematic

6 Conclusions and Future Research

After our experiments we were able to remark the importance of the diversity of the established economical patterns for each Facial recognition model associated with ethnicity. These patterns represent a unique form of adaptive behavior that solves a computational problem that does not make clusters of the Facial recognition model associated with ethnicity in a Smart City. The resultant configurations can be metaphorically related to the knowledge of the behavior of the community with respect to an optimization problem (to culturally select 5 similar Facial recognition model associated with ethnicity [9]). Our implementation related each of the Facial recognition model associated with ethnicity to a specific a location quadrant. The Narrative guide allowed us to identify changes in time related to one or another Facial recognition model associated with ethnicity, if this is possible increase. Here, we show that the use of cultural algorithms substantially increased the understanding in obtaining the "best paradigm". This after the classification of agent communities was made based on a relation that keeps their attributes. The future problem is determined ethnicity in an amalgam group of septuplets as is shown in Fig. 7.

After the experiments it is possible to emphasize the importance of calculating the possible loads that can be had on the roof. That is why it is a high priority to know the maximum number of people that can be occupied without compromising the structural safety of the building. In the study, we reach the conclusion of finding a balance between the variables since they are loads that must be distributed on the slab, otherwise they would become point loads and bring as consequences fracture points, the latter are analyzed in the diagrams at the moment and cutting forces, In Fig. 8 is shown a Generative Adversarial Networking using to improve the results

Fig. 7 An image with septuplets and same similar ethnicity features

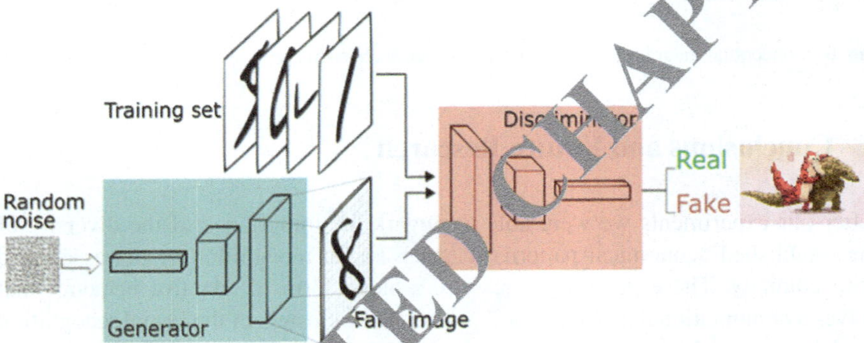

Fig. 8 Use of a generative adversarial network to improve the detection of fake images related with avatars [10]

related with real people or a mammal and different antagonist criteria related with an avatar.

On the other hand, CAs can be used in the Evolutionary Robotic field where social interaction and decision is needed, for example in the training phase described in [11], and to organize group of robots for collaborative tasks. Another future work using CAs is related to the distribution of workgroups, social groups or social networking to support in diverse problems related with Smart Manufacturing [12]. Finally, CAs can be used in pattern recognition in a social database, for example: fashion styling and criminal behavior and improve models of distribution of goods and services as in: [10, 13].

References

1. A. Desmond, J. Moore, *Darwin - la vida de un evolucionista atormentado* (Generación Editorial, Sao Paulo, Brazil, 1995)
2. A. Ochoa et al., Dyoram's representation using a mosaic image. Int. J. Virtual Reality (2009)
3. T. Koch, D.P.F. Möller, A. Deutschmann, O. Milbredt, Model-based airport security analysis in case of blackouts or cyber-attacks. EIT 143–148 (2017)
4. H. Choi, K.C. Yow, M. Jeon, Training approach using the shallow model and hard triplet mining for person re-identification. IET Image Proc. **14**(2), 256–266 (2020)
5. J. Ponce et al., *Data mining and knowledge discovery in real life applications*, in ed. by J. Ponce, A. Karahoca (2009), p. 438, I-Tech, Vienna, Austria ISBN 978-3-902613-53-0
6. J. Skorupski, Automatic verification of a knowledge base by using a multi-criteria group evaluation with application to security screening at an airport. Knowl.-Based Syst. **85**, 170–180 (2015)
7. J. Skorupski, P. Uchronski, Managing the process of passenger security control at an airport using the fuzzy inference system. Expert Syst. Appl. **54**, 284–293 (2016)
8. F. Waris, R.G. Reynolds, Optimizing AI pipelines: a game-theoretic cultural algorithms approach. CEC 1–10 (2018)
9. G. Suciu, A. Scheianu, A. Vulpe, I. Petre, V. Suciu, Cyber-attacks–The impact over airports security and prevention modalities. WorldCIST **3**, 154–162 (2018)
10. A.T. Arslan, E. Seke, Face depth estimation with conditional generative adversarial networks. IEEE Access **7**, 23222–23231 (2019)
11. S. Nolfi, D. Floreano, *Evolutionary Robotic: The Biology, Intelligence, and Technology of Self-Organization Machines* (MIT Press, Cambridge, MA, 2000)
12. R.G. Reynolds, W. Sverdlik, Problem solving using cultural algorithms, in *International Conference on Evolutionary Computation*, pp. 645–650 (1994)
13. Y.-N. Guo, Z. Yang, C. Wang, D. Gong, Cultural particle swarm optimization algorithms for uncertain multi-objective problems with interval parameters. Natural Comput. **16**(4), 527–548 (2017)
14. A. Ochoa et al., Baharastar – Simulador de Algoritmos Culturales para la Minería de Datos Social, in *Proceedings of COMCEV'2007* (2007)
15. C. Bassetti, R. Ferrario, C. Giorgetta, Work- and Job-related stress, emotions, and performance in critical situations, in *An Interdisciplinary Study in the Context of Airport Security, EAPCogSci* (2015)
16. Z.-J. Fan, Y.-J. Zheng, Evolutionary optimization of airport security inspection allocation. SEAL, 716–726 (2017)

References

Multi-level Image Thresholding Segmentation Using 2D Histogram Non-local Means and Metaheuristics Algorithms

Andrea A. Hernandez del Rio, Erik Cuevas and Daniel Zaldivar

Abstract One of the goals for the multi-level image thresholding segmentation is to divide the image into several homogeneous regions without overlapping. The performance of segmentation approaches when are used 1D histogram-based methods are unsatisfactory as they consider the gray level of an image only and do not deal with spatial correlation among the pixels. The alternative is to use a 2D histogram that permits to handle the situations described above. This chapter explains the use of PSO and SCA metaheuristics algorithms to find the best thresholds for images segmentation, using the two-dimensional (2D) histogram non-local means and Rényi entropy as an objective function. To compare the performance of the results it uses the method 2DNLMeKGSA propose by H. Mittal and M. Saraswat. The methods have tested on five images from the Berkeley Segmentation Dataset and Benchmark (BSDS300) in terms of subjective and objective evaluations.

Keywords 2D histogram · Image segmentation · Metaheuristic algorithms · Rényi entropy

1 Introduction

Segmentation is the process of subdividing an image into its constituent regions. The degree of detail to which the subdivision moved depends on the problem solved. More precisely, image segmentation is the process of assigning a tag to each pixel of the image so that pixels that share the same tag will also have specific similar characteristics. Several segmentation methods have proposed in the literature, which can be found in surveys [1, 2]. One of the simplest methods for segmentation is

A. A. Hernandez del Rio (✉) · E. Cuevas · D. Zaldivar
Universidad de Guadalajara, CUCEI, Guadalajara, Jalisco, Mexico
e-mail: AndreaA.Hernandez@alumno.udg.mx

E. Cuevas
e-mail: erik.cuevas@cucei.udg.mx

D. Zaldivar
e-mail: daniel.zaldivar@cucei.udg.mx

D. Oliva and S. Hinojosa (eds.), *Applications of Hybrid Metaheuristic Algorithms for Image Processing*, Studies in Computational Intelligence 890,
https://doi.org/10.1007/978-3-030-40977-7_6

Image Thresholding (TH) and is one of the most popular implemented in various applications areas for example medical [3–5].

The image threshold is divided into: bilevel and multi-level. In the former, one presume that the image is composed of a forefront and background, that have distinctively different gray-level distribution. In the multi-level, it is presuming that there are several components (segments) in the image, each of a similar value of gray level. One then try to locate the values of the thresholds that can separate the components of the image. As it can see, the multi-level situation is an extension of the bilevel one. Further, image thresholding segmentation can be categorized into six groups by Sezgin and Sankur [6] based on the information that is used by the methods:

- Histogram-based methods, where the crest, valleys, and curvatures of the smoothed histogram, are analyzed.
- Clustering-based methods, where the gray level samples clustered in two parts as background and forefront (object).
- Entropy-based methods result in algorithms that use the entropy of the forefront and background regions.
- Object attribute-based methods search a measure of similarity between the gray level and the binarized images.
- Spatial methods use higher-order probability distribution and a correlation between pixels.
- Local methods adapt the threshold value on each pixel to the local image characteristics.

The histogram-based methods most often use the one-dimensional histogram (1D), but the performance of 1D histogram only considers the gray level information of an image and does not take into consideration the spatial correlation between the pixels. In 1989, Abutaleb [7] proposed to extend the entropy-based thresholding algorithm to the 2-dimensional histogram (2D). In this approach, the original gray-level histogram is integrated with local averaging of pixels to form the gray-local 2D histogram. The experimental results of this method are attractive in comparison to 1D histogram-based methods. Therefore, to include greater post-filtering clarity, Mittal and Saraswat [8] in 2018, introduce a new 2D histogram based on non-local means filter. Non-local means filter [9] calculates the mean of all pixels in the image and is further weighted by the similarity of the pixels with the target pixel.

As it can see, the use of a 2D histogram for image thresholding segmentation has a high computational cost because of the exhaustive search for the thresholds [10], and it is better to consider some optimization algorithms to overcome this problem.

In general, optimization is the selection of the best solution from some set of available solutions. Many optimization algorithms have been proposed in the literature; they can be divided into two families: classic methods and metaheuristics methods.

Metaheuristics algorithms use natural phenomena to solve complex optimization problems [11], without the restrictions of classical optimization methods that require the objective function to be twice differentiable and unimodal. Metaheuristics algorithms have been employed by the researchers to solve the problem of image thresholding segmentation in several works [12, 13]. According to No Free Lunch theorem [14], no ideal metaheuristic algorithm exists for all optimization problems. That theorem is one of the reasons that in the literature exist many metaheuristic algorithms; some of these algorithms are:

- Differential Evolution (DE) development by Kennet Price and Rainer Storn in 1995 [15].
- Particle Swarm Optimization (PSO) introduced by Kennedy and Eberhart in 1995 [16].
- Genetic Algorithm (GA) introduced in 1962 by John Holland [17].
- Artificial Bee Colony (ABC) proposed by Dervis Karaboga in 2009 [18].
- Electromagnetism-Like Optimization (EMO) introduced in 2003 by Ilker Bibil and Shu-Chering Fang [19].
- Gravitational Search Algorithm (GSA) proposed by Esmat Rashedi, Hossein Nezamabadi-pour and Saeid Saryazdi in 2009 [20].
- Sine Cosine Algorithm (SCA) introduced by Seyedali Mirjalili in 2016 [21].
- Exponential Kbest Gravitational Search Algorithm designed in 2018 by H. Mittal and M. Saraswat [8].

Moreover, the list goes on because new or existing algorithms can outperform the other for the specific set of optimization problems as the image thresholding segmentation. This chapter focuses in explain the multi-level 2D histogram image thresholding segmentation using the PSO and SCA algorithms. PSO is one of the most popular algorithms in literature and has been used for multi-threshold image segmentation. SCA is a more recent algorithm that has also proven effective in finding the solution to different optimization problems, which is why they were selected to explain the multi-level 2D histogram image thresholding segmentation.

PSO was first intended for simulating social behavior, as a stylized representation of the movement of organisms in a bird flock or fish school. The algorithm was simplified and it was observed to be performing optimization. The PSO algorithm works by having a population of candidate solutions (called particles). These particles are moved around in the search-space according to a few simple formulae [22]. The movements of the particles are guided by their own best-known position in the search space as well as the entire population best-known position. When improved positions are being discovered these will then come to guide the movements of the population. The process is repeated and by doing, so it is hoped, but not guaranteed, that a satisfactory solution will eventually be discovered.

On the other hand, SCA creates multiple initial random candidate solutions and requires them to fluctuate outwards or towards the best solution using a mathematical model based on sine and cosine functions. Several random and adaptive variables also are integrated to this algorithm to emphasize exploration and exploitation of the search space in different milestones of optimization [21].

In the exploration phase, a metaheuristic algorithm combines the random solutions in the set of solutions with a high rate of randomness to find the promising regions of the search space. In the exploitation phase, however, there are gradual changes in the random solutions, exploitation is the mechanism of locally refining the best solutions previously found in the exploration phase.

Moreover, the solution of a metaheuristic algorithm is dependent on the selection of the objective function or cost function [23]. Entropy is a concept firstly used in the second law of Thermodynamics; it was introduced into physics by German physicist Rudolf Clausius in second half of eighteenth century. It represents a measure of disorder or randomness in system [24]. In an image, the high entropy of a segmented image represents better separation among regions. In general terms, in the image processing field it is possible to said that the Kapur entropy [25] is the most popular approach used for finding the best thresholds that segment a digital image. Kapur proposes the maximization of the entropy as a measure of the homogeneity among classes. The Kapur entropy has been used for different implementations, for example, for the segmentation of thermal images [26], breast cancer images [27] and breast histology images [28]. However, the use of Kapur entropy does not guarantee to obtain the best results in complex images. This fact occurs when the number of thresholds increases because each threshold increments the complexity to compute the entropy and decrease the accuracy of the segmentation.

Different entropies have been proposed to be used instead of Kapur; some examples are the Shannon entropy [29], Tsallis entropy [30], Rényi's entropy [31, 32], Cross-entropy [33] [MCET] and the generalized entropy [34]. In Information theory, Shannon's entropy measures the information contained in a dataset, but this information is not meaningful. Here is important to mention that Shannon is the base of all other entropies used for image segmentation. The Rényi's entropy is a modified version of the Shannon entropy that includes the maximum entropy sum method, and the entropic correlation method. The combination of entropies presented by Rényi provides better results that using them separately.

An image segmented technique based on 2D Rényi's entropy has been introduced by Sahoo and Arora [31]. Further, due to the complexity of 2D Rényi's entropy, Cheng et al. [35] introduced another image thresholding segmentation method based on 2D Rényi's gray entropy and fuzzy clustering where two 1D Rényi entropies were computed for forming 2D Rényi entropy. It has been observed from the literature that Rényi entropy on 2D histogram shows better performance [8, 36].

This chapter is organized as follows: Sect. 2 describes the main concepts of Non-local means; Particle swarm optimization; Sine–cosine algorithm and Rényi entropy. Meanwhile, Sect. 3 explain the methods and its mathematical formulation. In Sect. 4 presents the experimental results and the statistical analysis. Finally, Sect. 5 conclusions.

2 Preliminaries

In this section, a brief description of the non-local means algorithm, the particle swarm optimization algorithm, the sine–cosine algorithm and the entropy of Rényi is given.

2.1 2D Non-local Means

The two main limitations in image accuracy are categorized as blur and noise. Blur is intrinsic to image acquisition systems, as digital images have a finite number of samples and must satisfy the Shannon–Nyquist sampling conditions [37]. The second primary image perturbation is noise. Image noise is a random variation of brightness or color information in images. Several methods have been proposed to remove the noise and recover the correct image.

The principle of the first denoising methods was quite simple: replacing the value of a pixel with an average of the value of nearby pixels.

Non-local means [9] computes the weighted average of all the pixels in an image. Let $A(p)$ and $A(q)$ are the respective values of p an q pixel in an image A. Equation (1) can computed the non-local means of A.

$$B(p) = \frac{\sum_{q \in A} A(q)\omega(p, q)}{\sum_{q \in A} \omega(p, q)} \tag{1}$$

where, $B(p)$ is the output value of pixel p and $\omega(p, q)$ is the Gaussian weighting function defined by Eq. (2).

$$\omega(p, q) = \exp \frac{-|\mu(q) - \mu(p)|^2}{\sigma^2} \tag{2}$$

where, σ is the standard deviation, $\mu(p)$ and $\mu(q)$ are the local mean values at pixels p and q respectively and are calculated using Eqs. (3) and (4).

$$\mu(p) = \frac{1}{m \times m} \sum_{i \in F(p)} A(i) \tag{3}$$

$$\mu(q) = \frac{1}{m \times m} \sum_{i \in F(q)} A(i) \tag{4}$$

here, $F(p)$ is a square filter of size $m \times m$.

2.2 Particle Swarm Optimization

The PSO is an iterative algorithm formed by a group of particles in which each particle keeps track of its coordinates and shares them with the other particles.

In the application, each solution is considered as a particle with a best value or fitness that can be calculated using an objective function. All the particles preserve their personal best positions p, and there is a global best position g, for the entire group. They adjust their velocities by considering their personal best performances and the global best performance of the group and change their positions by adjusting the velocities [38].

The algorithm initialized with a random group of particles. The position of each particle is given in the first iteration by Eq. (5).

$$x_d^{i,t} = l_d + rand(u_d - l_d), \quad x_d^{i,t} \in \boldsymbol{x}^t \tag{5}$$

where $x_d^{i,t}$ is the ith particle of the population $\boldsymbol{x}^t$, i is the index that refers the number of particles which the maximum value is the size of the population ($i = 1, 2, \ldots, N$). The dimension of the problem is defined by the variable d and the number of iterations with the variable t. While l_d and u_d are the lower and upper limits, specifically for one of the dimensions of the search space and $rand$ is a random number between zero and one.

To obtain the new position of the particles is necessary to calculate the speed of each of them. In the first iteration, the speed is zero because the particles do not move, but Eq. (6) presents how the calculation of the velocity is carried out for the following iterations.

$$\boldsymbol{v}^{t+1} = v^t + rand1 \times \left(\boldsymbol{p} - \boldsymbol{x}^t\right) + rand2 \times \left(g - \boldsymbol{x}^t\right) \tag{6}$$

where v^{t+1} is the value of the velocity that is calculated for iteration $t + 1$, $\boldsymbol{v}^t$ is the velocity in the previous iteration t, $\boldsymbol{x}^t$ is the vector that contains the positions of each particle, $\boldsymbol{p}$ contains the best current positions associated with the vicinity of each particle, while g is the best current particle globally. On the other hand, $rand1$ and $rand2$ are random numbers usually evenly distributed in the range of zero to one [39].

After calculating the speed, the particles moved to new positions in the current iteration. This movement is described in Eq. (7).

$$\boldsymbol{x}^{t+1} = \boldsymbol{x}^t + \boldsymbol{v}^{t+1} \tag{7}$$

$\boldsymbol{x}^{t+1}$ is the vector where the new positions obtained in iteration $t + 1$ are stored, $\boldsymbol{x}^t$ corresponds to the previous positions of the particles. Finally, $\boldsymbol{v}^{t+1}$ is the velocity vector obtained using Eq. (6).

PSO is one of the most popular algorithms, and in the literature, it is possible to find some modifications that alter the way in how the particles are displaced or even changes in their initialization.

This chapter makes use of the basic PSO algorithm, and this version is the one used for the reported experiments.

The flow diagram in Fig. 1 shows in detail the basic structure of the PSO algorithm.

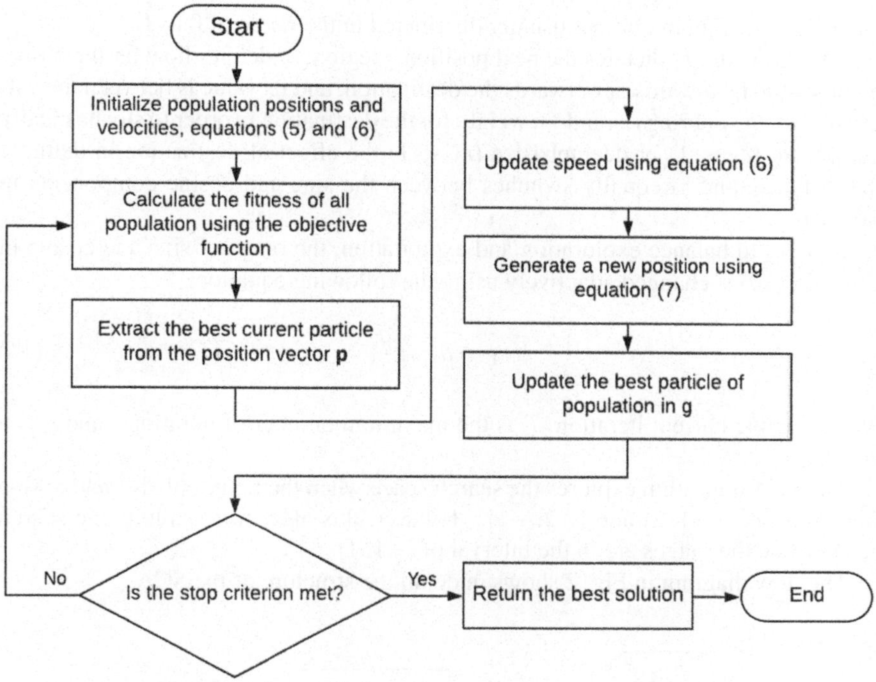

Fig. 1 Flowchart of the PSO algorithm

2.3 *Sine–Cosine Algorithm*

The Sine–Cosine Algorithm (SCA) creates multiple initial random candidate solutions (population) and requires them to fluctuate outwards or towards the best solution using a mathematical model based on sine and cosine functions. The following position updating equations are for exploration and exploitation phases, respectively:

$$X_i^{t+1} = X_i^t + r_1 \times \sin r_2 \times \left| r_3 P_i^t - X_i^t \right| \tag{8}$$

$$X_i^{t+1} = X_i^t + r_1 \times \cos r_2 \times \left| r_3 P_i^t - X_i^t \right| \tag{9}$$

where X_i^t is the position of the current position in ith dimension at ith iteration, $r_1/r_2/r_3$ are random numbers, P_i is position of the destination point in ith dimension, and || indicates the absolute value.

These two equations are combined to be used as follows:

$$X_i^{t+1} = \begin{cases} X_i^t + r_1 \times \sin r_2 \times \left| r_3 P_i^t - X_i^t \right|, & r_4 < 0.5 \\ X_i^t + r_1 \times \cos r_2 \times \left| r_3 P_i^t - X_i^t \right|, & r_4 \geq 0.5 \end{cases} \tag{10}$$

where r_4 is a random number usually distributed in the range of 0 to 1.

The parameter r_1 dictates the next position's region; r_2 defines how far the movement should be towards or outwards the destination, and its value is between the values of 0 to 2π; r_3 brings a random weight for the destination in order to stochastically emphasize ($r_3 > 1$) or deemphasize ($r_3 < 1$) the effect of destination in defining the distance, and r_4 equally switches between the sine and cosine components in Eq. (10).

In order to balance exploration and exploitation, the range of sine and cosine in Eqs. (8)–(10) is changed adaptively using the following equation:

$$r_1 = a - t\frac{a}{T} \tag{11}$$

where t is the current iteration, T is the maximum number of iterations, and a is a constant.

The SCA algorithm explores the search space when the ranges of sine and cosine functions are in $(1, 2]$ and $[-2, -1)$. However, this algorithm exploits the search space when the ranges are in the interval of $[-1, 1]$.

The flow diagram in Fig. 2 shows in detail the structure of the SCA.

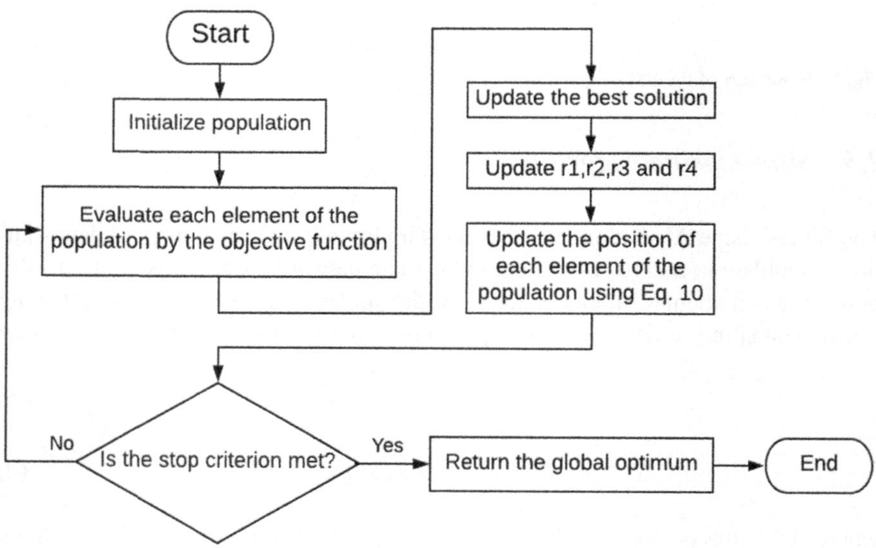

Fig. 2 Flowchart of the SCA

2.4 Rényi Entropy

Entropies quantify the diversity, uncertainty, or randomness of a system. It is also the amount of "noise" or "disorder" that a system contains or releases. In this way, we can talk about the amount of information a signal carry.

Shannon [37] states that the amount of information contained in a system A can be measure by Eq. (12) and is known as Shannon's entropy $S(A)$.

$$S(A) = -\sum_{i=1}^{E} p_i \log p_i \tag{12}$$

where E is the number of events with p_i probabilities, Rényi extended the Shannon entropy to define the Rényi entropy (R) according to Eq. (13).

$$R_\alpha(A) = \frac{1}{1-\alpha} \log \sum_{i=1}^{E} p_i^\alpha \tag{13}$$

where $\alpha > 0$ and $\alpha \neq 1$ is an arbitrary positive real number and is known as the order of the entropy. Here, A is a discrete random variable with possible outcomes $1, 2, \ldots, E$ and corresponding probabilities $p_i = \Pr(A = i)$ for $i = 1, 2, \ldots, E$. The logarithm is conventionally taken to be base two but if the probabilities are $p_i = 1/E$ for all $i = 1, 2, \ldots, E$, then all the Rényi entropies of the distribution are equal: $R_\alpha(A) = \log E$.

3 Image Thresholding Segmentation Method

Images contain different ambiguities generated by the scenes, for that reason is necessary to design robust image processing methods. In 2018, Mittal and Saraswat [8] based on the work of Abutaleb [7], Buades [9], Sahoo and Arora [31], proposed a new method of segmentation of images called 2DNLMeKGSA. This new method besides using the 2D histogram, the non-local means filter and Rényi entropy, it also modifies the GSA metaheuristic algorithm. In this chapter, the PSO and SCA algorithms are used instead of the modified eKGSA algorithm, but their results are compared with the previous one.

In Fig. 3 is illustrated the generic flow graph of the 2D non-local means using metaheuristic algorithm (eKGSA, PSO and SCA) method for multi-level image thresholding segmentation.

As can be appreciated in Fig. 3, the digital image is first converted into a gray-scale image A followed by its non-local means filtered image B and these two images are used to compute the 2D histogram. The generated 2D histogram is further given to one of the metaheuristic algorithms, eKGSA, PSO or SCA, to obtain multi-level

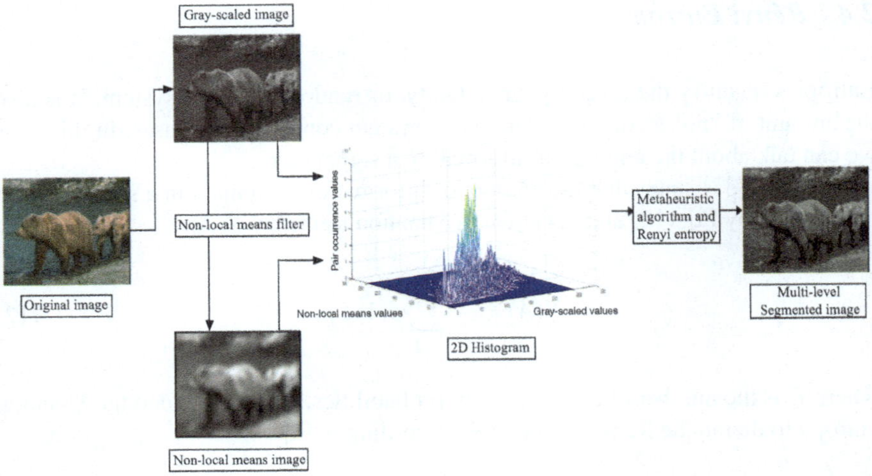

Fig. 3 Flow chart of the image segmentation methods

thresholds using the Rényi entropy as an objective function. The obtained thresholds are used to segment the image. The flow chart in Fig. 4 shows the generic structure for the methods.

3.1 Generate the Non-local Means 2D Histogram

Let $A(x, y)$ represents the gray level $([0 - L - 1])L = 256$, of a pixel at spatial coordinate (x, y) in an image of size $M \times N$. The $B(x, y)$ corresponds to the non-local means value of the pixel at (x, y), generated by applying the non-local means as described in Sect. 2.1. The non-local means 2D histogram is calculated by Eq. (14) using the two images (A and B).

$$H(i, j) = c_{ij} \tag{14}$$

where, $i = A(x, y)$, $j = B(x, y)$, and c_{ij} is the occurrence of the pair (i, j).

From the above constructed non-local means 2D histogram (H), the normalized is calculated as per Eq. (15)

$$P_{ij} = \frac{c_{ij}}{M \times N} \tag{15}$$

Figure 5 shows two selected images from Berkeley Segmentation Dataset and Benchmark (BSDS300) and their respective non-local means 2D histograms.

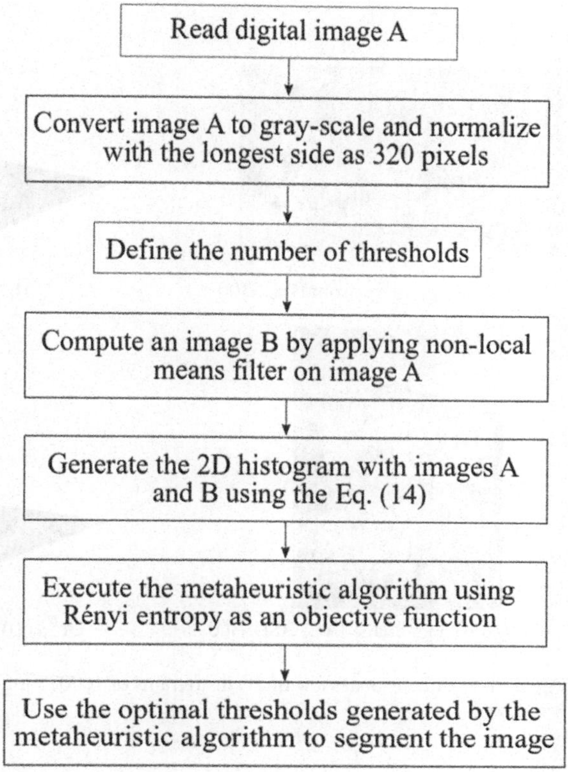

Fig. 4 Flow chart of the generic structure for the methods

3.2 *PSO and SCA*

For the experiments, the PSO and SCA metaheuristics algorithms were selected for the creation of the two segmentation methods based on the non-local means 2D histogram, also the eKGSA algorithm is used to compare the results. The names of the methods are: 2DNLMPSO, 2DNLMSCA and 2DNLMeKGSA. For the three algorithms (PSO, SCA and eKGSA) a population number (*NP*) of 50 was considered. In addition, each algorithm performs 1000 iterations (max_it). The parametric settings for each algorithm are presented in Table 1.

3.3 *Objective Function: Rényi Entropy*

The non-local means 2D histogram is subdivided into n^2 sub-regions, as shown in Fig. 6, where the division is made on the basis of number of gray-level thresholds $\{g_1, g_2, \ldots, g_{n-1}\}$ and non-local means thresholds $\{f_1, f_2, \ldots, f_{n-1}\}$.

Among the n^2 sub-regions, only diagonal sub-regions are considered since they contain maximum information of the images as apparent from Fig. 5 and are numbered as $\{1, 2, 3, \ldots, n\}$ in Fig. 6, these diagonal subregions are used to calculate the Rényi entropy for positive $\alpha \neq 1$, as depicted in Eq. (16).

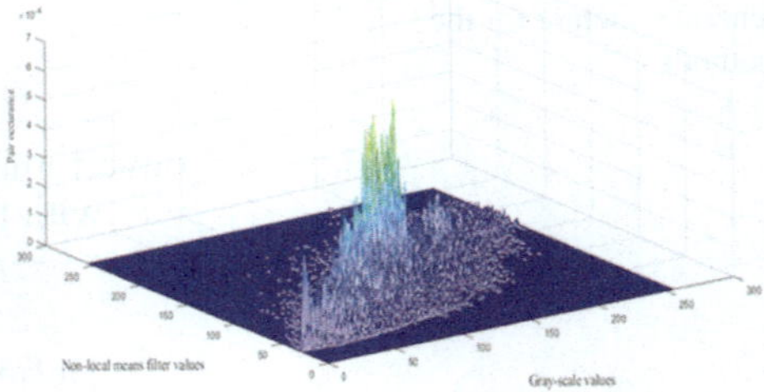

(1) 100075.jpg: image from BSDS300 (1a) 100075.jpg: 2D histogram

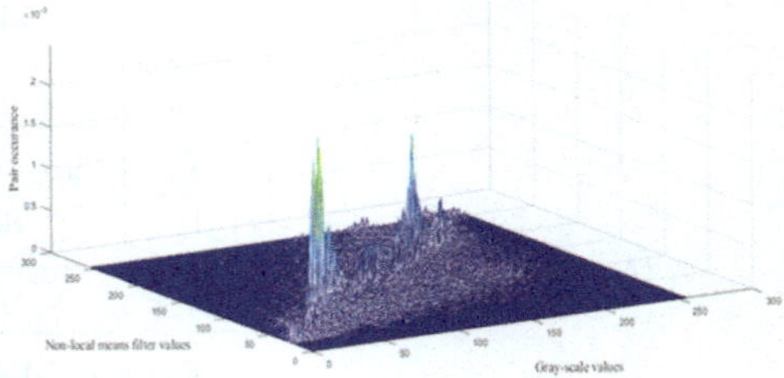

(2) 102061.jpg: image from BSDS300 (2a) 102061.jpg: 2D histogram

Fig. 5 Tree dimensional view of 2D histograms of 100075.jpg and 102061.jpg images taken from BSDS300

Table 1 Parametric settings

Algorithm	Parameter	Value
eKGSA [8]	Gconstant (Go)	100
	Beta (β)	20
	final_per	2
PSO [40]	Social coefficient c_1	2
	Cognitive coefficient c_2	2
	Velocity clamp	0
	Maximum inertia value (W_{min})	1
	Minimum inertia value (W_{max})	0.2–0.9
SCA [21]	No parameters	–

$$R^{\alpha}(g, f) = \sum_{k=1}^{n-1} R_k^{\alpha}(g_k, f_k) \tag{16}$$

where, $R_k^{\alpha}(g_k, f_k)$ is Rényi entropy of kth diagonal subregion $\{k = 1, 2, \ldots, n\}$ bounded by (g_k, f_k) and is defined by Eq. (17).

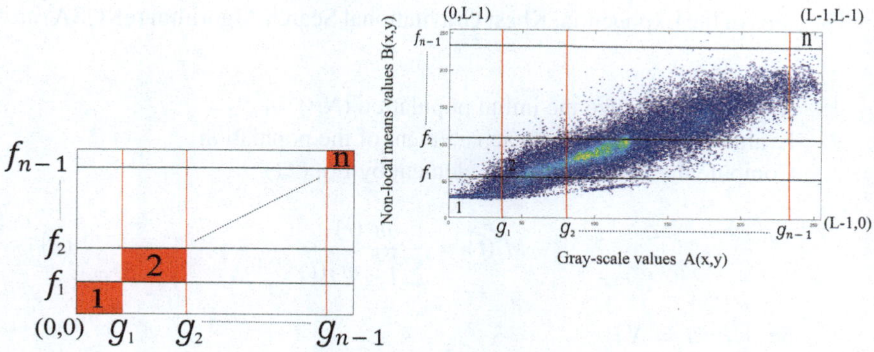

Fig. 6 Two-dimensional histogram non local means division for threshold

$$R_k^{\alpha}(g_k, f_k) = \frac{1}{1-\alpha} \log \sum_{i=t_{k-1}+1}^{g_k} \sum_{j=f_{k-1}+1}^{f_k} \left(\frac{P_{ij}}{P_k}\right)^{\alpha} \tag{17}$$

where, P_{ij} is the value of normalized 2D histogram at (i, j) location and P_k represents the total sum of values at kth diagonal subregion in the normalized 2D histogram as shown in Eq. (18).

$$P_k = \sum_{i=t_{k-1}+1}^{g_k} \sum_{j=f_{k-1}+1}^{f_k} P_{ij} \tag{18}$$

In this chapter, the methods use the Rényi entropy as an objective function which is maximized to obtain the optimal set of multi-level thresholds.

$$\text{Maximize: } R^{\alpha}(g, f) \quad \text{for } g, f \in [0, L-1] \tag{19}$$

It is reasonable to take the value of α in the range [0.1, 0.9]. Furthermore, the value of α is set to 0.45 for the experiments in Sect. 4.

3.4 The 2DNLMeKGSA Method

In 2018, Mittal and Saraswat [8] proposed a novel method for multi-level image thresholding segmentation using the two-dimensional non-local means histogram and a modified algorithm (eKGSA). In this new algorithm the authors propose an exponentially decreasing Kbest defined by Eq. (20).

$$Kbest\left(\frac{final\_per}{NP}\right)^{\frac{1}{max\_it}} \tag{20}$$

The steps of the Exponential Kbest Gravitational Search Algorithm (eKGSA) are:

1: Randomly initialize the initial population (*NP*)
2: Evaluate the fitness *fit* of each element of the population
3: Compute the mass *M* of each element by Eq. (21)

$$M_i(t) = \frac{m_i(t)}{\sum_{j=1}^{NP} m_j(t)} \tag{21}$$

4: Set $Kbest = NP$
5: while stopping criteria is not satisfied do
6: Compute the acceleration *a* by Eq. (22)

$$a_i^d(t) = \frac{F_i^d(t)}{M_i(t)} \tag{22}$$

7: Compute the velocity *v* of each element, using Eq. (23)

$$v_i^d(t+1) = rand \times v_i^d + a_i^d(t) \tag{23}$$

8: Update the position of each element by Eq. (24)

$$element_i^d(t+1) = element_i^d + v_i^d(t+1) \tag{24}$$

9: Evaluate the fitness fit for each element
10: Compute the mass *M* of each element by Eq. (21)
11: Update $Kbest$ using Eq. (20)
12: end while

In this chapter the 2DNLMPSO and 2DNLMSCA methods are based in the structure of the 2DNLMeKGSA method as show in the Sect. 3.

4 Experimental Results

The analysis has done on five images of Berkeley Segmentation Dataset and Benchmark (BSDS300) [41]; these images show in Table 2. The experiments were performed using Matlab 9.4 on an Intel® core i7-7700 processor with 16 GB of RAM.

Table 2 **a** 100075.jpg, **b** 100098.jpg, **c** 101085.jpg, **d** 101087.jpg, **e** 102061.jpg

(a)	(b)

(c)	(d)

(e)	

In BSDS300, each image is provided a set of ground truth images, compiled by human observers. The five images are normalized with the longest side as 320 pixels. The optimal multi-level thresholds in the 2D non-local means (2DNLM) image segmentation method are calculated using PSO, SCA and eKGSA metaheuristics algorithms. The number of selected thresholds for the images are 2, 3, 4 and 5. Each method is executing 30 times to minimize the interference.

4.1 Performance Evaluation Parameters

There are two kinds of quality assessments: subjective assessments and objective assessments. The first ones using methods where quality scores are evaluated by humans and the last ones are supported by measurements that can automatically estimate perceived image quality [42].

An objective image quality measure can be used in many important image and video processing applications. For that reason, the performance of the methods has been evaluated using six image segmentation performance parameters for objective assessment.

In addition, three images from BSDS300 along with the corresponding segmented images returned by each algorithm have been considered in this chapter for subjective analysis.

SSIM and FSIM

In digital images we can find that their pixels exhibit strong dependencies, especially when they are spatially proximate, and these dependencies carry information about the structure of the objects in the visual scene.

The structural similarity (SSIM) is used for measuring the similarity between two images. SSIM is a perception-based model that considers image degradation as perceived change in structural information, while also incorporating critical perceptual phenomena, including both luminance masking and contrast masking terms.

The SSIM index [43] calculated on various windows of an image. The measure between two windows A and D of standard size $m \times m$ is:

$$SSIM(A, D) = \frac{(2\mu_A\mu_D + C_1)(2\sigma_{AD} + C_2)}{\left(\mu_A^2 + \mu_D^2 + C_1\right)\left(\sigma_A^2 + \sigma_D^2 + C_2\right)} \tag{25}$$

where μ_A and μ_D are the mean value of the gray-scale image and the threshold image, respectively; σ_A^2 and σ_D^2 correspond to the standard deviation; σ_{AD} is the covariance of A and D; $C_1 = (k_1L)^2$ and $C_2 = (k_2L)^2$ are variables to stabilize the division with weak denominator, being L the dynamic range of the pixel values $\left(2^{No\,bits\,per\,pixel-1}\right)$, $k_1 = 0.01$ and $k_2 = 0.03$ by default.

A higher value of the SSIM indicates a better performance of the evaluated methodology.

The feature similarity Index (FSIM) defines the quality score, which reflects the significance of a local structure. In other words, FSIM establishes the similarity between two images [44], Eq. (26).

$$FSIM = \frac{\sum_{x\in\Omega} S_L(x) \cdot PC_m(x)}{\sum_{x\in\Omega} PC_m(x)} \tag{26}$$

where Ω represents the entire domain of the image:

$$S_L(x) = S_{PC}(x)S_G(x) \tag{27}$$

$$S_{PC}(x) = \frac{2PC_1(x)PC_2(x) + T_1}{PC_1^2(x) + C_2^2(x) + T_1} \tag{28}$$

$$S_G(x) = \frac{2G_1(x)G_2(x) + T_2}{G_1^2(x) + G_2^2(x) + T_2} \tag{29}$$

G is the gradient magnitude (GM) of an image and is defined as:

$$G = \sqrt{G_x^2 + G_y^2} \tag{30}$$

PC is the phase congruence:

$$PC(x) = \frac{E(x)}{(\varepsilon + \sum_n A_n(x))} \tag{31}$$

The magnitude of the response vector in x on n is $E(x)$ and $A_n(x)$ is the local amplitude of scale n. ε is a small positive number and $PC_m(x) = \max(PC_1(x), PC_2(x))$.

Table 3 reports the averages values of SSIM and FSIM of the three methods for each image and number of thresholds (Thr).

RMSE and PSNR

The root-mean-square error (RMSE) is a frequently used measure of the differences between values (sample or population values) predicted by a model or an estimator and the values observed. The RMSE represents the square root of the second sample moment of the differences between predicted values and observed values or the quadratic mean of these differences [45]. In general, a lower RMSE is better than a higher one and is defined as:

$$RMSE = \sqrt{\frac{\sum_{i=1}^{M}\sum_{j=1}^{N}(A(i,j) - D(i,j))}{M \times N}} \tag{32}$$

where A is the gray-scale image and D is the segmented image; the total number of rows is M, and the total number of columns is N.

The peak signal-to-noise ratio (PSNR), is an engineering term for the ratio between the maximum possible power of a signal and the power of corrupting noise that affects the fidelity of its representation. Because many signals have an extended dynamic range, PSNR usually expressed in terms of the logarithmic decibel scale.

In the case of images, PSNR compares the similarity of the original image against the segmented [46], Eq. (33).

$$PSNR = 20\log_{10}\left(\frac{255}{RMSE}\right)\text{dB} \tag{33}$$

Table 3 Quality results SSIM and FSIM of the segmented images for 2DNLMeKGSA, 2DNLMPSO and 2DNLMSCA

Parameters		SSIM			FSIM		
Image	Thr	eKGSA	PSO	SCA	eKGSA	PSO	SCA
(a)	2	**0.2741**	0.1361	0.1488	**0.5524**	0.5274	0.5307
(a)	3	0.4036	0.4511	**0.4648**	0.6118	0.6319	**0.6359**
(a)	4	0.4810	0.5291	**0.5437**	0.6535	0.6787	**0.6808**
(a)	5	0.5628	0.6076	**0.6381**	0.7042	0.7167	**0.7305**
(b)	2	0.4336	**0.4877**	0.4861	0.6443	**0.6882**	0.6835
(b)	3	0.5660	**0.6205**	0.6148	0.7292	**0.7595**	0.7488
(b)	4	0.6512	**0.7203**	0.7081	0.7891	**0.8089**	0.7935
(b)	5	0.6897	0.7587	**0.7806**	0.8011	**0.8466**	0.8278
(c)	2	**0.3918**	0.3754	0.3744	0.6319	**0.6515**	0.6509
(c)	3	0.5083	0.5383	**0.5556**	0.7105	0.7288	**0.7346**
(c)	4	0.6180	0.6262	**0.6616**	0.7595	0.7807	**0.7943**
(c)	5	0.6506	0.7082	**0.7427**	0.7812	0.8243	**0.8372**
(d)	2	0.6036	**0.6504**	**0.6504**	0.7028	0.7308	**0.7317**
(d)	3	0.6751	**0.7237**	0.7187	0.7410	**0.7789**	0.7756
(d)	4	0.7400	**0.7748**	0.7723	0.7791	0.8078	**0.8149**
(d)	5	0.7613	0.8121	**0.8153**	0.7962	0.8372	**0.8388**
(e)	2	0.4449	0.4573	**0.4672**	0.6726	0.6818	**0.6823**
(e)	3	0.5950	0.6031	**0.6411**	0.7127	0.7235	**0.7283**
(e)	4	0.6395	0.7016	**0.7273**	0.7433	0.7628	**0.77**
(e)	5	0.6816	0.7222	**0.7591**	0.7594	0.7832	**0.793**

Source: Bold values highlight the best result among the algorithms with respect to the performance evaluation parameter and the number of thresholds used in the image

Table 4 shows the averages of the RMSE and PSNR metrics.

NCC and AD

Correlation is an essential tool in image processing, pattern recognition, and other fields. The correlation between two signals is a standard approach to feature detection as well as a building block for more sophisticated recognition techniques.

The normalized cross-correlation (NCC) is used for template matching, where images are first normalized due to lighting and exposure conditions [47].

$$NCC = \frac{1}{n}\sum_{i,j}\frac{1}{\sigma_A \sigma_D}\left(A(i,j) - \overline{A}\right)\left(D(i,j) - \overline{D}\right) \tag{34}$$

It considered that a higher value of the NCC indicates a better performance of the evaluated methodology.

Table 4 Quality results RMSE and PSNR of the segmented images for 2DNLMeKGSA, 2DNLMPSO and 2DNLMSCA

Parameters		RMSE			PSNR		
Image	Thr	eKGSA	PSO	SCA	eKGSA	PSO	SCA
(a)	2	**59.6641**	74.6556	73.2235	**12.8821**	10.6744	10.9047
(a)	3	47.7770	41.7095	**40.676**	14.8958	15.7978	**15.9984**
(a)	4	40.9691	35.8857	**34.5551**	16.1942	17.11	**17.4397**
(a)	5	33.6769	29.8144	**27.9458**	17.9615	18.7375	**19.2859**
(b)	2	64.7034	**58.467**	58.7482	11.9554	**12.7941**	12.7518
(b)	3	50.8178	**44.322**	44.6076	14.0745	**15.2043**	15.1494
(b)	4	40.4858	**36.3913**	38.2784	16.0660	**16.9366**	16.4949
(b)	5	37.5271	**30.8445**	32.4338	16.7512	**18.3928**	17.9514
(c)	2	52.2754	**50.4221**	50.4987	13.8102	**14.0792**	14.0662
(c)	3	41.5510	38.3491	**37.368**	15.8464	16.4719	**16.696**
(c)	4	34.3590	31.9081	**30.4265**	17.4762	18.0833	**18.486**
(c)	5	31.3554	27.0636	**25.9038**	18.2716	19.5306	**19.8723**
(d)	2	81.4906	**81.2042**	81.9092	**10.0228**	9.9416	9.8649
(d)	3	**58.3110**	67.7963	71.1937	**13.1247**	11.5182	11.0885
(d)	4	49.1715	**48.3497**	60.6356	14.5051	**14.5124**	12.5385
(d)	5	40.6227	**39.1502**	43.4206	16.2890	**16.3759**	15.4761
(e)	2	57.1889	49.9405	**49.2022**	13.0303	14.1639	**14.2925**
(e)	3	43.4168	37.9631	**36.695**	15.4946	16.5727	**16.857**
(e)	4	38.2771	31.0024	**29.1761**	16.5981	18.3369	**18.8442**
(e)	5	33.7878	27.9589	**25.8339**	17.6893	19.2461	**19.9208**

Source: Bold values highlight the best result among the algorithms with respect to the performance evaluation parameter and the number of thresholds used in the image

Average Difference (AD) is the average difference between the pixel values from the reference signal and test image [48]. The Eq. (35) gives it.

$$AD = \frac{1}{MN} \sum_{i=1}^{M} \sum_{j=1}^{N} (A(i, j) - D(i, j)) \tag{35}$$

In the AD metric if the evaluated methodology gets lower value is better than the higher one.

The results for the two metrics (NCC and AD) are presented in Table 5, listed by each image and number of thresholds (Thr) obtained by each of the methods evaluated.

Table 6 lists the best threshold (Thr) set of each method that is obtained for each image using the best fitness value as a reference. In maximization algorithms, the higher the value (fitness) of the objective function better is the solution. For subjective

Table 5 Quality results NCC and AD of the segmented images for 2DNLMeKGSA, 2DNLMPSO and 2DNLMSCA

Parameters			NCC			AD	
Image	Thr	eKGSA	PSO	SCA	eKGSA	PSO	SCA
(a)	2	**0.5143**	0.3938	0.4055	**52.2969**	67.6651	66.2388
(a)	3	0.6228	0.6745	**0.6804**	40.8805	35.5196	**34.6613**
(a)	4	0.6794	0.7221	**0.7297**	34.5219	30.3047	**29.0105**
(a)	5	0.7419	0.7597	**0.7723**	28.1163	25.1057	**23.3593**
(b)	2	0.6243	**0.6594**	0.6555	55.3140	**51.3983**	51.4596
(b)	3	0.7245	**0.7493**	0.7404	43.1060	38.536	**38.4851**
(b)	4	0.7862	**0.7915**	0.7737	34.4242	**31.0673**	31.9568
(b)	5	0.8071	**0.8297**	0.8111	30.9721	**26.1659**	26.8998
(c)	2	0.6189	**0.6483**	0.6478	43.2892	**43.7838**	43.8368
(c)	3	0.7127	0.7383	**0.7402**	34.6686	31.5623	**30.6647**
(c)	4	0.7591	0.7867	**0.7888**	27.4371	25.9436	**24.5671**
(c)	5	0.7857	0.8169	**0.8173**	24.9343	21.6623	**20.4588**
(d)	2	0.5452	**0.5483**	0.5441	66.1854	**62.4075**	62.8271
(d)	3	**0.6889**	0.6267	0.6069	**48.8345**	51.4069	53.4106
(d)	4	**0.7360**	0.7344	0.6679	40.7086	**39.0205**	45.0476
(d)	5	**0.7967**	0.7858	0.7597	32.9661	**31.7275**	34.4248
(e)	2	0.6372	0.6871	**0.689**	49.1088	43.7285	**42.8636**
(e)	3	0.7296	0.7644	**0.7652**	36.4256	31.6788	**30.7595**
(e)	4	0.7673	0.8068	**0.8106**	31.6512	25.9032	**24.0348**
(e)	5	0.7980	0.8324	**0.834**	27.9060	23.3169	**21.4133**

Source: Bold values highlight the best result among the algorithms with respect to the performance evaluation parameter and the number of thresholds used in the image

analysis, Figs. 7, 8 and 9 shows the segmented images with the thresholds reported in Table 6, obtained by the 2DNLMeKGSA, 2DNLMPSO and 2DNLMSCA methods.

From Figs. 7, 8, 9 and 10, the three selected images are divided for method and the number of thresholds; also, the last column shows the original gray-scale image.

4.2 Experimental Analyses

As mentioned above, 30 runs performed for each of the images for each threshold number and each method. Tables 3, 4 and 5 where the results of the metrics are reported only show the averages of the experiments performed. When analyzing Tables 3, 4 and 5, the 2DNLMSCA is the method with higher values as can be seen in the results for SSIM, FSIM, PSNR, and NCC for the images (a), (c) and (e).

Table 6 Best threshold set and best fitness of each image

Parameters		Thresholds			Fitness		
Image	Thr	eKGSA	PSO	SCA	eKGSA	PSO	SCA
(a)	2	100, 179	124, 191	124, 193	23.9565	**24.1583**	24.1469
(a)	3	81, 151, 187	50, 122, 188	45, 119, 201	30.3663	30.8791	**30.8879**
(a)	4	62, 106, 134, 194	44, 107, 152, 189	38, 88, 161, 200	35.6759	36.8224	**36.9744**
(a)	5	7, 45, 84, 145, 200	31, 71, 120, 167, 217	31, 82, 127, 183, 207	40.354	**42.4871**	42.2743
(b)	2	101, 164	81, 159	82, 155	24.6195	24.6746	**24.6749**
(b)	3	87, 126, 178	62, 117, 170	52, 103, 180	30.7509	31.0355	**31.0374**
(b)	4	66, 121, 161, 196	41, 79, 130, 174	41, 86, 136, 188	35.9847	**36.991**	36.9568
(b)	5	40, 96, 132, 167, 186	43, 64, 91, 147, 191	29, 58, 101, 137, 196	41.9522	42.2835	**42.4425**
(c)	2	72, 151	82, 148	82, 149	24.4037	**24.4909**	24.4894
(c)	3	54, 96, 164	65, 109, 159	59, 103, 161	30.3032	30.7893	**30.8011**
(c)	4	54, 104, 150, 196	52, 89, 125, 172	52, 86, 130, 166	36.0174	36.5225	**36.6137**
(c)	5	38, 83, 116, 166, 181	54, 82, 116, 149, 185	52, 81, 108, 147, 182	41.2841	41.9202	**41.9618**
(d)	2	78, 112	62, 117	62, 113	22.41	22.7258	**22.7285**
(d)	3	21, 79, 121	50, 92, 132	44, 79, 127	28.0479	28.4909	**28.538**
(d)	4	37, 68, 134, 170	29, 66, 108, 137	43, 83, 116, 155	32.9423	33.655	**33.7915**
(d)	5	55, 85, 110, 160, 179	39, 82, 108, 138, 177	35, 74, 116, 148, 179	37.4668	38.7609	**38.7919**
(e)	2	96, 168	79, 156	79, 152	23.6489	**23.8843**	23.8797
(e)	3	43, 85, 165	59, 118, 167	54, 111, 171	29.7943	**30.2992**	30.2781
(e)	4	48, 86, 132, 194	46, 85, 127, 167	53, 83, 121, 182	35.1997	35.8733	**35.9853**
(e)	5	34, 109, 149, 182, 216	56, 84, 122, 159, 193	26, 72, 123, 164, 203	40.1334	**41.2232**	41.189

Source: Bold values highlight the best result among the algorithms with respect to the performance evaluation parameter and the number of thresholds used in the image

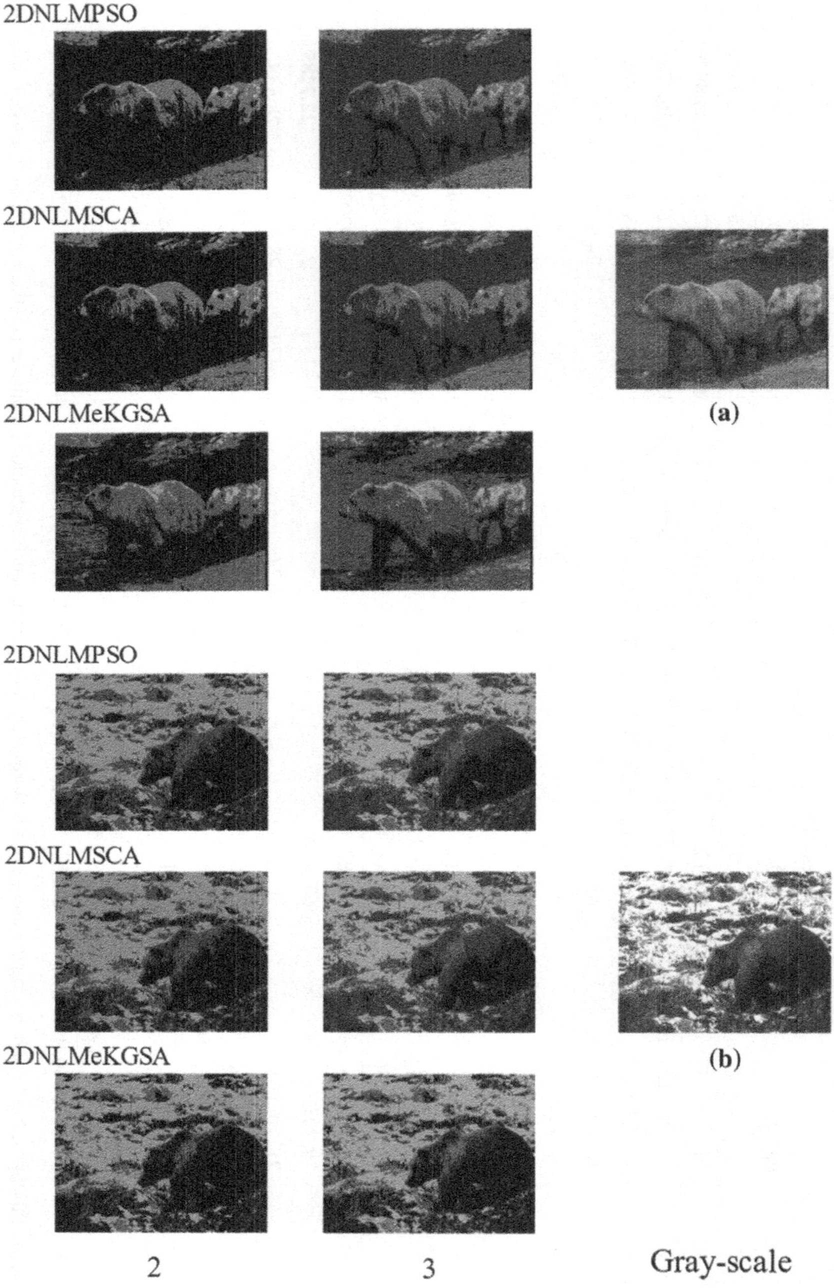

Fig. 7 Visual quality comparison between gray-scale image and segmented images obtained by 2DNLMeKGSA, 2DNLMPSO and 2DNLMSCA multilevel thresholding methods for 2 and 3 thresholds

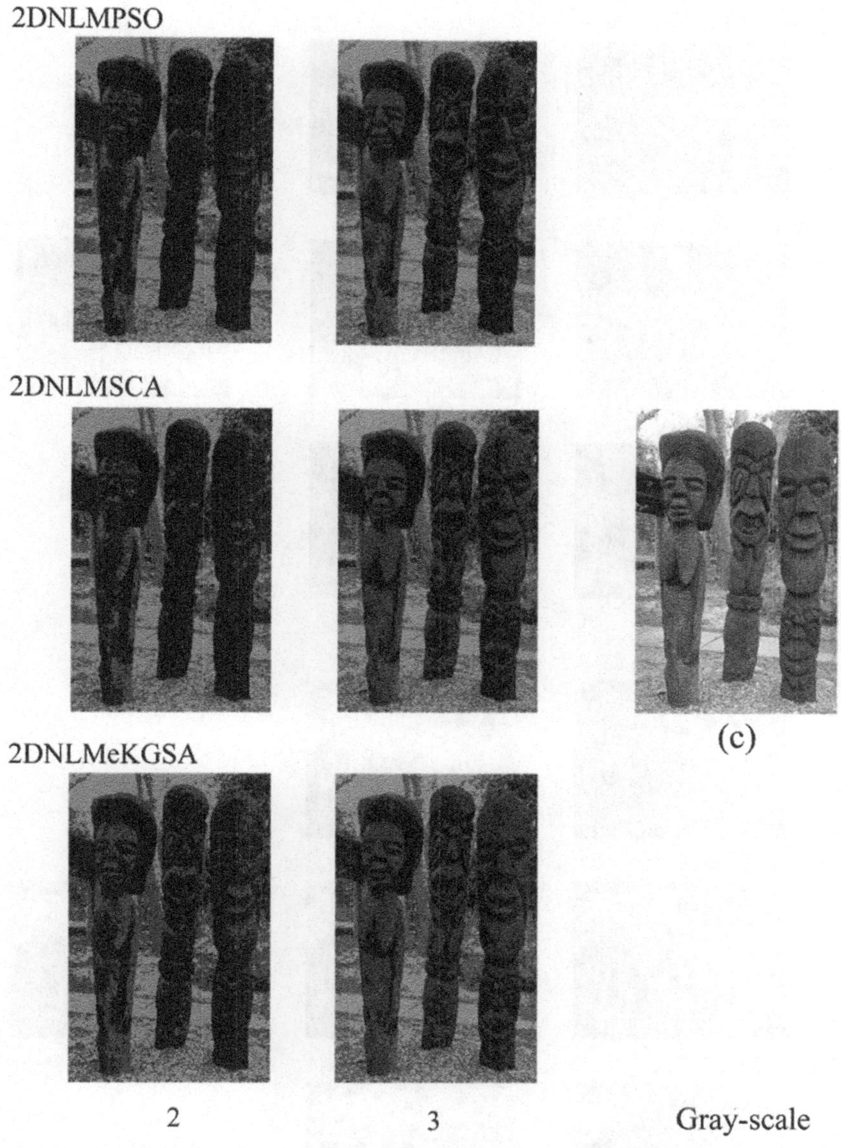

Fig. 8 Visual quality comparison between gray-scale image and segmented images obtained by 2DNLMeKGSA, 2DNLMPSO and 2DNLMSCA multilevel thresholding methods for 2 and 3 thresholds

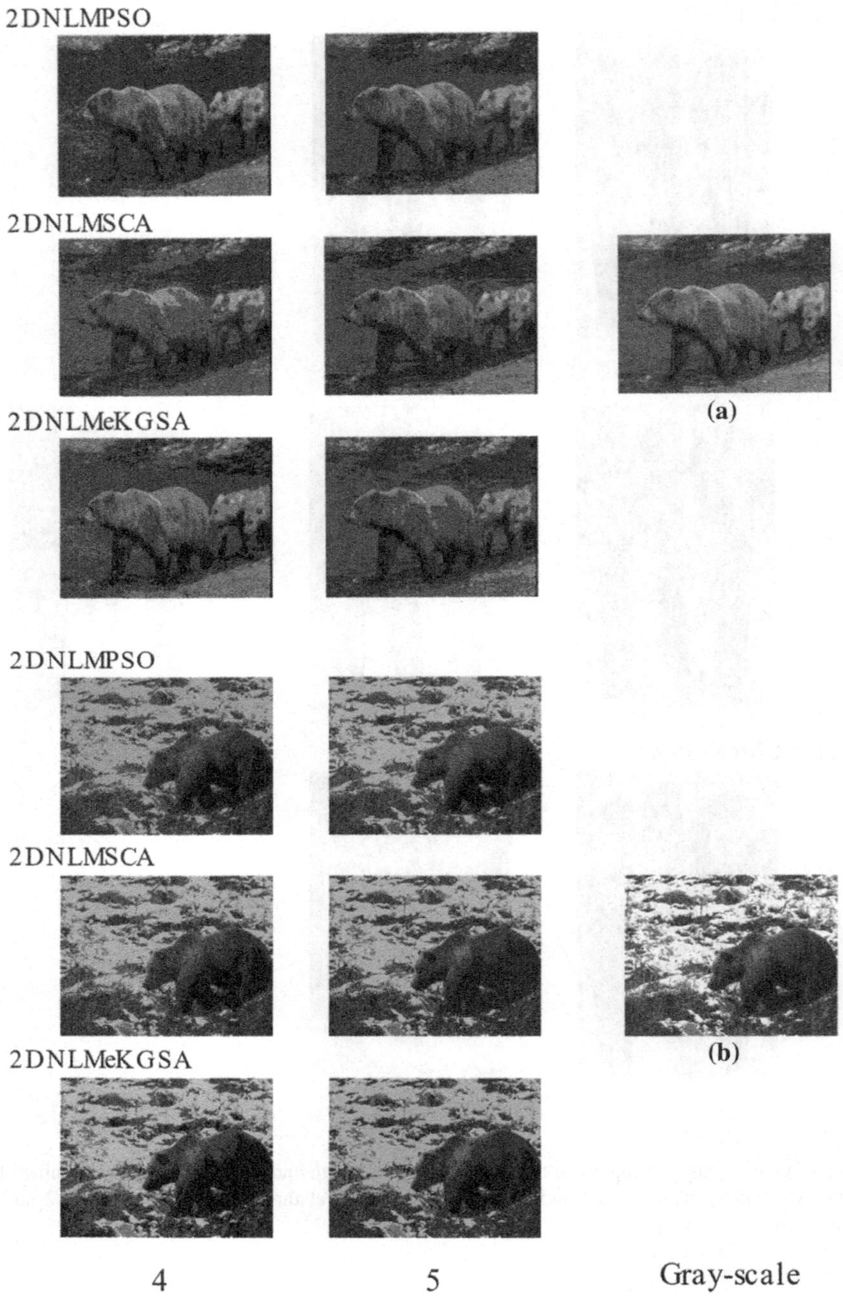

Fig. 9 Visual quality comparison between gray-scale image and segmented images obtained by 2DNLMeKGSA, 2DNLMPSO and 2DNLMSCA multilevel thresholding methods for 4 and 5 thresholds

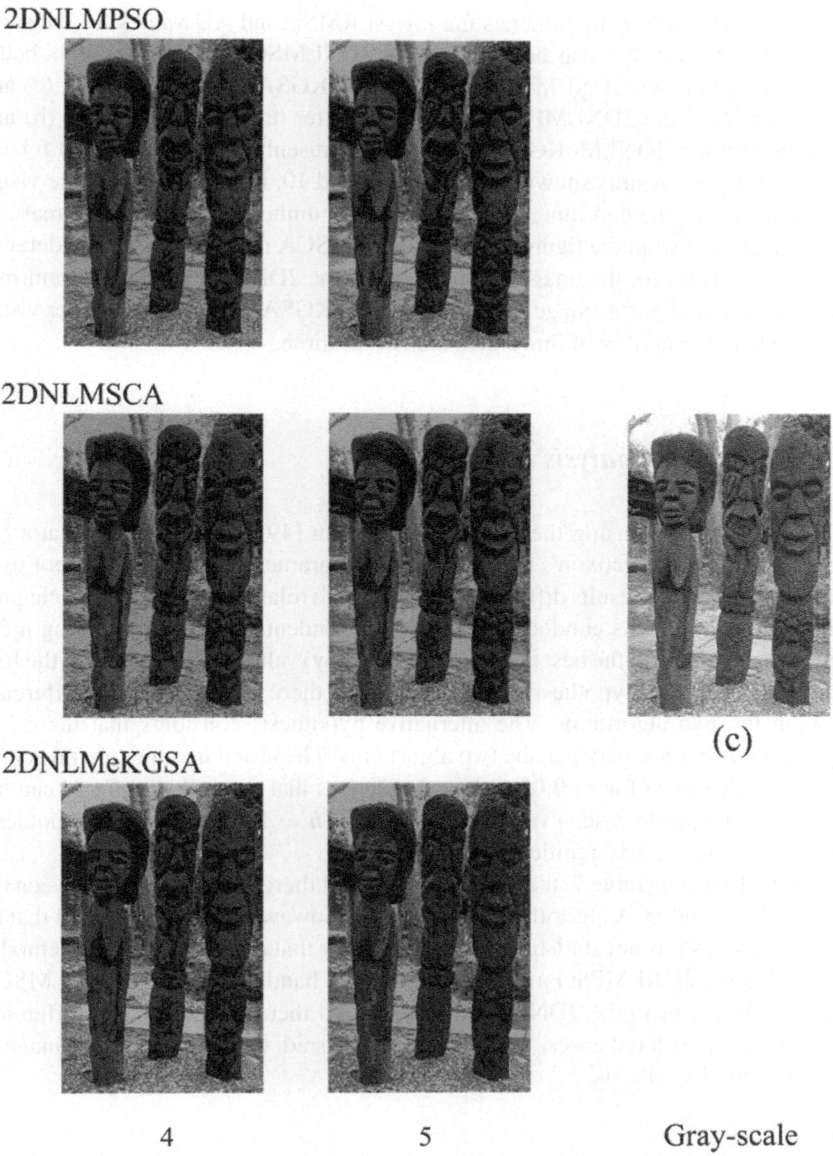

Fig. 10 Visual quality comparison between gray-scale image and segmented images obtained by 2DNLMeKGSA, 2DNLMPSO and 2DNLMSCA multilevel thresholding methods for 4 and 5 thresholds

Also, the 2DNLMSCA method has the lowest RMSE and AD values for the same images, for the above it can be said that the 2DNLMSCA method presents better performance than the 2DNLMPSO and 2DNLMeKGSA in the images (a), (c) and (e); nonetheless, the 2DNLMPSO method has better results for the image (b) and (d), although the 2DNLMeKGSA method also presents promising results for the image (d). By the results shown in Figs. 7, 8, 9 and 10, it is detected that the visual quality of the segmented image improves as the number of thresholds increase. It can be discerned from the figures that the 2DNLMSCA method returns fine detailed segmented images for the images (a) and (c) and the 2DNLMPSO method returns a higher visual quality for image (b). The 2DNLMeKGSA method gives better visual results when the number of thresholds are two or three.

4.3 Statistical Analysis

A statistical analysis using the Wilcoxon rank sum [49] test is performed at a 5% significance level. Wilcoxon's rank test is a non-parametric significance proof used to assess statistically results differences between two related methods. In this chapter, Wilcoxon's rank test is conducted with 30 independent samples considering a 5% significance level over the best fitness (Rényi entropy) value corresponding to the four thresholds. The null hypothesis is constructed as: there is no significant difference between the two algorithms. The alternative hypothesis considers that there is a significant difference between the two algorithms. The $\boldsymbol{w}$ and $\boldsymbol{h}$ values are presented in Table 7. A value of $\boldsymbol{w} > \mathbf{0.05}$ or $\boldsymbol{h} = \mathbf{0}$ indicates that the null hypothesis can not be rejected. Opposite case, a value of $\boldsymbol{w} < \mathbf{0.05}$ or $\boldsymbol{h} = \mathbf{1}$ means the null hypothesis can be rejected at a 5% significance level.

When observing Table 7, the results suggest that there is a difference between the PSO, eKGSA and SCA algorithms in most cases, however, it can be detected that on seven occasions it is not statistically possible to say that the 2DNLMSCA method is different to the 2DNLMPSO method. On the other hand, between the 2DNLMSCA method is superior to the 2DNLMeKGSA method there is a significant difference in the twenty registered cases. Therefore it can be said, that the methods manage to reject the null hypothesis.

5 Conclusion

In this chapter, three multi-level image thresholding segmentation methods have been explained. These methods use the 2D histogram non-local means and Rényi entropy as an objective function for the metaheuristic algorithms (eKGSA, PSO and SCA) in order to obtain the optimal thresholds that segment the images more precisely. The experiments are conducted on five images from the Berkeley Segmentation Dataset

Table 7 Statistical analysis (Wilcoxon rank sum test) of the metaheuristic algorithm based multilevel thresholding methods. w = probability of the statistic, h = 1 means the null hypothesis can be rejected at 5% level of significance

Image	Thr	2DNLMSCA versus 2DNLMPSO		2DNLMSCA versus 2DNLMeKGSA	
		w	h	w	h
(a)	2	**1.29E−06**	1	**3.02E−11**	1
(a)	3	0.4204	0	**3.02E−11**	1
(a)	4	0.6204	0	**3.02E−11**	1
(a)	5	**0.0371**	1	**3.02E−11**	1
(b)	2	0.9941	0	**3.02E−11**	1
(b)	3	**0.0303**	1	**3.02E−11**	1
(b)	4	**0.0122**	1	**3.02E−11**	1
(b)	5	**1.34E−05**	1	**4.08E−11**	1
(c)	2	0.9823	0	**3.02E−11**	1
(c)	3	**0.0156**	1	**3.02E−11**	1
(c)	4	**0.0406**	1	**3.02E−11**	1
(c)	5	0.112	0	**3.02E−11**	1
(d)	2	**0.0011**	1	**3.02E−11**	1
(d)	3	**4.12E−06**	1	**3.02E−11**	1
(d)	4	**2.15E−06**	1	**3.02E−11**	1
(d)	5	0.085	1	3.02E−11	1
(e)	2	0.4247	0	3.02E−11	1
(e)	3	**0.0215**	1	3.02E−11	1
(e)	4	**3.37E−04**	1	3.02E−11	1
(e)	5	0.2009	0	3.02E−11	1

Source: Bold values highlight the best result among the algorithms with respect to the performance evaluation parameter and the number of thresholds used in the image

and Benchmark (BSDS300) for subjective as well objective assessment on 2-level, 3-level, 4-level, and 5-level image thresholding segmentations. The objective analysis is done on six measures.

The experimental results suggest that the 2DNLMSCA method yields solutions of higher quality in the images (a), (c) and (e). However, the objective and subjective analysis of the 2DNLMPSO method exhibits better results for images (b) and (d). Therefore, the three methods can be used to segment different types of images. For future work, it is suggested to use some spatial properties of the images to improve thresholding-based image segmentation methods. Also, the use of new metaheuristic algorithms as well as their modification to improve their performance and obtain more accurate results are suggested.

References

1. N.M. Zaitoun, J. Aqel, Survey on image segmentation techniques. Procedia Procedia Comput. Sci. **65**, 797–806 (2015)
2. Y.J. Zhang, A survey on evaluation methods for image segmentation. Pattern Recognit. **29**(8), 1335–1346 (1996)
3. D. Oliva, S. Hinojosa, E. Cuevas, G. Pajares, O. Avalos, J. Gálvez, Cross entropy based thresholding for magnetic resonance brain images using crow search algorithm. Expert Syst. Appl. **79**, 164–180 (2017)
4. O. Tarkhaneh, H. Shen, An adaptive differential evolution algorithm to optimal multi-level thresholding for MRI brain image segmentation. Expert Syst. Appl. **138**, 112820 (2019)
5. S. Kotte, R.K. Pullakura, S.K. Injeti, Optimal multilevel thresholding selection for brain MRI image segmentation based on adaptive wind driven optimization. Measurement **130**, 340–361 (2018)
6. M. Sezgin, B. Sankur, Survey over image thresholding techniques and quantitative performance evaluation. J. Electron. Imaging **13**(1), 146 (2004)
7. A.S. Abutaleb, Automatic thresholding of gray-level pictures using two-dimensional entropy. Comput. Vision Graph. Image Process. **47**(1), 22–32 (1989)
8. H. Mittal, M. Saraswat, An optimum multi-level image thresholding segmentation using non-local means 2D histogram and exponential Kbest gravitational search algorithm. Eng. Appl. Artif. Intell. **71**, 226–235 (2018)
9. A. Buades, B. Coll, J.-M. Morel, A non-local algorithm for image denoising, in *2005 IEEE Computer Society Conference on Computer Vision and Pattern Recognition* (*CVPR'05*), vol. 2 (2005), pp. 60–65.
10. N. Otsu, A threshold selection method from gray-level histograms. IEEE Trans. Syst. Man. Cybern. **9**, 62–66 (1979)
11. X.-S. Yang, *Nature-Inspired Optimization Algorithms* (Elsevier, Amsterdam, 2014), p. iii
12. X. Zhao, M. Turk, W. Li, K. Lien, G. Wang, A multilevel image thresholding segmentation algorithm based on two-dimensional K–L divergence and modified particle swarm optimization. Appl. Soft Comput. **48**(C), 151–159 (2016)
13. S. Hinojosa et al., Unassisted thresholding based on multi-objective evolutionary algorithms. Knowledge-Based Syst. **159**, 221–232 (2018)
14. D.H.H. Wolpert, W.G.G. Macready, No free lunch theorems for optimization. IEEE Trans. Evol. Comput. **1**(1), 67–82 (1997)
15. K. Price, R. Storn, Differential evolution—a simple and efficient heuristic for global optimization over continuous spaces. J. Glob. Optim. **11**, 341–359 (1997)
16. B. Chopard, M. Tomassini, Particle swarm optimization, in *Natural Computing Series*, vol. 4 (2018), pp. 97–102
17. J.H. Holland, Outline for a logical theory of adaptive systems. J. ACM **9**(3), 297–314 (1962)
18. D. Karaboga, B. Akay, A comparative study of artificial bee colony algorithm. Appl. Math. Comput. **214**(1), 108–132 (2009)
19. Ş.İ. Birbil, S.-C. Fang, An electromagnetism-like mechanism for global optimization. J. Glob. Optim. **25**(3), 263–282 (2003)
20. E. Rashedi, H. Nezamabadi-pour, S. Saryazdi, GSA: a gravitational search algorithm. Inf. Sci. (Ny) **179**(13), 2232–2248 (2009)
21. S. Mirjalili, SCA: a sine cosine algorithm for solving optimization problems. Knowledge-Based Syst. **96**, 120–133 (2016)
22. Y. Zhang, S. Wang, G. Ji, A comprehensive survey on particle swarm optimization algorithm and its applications. Math. Probl. Eng. **2015**, 1–38 (2015)
23. S. Sarkar, S. Das, S.S. Chaudhuri, A multilevel color image thresholding scheme based on minimum cross entropy and differential evolution. Pattern Recognit. Lett. **54**, 27–35 (2015)
24. J.D. Bekensteing, Black holes and entropy. General relativity's centinnia. Phys. Rev. D. **7**, 23333 (1973)

25. J.N. Kapur, P.K. Sahoo, A.K.C. Wong, A new method for gray-level picture thresholding using the entropy of the histogram. Comput. Vision Graph. Image Process. **29**(3), 273–285 (1985)
26. S. Hinojosa, G. Pajares, E. Cuevas, N. Ortega-Sanchez, Thermal image segmentation using evolutionary computation techniques. Stud. Comput. Intell. **730**, 63–88 (2018)
27. M.A. Díaz-Cortés et al., A multi-level thresholding method for breast thermograms analysis using Dragonfly algorithm. Infrared Phys. Technol. **93**, 346–361 (2018)
28. S. Hinojosa, K.G. Dhal, M.A. Elaziz, D. Oliva, E. Cuevas, Entropy-based imagery segmentation for breast histology using the stochastic fractal search. Neurocomputing **321**, 201–215 (2018)
29. R. Benzid, D. Arar, M. Bentoumi, A fast technique for gray level image thresholding and quantization based on the entropy maximization, in *5th International Multi-Conference on Systems, Signals and Devices*, vol. 2, no. 1 (2008), pp. 1–4
30. S. Sarkar, S. Das, S.S. Chaudhuri, Multilevel image thresholding based on Tsallis entropy and differential evolution, in *Swarm, Evolutionary, and Memetic Computing, SEMCCO 2012*, vol. 7677 (2012)
31. P.K. Sahoo, G. Arora, A thresholding method based on two-dimensional Renyi's entropy. Pattern Recognit. **37**, 1149–1161 (2004)
32. S. Lan, L.I.U. Li, Z. Kong, J.G. Wang, Segmentation approach based on fuzzy Renyi entropy. *Chinese Conference on Pattern Recognition (CCPR)* (2010)
33. N.R. Pal, On minimum cross-entropy thresholding. Pattern Recognit. **29**(4), 575–580 (1996)
34. M. Masi, A step beyond Tsallis and Rényi entropies. Phys. Lett. Sect. A Gen. At. Solid State Phys. **338**(3–5), 217–224 (2005)
35. C. Cheng, X. Hao, S. Liu, Image segmentation based on 2D Renyi gray entropy and fuzzy clustering, in *2014 12th International Conference on Signal Processing (ICSP)* (2014), pp. 738–742
36. X.-F. Li, H.-Y. Liu, M. Yan, T.-P. Wei, Infrared image segmentation based on AAFSA and 2D-Renyi entropy threshold selection. DEStech Trans Comput Sci Eng, **o**, aice–ncs (2016)
37. C.E. Shannon, A mathematical theory of communication. ACM SIGMOBILE Mob. Comput. Commun. Rev. **5**(1), 3 (2001)
38. S. Borjigin, P.K. Sahoo, Color image segmentation based on multi-level Tsallis–Havrda–Charvát entropy and 2D histogram using PSO algorithms. Pattern Recognit. **92**, 107–118 (2019)
39. E.V. Cuevas Jimenez, J.V. Osuna Enciso, D.A. Oliva Navarro, M.A. Diaz Cortez, *Optimizacion: Algoritmos Programados Con MATLAB* (Alfaomega, Mexico, 2016)
40. R. Eberhart, J. Kennedy, Particle swarm optimization, in *Proceedings of the IEEE International Conference on Neural Networks* (Citeseer)
41. The Berkeley segmentation dataset and benchmark. [Online]. Available: https://www2.eecs.berkeley.edu/Research/Projects/CS/vision/bsds/. Accessed: 11 Jun 2019
42. A. Tanchenko, Visual-PSNR measure of image quality. J. Vis. Commun. Image Represent. **25**(5), 874–878 (2014)
43. Z. Wang, A.C. Bovik, H.R. Sheikh, E.P. Simoncelli, Image quality assessment: from error visibility to structural similarity. IEEE Trans. Image Process. **13**(4), 600–612 (2004)
44. L. Zhang, L. Zhang, X. Mou, D. Zhang, FSIM: a feature similarity index for image quality assessment. IEEE Trans. Image Process. **20**(8), 2378–2386 (2011)
45. R.J. Hyndman, A.B. Koehler, Another look at measures of forecast accuracy. Int. J. Forecast. **22**(4), 679–688 (2006)
46. Q. Huynh-Thu, M. Ghanbari, Scope of validity of PSNR in image/video quality assessment. Electron. Lett. **44**(13), 800 (2008)
47. J.P. Lewis, Fast template matching template. Pattern Recognit. **10**(11), 120–123 (1995)
48. C.S. Varnan, A. Jagan, J. Kaur, D. Jyoti, D.S. Rao, Image quality assessment techniques in spatial domain. Int. J. Comput. Sci. Technol. **2**(3), 177–184 (2011)
49. F. Wilcoxon, Individual comparisons by ranking methods. Biometrics Bull. **1**(6), 80 (1945)

Hybrid Metaheuristics and Other Image Processing Tasks

Comparison of Metaheuristic Methods for Template Matching

Gemma Corona, Marco Pérez-Cisneros, Oscar Maciel-Castillo, Adrián González and Fernando Fausto

Abstract In many image processing applications, the Template Matching (TM) technique plays an important role in recognizing and locating patterns or objects within a digital image. The main task of the TM is to seeks and find a position within an original image that resembles a predetermined sub-image (template) and a corresponding region of the original image as much as possible. TM involves two main aspects: similarity measurement and search strategy. The simplest method applied to TM involves a comprehensive calculation of the value of the Normalized Cross-Correlation (NCC) over all the pixel locations of the source image (search strategy). Unfortunately, the high computational cost that implies the evaluation of the NCC coefficient makes this approach restricted. We propose several TM methods based on evolutionary approaches as an alternative to reduce the number of search locations in the TM process. We have conducted a comparison of several evolutionary methods, to obtain which of these is the optimum to perform the TM task. Experimental results of this comparison show us which methods achieve the best balance between estimation and computational cost.

1 Introduction

Artificial vision is a broad field within research that aims to create machines capable of understanding their perceptible structure. TM techniques have proven to be a very useful tool for this process of intelligent perception and have led machines to a superhuman performance in tasks such as the location and recognition of objects or patterns in digital images, which represents one of the most important tasks for several applications of image processing and computer vision, such as facial recognition, industrial inspection, classification of objectives, digital photometry, remote sensors among others [1].

G. Corona (✉) · M. Pérez-Cisneros · O. Maciel-Castillo · A. González · F. Fausto
Departamento de Computación, Universidad de Guadalajara, CUCEI, Guadalajara, Jalisco, México
e-mail: gemma.cnunez@academicos.udg.mx

D. Oliva and S. Hinojosa (eds.), *Applications of Hybrid Metaheuristic Algorithms for Image Processing*, Studies in Computational Intelligence 890,
https://doi.org/10.1007/978-3-030-40977-7_7

Template Matching (TM) is a computer vision tool which allows us to find objects or their parts by looking for a sample sub-image (known as a template) or the most similar region in an image. In a typical TM procedure, the similarity between a template and a region is determined by applying a formula to measure specific similarity on a neighborhood around a particular pixel location within the image. Among the most common metrics used to evaluate the similarity between the template and the source image are the sum of the absolute differences (SAD), the sum of the squared differences (SSD) and the normalized cross-correlation (NCC). The calculation of these similarity measures has a high computational cost, and it represents a slow operation in a TM process. To this we add that traditional TM algorithms perform exhaustive searches in each pixel location of the original image and this limits the use of those recognition techniques in most artificial vision applications in real time [2].

We proposed several TM algorithms based on optimization techniques as an alternative to reduce the computational cost related to the pairing process, said techniques use search strategy to get a specific location on a limited subset of pixels within the image plane, this allows a noticeable reduction in the number of function evaluations required in this process.

In this chapter we present a comparative study related to various metaheuristic optimization methods that we implemented to solve the TM problem, including Artificial Bee Colony (ABC) [3], The Ant Lion Optimizer (ALO) [4], Bat-Inspired Algorithm (BA) [5], Covariance Matrix Adaptation Evolution Strategy (CMAES) [6], Cuckoo Search Algorithm (CS) [7], Crow Search algorithm (CSA) [8], Differential Evolution (DE) [9], Firefly Algorithm (FA) [10], Fuzzy Adaptive Particle Swarm Optimization (FUZZY) [11], Grey Wolf Optimizer (GWO) [12], Harmony Search (HS) [13], Locust Swarm-II (LSII) [14], Moth-flame optimization (MFO) [15], Particle Swarm Optimization (PSO) [16], Simulated Annealing (SA) [17], Selfish Herd Optimizer (SHO) [18], The Social Spider Optimization (SSO) [19], The Whale Optimization Algorithm (WOA) [20] and Yellow Saddle Goatfish Algorithm (YSGA) [21]. For this comparison, we have considered a set of test images with specific characteristics and difficulties in relation to the TM task.

For all cases, the objective function considered is the "normalized cross-correlation". This chapter is organized as follows: in Sect. 2, We illustrate the TM process, emphasizing on its most important characteristics; in Sect. 3, we present a general scheme of operation that governs most metaheuristic optimization algorithms, In Sect. 4, we discuss the general procedure of TM algorithms based on metaheuristic optimization; In Sect. 5, we introduce a series of comparative results between different TM methods based on metaheuristics, and a discussion regarding the results; finally, in Sect. 6 we provide the conclusions got after analyzing the comparison.

2 Template Matching Process

To illustrate the process known as Template Matching (TM), let I denote an intensity image with a certain spatial resolution (size) $M \times N$ and let R represent a reference image (or image template) of size m × n. Furthermore, let (x, y) denote a coordinate pair of positions of R. If we consider the shifted reference image $R_{u,v}(x, y) = R(x - u, y - v)$ such that u and v represent and horizontal and vertical displacement over the source image I respectively, then the matching problem may be summarized as follows: Given a source image I and a reference image R, find the offset (uv) within a search region $S \in I$ such that the similarity between the shifted reference image $R_{u,v}(x, y)$ and the corresponding subimage of I is maximum. To solve such problem, first we must address two important issues: (1) determining an appropriate similarity measurement to validate a match occurrence, and (2) developing an efficient search strategy to find the optimal template displacement (uv) (Fig. 1).

Although there are several metrics used to evaluate the similarity between two images, in the TM process the most commonly used measurements include the Sum of Absolute Differences (SAD), the Sum of Squared Differences (SSD), and the Normalizes Cross-Correlation (NCC). However, the calculation of such similarity measurements demands a high computational cost, and, it represents the most time-consuming operation on a TM process [22]. Although such metrics allow an adequate measurement of the similarity between a pair of images, the NCC coefficient is considered being the most robust measurement among them, and as such is most commonly used [22]. The NCC value between a source image I of size $N \times M$ and an image template R of size $m \times n$, at a given image displacement (uv), is given as follows:

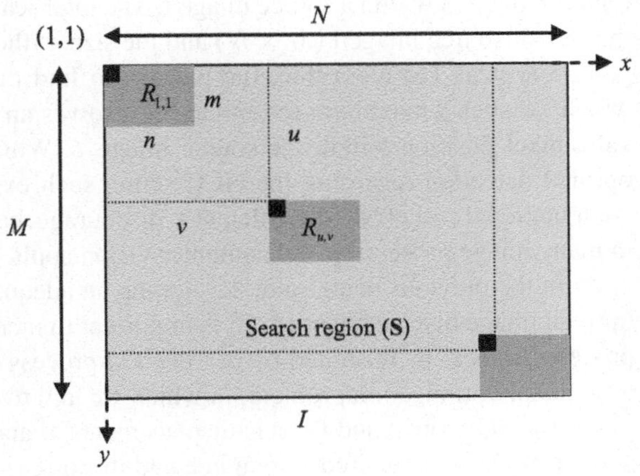

Fig.1 Template matching process

$$NCC = \frac{\sum_{i=1}^{m}\sum_{j=1}^{n}[I(u+i, v+j) - \overline{I}(u,v)] * [R(i,j) - \overline{R}]}{\sqrt{\left[\Sigma_{i=1}^{m}\Sigma_{j=1}^{n}\left[I(u+i, v+j) - \overline{I}(u,v)\right]\right]^2 * \left[\Sigma_{i=1}^{m}\Sigma_{j=1}^{n}\left[R(i,j) - \overline{R}\right]\right]^2}} \tag{1}$$

where $\overline{I}(uv)$ denotes the gray-scale average intensity of the source image I for the coincident region of the image template R, while $\overline{R}$ stand for the gray-scale average intensity of the image template R. Such values are given as follows:

$$\overline{I}(uv) = \frac{1}{m \cdot n}\sum_{i=1}^{m}\sum_{j=1}^{n} I(u+i, v+j)\overline{R} = \frac{1}{m \cdot n}\sum_{i=1}^{m}\sum_{j=1}^{n} R = (i, j) \tag{2}$$

The NCC computation yields to values defined between the interval of $[-1, 1]$, where a score of NCC = 1 implies the best similarity between the image template R and its corresponding sub-image I, whereas a value of NCC = −1 means that both of such images are entirely different.

In TM process, let $(\hat{u}, \hat{v})$ denote the position within the source image I in which is the best resemblance (maximum NCC value) between R and I is found. We can define such image position as follows:

$$(\hat{u}, \hat{v}) = \arg\max_{(u,v)\epsilon \mathbf{S}} NCC(\hat{u}, \hat{v}) \tag{3}$$

where $\mathbf{S} = \{(u, v)|1 \leq u \leq M - m, 1 \leq v \leq N - n\}$, as previously stated, denotes a search region defined within the source image I.

Figure 1 shows, the reference image R shifted by an offset (uv) across the search region S defined within a source image I. The total search region S depends on both; the size of source image I $(M \times N)$ and the size of the reference image R $(m \times n)$.

In a typical TM algorithm, the process to find the image displacement $(\hat{u}, \hat{v})$ which satisfies a maximum resemblance, involves an exhaustive search over every valid pixel position within the source image I. While this approach yields to an optimal detection regarding the NCC score, such exhaustive search and the high computational cost of NCC coefficient's, discourage the use of classic TM approaches in many image processing and computer vision applications.

With the previous being said, developing an adequate search strategy to find the optimal image displacement $(\hat{u}, \hat{v})$ is important to increase the efficiency of the TM process. In fact, as illustrated by (4), the TM process could be effectively modeled as a global optimization problem, in which we aim to find the optimal combination of discrete horizontal and vertical displacements, $\hat{u}$ and $\hat{v}$ respectively, such that the similarity between the image template and its corresponding sub-image in position $(\hat{u}, \hat{v})$ yields to a maximum NCC value. For this reason, the use of optimization techniques, as an alternative to solve efficiently the problem of TM, become intuitive [23].

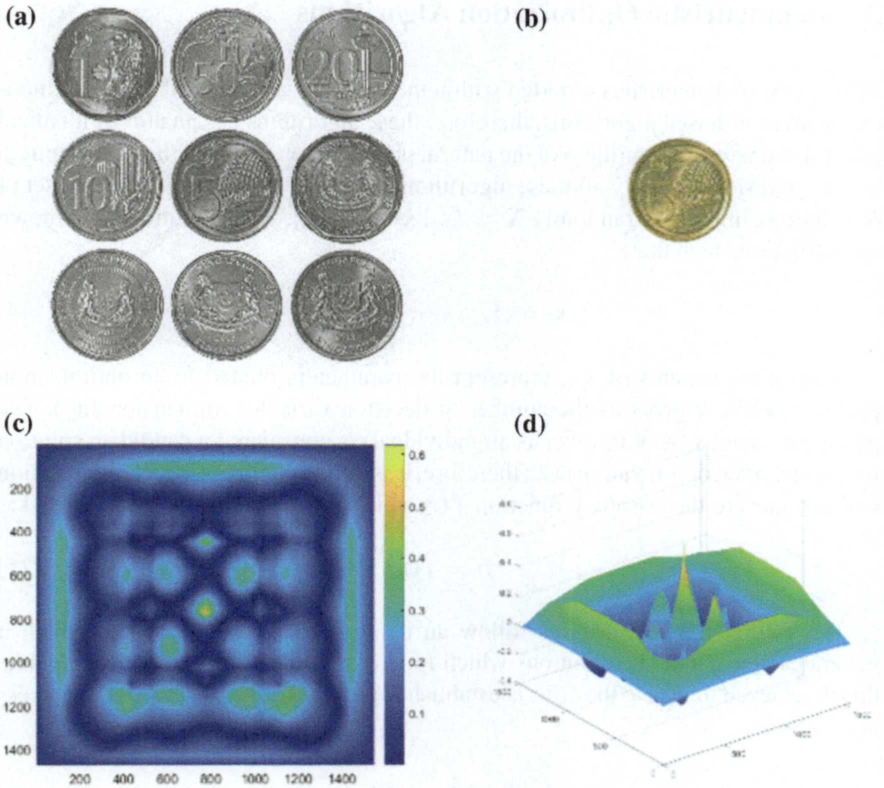

Fig. 2 **a** Example of a source image, **b** a reference image (image template), **c** color-encoded NCC values corresponding to the template matching process between (**a**) and (**b**), and **d** the NCC multimodal surface of (**c**)

Figure 2 illustrates the TM process regarding to the NCC coefficient: Fig 2a, b illustrate both, an example source image and an image template, respectively; Fig. 2c shows the color-encoded NCC values corresponding to all locations within the valid search region **S** on the source image (full search strategy); finally, Fig. 2d presents the NCC surface, which exhibits the highly multimodal nature of a typical TM problem. By observing both, Fig. 2c, d, the surface generated by the NCC values has several local maxima positions, while it only has a single global maximum. In such situations, classical optimization methods (particularly those founded on gradient-based techniques) get stuck in local optimal values, which makes them inadequate for solving this kind of optimization problems.

3 Metaheuristic Optimization Algorithms

Most of the metaheuristics reported within the literature are inspired by nature models population-based algorithms, therefore, these algorithms use an almost identical general framework, regardless of the natural phenomenon in which they are inspired.

The first step of nearly all these algorithms is similar, it starts by defining a set of N solutions initialized randomly $\mathbf{X} = \{\mathbf{x}_1, \mathbf{x}_2, \ldots, \mathbf{x}_N\}$, these solutions are known as population, such that:

$$\mathbf{x}_i = [x_{i,1}, x_{i,2}, \ldots, x_{i,d}] \tag{4}$$

where the elements of $x_{i,n}$ represent the parameters related to an optimization problem, while d gives us the number of decision variables (dimensionality). The parameter array $\mathbf{x}_i \in \mathbf{X}$ (known as an individual) is considered a candidate solution for the specified optimization task; therefore, it assigns a quality value to each solution which relates to the objective function $f(.)$ which describes the optimization task:

$$f_i = f(\mathbf{x}_i) \tag{5}$$

The methods nature inspires follow an iterative search script, within which it generates new candidate solutions which modify the individuals that are available, this is achieved thanks to the criteria established previously. In most cases the process is as follows:

$$\mathbf{x}_i^{'} = \mathbf{x}_i + \Delta\mathbf{x}_i \tag{6}$$

where $\mathbf{x}_i^{'}$ tells us about the candidate solution which is generated by adding a specific vector updated with the reference $\mathbf{x}_i$ to $\Delta\mathbf{x}_i$.

Most nature-inspired algorithms include some kind of selection, to compare the solutions generated to the current population $\mathbf{x}^k$ (where k represents the current iteration), the purpose is to choose the best individuals among them. The result is a new set of solutions $\mathbf{X}^{k+1} = \{\mathbf{x}_1^{k+1}, \mathbf{x}_2^{k+1}, \ldots, \mathbf{x}_n^{k+1}\}$, corresponding to the next iteration "$k + 1$".

This process repeats iteratively until it reaches the maximum number of iterations, and it met the determined stopping criterion. Once finished the process, the algorithm shows the best approximation for the global optimum (best solution).

4 The General Procedure Followed by TM Algorithms Based on Metaheuristic Optimization Methods

In Sect. 2 we explain a typical TM method which can find the precise position (uv) to calculate a certain similarity metric on all valid pixel locations within a specific search

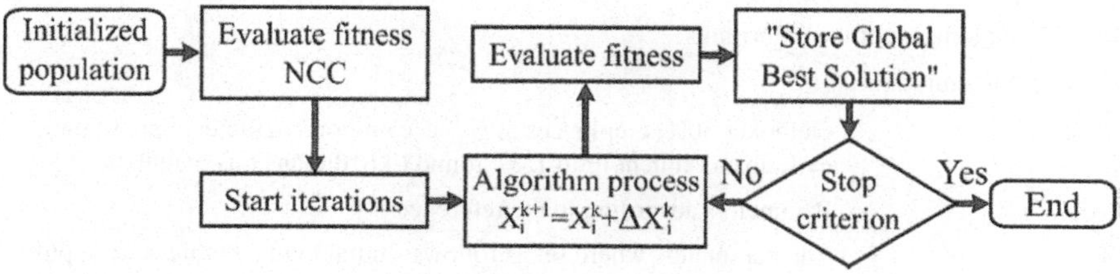

Fig. 3 The general procedure by TM algorithms based on metaheuristic optimization methods

region *S* of an image. However, these approaches generate a high computational cost [1]. Because of this, there are different TM algorithms [2, 22, 24] that have been proposed to speed up the search process allowing to compute only a limited set of search locations. These methods allow to reduce the computational cost of the TM technique, but lack the ability to explore the search region *S* effectively, and suffer premature convergence, resulting in values with sub-optimal solutions. These problems relate to the operators used when modifying particle positions during the evolutionary process, as shown in Fig. 3. In most of these methods, the position of each search agent for the next iteration is updated considering an attraction to the best individual information so far [25]. Because of this, the whole population tends to concentrate around the most known location in the source image, which favors a premature convergence towards a local optimum of the multimodal surface of the image.

5 Experimental Results

In this comparison study, we used a set of 19 algorithms in the TM method, to test their performance. The algorithms that we used in the comparison are: ABC, ALO, BA, CMAES, CS, CSA, DE, FA, FUZZY, GWO, HS, LSII, MFO, PSO, SA, SHO, SSO, WOA, YSGA. In Table 1, the sets of parameters applied to each algorithm are show; the parameter configuration applied to each algorithm correspond to those recommended by their respective authors:

For each algorithm, the population size is 50 ($N = 50$) while the maximum number of iterations is 300 ($ItersN = 300$). In n TM tests. Each test represents a unique problem since the dimensions used are different for each image.

We reported the evaluation of the performance on 30 different executions in Table 1 considering the following indices: Average f_{AVG}, Median f_{MEDIAN} and Standard deviation f_{STD}, which represent the percentage of all test in which a given image template was successfully detected (view Table 2).

From Table 2 we were able to see that LS-II obtained the best AVG and MEDIAN from every image, and with a small STD which translates to obtain consistent better

Table 1 Parameters of the algorithms

Setting configuration	
ABC	Onlooker 50%, employees 50%, acceleration coefficient upper bound = 1, abandonment limit $L =$ round (0.6*dimensions*population) [3]
ALO	Parameters according to the reference [4]
BA	The parameters where set as follows: Initial loudness rate $A = 2$, pulse emission rate $r = 0.9$, minimum frequency $f_{\min} = 0$ and maximum frequency $f_{\max} = 1$, respectively [5]
CMAES	The ones proposed by the author [6]
CS	The ones proposed by the author [7]
CSA	The awareness probability is set to $AP = 0.1$, while the flight length is given as $fl = 2$ [18]
DE	The crossover rate is set to $CR = 0.5$, while the differential weight is given as $F = 0.2$ [9]
FA	The randomness factor and the light absorption coefficient are set to $\alpha = 0.2$ and $\gamma = 1.0$, respectively [10]
FUZZY	The ones proposed by the author [11]
GWO	The algorithm's parameter a is set to decrease linearly from 2 to 0 [12]
HS	$HMCR = 0.7$ Harmony Memory Considering Rate finds notes randomly within the possible playable range. $PAR = 0.3$ Pitch Adjusting Rate [13]
LSII	The number of best individuals considered for the social phase operator is set to q= 10 [14]
MFO	The number of flames is set as $N_{\text{flames}} = \text{round}\left((N_{\text{pop}} - k) * (N_{\text{pop}} - 1)/k_{\max}\right)$, where N_{pop} is the population, $k_{\max}$ the maximum number of iterations and k the actual iteration [15]
PSO	The cognitive and social coefficients are set to $c_1 = 2.0$ and $c_2 = 2.0$, respectively. Also, the inertia weight factor ω is set to decreases linearly from 0.9 to 0.2 as the search process evolves [16]
SA	The algorithm's initial temperature is set to $T^0 = 1$, while the cooling schedule employed correspond to a geometrical cooling scheme considering a cooling rate of $\beta = 0.98$ [17]
SHO	The proposed method is tested by considering a herd population proportion randomly chosen from between a 70 and 90%, with the remaining individuals assigned as predators [18]
SSO	The female attraction probability parameter is set as $PF = 0.7$ [19]
WOA	The internal parameters A and C are set to decreases linearly from 2 to 0, and -1 to -2, respectively [20]
YSGA	Cluster number = 4 [21]

Table 2 Performance comparison of algorithms proposed for the experimental set show

		$IM1$	$IM2$	$IM3$	$IM4$	$IM5$	$IM6$
ABC	f_{AVG}	0.8200	0.6527	0.6424	0.6944	0.7186	0.4524
	f_{MEDIAN}	0.8639	0.6647	0.6230	0.7053	0.7270	0.4311
	f_{STD}	0.1359	0.2029	0.0628	0.2052	0.1351	0.1860
ALO	f_{AVG}	0.4757	0.2962	0.4783	0.2727	0.4139	0.1594
	f_{MEDIAN}	0.4484	0.2796	0.4910	0.2513	0.4001	0.1592
	f_{STD}	0.1219	0.0617	0.0490	0.0538	0.1116	0.0290
BA	f_{AVG}	0.5966	0.3520	0.5344	0.3538	0.5288	0.2408
	f_{MEDIAN}	0.5636	0.3196	0.5344	0.2591	0.4760	0.2050
	f_{STD}	0.1699	0.1487	0.0338	0.2112	0.1709	0.1118
CMAES	f_{AVG}	0.7834	0.6855	0.6362	0.6123	0.5487	0.2620
	f_{MEDIAN}	0.7797	0.7642	0.6267	0.5081	0.5249	0.2341
	f_{STD}	0.1861	0.2538	0.0654	0.3289	0.1327	0.1047
CS	f_{AVG}	0.7475	0.4646	0.6344	0.5184	0.6868	0.3327
	f_{MEDIAN}	0.7562	0.4212	0.6219	0.4711	0.6487	0.2891
	f_{STD}	0.1340	0.1353	0.0598	0.1607	0.1707	0.1326
CSA	f_{AVG}	0.7881	0.6363	0.5927	0.5981	0.7639	0.3150
	f_{MEDIAN}	0.7982	0.6602	0.5874	0.5636	0.7991	0.3160
	f_{STD}	0.1363	0.1977	0.0527	0.1803	0.1183	0.1021
DE	f_{AVG}	0.8246	0.5689	0.6197	0.6397	0.7082	0.4062
	f_{MEDIAN}	0.8014	0.5382	0.6055	0.5870	0.7132	0.4035
	f_{STD}	0.1154	0.2126	0.0627	0.2145	0.1268	0.1368
FA	f_{AVG}	0.8395	0.6065	0.6239	0.5546	0.7225	0.3781
	f_{MEDIAN}	0.8796	0.6017	0.6065	0.4976	0.7304	0.3269
	f_{STD}	0.1241	0.1824	0.0781	0.1585	0.1302	0.1833
FUZZY	f_{AVG}	0.7096	0.3665	0.5799	0.4402	0.5984	0.2326
	f_{MEDIAN}	0.6973	0.3530	0.5696	0.4051	0.5912	0.2013
	f_{STD}	0.1443	0.0695	0.0739	0.1277	0.1432	0.0566
GWO	f_{AVG}	0.9910	0.8477	0.7059	0.7482	0.7752	0.3197
	f_{MEDIAN}	0.9990	0.9885	0.6340	0.9988	0.8309	0.2341
	f_{STD}	0.0107	0.2500	0.1495	0.2888	0.2139	0.2138
HS	f_{AVG}	0.8208	0.5578	0.5999	0.8736	0.8373	0.4471
	f_{MEDIAN}	0.9026	0.4687	0.5762	0.9988	0.9092	0.3496
	f_{STD}	0.1880	0.2215	0.0659	0.2314	0.1712	0.2313
LS-II	f_{AVG}	0.9989	0.9879	0.9875	0.9984	0.9992	0.9932

(continued)

Table 2 (continued)

		$IM1$	$IM2$	$IM3$	$IM4$	$IM5$	$IM6$
	f_{MEDIAN}	0.9989	0.9879	0.9973	0.9984	0.9992	0.9932
	f_{STD}	5.6×10^{-16}	3.3×10^{-16}	0.0536	4.5×10^{-16}	4.5×10^{-16}	3.3×10^{-16}
MFO	f_{AVG}	0.8975	0.7291	0.6908	0.7340	0.8043	0.4593
	f_{MEDIAN}	0.9990	0.9330	0.6288	0.8480	0.9615	0.2641
	f_{STD}	0.1722	0.3005	0.1626	0.2662	0.2182	0.3320
PSO	f_{AVG}	0.9513	0.6994	0.6559	0.6980	0.8239	0.5184
	f_{MEDIAN}	0.9990	0.8762	0.5698	0.5247	0.9903	0.2701
	f_{STD}	0.1463	0.3070	0.1347	0.2802	0.2258	0.3418
SA	f_{AVG}	0.1241	0.1560	0.0925	0.1774	0.0332	0.0210
	f_{MEDIAN}	0.1711	0.1544	0.1113	0.1798	0.0199	0.0205
	f_{STD}	0.2410	0.1065	0.2371	0.1094	0.1652	0.0618
SHO	f_{AVG}	0.8802	0.7028	0.6494	0.8071	0.6960	0.7106
	f_{MEDIAN}	0.9990	0.9885	0.5698	0.9988	0.5538	0.9932
	f_{STD}	0.2025	0.3119	0.1616	0.3032	0.2358	0.3781
SSO	f_{AVG}	0.8640	0.7301	0.6829	0.8721	0.7945	0.4841
	f_{MEDIAN}	0.9990	0.9885	0.5698	0.9988	0.9992	0.2341
	f_{STD}	0.2019	0.3082	0.1934	0.2358	0.2386	0.3667
WOA	f_{AVG}	0.8290	0.4634	0.6129	0.5125	0.6295	0.2568
	f_{MEDIAN}	0.9932	0.3595	0.5686	0.3984	0.5019	0.2168
	f_{STD}	0.2295	0.2480	0.1263	0.2624	0.2392	0.1482
YSGA	f_{AVG}	0.8551	0.5150	0.6708	0.6314	0.5870	0.2968
	f_{MEDIAN}	0.9239	0.4000	0.6334	0.5534	0.5282	0.2439
	f_{STD}	0.1608	0.2343	0.0927	0.2592	0.1538	0.1471

results than the other algorithms, some algorithms that obtained good results obtaining second place were GWO in images 1–3, HS in images 4 and 5 and SHO in IMG6.

Table 3 represents the success rate of the algorithms; this is the number of times they placed the template inside their correct reference in the search image, LSII obtains and average of 100% which means that it was able to find the template in each of the 30 executions, GWO was able to obtain a 100% in IM1, but as we see in Table 2 the AVG of LSII got better results making it the best option even in I.

Table 3 Results of success rate

	*IM*1 (%)	*IM*2 (%)	*IM*3 (%)	*IM*4 (%)	*IM*5 (%)	*IM*6 (%)	*IM*7 (%)
ABC	93.33	86.67	50.00	90.00	16.67	63.33	66.67
ALO	3.33	10.00	6.67	30.00	30.00	10.00	15.00
BA	60.00	10.00	16.67	26.67	20.00	20.00	25.56
CMAES	76.67	73.33	73.33	73.33	60.00	20.00	62.78
CS	96.67	56.67	53.33	76.67	23.33	43.33	58.33
CSA	53.33	23.33	30.00	56.67	0.00	20.00	30.56
DE	96.67	76.67	36.67	90.00	6.67	60.00	61.11
FA	93.33	90.00	46.67	73.33	10.00	46.67	60.00
FUZZY	93.33	26.67	36.67	70.00	23.33	23.33	45.56
GWO	100.00	80.00	46.67	63.33	30.00	16.67	56.11
HS	76.67	53.33	30.00	93.33	16.67	53.33	53.89
LSII	100.00	100.00	10.00	100.00	100.00	100.00	100.00
MFO	76.67	63.33	40.00	70.00	33.33	36.67	53.33
PSO	93.33	56.67	30.00	53.33	30.00	40.00	50.56
SA	13.33	0.00	6.67	23.33	23.33	20.00	14.44
SHO	80.00	56.67	23.33	76.67	40.00	63.33	56.67
SSO	76.67	60.00	36.67	83.33	33.33	40.00	55.00
WOA	86.67	23.33	40.00	63.33	36.67	10.00	43.33

6 Conclusions

In this chapter, a comparative study between 19 metaheuristics-based Template Matching (TM) algorithms has been presented. Overall, the TM problem could be quite difficult to solve; this is because the NCC function evaluation models multimodal surfaces which are difficult to handle with traditional optimization algorithm, hence making metaheuristic search algorithms a good alternative to tackle this problem. On each of the compared metaheuristic methods, solutions are represented as 2-D coordinates within a search region over a given source image. The NCC coefficient (represented as a fitness value) is computed for each individual search agent, and then, used to evaluate the matching quality between a given image template and a coincident region of the source image. While all of the compared methods yield to a notorious reduction on the computational cost related to traditional exhaustive TM approaches, the overall performance of these techniques is not equal; this difference in performance is related to the different search strategies and mechanisms implemented by each individual method. Our comparative results suggest that methods such as HS and LSII are able to solve the TM much more competitively than all other techniques. This notorious performance may be attributed to the dedicated exploration and exploitation operators implemented by this algorithm, namely the solitary and social phase operators.

Appendix

Images dataset use on our experimental setup.

	Image	Template	NCC surface	Image resolution
$IM1$			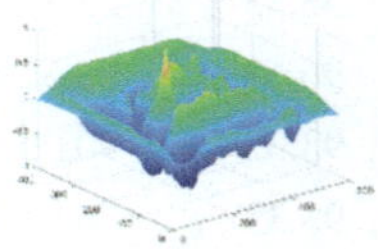	447 × 344
$IM2$		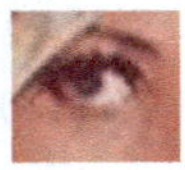	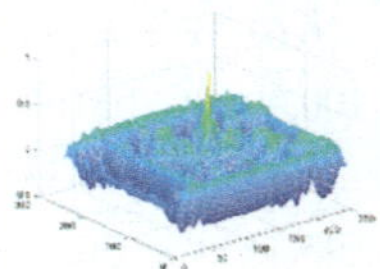	220 × 220
$IM3$			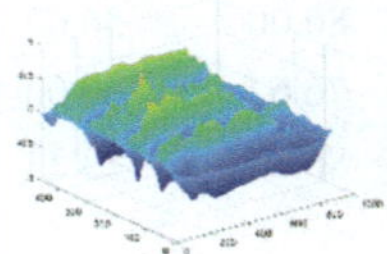	747 × 400
$IM4$			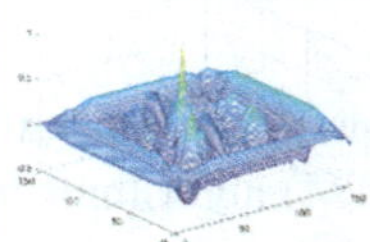	120 × 120
$IM5$	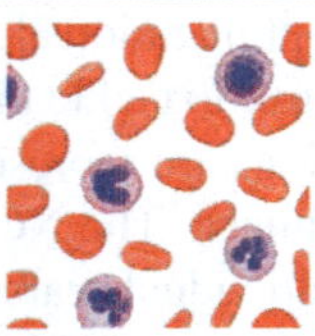	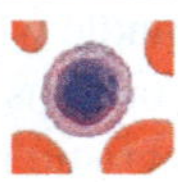	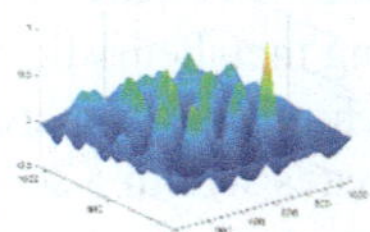	800 × 800
$IM6$			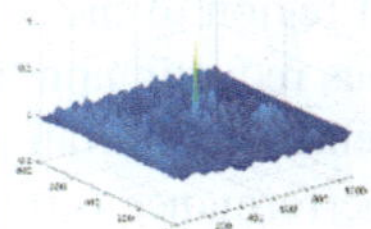	880 × 651

References

1. A. González, E. Cuevas, F. Fausto, A. Valdivia, R. Rojas, A template matching approach based on the behavior of swarms of locust. *Appl. Intell.***47**(4), 1087–1098 (2017)
2. H. Grailu, M. Lotfizad, H. Sadoghi-Yazdi, An improved pattern matching technique for lossy/lossless compression of binary printed Farsi and Arabic textual images. Int. J. Intell. Comput. Cybern. **2**(1), 120–147 (2009)
3. D. Karaboga, B. Basturk, On the performance of artificial bee colony (ABC) algorithm. Appl. Soft Comput. J. **8**(1), 687–697 (2008)
4. S. Mirjalili, The ant lion optimizer. Adv. Eng. Softw. **83**, 80–98 (2015)
5. X.-S. Yang, *A New Metaheuristic Bat-Inspired Algorithm* (Springer, Berlin, Heidelberg, 2010), pp. 65–74
6. P.K. Nikolaus Hansen, S.D. Müller, Reducing the time complexity of the derandomized evolution strategy with covariance matrix adaptation (CMA-ES). *Evol. Comput.* **149**(1), 1–18 (2003)
7. A.A. El-Fergany, A.Y. Abdelaziz, Cuckoo search-based algorithm for optimal shunt capacitors allocations in distribution networks. *Electr. Power Compon. Syst.***41**(16), 1567–1581 (2013)
8. A. Askarzadeh, A novel metaheuristic method for solving constrained engineering optimization problems: Crow search algorithm. Comput. Struct. **169**, 1–12 (2016)
9. K.V. Price, *Differential Evolution* (Springer, Berlin, Heidelberg, 2013), pp. 187–214
10. X.-S. Yang, *Firefly Algorithm, Stochastic Test Functions and Design Optimisation*, pp. 1–12 (2010)
11. Y. Shi, R.C. Eberhart, Fuzzy adaptive particle swarm optimization, in *Proceedings of the 2001 Congress on Evolutionary Computation(IEEE Cat. No.01TH8546)*, vol. 1 (1997)
12. S. Mirjalili, S.M. Mirjalili, A. Lewis, Grey Wolf optimizer. Adv. Eng. Softw. **69**, 46–61 (2014)
13. Z. Woo, J. Hoon, G.V. Loganathan, A new heuristic optimization algorithm: harmony search. *Simulation* (2001)
14. O. Camarena, E. Cuevas, M. Pérez-Cisneros, F. Fausto, A. González, A. Valdivia, Ls-II: an improved locust search algorithm for solving optimization problems. Math. Probl. Eng. **2018**, 1–15 (2018)
15. S. Mirjalili, Moth-flame optimization algorithm: a novel nature-inspired heuristic paradigm. Knowl.-Based Syst. **89**, 228–249 (2015)
16. J. Kennedy, R. Eberhart, B. Gov, Particle swarm optimization. *Encycl. Mach. Learn.*, 760–766 (2010)
17. S. Kirkpatrick, C.D. Gelatt, M.P. Vecchi, Optimization by simulated annealing. *Science***220**(4598), 671–680 (1983)
18. F. Fausto, E. Cuevas, A. Valdivia, A. González, A global optimization algorithm inspired in the behavior of selfish herds. BioSystems **160**, 39–55 (2017)
19. E. Cuevas, M. Cienfuegos, D. Zaldívar, M. Pérez-Cisneros, A swarm optimization algorithm inspired in the behavior of the social-spider. Expert Syst. Appl. **40**(16), 6374–6384 (2013)
20. S. Mirjalili, A. Lewis, The whale optimization algorithm. Adv. Eng. Softw. **95**, 51–67 (2016)
21. D. Zaldívar, B. Morales, A. Rodríguez, A. Valdivia-G, E. Cuevas, M. Pérez-Cisneros, A novel bio-inspired optimization model based on Yellow Saddle Goatfish behaviour. *Biosystems***174**, 1–21 (2018)
22. A.H. Gandomi, A.H. Alavi, Krill herd: a new bio-inspired optimization algorithm. Commun. Nonlinear Sci. Numer. Simul. **17**(12), 4831–4845 (2012)
23. E. Cuevas, A. Echavarría, D. Zaldívar, M. Pérez-Cisneros, A novel evolutionary algorithm inspired by the states of matter for template matching. Expert Syst. Appl. **40**(16), 6359–6373 (2013)

24. N. Dong, C.H. Wu, W.H. Ip, Z.Q. Chen, C.Y. Chan, K.L. Yung, An improved species based genetic algorithm and its application in multiple template matching for embroidered pattern inspection. Expert Syst. Appl. **38**(12), 15172–15182 (2011)
25. G. Chen, C.P. Low, Z. Yang, Preserving and exploiting genetic diversity in evolutionary programming algorithms. IEEE Trans. Evol. Comput. **13**(3), 661–673 (2009)

Novel Feature Extraction Strategies Supporting 2D Shape Description and Retrieval

P. Govindaraj and M. S. Sudhakar

Abstract Acute shape characterization in image retrieval tasks remain a persisting issue in computer vision determining their retrieval performance. This chapter contributes and relatively compares three such descriptors that are further tested for shape classification by employing a supervised machine learning mechanism. The core objective of this chapter is the effective exploitation of simple computing concepts for realizing shape descriptors aiding retrieval. Accordingly, simple and novel shape descriptors with its performance analysis are presented in this chapter. The potency of these methods is investigated using the Bull's Eye Retrieval (BER) rate on benchmarked datasets such as the Kimia, MPEG-7 CE Shape-1 part B and Tari-1000. Consistent BER greater than 90% attained across the diverse datasets affirms the descriptors efficacy, consequently signifying the robustness of these descriptors towards diverse affine transformations thereby, making it suitable for dynamic CBIR applications.

Keywords Feature extraction · Hexagonal imaging · Phase congruency · Laws texture energy measures

1 Introduction

Shape is considered as a vital feature for object recognition, classification and retrieval in computer vision and multimedia processing. Moreover, it reduces the feature size particularly when characterizing objects. Finding good shape descriptors and similarity measures are the central issues in these applications. Numerous

P. Govindaraj
Department of Electronics & Electrical Engineering, Indian Institute of Technology, Guwahati, Guwahati, Assam, India
e-mail: govindaraj.p@iitg.ac.in

M. S. Sudhakar (✉)
Vellore Institute of Technology, Vellore, India
e-mail: sudhakar.ms@vit.ac.in

D. Oliva and S. Hinojosa (eds.), *Applications of Hybrid Metaheuristic Algorithms for Image Processing*, Studies in Computational Intelligence 890,
https://doi.org/10.1007/978-3-030-40977-7_8

shape descriptors available in literature abiding invariance are deemed fit for image segmentation and retrieval [1] and mainly focus on improving their retrieval performance neglecting their higher computational aspects. This necessitates the need for computationally simple shape-based feature characterization schemes supporting image storage and retrieval.

Generally, shape characterization methods are mainly categorized into contour and region-based methods [2]. Subsequently, they are subdivided into global and structural schemes based on regional approach. Shape features falling under global category capture the integral boundary information for shape description. Rather the structural approach segments the boundary information for feature characterization. A brief review of the prevailing descriptors with its merits and demerits are chronologically listed below.

The widespread contour-based global descriptor, shape context (SC) [3] spatially correlates each point with all other points for representing the contour. Then, for each point in the contour, a sixty dimensional histogram is formulated using the spatial distribution of points in the contour. Ling et al. [4] built an articulation invariant feature for shape images by combining the inner-distance and multidimensional scaling (MDS). The approach in [3] evaluates and selects the shortest Euclidean distance between all points within the contour to formulate the feature descriptor followed by dynamic programming for shape matching and retrieval. Global descriptors [5, 6] utilize triangle area representation (TAR) configure boundary points in the contour at different scales with signed areas of triangles. The triangle area measures the curvature of the localized points and their concavity or convexity is signified by the sign of area value. Souza et al. [7] proposed a new shape descriptor termed as Hough Transform Statistics neighbourhood (HTSn) based on the Hough transform statistics to characterize object shapes. Irrespective of the image size, the feature vector size remains constant because of Hough space. Piecewise approximation of the shape information into conic-section coefficients was performed in [8]. Projective conic invariants with their pairs were used for generating the feature vectors. These feature extraction methods consider the overall shape information to construct their feature vectors.

Wu et al. [9] employed Tchebichef moments for shape description by relating the lower order invariants. Likewise, the Common Base Triangle Area (CBTA) of Hu et al. [1] extracted features from common base triangles. The resulting features were later modified by dynamic programing for shape matching and retrieval. For identifying shape boundaries, Fourier components of Circle Views (CVs) signatures were used in [10] for formulating the feature vectors. The invariant multi-scale descriptor [11] compact the shape contour into feature points using dynamic programing. Kaothanthong et al. [12] generated shape features using the intersection pattern of the distribution of line segments.

To achieve good retrieval rate, the above schemes depend heavily on dynamic programming that adds to the complexity of the overall retrieval process and restricts its usage in critical real-time applications. Hence, prospective hardware friendly schemes offering high retrieval accuracy is required for shape-based applications in image processing and computer vision problems. Thus, this chapter presents and discusses three shape description schemes that are realized using hybrid metaheuristic algorithms and its application to shape retrieval. Algorithmic hybridization is accomplished in these schemes by cross linking trigonometrical concepts with point-wise pixel processing to render acute and precise shape descriptors yielding higher retrieval rate. The presented light-weight shape descriptors aim to maximize the matching accuracy by strictly enforcing localization on shape characterization.

The rest of the chapter is ordered as follows: Sect. 2 details the General shape retrieval framework with three different edge detection and feature extraction techniques for binary images. The performance of the proposed descriptors on different shape datasets are discussed in Sect. 3. Finally, Sect. 4 concludes the presented works.

2 General Shape Retrieval Framework

A retrieval framework can be classified into four main subgroups, namely, input, feature extraction, matching and display. Figure 1 outlines the background process of a retrieval framework. User presents the query image to the framework, from which the features are extracted and then matched with the feature database. Finally, the corresponding matches for the given input query are displayed according to their matching ranks.

Aligning with the above methodology three shape descriptors merged with the corresponding classification scheme is rendered in this chapter supporting retrieval. The details of which are discussed under relevant sub-headers.

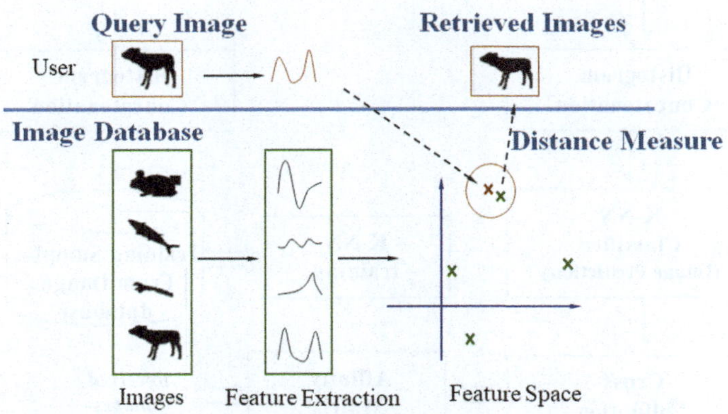

Fig. 1 General shape retrieval framework

2.1 Shape Characterization Using Laws of Texture Energy Measures

Feature characterization plays an important role for rendering highly precise feature descriptor that determines the efficiency of the shape retrieval methods. This sub-section details a novel retrieval scheme employing the Laws of Texture Energy Measures (LTEM). An outline of the proposed intention is displayed in Fig. 2. At the onset, each shape (target) image is processed by the LTEM operator for extracting the features that are then fabricated into shape histograms. This process is then repeated for the database images and the resulting shape histograms are accumulated into a feature matrix. Then, the k-Nearest Neighbour (k-NN) algorithm spatially classifies the feature database into respective clusters. In the retrieval phase, the LTEM features of the querying shape are extracted and then matched with the feature database. The images that closely map with the input are retrieved and displayed on the output window.

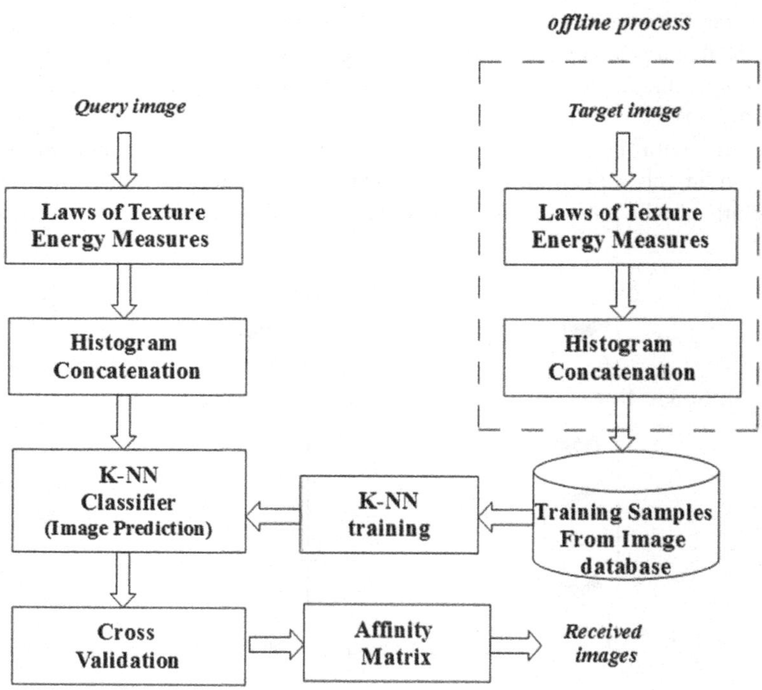

Fig. 2 Block-diagram of shape retrieval scheme using LTEM

2.1.1 Laws of Texture Energy Measures

LTEMs simplicity in feature extraction makes it widely popular across content based image retrieval, texture analysis [13, 14] and iris recognition [15]. It accomplishes the same by employing fixed window size diverse filter masks capable of extracting level, edge, spot and ripple features of an image. These four LTEM masks are presented in Eqs. (2.1)–(2.4).

$$L5(level) = [1 \quad 4 \quad 6 \quad 4 \quad 1] \tag{2.1}$$

$$E5(edge) = [-1 \quad -2 \quad 0 \quad 2 \quad 1] \tag{2.2}$$

$$S5(spot) = [-1 \quad 0 \quad 2 \quad 0 \quad -1] \tag{2.3}$$

$$R5(ripple) = [1 \quad -4 \quad 6 \quad -4 \quad 1] \tag{2.4}$$

This approach introduces a novel combination of the aforementioned masks to construct a 2-D filter for feature extraction. The combinations are listed below.

L5E5/E5L5	L5S5/S5L5	L5R5/R5L5
E5S5/S5E5	E5R5/R5E5	E5E5
S5R5/R5S5	S5S5	R5R5

From these 16 masks, L5L5 mask is neglected owing to its zero mean nature. Here, to obtain the shape boundary information, L5E5 and E5L5 is used in vertical and horizontal directions. Then the obtained information in both the directions is merged using logical OR operation to obtain single feature extracted image.

2.1.2 Histogram Formulation

The resulting LTEM features are subsequently transformed into histograms by locally dividing them into non-overlapping regions of size $r \times r$ and then concatenated to get the final shape feature vector (h) as denoted in Eq. (2.5). The histogram fabrication process is portrayed in Fig. 3.

$$h = [h_1\, h_2\, h_3 \ldots h_i] \tag{2.5}$$

2.2 *Shape Characterization Using Hexagonal Grid-Based Triangular Tessellation (HGTT)*

The second method organises each input image into overlapping hexagonal grids from 5×5 image sub-region and then tessellating them into triangles for extraction of localized features. This process is then repeated on the entire image and the

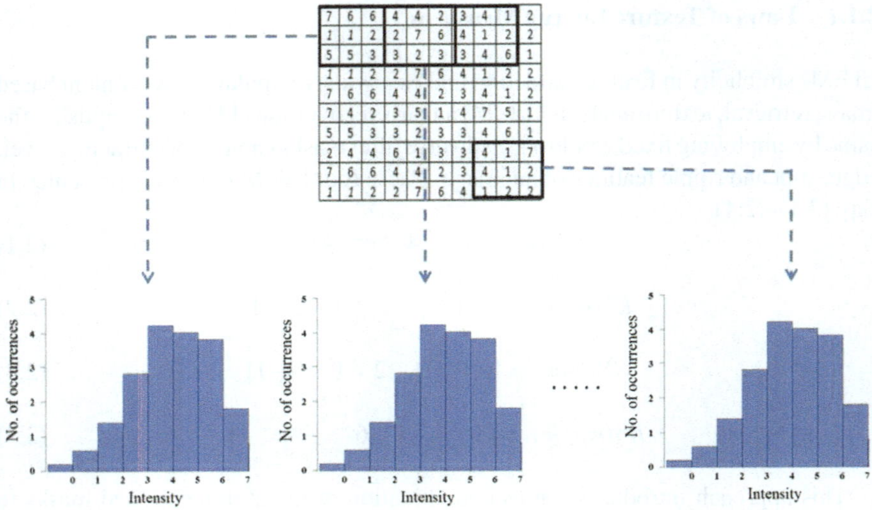

Fig. 3 Formulating feature histograms

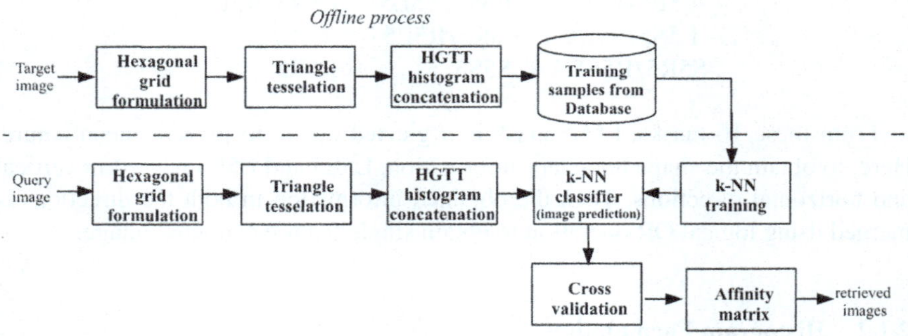

Fig. 4 Block-diagram of HGTT based shape retrieval

yielded features are transformed to produce the shape histogram. Similar process is undergone by the images present in the master database and the resulting features are stored into a feature matrix. Feature categorization is then performed by the k-Nearest Neighbour (k-NN) algorithm. During retrieval, the HGTT features of the query are extracted and compared against the accumulated features. Shapes matching with the given query are retrieved and displayed based on their ranks. An outline of the aforesaid discussion is depicted in Fig. 4, and the details pertaining to the individual blocks are elaborated under the following sub-sections.

This scheme initially resizes the input image into fixed size for attaining constant length features from varying sized input images using bilinear interpolation. The uniform sized images are then partitioned into 5 × 5 sub-regions and organized into hexagonal grids. The hexagon frames are further sub-divided into six equilateral triangles owing to the inherent nature of hexagonal arrangement. Later, the side

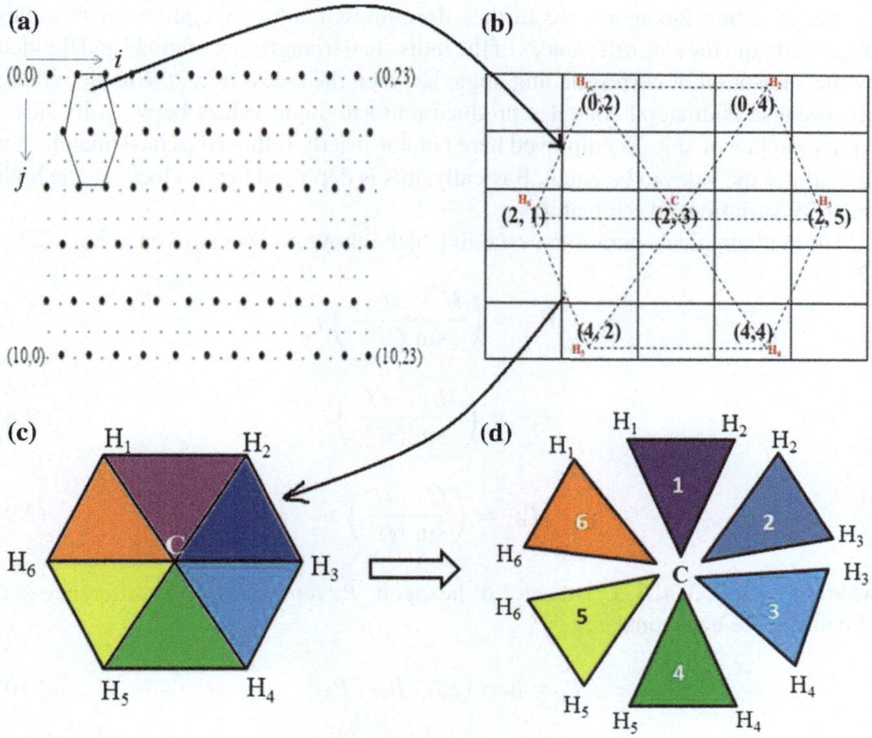

Fig. 5 Formulation of hexagons and triangles from an image sub-region

differences of each triangle are determined. The maximum amongst them is selected and divided by the sinusoidal angle value (adhering to law of sines) to produce the localised shape feature. The remaining triangles in the hexagon undergo the similar process and the resulting maximum features from the six triangles finally replace the edges of the hexagon. This geometric organization enforces congruence amongst the pixel values and enriches the high frequency content present in each sub-region. Furthermore, it heightens the interaction amongst the pixels in the local neighbourhood to yield a strongly linked feature map. This process is detailed in Fig. 5.

The mechanism initially treats each image as a square geometrical object. Firstly, each image is divided into 5 × 5 overlapping sub-regions. Then, each region is hexagonally sampled by alternate suppression of rows and columns present in the grid [16]. The sub sampling process is carried out using Eq. (2.6) with the related operations outlined in Fig. 5a, b.

$$
\begin{aligned}
H(i, j) &= I(2i, 2j) \qquad \textit{if } i \textit{ is even} \\
&= I(2i, 2j+1) \quad \textit{if } i \textit{ is odd,}
\end{aligned} \tag{2.6}
$$

The resulting hexagons are further decomposed into six equilateral triangles (Fig. 5c, d) and the side differences of the individual triangles are attained and divided by the sinusoidal of corresponding angle between the sides. Here, the angle will be 60° because equilateral triangles produce uniform angle values between its sides. Although, law of sines is employed here but not strictly followed as its demands that the ratio of the sides to be equal. Basically, this is deployed here to localize the high intensity variations of the features.

The mathematical operations pertaining to the above process is given in Eqs. (2.7)–(2.10).

$$P_{d1} = \left(\frac{H_i - H_{i+1}}{\sin 60^\circ}\right), \tag{2.7}$$

$$P_{d2} = \left(\frac{H_{i+1} - C}{\sin 60^\circ}\right), \tag{2.8}$$

$$P_{d3} = \left(\frac{C - H_i}{\sin 60^\circ}\right), \tag{2.9}$$

where, $i = 1, 2, 3, 4, 5$. C is center of hexagon, P_d represents pixel difference and P denotes the hexagonal edges.

$$P_i = \max\left(P_{d1}, P_{d2}, P_{d3}\right), \tag{2.10}$$

To further enhance the pixel interactions, the attained decimal point-wise responses are binary transformed thereby, boosting the binary interactions amongst the pixels and aid them in enhancing the shape boundaries. The binary image is re-framed by considering the six hexagonal pixels bits as shown in Fig. 6 and mapped into corresponding decimal levels from 0 to 7. The binary to decimal (D) conversion is given in Eq. (2.11).

$$D_i = bin2dec\,(B_i\ B_{i+1}\ B_c), \tag{2.11}$$

For $i = 1, 2, 3, 4, 5$ and B_c is center pixel.

Here, the acronym *bin2dec* signifies the binary to decimal conversion process. This conversion results in normalized point values to yield fewer bin shape histograms. The arranged binary processing bestows efficient shape matching and warrants increased retrieval accuracy.

Finally the feature vectors are obtained by histogram concatenation as given in Sect. 2.1.2.

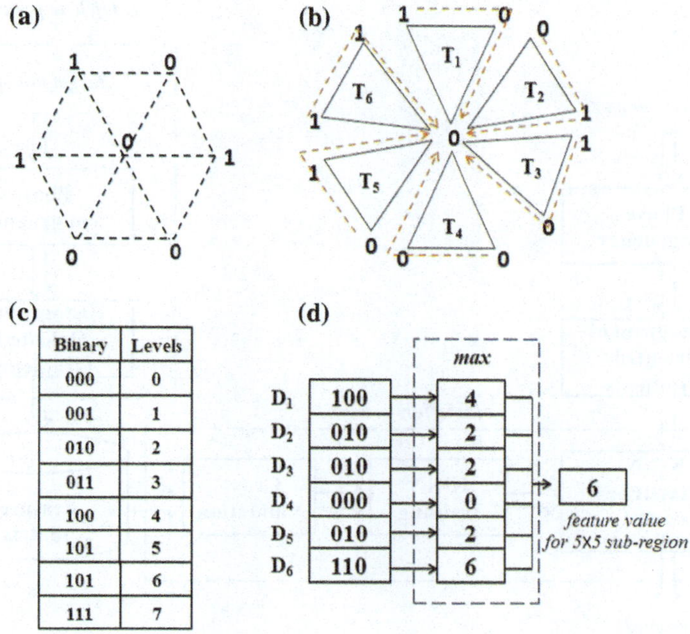

Fig. 6 Binary feature arrangement

2.3 *Shape Description by Blending Phase Congruency and Histogram of Oriented Gradients*

The final mechanism processes each shape image using the phase congruency (PC) operator. The resulting PC features are then fabricated into shape histograms using histogram of oriented gradients (HOG). This process is then applied to every other image in the database and the resulting feature vectors are systematized into a matrix. Then, the k-NN classifier operates on the feature database to spatially categorize the features into separate clusters. In the retrieval stage, an input image is processed by the PC-HOG feature to attain the feature vectors that are relatively compared against the stored features for identifying the matched images. Shapes mapping with the given query are retrieved and displayed on the output window according to their ranks. The block diagram of PC-HOG based shape retrieval is given in Fig. 7.

PC is capable of detecting high frequency features and represents the measure of feature significance. PC features correspond to maximally in-phase Fourier components of the given image. These features denote the ratio between the energy to the locally summed up Fourier components and represent their principal components that dually determine the presence of edges and corner. Also, the capitulated large minimum and maximum moment points assists in distinguishing the edges and the corners post feature extraction. This inherent feature arrangement makes them robust to magnification and illumination.

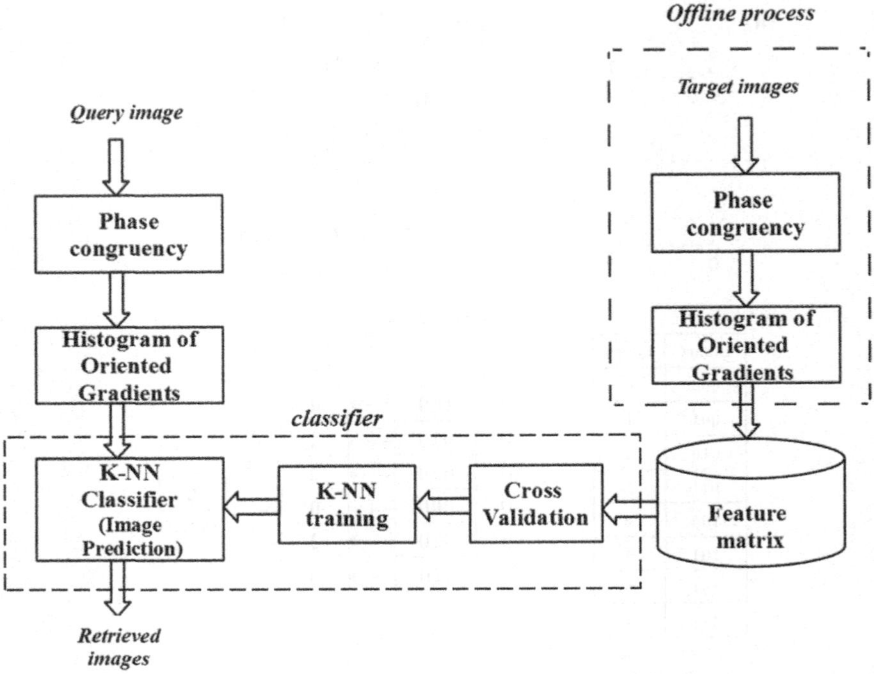

Fig. 7 Block diagram of PC-HOG based shape retrieval

2.3.1 Feature Extraction Using 2D Phase Congruency

The 2-D phase congruency [17] at a spatial location (i, j) is defined in (2.12),

$$PC_2(i, j) = \frac{\sum_{\Theta} \sum_n W_o(i, j) \lfloor A_{n\Theta}(i, j) \Delta\varphi_{n\Theta}(i, j) - T_o \rfloor}{\sum_{\Theta} \sum_n A_{n\Theta}(i, j) + \varepsilon} \tag{2.12}$$

where, $A_{n\Theta}(i, j)$ represents the image energy at location (i, j), computed using a 2D log-Gabor filter with $\Delta\varphi_{n\Theta}(i, j)$ corresponding to the phase deviation function and the threshold T_o controls the noise level of the image energy map. This operation is encapsulated with $\lfloor . \rfloor$ that indicates the enclosed quantity is equal to it when its value is positive and zero otherwise. And ε introduced in (2.12) represents a small constant to avoid division by zero. Finally, the sigmoidal Weighting function, $W_o(i, j)$ defined in (2.13) penalises the frequency distributions that are particularly narrow.

$$W_o(i, j) = \frac{1}{1 + e^{\gamma(c - s(i, j))}} \tag{2.13}$$

The parameters γ and c corresponds to gain and cut-off value respectively, and

$$s\,(i,j) = \frac{1}{N_s}\left(\frac{\sum_n A_{n\Theta}\,(i,j)}{A_{max}\,(i,j) + \varepsilon}\right) \tag{2.14}$$

Denotes the filter response spread computed as the sum of the amplitudes of the responses ($A_{n\Theta}$) divided by the highest individual response (A_{max}) to obtain the notional "width" of the distribution. N_s is total number of scales being considered. The phase deviation function $\Delta\varphi_{n\Theta}\,(i,j)$ is mathematically defined as,

$$\begin{aligned}\Delta\varphi_{n\Theta}\,(i,j) &= \cos(\varphi_{n\Theta}\,(i,j) - \overline{\varphi_{n\Theta}}\,(i,j)) - |\sin\,(\varphi_{n\Theta}\,(i,j) - \overline{\varphi_{n\Theta}}\,(i,j))| \\ &= e_{n\Theta}\,(i,j)\,\overline{\varphi_e}\,(i,j) + o_{n\Theta}\,(i,j)\,\overline{\varphi_o}\,(i,j) \\ &\quad - |e_{n\Theta}\,(i,j)\,\overline{\varphi_o}\,(i,j) + o_{n\Theta}\,(i,j)\,\overline{\varphi_e}\,(i,j)|\end{aligned} \tag{2.15}$$

where, $\overline{\varphi_{n\Theta}}$ mean phase response. $\overline{\varphi_e}$ and $\overline{\varphi_\Theta}$ is even and odd mean phase angle respectively.They are accordingly defined in (2.16) and (2.17),

$$\overline{\varphi_e}\,(i,j) = \frac{\sum_n e_{n\Theta}\,(i,j)}{E\,(i,j)} \tag{2.16}$$

$$\overline{\varphi_o}\,(i,j) = \frac{\sum_n o_{n\Theta}\,(i,j)}{E\,(i,j)} \tag{2.17}$$

The local energy ($E(i,j)$) pertaining to the evaluation of even and odd phase response is defined in (2.18)

$$E\,(i,j) = \sqrt{\left(\sum_n e_{n\Theta}\,(i,j)\right)^2 + \left(\sum_n o_{n\Theta}\,(i,j)\right)^2} \tag{2.18}$$

$e_{n\Theta}$ and $o_{n\Theta}$ are the convolution results of the input image, $I\,(i,j)$ with quadrature even and odd 2D log-Gabor filters $M^e_{n\Theta}$ and $M^o_{n\Theta}$ respectively, at orientation Θ and scale n as given in (2.19)

$$[e_{n\Theta}\,(i,j)\,,\, o_{n\Theta}\,(i,j)] = \left[I\,(i,j) * M^e_{n\Theta},\, I\,(i,j) * M^o_{n\Theta}\right] \tag{2.19}$$

The amplitude response ($A_{n\Theta}$) and phase response ($\varphi_{n\Theta}$) is given (2.20) and (2.21) respectively.

$$A_{n\Theta}\,(i,j) = \sqrt{e^2_{n\Theta}\,(i,j) + o^2_{n\Theta}\,(i,j)} \tag{2.20}$$

$$\varphi_{n\Theta} = atan\left(\frac{e_{n\Theta}\,(i,j)}{o_{n\Theta}\,(i,j)}\right) \tag{2.21}$$

The threshold (T_o) administering the noise level is given by,

$$T_o = k\overline{A_{n\Theta}}' \sum_{n=1}^{N_s} \frac{1}{m^n} \tag{2.22}$$

where, $\overline{A_{o\Theta}}' = e^{\overline{\log A_{n\Theta}(i,j)}}$ estimates the mean noise response of the smallest scale 2D filter pair over the image and m is the scaling factor between the successive filters. The 2D log-Gabor filter is designed as in Eq. (2.23).

$$G(\omega, \theta_m) = \exp\left(-\frac{\left(\log\left({}^{\omega}/_{\omega_o}\right)\right)}{2\sigma_n^2} - \frac{(\theta - \theta_o)^2}{2\sigma_\theta^2}\right) \tag{2.23}$$

Here, θ_0 is the orientation angle of the filter. ω_o is the center frequency of the filter, σ_n and σ_θ are the control parameters for bandwidth and angle range of the filter.

The raw phase congruency images were obtained by applying Eq. (2.12) to the images with the following parameters. Local frequency information was obtained using two-octave bandwidth filters over four scales and six orientations. The wavelength of the smallest scale filters was 3 pixels; the scaling between successive filters was 2. Thus, over the four scales, the filter wavelengths were 3, 6, 12, and 24 pixels. The filters were constructed directly in the frequency domain as polar-separable functions: a logarithmic Gaussian function in the radial direction and a Gaussian in the angular direction. In the angular direction, the ratio between the angular spacing of the filters and angular standard deviation of the Gaussians was 1.2. This results in coverage of the spectrum that varies by less than 1%. A noise compensation k value of 2 was used. The frequency spread weighting function cut-off fraction, c, was set at 0.4. The value of ε, the small constant used to prevent division by zero in the case where local energy in the image becomes very small, was set at 0.01.

2.3.2 Histogram Fabrication Using HOG

The occurrences of localized gradient orientations in an image are characterized using the HOG. They are obtained by calculating the distribution of gradient directions. Magnitude of gradients is large around edges and corners of the image that carry more shape information than the other regions; hence, making them useful. HOGs acute discerning ability is well demonstrated in several computer vision applications [18]. Attaining HOG feature vector is outlined below. In the first step, the horizontal gradients (g_x) and vertical gradients (g_y) are calculated using filter kernels $[-1\ 0\ 1]$ and $[-1\ 0\ 1]^T$ respectively. Next, the magnitude and the direction of gradient is obtained using (2.24) and (2.25),

$$magnitude,\ g = \sqrt{{g_x}^2 + {g_y}^2} \tag{2.24}$$

$$direction, \theta = \arctan\left(\frac{g_y}{g_x}\right) \tag{2.25}$$

Here, range of direction (θ) is 0 to 180 because calculated gradients are unsigned. In this approach, the wrap-around processed image is further divided into 8×8 sub-regions and gradients of these regions are obtained using (2.24) and (2.25). Each sub-region contains a total of 128 numbers attaining by combining direction (64 numbers) and magnitude (64 numbers) that is further converted into histogram of a 9-bin. The bins of histogram is an array of 9 numbers corresponding to angles 0, 20, 40, 60, 80, 100, 120, 140, 160. If the direction of the gradients exactly matches the array value, then magnitude of the gradient is mapped to the corresponding value. Else, the values are splitted between two neighbouring bins according to the distance from the bin. An illustration of the above process is shown in Fig. 8.

Thus, the feature formulation and representation mechanism produces a 9-bin histogram for each sub-region of the image. This process is reiterated on the remaining sub-regions to finally formulate the shape histogram that represents the feature descriptor of the given image.

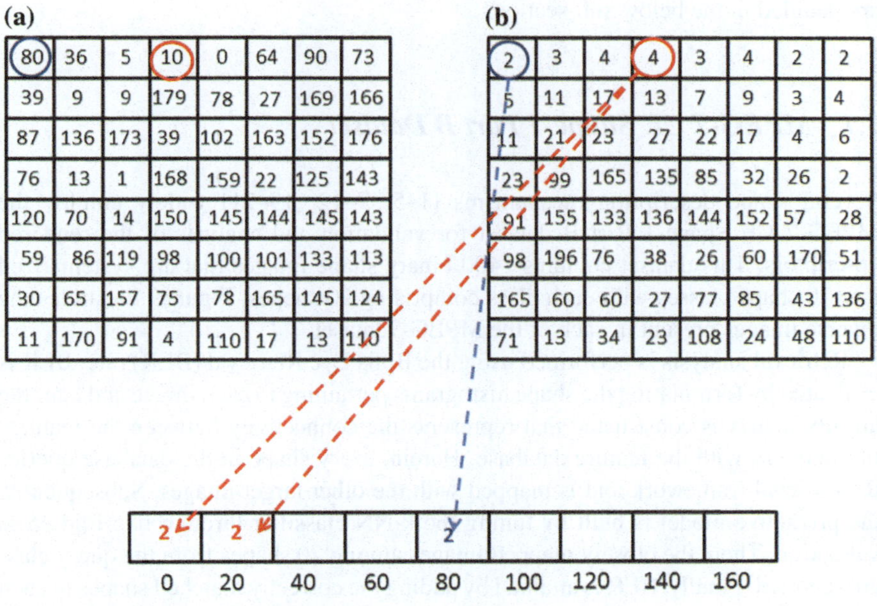

Fig. 8 Formulation of histogram of oriented gradients

3 Evaluation

Efficiency of these methods is effectively investigated with the relevant performance scores and is elaborated in subsequent sections. Widespread shape databases MPEG-7 CE Shape-1 Part B, TARI-1000 and Kimia's 99 datasets are utilized for effective investigation of various descriptors covering diverse shape characteristics. Herein, Bull's Eye Retrieval (BER) is used to analyse the descriptors efficacy. For analysing the descriptor characteristics, a shape retrieval framework is realized by merging the diverse feature extraction mechanism with k-Nearest Neighbour (k-NN) classifier for feature categorization. Later, cross validation on the individual image dataset is performed by dividing them into testing and training shapes depending upon the fold number associated with it.

The k-NN classifier was tuned using the parameters given below: the number of observations (N) is assigned the dataset size. Euclidean distance metric is chosen for matching along with the k taking the value of 2 representing the neighbours. Also, the feature values along with the corresponding labels are loaded into X and Y respectively. The framework is realized in MATLAB R2015a on the Windows7 platform that runs on an $i3$ CPU with 2 GHz clock speed. The experiments performed are detailed in the below sub-sections.

3.1 MPEG-7 CE Shape-1 Part B Dataset

Several shape description mechanisms [1–5, 7–12, 19–21] widely employ the MPEG-7 CE Shape-1 Part B dataset for validation and analysis of the rendered descriptors. This dataset contains 1400 binary shape images that are systematized into 70 shape classes with each class comprising 20 shapes. Figure 9 illustrates the diverse images present in each of the MPEG-7dataset.

Retrieval analysis is performed using the Bulls Eye Retrieval (BER) rate. BER is evaluated by formulating the shape histograms pertaining to each image and later the affinity matrix is constructed that represents the connectivity between the features of the query with the feature database. Herein, every shape in the database queries the retrieval framework and is mapped with the other target images. Subsequently, the predictive model is built by tuning the k-NN classifier through five-fold cross validation. Then, the closely mapped images among 20 shapes from the query class are reported. Finally, BER is attained by adding the correctly matched shapes in each class and dividing them by the dimensions of the feature matrix (Table 1).

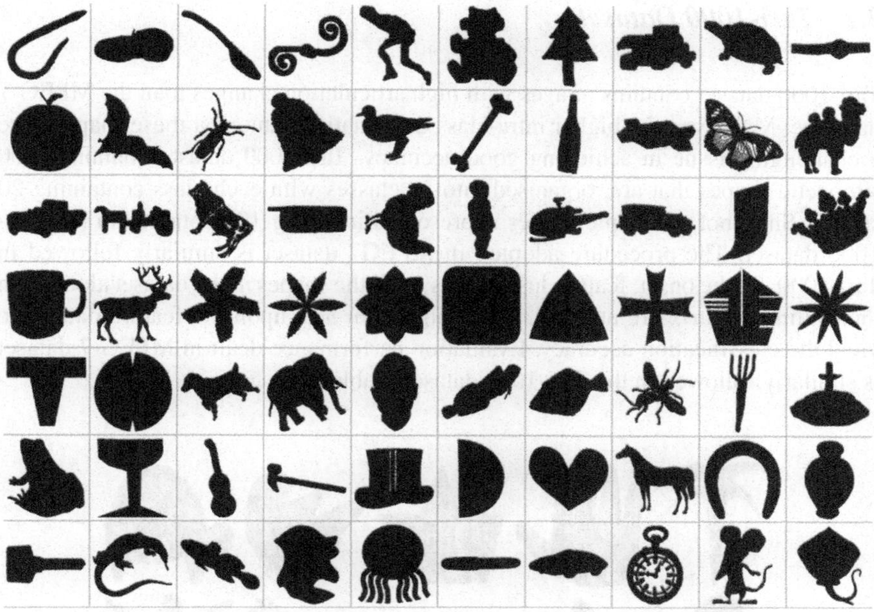

Fig. 9 Example images of MPEG-7 dataset

Table 1 BER comparison of different shape retrieval methods on the MPEG-7 dataset [19]

Method	BER rate (%)
ECCobj2D	54.56
SC	64.59
IDSC	68.63
Mean-EuDist	77.69
IDSC+DP	85.40
SC+DP	86.60
Shape tree	87.70
CBTA	89.45
Height functions	89.66
Shape vocabulary	90.41
IMD	91.25
CBTA+SC	93.65
LTEM+hist	94.64
HGTT	95.05
PC-HOG	96.97

3.2 Tari-1000 Dataset

Tari-1000 dataset contains images with high articulation changes than the MPEG-7 database. Moreover, the higher intra-class deformations amongst these shapes pose a challenging issue in achieving good accuracy. Tari-1000 dataset contains 1000 silhouette shapes that are, organised into 50 classes with each class containing 20 shapes. Snapshot of sample shapes is presented in in Fig. 10 contained in the Tari-1000 dataset. The procedure adopted in MPEG7 dataset is similarly followed in Tari-1000. At the onset, feature histograms of all the shapes in the dataset along with their affinity matrix are attained. k-NN algorithm acts upon the feature dataset to yield the classification accuracy. Evaluation performance dealt in MPEG-7 dataset is similarly followed in the Tari-1000 dataset (Table 2).

Fig. 10 Sample images of Tari-1000 dataset

Table 2 Comparison of BER rates of different shape retrieval methods on the Tari-1000 dataset [20]

Method	BER (%)
LTEM+hist	91.70
PC-HOG	93.67
HGTT	93.87
SC	94.17
IDSC	95.33
ASC	95.44
SC+LP	97.79

3.3 Kimia's 99 Dataset

Another benchmark dataset that is commonly used for shape retrieval experiments is the Kimia's dataset. Despite the availability of Kimia's 25, Kimia's 256 dataset here Kimia's 99 dataset is adopted as Kimia's 25 correspond to a small dataset and several shapes of Kimias 256 are covered in the Tari-1000 dataset. Figure 11 illustrates the shape samples of the Kimia's 99 dataset.

The number 99 suffixing Kimia's dataset represent the total 99 shapes that are systematized into 9 classes with each class consisting of 11 images. Each shape is treated as the query and the corresponding matched shapes are tabulated as the number of shapes from the same class ranked from top 1 to top 10.

From the above relative discussions performed on the three diverse image datasets it is evident the presented three shape descriptors is found to excel its predecessors in MPEG7 dataset and relatively compete with them in the other two datasets. This enunciates the fact that the presented descriptors are light weight in nature and aims to offer improved accuracy when compared with its counter parts that are realized using complex computational modules (Table 3).

Fig. 11 Shape images of Kimia's 99 dataset

Table 3 Recognition result of Kimia's 99 dataset [21]

Method	1st	2nd	3rd	4th	5th	6th	7th	8th	9th	10th
ECCobj2D	94	85	81	73	81	73	64	59	56	35
SC	97	91	88	85	84	77	75	66	56	37
Mean-DIR Euclidean	97	92	89	86	81	80	74	74	70	52
CPDH	98	94	95	92	90	88	85	84	71	52
LTEM+hist	99	99	96	94	87	86	85	80	82	78
IDSC+DP	99	99	99	98	98	97	97	98	94	79
HGTT	99	98	98	94	95	94	94	93	91	87
PC-HOG	99	97	92	92	86	85	84	78	74	77

4 Conclusion

This chapter presents three novel and light shape descriptors offering effective retrieval performance in shape retrieval. Experiments performed on well-established datasets prove that the descriptors are robust against affine transformations and the yielded retrieval accuracy using BER affirms the same. Consistent BER and performance results recorded across diverse dataset indicate that the presented descriptors are suitable for fast retrieval from large image database without compromising on the retrieval accuracy.

References

1. D. Hu, L. Shang, Z. Zhu, J. Yang, W. Huang, J. Yang, L. Shang, Z. Zhu, J. Yang, W. Huang, Shape matching and object recognition using common base triangle area. IET Comput. Vis. **9**, 769–778 (2015)
2. D. Zhang, G. Lu, Review of shape representation and description techniques. Pattern Recogn. **37**(1), 1–19 (2004)
3. S. Belongie, J. Malik, J. Puzicha, Shape matching and object recognition using shape contexts. IEEE Trans. Pattern Anal. Mach. Intell. **24**, 509–522 (2002)
4. H. Ling, D.W. Jacobs, Shape classification using the inner-distance. IEEE Trans. Pattern Anal. Mach. Intell. **29**, 286–299 (2007)
5. N. Alajlan, G. Freeman, I.El Rube, M.M.S. Kamel, G. Freeman, Shape retrieval using triangle-area representation and dynamic space warping. Pattern Recognit. **40**, 1911–1920 (2007)
6. N. Alajlan, M.S. Kamel, G.H. Freeman, Geometry-based image retrieval in binary image databases. IEEE Trans. Pattern Anal. Mach. Intell. **30**, 1003–1013 (2008)
7. G.B. de Souza, A.N. Marana, HTS and HTSn: New shape descriptors based on Hough transform statistics. Comput. Vis. Image Underst. **127**, 43–56 (2014)
8. P. Srestasathiern, A. Yilmaz, Planar shape representation and matching under projective transformation. Comput. Vis. Image Underst. (2011)
9. H. Wu, S. Yan, Computing invariants of Tchebichef moments for shape based image retrieval. Neurocomputing **215**, 110–117 (2016)
10. H.H.D. Jomma, A.I.A. Hussein, Circle views signature: a novel shape representation for shape. Recogn. Retrieval **39**, 274–282 (2016)

11. J. Yang, H. Wang, J. Yuan, Y. Li, J. Liu, Invariant multi-scale descriptor for shape representation, matching and retrieval. Comput. Vis. Image. **145**, 43–58 (2016)
12. N. Kaothanthong, J. Chun, T. Tokuyama, Distance interior ratio: a new shape signature for 2D shape retrieval. Pattern Recognit. Lett. **78**, 14–21 (2016)
13. P. Howarth, S. Rger, Evaluation of texture features for content-based image retrieval. Int. Conf. Image Video. (2004)
14. E. Acar, M. Özerdem, An iris recognition system by laws texture energy measure based k-NN classifier. Signal Process. Commun. (2013)
15. A. Setiawan, J. Wesley, Y. Purnama, Mammogram classification using laws texture energy measure and neural networks. Proc. Comput. Sci. (2015)
16. K. Sankar, T. Sanjay, E. Rajan, Hexagonal pixel grid modelling and processing of digital images using CLAP algorithms (2004)
17. P. Kovesi, Edges are not just steps, in *Proceedings of the Fifth Asian Conference on Computer Vision* Melbourne, vol. 8, pp. 22–8 (2002)
18. A. Gudigar, S. Chokkadi, A review on automatic detection and recognition of traffic sign. Multimed. Tools. **75**, 333 (2016)
19. P. Govindaraj, M.S. Sudhakar, Hexagonal grid based triangulated feature descriptor for shape retrieval. Pattern Recognit. Lett. **116**, 157–163 (2018)
20. P. Govindaraj, M.S. Sudhakar, A new 2D shape retrieval scheme based on phase congruency and histogram of oriented gradients. Signal, Image Video Process **13**(4), 771–778 (2019)
21. P. Govindaraj, M.S. Sudhakar, Shape characterization using laws of texture energy measures facilitating retrieval. Imaging Sci. J. **66**, 98–105 (2018)

Clustering Data Using Techniques of Image Processing Erode and Dilate to Avoid the Use of Euclidean Distance

Noé Ortega-Sánchez, Erik Cuevas, Marco A. Pérez and Valentín Osuna-Enciso

Abstract Clustering is one of the most popular methods of machine learning. The process of clustering involves the division of a set of abstract objects into a certain number of groups which integrated with objects of similar characteristics. Therefore, a cluster integrates objects which are similar to them, but dissimilar to the elements that belong to the rest of the clusters. Several clustering methods have proposed in the literature with different performance levels. All these techniques use as similarity criterion the Euclidean distance among cluster elements. However, there exist diverse scenarios where the Euclidean distance cannot be utilized appropriately to separate the elements in groups. Under such conditions, traditional cluster methods cannot directly apply. On the other hand, the operations of dilate and erode are a set of non-linear operators that modify the shape of a data group in the feature space, to obtain a monolithic object. Although morphological operations have demonstrated its importance in several engineering fields as image processing, its use as a clustering technique has been practically overlooked. In this work, an alternative clustering algorithm is proposed to group elements without considering the distance as a similarity criterion. In our approach, the data were separated into different groups by considering morphological operations. Under this scheme, the procedure allows the integration of data points, which present a spatial connection. Since the proposed algorithm does not use the distance in its functioning, it solves complex clustering problems which traditional clustering algorithms cannot.

Keywords Clustering · Non-Euclidean distance · Similarity criteria

N. Ortega-Sánchez (✉) · E. Cuevas · M. A. Pérez · V. Osuna-Enciso
Universidad de Guadalajara, CUCEI, Guadalajara, Jalisco, Mexico
e-mail: noe.ortega@academicos.udg.mx

E. Cuevas
e-mail: erik.cuevas@cucei.udg.mx

M. A. Pérez
e-mail: marco.perez@cucei.udg.mx

V. Osuna-Enciso
e-mail: valentin.osuna@cucei.udg.mx

D. Oliva and S. Hinojosa (eds.), *Applications of Hybrid Metaheuristic Algorithms for Image Processing*, Studies in Computational Intelligence 890,
https://doi.org/10.1007/978-3-030-40977-7_9

1 Introduction

The aim of analyzing the data in any area is to show the tendency and predict some trends, based on results. In the literature has been developed a variety of techniques and methods to clustering data. In this context, clustering is [1] one of the top-rated techniques for getting information. In this chapter will study the popular and most representatives' algorithms to clustering.

The clustering techniques have a proposed create a set of objects which are at the same characteristics. The set of C in n elements it is subletting into groups based on the distribution. In a past way, it has been techniques like hierarchical [2] and partitional [3], and the specification of the Euclidean distance was used to create subsets. In this sense, despite being a good strategy has a weakness in different cases where the elements have distribution amorphous and irregular. The process of clustering involves the division of set in an abstract object into groups with similar characteristics; it is a critical phase, dividing the set of elements in subgroups that share characteristics. In other words, the set is grouped in with the similarity of characteristics to be discriminated to the others set of data. Clustering is used as a preprocessing task to build sophisticated machine learning systems with applications in data mining such as cancer detection [4], search engine development [5] and large spatial databases [6]. In the same way, big data has been studied by researchers [7, 8], several clustering algorithms have been proposed in the literature with different performance levels. Some examples of clustering approaches include the K-means [9], K-medoids [10], cluster center fast [9], and genetic algorithms (GA) [11].

Traditional clustering algorithms present interesting results. However, they maintain critical flaws such as the distance-inconsistency, high computational cost, and the a priori determination of the cluster number k.

In this chapter, an alternative clustering algorithm is proposed to group elements without considering the distance as a similarity criterion. In our approach, the data are separated into different groups by considering morphological operations. Under this scenario, allows the integration of data points, which present a spatial connection. Since the proposed algorithm does not use the distance in its functioning, it solves complex clustering problems which traditional clustering algorithms cannot.

The work was organized in the following sections. The second section will explain the context of data, operations of erode and dilate, the popular techniques for clustering. The third section will study three metrics to validate the quality of the clustering. The fourth section is devoted to explaining the integration of the concepts and the approach. The fifth section was reserve to show the results of the experiments and offer a visual and quantitative comparative. On the last section it concluded and discussed the results.

2 Data Representation and Clustering Strategies

The Euclidean distance is a tool that helps at the cluster algorithms, and present strong performance in some cases, but it has some deficient when the data representation is not well distributed. In the case of the set create an irregular shape, the effectiveness of the Euclidean distance is lower, then the processes of clustering are not worn correctly. Another example of failure is in the sets where the center of the data it is inexistent. The proposal idea is using a technique of image processing to represent the data in an image. With them can extract similar characteristics and recognizing visual patterns.

This section is devoted to explanted the concepts which are involved in this work. It is necessary to see three concepts that are mixed to obtain the algorithm. Hierarchical [2], partitional [3], and grid-based [6] clustering techniques use as clustering criterion the Euclidean distance among the elements in the groups.

It will study lightly the data representation, the operations of dilate and erode, and the generalities of K-means and GA. It is necessary to understand how they evaluate the elements to confirm the clusters in a base of Euclidean distance, like a measure of dissimilarity. Under those circumstances, K-means and GA show lightly, we know two ways of clustering: K-means and GA, is evident this kind of strategies have a good performance, but this does not mean they could be the solution to all problems.

2.1 Data Representation

The source of data comes from all human activities or natural elements; the attempt to classified the data it shows in Fig. 1. It is a hierarchical classification form source (if is Discrete or Continuos) and the possible representation (nominal or binary). The binary is divided into Symmetrical, which means the representation as well distributed and asymmetrical, which means it is loaded more into an area. In this chapter it uses data that come from different sources of information to challenge the

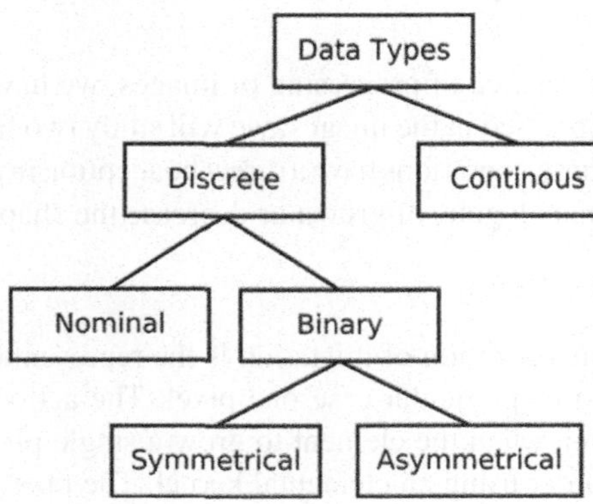

Fig. 1 Classification of data

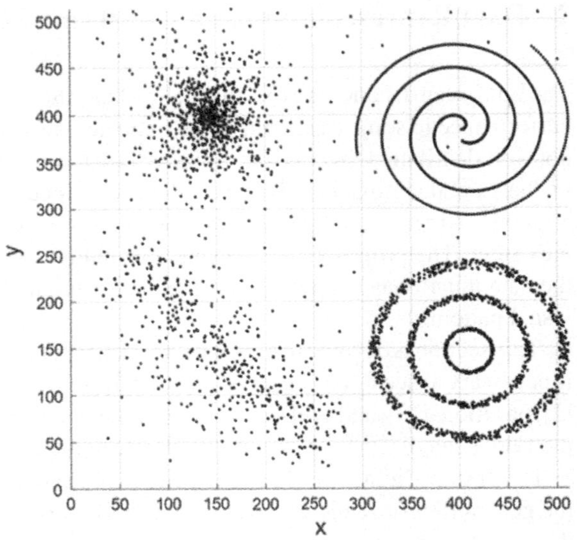

Fig. 2 Example of clustering problems whose solution cannot be found by approaches based on distance algorithms. One of the most difficult images is synthetic because they were designed to prove and defy.

The first requisite to clustering is the conception of the representation of data. In this context, the source of data is variety and inspired. It has factors to interpret the data type. The initial conception of data has represented as a visualization in a two-dimensional plot, and each point represents an object, but this representation is some unfair nevertheless is an excellent description to decide in respect to the clustering. In Fig. 2, it is possible to differentiate the cluster due to a kind of aggrupation.

There existed some representation of data type to guide a study of clustering in a set which has non-usual types the measures take a particular relevance, and it will be defined in the meaning of context work.

2.2 *Operation of Dilatation and Erosion*

In the area of processing of images, we have techniques to deal with some features expressed in the images; we will study two fundamental operations: dilate and erode. These operations have a direct conception in a physical phenomenon, where they have an analogous of grown or decrease the shape or area of the data representation.

(a) *Dilate*

The operation of dilatation is the representation to grow up the space of element or in this particular case one pixel. The action is creating another layer to the shape, even when the element to grow a single pixel; in this sense, the operation creates a border using an elemental kernel. The operation is defined as follows:

$$I \oplus H = \{(x', y') = (x + i, y + j) | (x', y') \in P_I(i, j) \in P_H\} \quad (1)$$

The explanation is simple when we identify a pixel to grown up the boundary is operating with a kernel based on the four cardinal points showed in Fig. 3b. (4 connected), also it is possible to use the surrounding eight pixels (8 connected).

The H is the representation of the structure where is applied to create the layer in the interest pixel, with this in mind, each pixel is reflected has a new area of pixels.

(b) *Erode*

The operation of erode is the elimination of the layer from on the objects, in this sense is notorious the object is devasted Eq. 2, show how to decrease the numbers of pixel to take a lower volume than the original form.

$$I \ominus H = \{(x', y') | (x' + i, y' + j) \in P_I(i, j) \forall (i, j) \in P_H\} \quad (2)$$

In other words, the elimination of the layer depends directly on the kernel (Fig. 4b), if this condition is met the pixel is respected otherwise it is eliminated, resulting in smaller artifact showed in Fig. 4c.

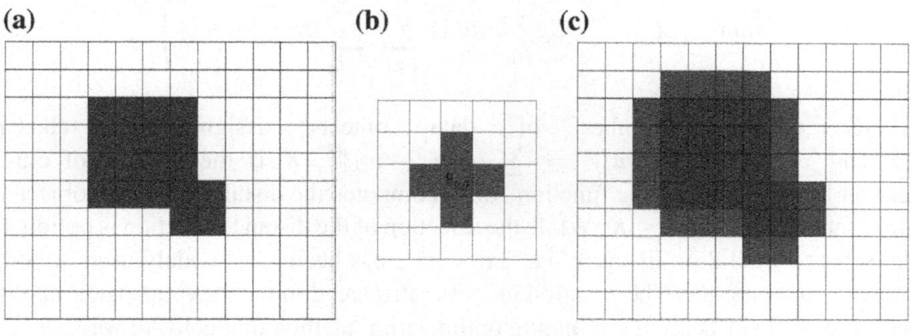

Fig. 3 **a** Original image, **b** kernel H to operate the dilate 4-connected, **c** the image resulting applying the operation with H

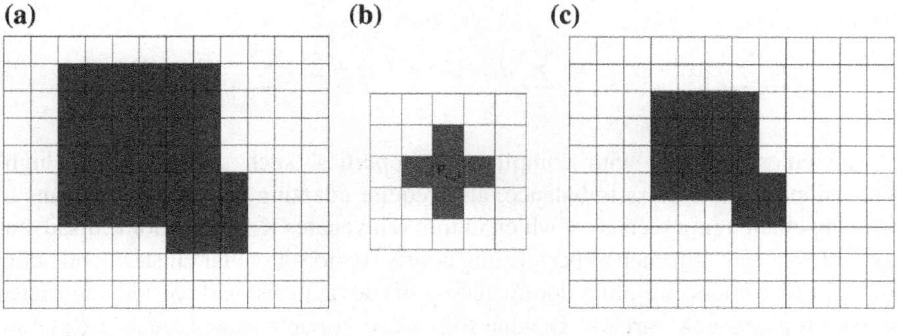

Fig. 4 **a** Original image, **b** kernel H to operate the erode 4-connected, **c** the image resulting applying the operation with H

The first thought about the previous operations is they are operation complementary, but is entirely false; the behavior is not reversible. The use of this operation in image processing is used to connect near objects and define the neighborhood.

2.3 K-means

One of the more popular algorithm to clustering is K-means, and also is first [12–14] and proposals with notable performance in this practice, result in several variations [15]. It shows remarkable results in the different application area; this is the main reason to compare with the novel approach in the investigation.

The behavior of K-means is defined regarding points [11, 16], where each point is this terms the representation of data. In this way come the next definitions to understand the algorithm $X = \{x_1, x_2, ..., x_n\}$ and variant coefficient. Delving into the mathematics behind K-means is beneficial. Suppose $D = \{x_1, \ldots, x_n\}$ is the data set to be clustered. K-means can be expressed by an objective function that depends on the proximities of the data points to the cluster centroids as follows:

$$\min(C_1, C_2, \ldots, C_K) = \min\left(\sum_{i=1}^{K} \sum_{x_j \in C_k} dist\left(x_j, m_i\right)\right) \tag{3}$$

where x_j the number of data objects assigned to cluster C_k, and $\sum_{x_j \in C_k} dist\left(x_j, m_i\right) = \sum_{x \in C_k} \|x - m_k\|^2$, K is the number of clusters set by the user, and the function "dist" computes the distance between object x and centroid m_k, $1 \le k \le K$. While the selection of the distance function is optional, the squared Euclidean distance, i.e. $\|x - m\|^2$, has been most widely used in both research and practice. The iteration process introduced in the previous paragraph is indeed a gradient-descent alternating optimization method that helps to solve.

In this way, it is one of their limitations, in the case of two dimensions, is stable, but when it increased the complexity of calculus is overcoming [17]. Showing in Eq. 4.

$$D_k = \sum_{i=1}^{k} \sum_{j=1}^{k} \left\|x_i - x_j\right\|^2 = \sum_{l=1}^{k} dist(C_l + C_l) + 2 \sum_{1 \le i \le j \le k} dist(C_i + C_j) \tag{4}$$

The emerging data with complicated properties, such as large-scale, high-dimensionality, and class imbalance, also require adapting the classic K-means to different challenging scenarios, which in turn rejuvenates K-means. Some disadvantages of K-means [8], such as performing poorly for non-globular clusters, and being sensitive to outliers, are often dominated by the advantages, and partially corrected by the proposed new variants. In what follows, we review some recent research on K-means from both the theoretical perspective and the data-driven perspective. Note that we here do not expect to cover all the works of K-means, but would instead introduce some aspects that relate to the central theme.

2.4 GA

Genetic algorithms (GAs) [18–20] are randomized search and optimization techniques guided by the principles of evolution and natural genetics, having an amount of implicit parallelism. A collection of such strings is called a population [20–22]. General developed the algorithm is characterized in the next stages: first made a randomized set of elements, where it selected the best of them in based on the fitness. The second stage, all elements are allowed to create a new generation based on their characteristics with certain modifications. Third, is made an evaluation, where the fitness check is performed to allow to elements advance or die. In the last stage at the end of the previous steps, the items that have shown the best results are replaced. The cycle is become initiated to reach the best performance.

The algorithm can recombine different solutions to get better ones, and so, it can use the benefits of assortment. The robustness of the algorithm should mention as something essential for the algorithm success, refers to the ability to perform consistently well on a broad range of problem types. There is no particular requirement on the problem before using GAs, so it can be applied to resolve any issue. Nature has been intelligent when it comes to evolution, and the following strategy by herself, trying, and fail, with this was capable of making significant advances to create functional ecosystems [19, 23, 24]. With this in mind, it is possible to imagine a synthetic system where are capable of solving different types of problems. It is simple in the design, and it is capable of finding a better solution in each generation, in other words, a set of elements was ranked based on their response to a fitness function.

3 Metric to Validate the Comparison of Clusters

The problems of testing for clustering tendency it could describe by an optimization way, in this terms are necessary a way to evaluate to give a measure, with this is possible compare the method with other. On this developed is highly necessary. In this sense, the evaluation provides evidence to accept or reject.

The question, in this case, is how to evaluate, and the answer is: with metrics to indicate the quality of the response. The metrics to evaluate are: Dunn Index [25], Davies-Bouldin Index [26] and Hopkins Statistics [27]; the first two are popular, and last is a method used in recent time. The complexity of evaluating a cluster it is by self a hard work [28], and complicated [29, 30], due to the lack of a standardized criterion. On state of the art have different studies, and they try to solve the dilemma, they based in the internal or external quality. The goal of clustering is to group data points that are similar according to nearness and give a decision about how good is the clustering [30–32].

(A) *Dunn Index*

It is one of the most used but also is one of the oldest in the cluster literature. It is the duty of classification of clusters said how acceptable is based on how well is separate and compact [25]. With this in mind, the evaluation of inter-cluster is maximized, and intra-cluster is minimized, in described in the Eq. 5.

$$DU_k = \min_{i=1,\dots,k}\left\{\min_{j=1+1,\dots,k}\left(\frac{diss(c_i, c_j)}{\max_{m=1,\dots,k}} diam(c_m)\right)\right\} \tag{5}$$

(B) *Davies-Bouldin Index*

This index was designed to find a general separation measure and decided without user interaction [26]. In this sense the metric has the same properties and qualities than Dunn Index, in other words, both find a well separate and compact cluster, the main difference in this index it is looking a maximized the inter-cluster distance. A minimized index gives the best separation of the cluster in automatic. The definition of this behavior is the following:

$$DB_k = \frac{1}{k}\sum_{x=1}^{k} \max_{j=1,\dots,k,i\neq j}\left\{\frac{diam(c_i) + diam(c_j)}{d(z_i, zj)}\right\} \tag{6}$$

(C) *Hopkins Statistics*

In nature, some phenomena tend to generate their patterns. One of the proposals for resolving is the Hopkins Statistics [27] is described in Eq. 7, where is prove a cluster with itself and synthetic. A result is usually tested by building an alternative hypothesis and comparing it against a null hypothesis. The hypothesis is a statement of randomness and could be based on the random label.

$$H = \frac{\sum_{j=1}^{m} u_j^d}{\sum_{j=1}^{m} u_j^d + \sum_{i=1}^{m} w_i^d} \tag{7}$$

4 The Approach of Images Techniques to Clustering

It is necessary to connect each pixel to represent the data in subsets, and the algorithm used was presented by Haralick [33]. The pixel connected were labeling include the region which is in the same. Therefore, all pixels connected represent the same subset. It should be mentioned connect components is one image processing techniques that can make a unit change from pixel to region, to understand this asseveration we see the region of pixels like a complex unit of elements. Another property of the technique is the use of filtering to manipulate the image and propitiate results with better characteristics in the labeling. Therefore, it is possible to build an N-tuple, and it used to know some properties like several objects, measures, and regions.

Before using the Haralick Algorithm, it is necessarily a preprocessing to enhance the image to get an optimal result; this means that it was used the operation of dilate and erode before the algorithm. It studies lightly in section three, and the general implementation of this approach is the following algorithm:

```
procedure iterative;
Initialize of each one-pixel to unique label
for L← 1 to nLines do
    for P←1to nPixels do
        if I(L,P)==1
            then label(L,P)←newLabel()
        else label(L,P)←0
    end for
end for
    Iteration of top-down followed by bottom-up passes
repeat
```

In the images, we have some components that in nature have nearly boundary with other components, and the associated operator is called a connected component's operator, their function is defined as the neighborhood between elements.

The algorithm selects the minimum of labels of its neighbors to assign to the pixel. It does not directly keep track of equivalences but instead uses a number of passes through the image to complete the labeling. In definition the connected component of neighbor is two-pixel p and q belong to the same connected component C if there is sequence of one pixel (p_0, p_1, p_2, …, p_n) of C where $p_0 = p$, $p_n = q$, and p_i is a neighbor of p_{I-1} for $I = 1, \ldots, n$, then the result connected regions are called 8 connected, as a result of this operation, the labeling of Fig. 5a is grouping with numbers as showing in Fig. 5b.

The strategy of a component operator is used in applicated in images which consist to a small number of objects against a contrasting black ground, perhaps the use of the strategy has shown a good performance to create a monolithic structure.

Fig. 5 **a** Original image, **b** image result after labeling

(a)

0	1	1	0	1	1	0
1	1	1	0	1	0	0
1	0	0	0	1	0	0
1	1	0	1	1	0	1
0	0	0	0	0	0	1
0	0	0	1	0	0	1
0	0	1	1	1	1	1

(b)

0	1	1	0	2	2	0
1	1	1	0	2	0	0
1	0	0	0	2	0	0
1	1	0	2	2	0	3
0	0	0	0	0	0	3
0	0	0	3	0	0	3
0	0	3	3	3	3	3

The visualization of data as we see is important in cluster analyses and is an essential step in the application. Another important aspect is the viewing is vital and avoid misunderstanding in the input and output information. The fusion of techniques and a different focus is presented in this chapter provide a capability to mix the operations of dilate and erode with the Haralick Algorithm to create an improved way to generate clustering in a fast way.

5 Experimental Results

The experiments were carried out as follows, a set of test images will be used, with which the capacities and weaknesses of each of the algorithms can be dismantled. Moreover, it will take the tests with the index that have been explained briefly.

Some of the representative clusters are showed in Fig. 6 were the level of complexity proportional to the shape of the data. It is appreciable in Fig. 6d exist two spiral and three rings that represent a challenge to the algorithms based on Euclidean distance, in this approach using operator of images and labeling algorithm, the process is fast and precise. Nevertheless, even this approach is not the best and some problems with how it is Fig. 6f has difficulty in creating the cluster, but as is known, all algorithms have a weakness and cannot solve everything.

It could see relevant information in Table 1. It has a smaller value in the banana test, (on With Dunn index) which indicate that it is being clustered effectively even in not a regular shape. In this case, the cloud of information does not have a consistent form, and the analyses throw a high number in comparison to the other metrics. Another example is the gaussinas1 that loses but not by much. It hopes the proposed approach does not win all the tests; in the rest of the test won and showed a good performance. Continuing with the image banana It can create clusters in such a way that there is no standard deviation with the Davies-Bouldin metric. But the value is higher, which means that despite not having a regular shape can be formed as a cluster, and the other test has the same tendency. Lastly, with the Hopkins Statistics it gets a higher value, the desired would be a lesser value in spite of them, in the other experiments, the values are competitive.

Table 2 shows the results of the different algorithms, showing that our algorithm has high capacity when recognizing cluster as in the banana or 2sp2glob. On the other hand, it is observed that it has difficulties to make the cluster in data groups that are dispersed, where it is deficient in the action of grouping but even in this case create a tentative cluster.

Table 2 shows the clusters it has been selected by algorithms and how the performance is variable. It should be mentioned that the algorithms have strengths and weaknesses. In general, we could see that our proposal stands out for the clusters that have irregular shapes because it does not depend on the Euclidean distance. It is one of the reasons that where the cluster is well defined is that it has a suboptimal performance. Only to mention, the image of 2sp2glob shows a cluster in the spiral than the

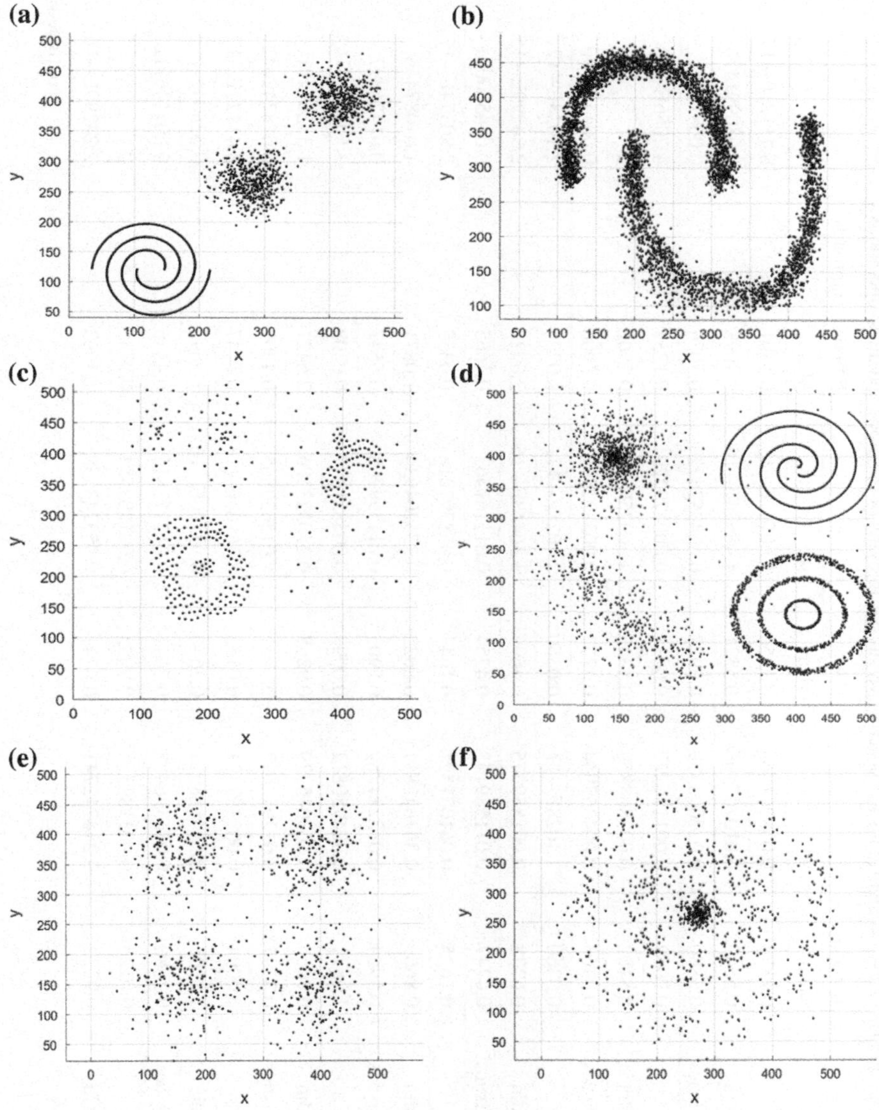

Fig. 6 Images that represent a challenge to the different algorithms **a** 2sp2glob, **b** banana, **c** compound, **d** impossible, **e** square1, **f** rings

other algorithms could not recognize; another remarkable is the image impossible where K-means and GA did not recognize the nest ring.

Table 1 Comparative in three indexes and two classical algorithms to clustering versus the applying of the erode and dilate operations

	Index	K-means		GA		Proposed approach	
		Mean	sd	Mean	sd	Mean	sd
2sp2glob	Dunn	0.1872	0.038261863	0.1950	5.60747E−17	0.1301	8.41121E−17
	Davies-Bouldin	0.5542	0.010207306	0.5521	4.48598E−16	2.4098	0.030991476
	Hopkins Statistics	0.3880	0.022005339	0.3927	0.004852472	0.3076	0.008193
Banana	Dunn	0.0039	0.001186277	0.0021	0.000279729	0.1235	4.2056E−17
	Davies-Bouldin	0.5402	0.003868859	0.5343	4.56925E−05	0.7502	0
	Hopkins Statistics	0.3763	0.004357954	0.3767	0.003792596	0.3906	0.004450796
Blobs	Dunn	0.0360	2.1028E−17	0.0360	2.1028E−17	0.0225	2.1028E−17
	Davies-Bouldin	0.6738	0.001859275	0.6699	3.36448E−16	2.0922	2.24299E−15
	Hopkins Statistics	0.4747	0.011589148	0.4754	0.011161363	0.4665	0.014550091
Compound	Dunn	0.1625	0.061031521	0.1643	0.058797218	0.0355	2.80374E−17
	Davies-Bouldin	0.8669	0.361642811	0.8622	0.373204934	1.0871	1.01087446
	Hopkins Statistics	0.5904	0.013211789	0.5903	0.011599708	0.5547	0.01426897
Impossible	Dunn	1.0871	0.004950812	0.0098	0.004851697	0.0308	3.50467E−17
	Davies-Bouldin	0.9666	0.263045279	0.9268	0.250523106	2.6421	1.35269491
	Hopkins Statistics	0.3797	0.012912692	0.3781	0.015039865	0.4309	0.018742042
Smile2	Dunn	0.0094	0.003307491	0.0113	8.76168E−18	0.1447	8.41121E−17
	Davies-Bouldin	1.0108	0.384818901	0.7646	8.97196E−16	1.5696	0.02504512
	Hopkins Statistics	0.2680	0.016396425	0.2641	0.017172815	0.4346	0.007446712
Rings	Dunn	0.0128	0.003189358	0.0085	0.001809554	0.0235	0
	Davies-Bouldin	0.9222	0.214742741	0.6741	0.053243558	1.4063	2.24299E−16

(continued)

Table 1 (continued)

	Index	K-means		GA		Proposed approach	
		Mean	sd	Mean	sd	Mean	sd
	Hopkins Statistics	0.4255	0.013041702	0.4069	0.011662421	0.4460	0.020837296
Gaussians1	Dunn	1.3455	0	1.3455	0	0.0289	2.1028E−17
	Davies-Bouldin	0.0811	0	0.0811	0	0.6185	6.72897E−16
	Hopkins Statistics	0.4801	0.017965326	0.4814	0.023202609	0.4260	0.035175443
Long1	Dunn	0.0095	0.016356342	0.0066	8.76168E−19	0.0247	1.40187E−17
	Davies-Bouldin	0.3625	0.096518173	0.3371	2.24299E−16	1.2166	8.97196E−16
	Hopkins Statistics	0.4083	0.01236654	0.4069	0.008204147	0.4235	0.030613195
Elliptical_10_2	Dunn	0.0340	0.013205005	0.0299	0.01653731	0.0074	0
	Davies-Bouldin	0.7993	0.263154777	0.7969	0.22556434	1.8334	0.468453881
	Hopkins Statistics	0.4127	0.022994547	0.4040	0.031292038	0.4367	0.019976935

Table 2 Clustering results of data sets

	Proposed approach	K-means	GA
2sp2glob	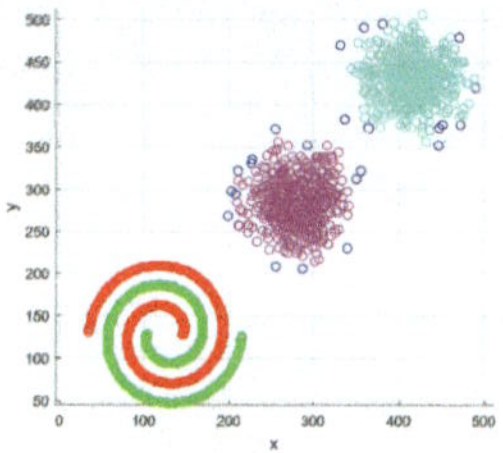	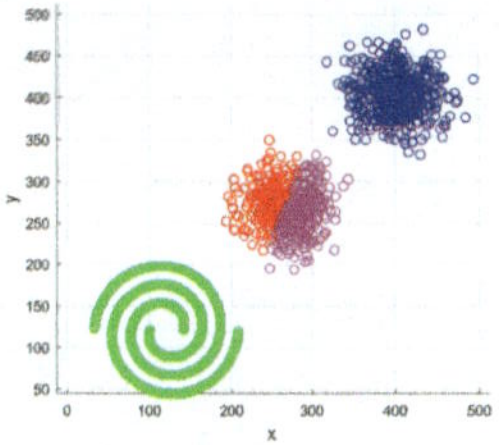	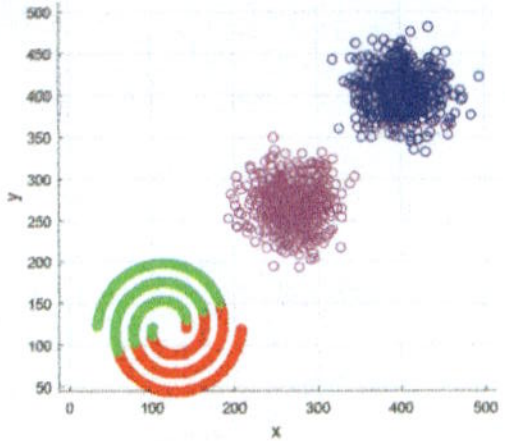
Banana	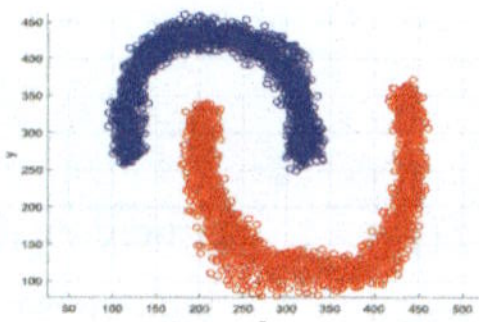	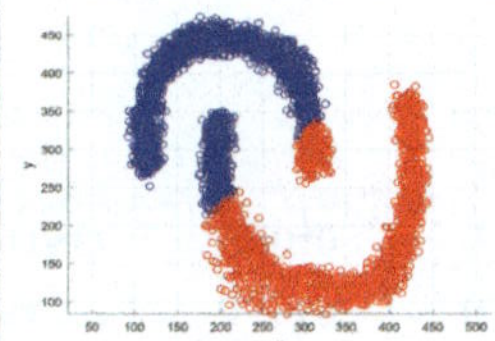	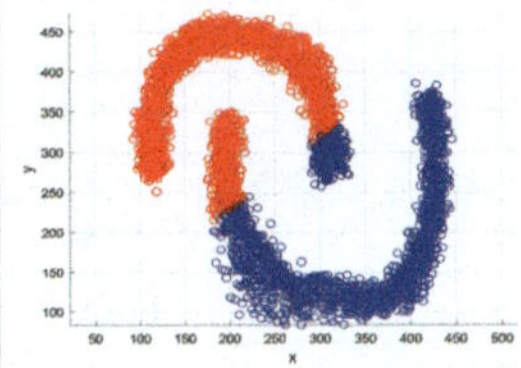
Compound	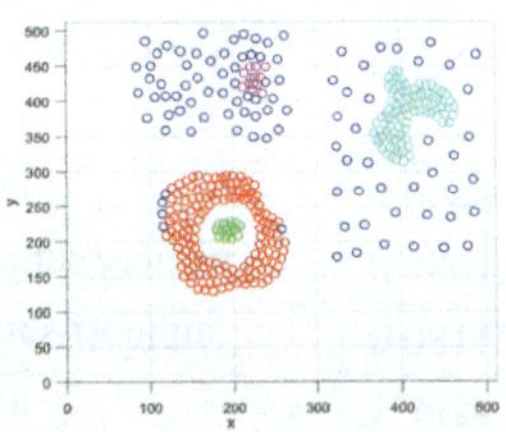	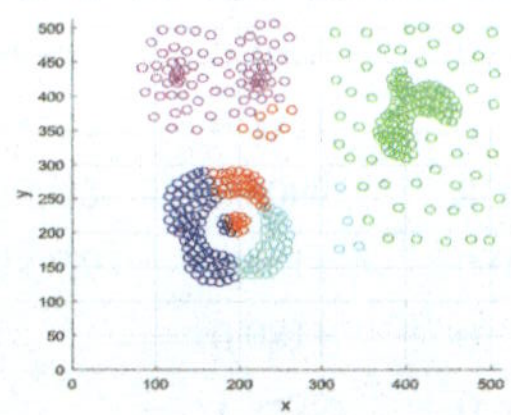	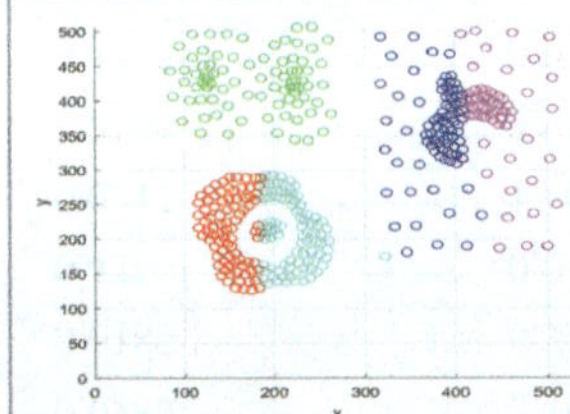

(continued)

Table 2 (continued)

	Proposed approach	K-means	GA
Impossible		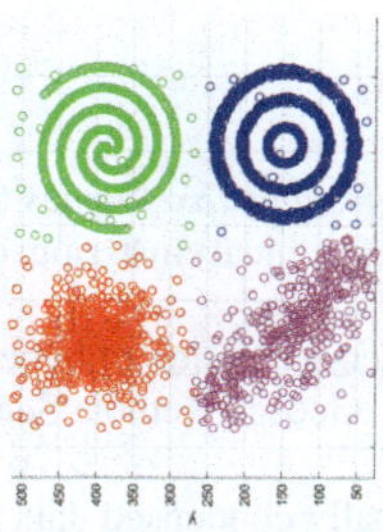	
Rings	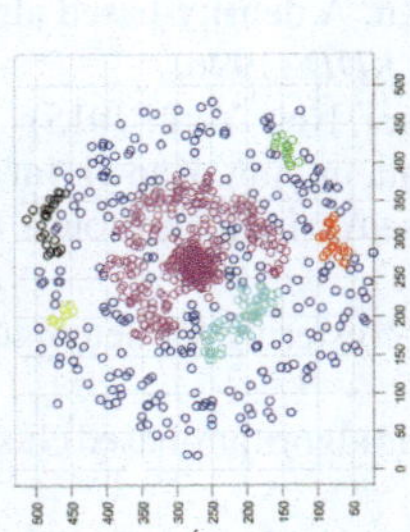	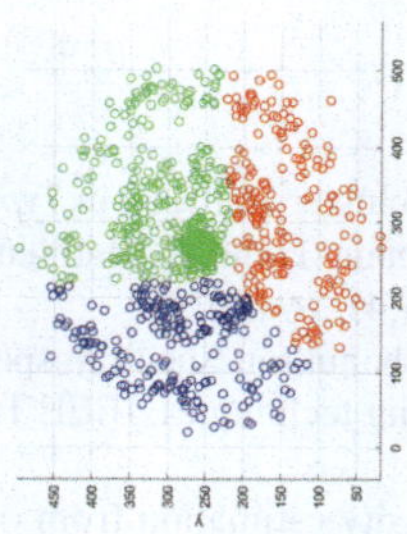	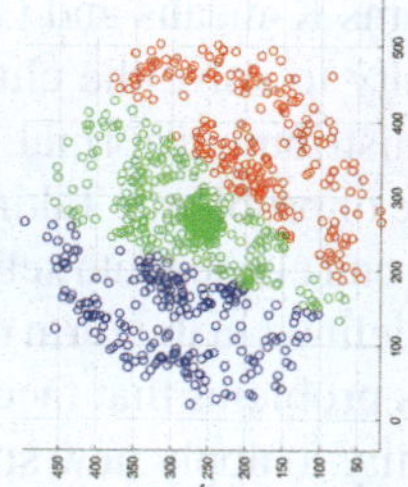
Square1	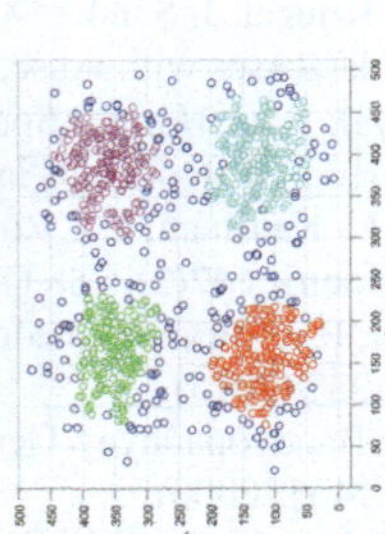	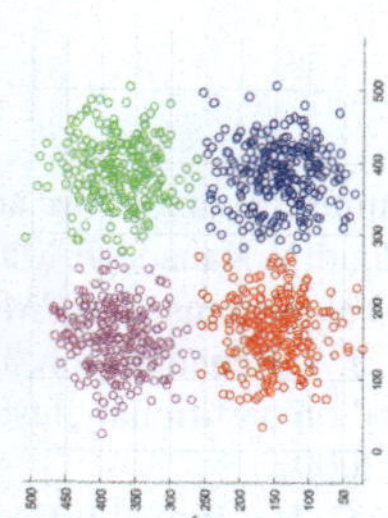	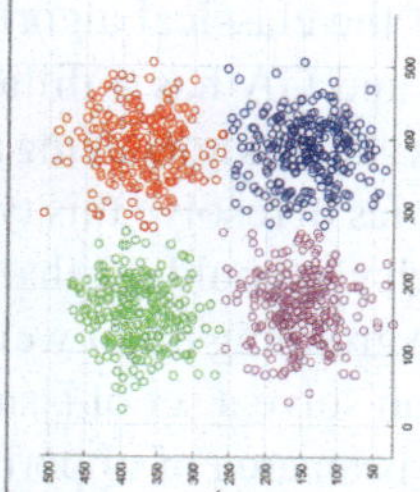

6 Conclusion and Discussion

In this chapter, we show an unusual technique to separate complex data set by using the operations of dilate and erode in combination with the Haralick Algorithm, the comparison of the classical algorithms K-means and GA. It has been demonstrated that K-means and GA has a difficulty to solve the clustering sets, due to irregular characteristics, such as not having a distribution normal. It shows that the new strategy has the capacities to resolve this type of problems. Taking into account the properties of our approach you could see that it can cluster data sets that have an irregular shape, where the data grouping is not well defined in the form of a cloud or a stack. With our approach, it can solve a set of issues problems that faced a challenge to the classical algorithms. It is an area of opportunity to apply new strategies that could resolve. It showed in the experiments the advantages in the sets but had the chance to improve this approach in another kind of images.

References

1. K. Bailey, Numerical taxonomy and cluster analysis, in *Typologies and Taxonomies* (1994)
2. V. Cohen-Addad, V. Kanade, F. Mallmann-Trenn, C. Mathieu, Hierarchical clustering: objective functions and algorithms. J. ACM **66**(4), 26 (2019)
3. Y. Tarabalka, J.A. Benediktsson, J. Chanussot, Spectral–spatial classification of hyperspectral imagery based on partitional clustering techniques. IEEE Trans. Geosci. Remote Sens. **47**(8), 2973–2987 (2009)
4. M. Girolami, C. He, Probability density estimation from optimally condensed data samples. IEEE Trans. pattern Anal. (2003)
5. T. Liu, C. Rosenberg, H.A. Rowley, Clustering billions of images with large scale nearest neighbor search, in *Proceedings—IEEE Workshop on Applications of Computer Vision, WACV 2007* (2007)
6. M. Ester, H. Kriegel, J. Sander, X. Xu, A density-based algorithm for discovering clusters in large spatial databases with noise, in *KDD* (1996)
7. C.C. Aggarwal, *Data Mining* (Springer, New York, 2015)
8. X. Wu et al., Top 10 algorithms in data mining. Knowl. Inf. Syst. **14**(1), 1–37 (2008)
9. J.E. Gentle, L. Kaufman, P.J. Rousseuw, Finding groups in data: an introduction to cluster analysis. Biometrics **47**(2), 788 (1991)
10. S. Harikumar, P.V. Surya, K-medoid clustering for heterogeneous datasets. Procedia Comput. Sci. **70**, 226–237 (2015)
11. U. Maulik, S. Bandyopadhyay, Genetic algorithm-based clustering technique. Pattern Recognit. **33**(9), 1455–1465 (2000)
12. G.H. Ball, D.J. Hall, ISODATA, a novel Method of data analysis and pattern classification (1965)
13. S. Lloyd, Least squares quantization in PCM. IEEE Trans. Inf. Theory **28**(2), 129–137 (1982)
14. J. MacQueen, Some methods for classification and analysis of multivariate observations, in *Proceedings of the Fifth Berkeley Symposium on Mathematical Statistics and Probability* (1967)
15. R.M. Gray, D.L. Neuhoff, Quantization. IEEE Trans. Inf. Theory **44**(6) (1998)
16. S.Z. Selim, M.A. Ismail, K-means-type algorithms: a Generalized convergence theorem and characterization of local optimality. IEEE Trans. Pattern Anal. Mach. Intell. **PAMI-6**(1), 81–87 (1984)

17. H. Xiong, J. Wu, J. Chen, K-means clustering versus validation measures: a data-distribution perspective. IEEE Trans. Syst. Man, Cybern. Part B **39**(2), 318–331 (2009)
18. S. Bandyopadhyay, S.K. Pal, B. Aruna, Multiobjective GAs, quantitative indices, and pattern classification. IEEE Trans. Syst. Man Cybern. Part B **34**(5), 2088–2099 (2004)
19. S. Bandyopadhyay, U. Maulik, Nonparametric genetic clustering: comparison of validity indices. IEEE Trans. Syst. Man Cybern. Part C (Applications Rev.) **31**(1), 120–125 (2001)
20. D.E. Goldberg, J.H. Holland, Genetic algorithms and machine learning. Mach. Learn. **3**(2/3), 95–99 (1988)
21. L. Davis, Handbook of genetic algorithms (1991)
22. Z. Michalewicz, *Genetic Algorithms + Data Structures = Evolution Programs* (Springer, Berlin, Heidelberg, 1996)
23. W. Song, S.C. Park, *Genetic Algorithm-Based Text Clustering Technique* (Springer, Berlin, Heidelberg, 2006), pp. 779–782
24. W. Song, J.Z. Liang, S.C. Park, Fuzzy control GA with a novel hybrid semantic similarity strategy for text clustering. Inf. Sci. (Ny) **273**, 156–170 (2014)
25. J.C. Dunn†, Well-separated clusters and optimal fuzzy partitions. J. Cybern. **4**(1), 95–104 (1974)
26. D.L. Davies, D.W. Bouldin, A cluster separation measure. IEEE Trans. Pattern Anal. Mach. Intell. **PAMI-1**(2), 224–227 (1979)
27. A. Banerjee, R.N. Dave, Validating clusters using the Hopkins statistic, in *2004 IEEE International Conference on Fuzzy Systems* (IEEE Cat. No. 04CH37542), vol. 1, pp. 149–153
28. W.S. Sarle, A.K. Jain, R.C. Dubes, Algorithms for clustering data. Technometrics **32**(2), 227 (1990)
29. A.F. Famili, G. Liu, Z. Liu, Evaluation and optimization of clustering in gene expression data analysis. Bioinformatics **20**(10), 1535–1545 (2004)
30. M. Halkidi, Y. Batistakis, M. Vazirgiannis, Cluster validity methods : part I
31. M. Halkidi, Y. Batistakis, M. Vazirgiannis, Clustering validity checking methods: part II
32. M. Kim, R.S. Ramakrishna, New indices for cluster validity assessment (2005)
33. R.M. Haralick, L.G. Shapiro, Computer and robot vision, vol. 1 (Addison-Wesley Reading, 1992)

Estimation of the Homography Matrix to Image Stitching

Cesar Ascencio

Abstract In many problems of computer vision, there are some problems for the estimation of homography matrix between images, the homography matrix is necessary to solve different problems in this area of science, one of them is Stitching, which in simple words is the process by several images are combined to produce a panoramic image or a high resolution image, usually through a computer program. The homography matrix is a fundamental basis to perform the Stitching on the images, there are many methods to calculate the homography, the most common to find this estimation is the random sampling consensus (RANSAC). But there are some works that consider the estimation process in a different way, the way in which these works deal with the problem is taking the problem as a multi-objective estimation process, with this approach it is possible to facilitate the calculation of multidimensional problems. In order to solve the multi-objective formulation, many different evolutionary algorithms have been explored, obtaining good results in their tests. In this chapter the problem of the estimation of the homography matrix is considered as a problem of multi-objective optimization and will be faced with the evolutionary algorithm ABC.

Keywords Stitching · Homography matrix · Artificial Bee Colony (ABC)

1 Introduction

With the development of digital photography and according to the cameras have been available to anyone and anywhere, have presented new problems that demand new requirements in the area of image processing, in order to solve this, a tool called stitching was developed.

Stitching is the technique that can be implemented in different problems, for example, one of these problems is the need to portray a panoramic shot without a wide-angle lens, another is when you have a focal length too long and we cannot place ourselves in a position optimal for the photograph to cover what we need, or

C. Ascencio (✉)
Universidad de Guadalajara, CUCEI, Guadalajara, Jalisco, Mexico
e-mail: cesar.ascencio@alumno.udg.mx

D. Oliva and S. Hinojosa (eds.), *Applications of Hybrid Metaheuristic Algorithms for Image Processing*, Studies in Computational Intelligence 890,
https://doi.org/10.1007/978-3-030-40977-7_10

in case of needing to create a cylindrical or spherical environment for a 3D job, it is also used when we want to reconstruct a mural (or something similar) from photos taken in parallel, between others.

To be able to perform image stitching it is necessary to perform an image registration, which is the process of transforming different data sets into a coordinate system, in this case it involves multiple photographs. Image registration is used to look for alignments that minimize the sum of the absolute differences between the superimposed pixels [1].

After the registration of the images, to match the images from the standpoint in which they overlap, both are transformed with the alignment. With simpler words, the alignment is a change in the coordinate system that results in the system adopting an optimal posture, thus, the image coincides with a required point of view. There are different types of transformations through which an image can be submitted: pure translation, pure rotation, a transformation of similarity that includes translation, rotation and scaling of the image to be transformed, affine or projective transformation.

The projective transformation can be described mathematically as:

$$x' = H \cdot x,$$

where X are the points in the old coordinate system, X' are the corresponding points in the transformed image and H is the homography matrix.

The homography matrix is geometrically defined as a projective transformation that determines a correspondence between two flat geometric figures that means that a homography is the relation of two images of the same surface flat in space. So you have to find the matrix of homography of the images to which you want to apply the stitching.

Currently some standard methods are used for the calculation of homography matrix, one of the most used is RANSAC which works by trying to fit several models using some of the point pairs detected in the images, then checking if the models were able to relate most of the points.

But there are also other ways of calculating the homography matrix the most common way with RANSAC which is a non-deterministic algorithm in the sense that it produces a reasonable result only with a certain probability, higher as more iterations are allowed. In this case the value of the homography matrix will be estimated using an evolutionary algorithm (EA), that is a generic metaheuristic optimization algorithm based on population and uses mechanisms inspired by biological evolution, such as reproduction, mutation, recombination and selection. Each possible solution is an individual in a population of the optimization problem. More specifically, the artificial colony of bees (ABC) will be used.

The Artificial Bee Colony (ABC) is a bio-inspired optimization algorithm that is based on the behavior of bees to optimize a certain function, the bees follow certain patterns to find their food and their method works well, this algorithm imitates that behavior to achieve finding the optimization of the determined function.

The estimation of the homography matrix will be carried out as a multiobjective optimization problem that will be optimized with the Bee Colony Artificial Evolutionary Algorithm, the objective of this algorithm is to find the minimum error between each individual (possible solution matrix) until finding the homographic matrix with the correct values (matrix that returns the minimum error) to perform the stitching. Based on the two images in which you want to do stitching, points are assigned in parts of the images that can be seen in both, this is the basis for the calculation of all operations.

The results obtained from stitching calculation with both programs (the program where the automatic calculation is made with the openCV libraries and the program where the evolutionary algorithm is used) will be compared at the end of the chapter. Both errors resulting from the homography matrices, among other parameters, will be taken into account.

2 Stitching

Stitching is a process by which two or more images are combined by superimposing parts of the images that are identical. All approaches to image stitching require that the superimposed parts are identical to have good results [2].

Image stitching is commonly used in the following applications:

- Multiple image super-resolution (Fig. 1).
- Create a cylindrical or spherical environment for a 3D job.
- Medical imaging.
- High-resolution photomosaic in digital maps.

Stitching process can involve some issues, the illumination of the images can vary even if they are taken almost at the same moment, which can cause the stitching of the images to be noticed, other problems to be treated are: lens distortion, presence of parallax, movement of the scene and the differences in exposure or when set of images will not have a reasonable amount of overlap (at least 15–30%) to overcome lens distortion and have enough detectable features.

The stitching process can be divide into:

Fig. 1 Panoramic image created by multiple high resolution images with stitching

- Key point detection (feature detection).
- Image registration.
- Alignment.

As in this chapter you only want to calculate the homography matrix to perform the Stitching, only the steps that make up the Stitching are omitted and instead of these the key points in each image are marked with an automatic algorithm. Even so, each point of the Stitching process is briefly described.

2.1 *Feature Detection*

Feature detection is necessary to automatically find correspondences between images. It is necessary to have solid correspondences to correctly align one image with the other that you want to overlap. Corners, blobs, Harris corners, and difference of gaussian (DoG) are good features easy to detect by computer, since they are different and appear in images repeatedly.

The features should be distinctive objects, which must be easily identifiable, they are recognized throughout the image. In general, a physical interpretation of the characteristics is very necessary. The images must have a set of detected characteristics with the greatest number of common elements, considering the images that do not cover exactly the same panorama, when there are objects or when other changes occur.

There may be problems when the features are detected, some of these problems are derived from an incorrect detection of characteristics that in turn can be the result of a degraded or poor quality image. There are many factors that can interfere in that the corresponding physical characteristics are different, a factor can be the spectral sensibility of the sensors, so it is necessary to choose a robust feature descriptor that considers these circumstances. Simultaneously, these descriptors must have some stability so that they perform well even with certain variations or noise and discreet enough to distinguish between the characteristics. In Fig. 2 you can see two images with visible feature detection.

During an automation of this registration, two main approaches to feature understanding have been formed.

1. Area methods are area-based methods does not focus on the detection of features but focuses more on the feature matching.
2. Feature methods are feature-based methods that focuses on the extraction of characteristic structures outstanding in images [3–5].

2.2 *Image Registration*

Image registration is the process of overlaying two or more images of the same scene taken in different conditions, such as images taken at different times, from

Fig. 2 Example feature detection

different viewpoints, and/or by different sensors, geometrically aligns two images. Image registration is a very important step in all image analysis tasks in which the final information is obtained from the combination of several data sources such as the fusion of images, change detection and the restoration of multichannel images [3–5], with others words, image registration involves matching features in a set of images or using direct alignment methods to search for image alignments that minimize the sum of absolute differences between overlapping pixels.

The applications that can be given to image registration can be divided into 4 groups according to the image:

- Different viewpoints (multiview analysis). Images of the same scene taken from different angles.
- Different times (multitemporal analysis). Images of the same scene are taken from different times.

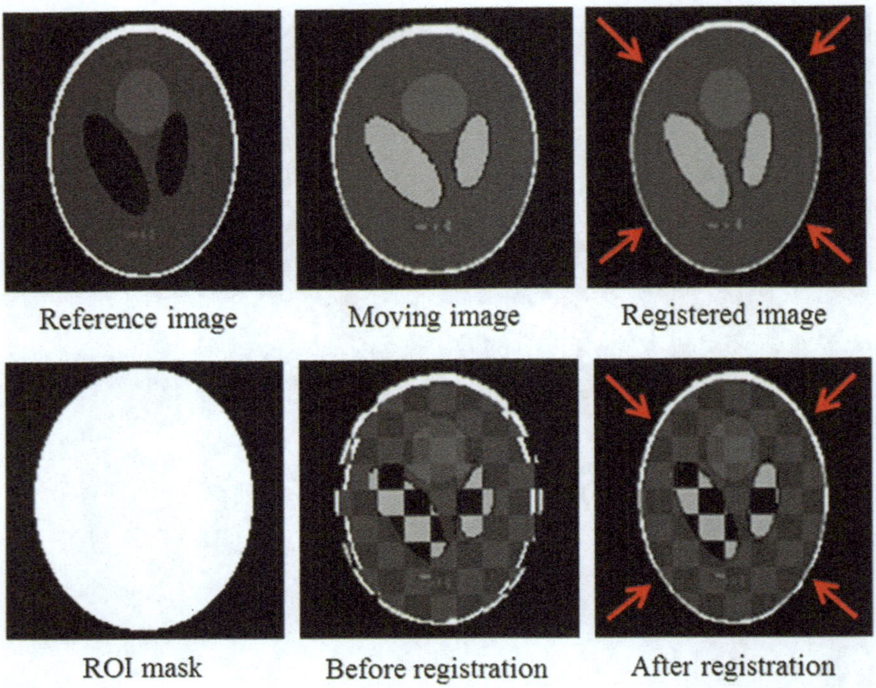

Fig. 3 Example of image registration. Image from https://www.uniklinik-freiburg.de/mr-en/research-groups/postprocessing/dce-mri-registration.html

- Different sensors (multimodal analysis). Images of the same scene are taken from different sensors.
- Scene to model registration. Images of a scene and a model of the scene are registered [3–5].

Figure 3 shows an example of image registration.

2.3 *Alignment*

Image alignment algorithms is able to find the correspondence relationships between images even though they have varying degrees of overlap. Alignment in other words, the coordinate system is changed so that it adopts another coordinate system that allows the image to be matched in the required viewpoint. The types of transformation to which an image can be submitted are: pure rotation, pure translation, rotation and scaling of the image which needs to be transformed, a similarity transform which includes translation or projective transform [6].

An essential problem in image alignment is first you must determine a mathematical model that can relate the coordinates of each pixel of an image with each pixel coordinate of the other image.

Projective transformation can be mathematically described as Eq. 1.

$$x' = H \cdot x \tag{1}$$

where x are the points in the old coordinate system, x′ is the corresponding points in the transformed image and H is the homography matrix.

Figure 4 is divided by 3 sections *A*, *B* and *C* where the complete process is observed by parts to perform a stitching, in section *A*, the feature detection of the two images is shown. In section *B* the matching detection between the images was made, in this

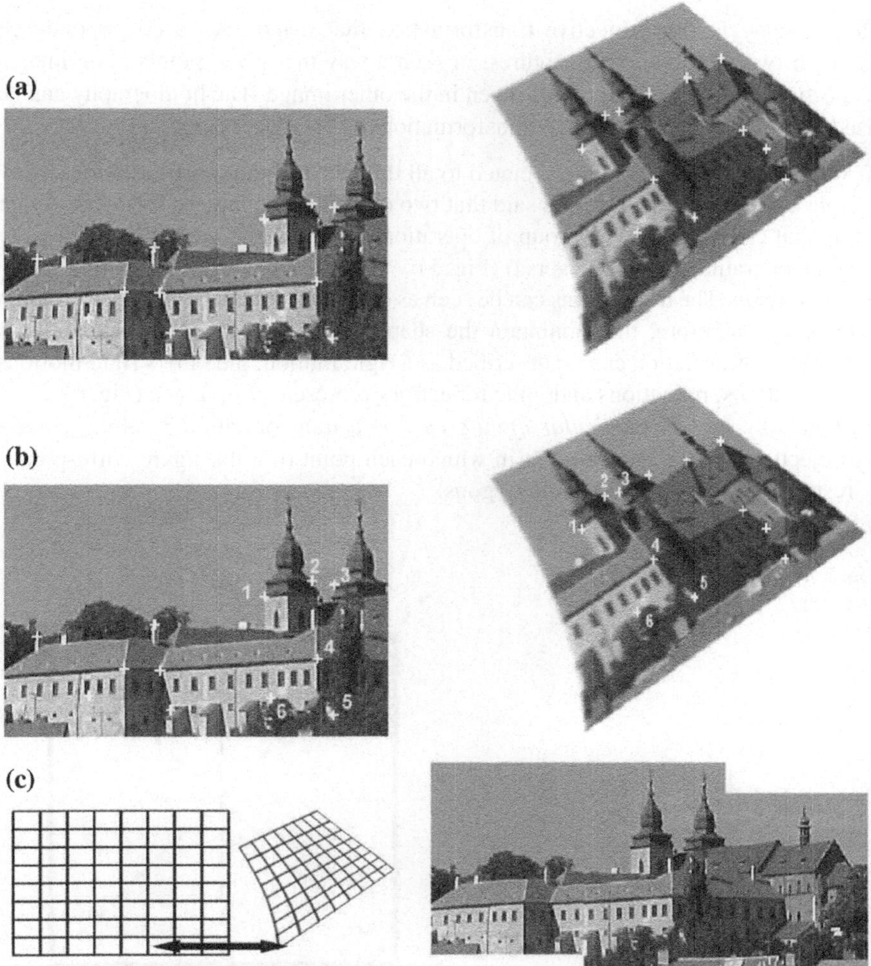

Fig. 4 Complete process to stitch images (image registration methods, https://doi.org/10.1016/S0262-8856(03)00137-9)

step each point identified by the feature detection corresponding to its counterpart of the other image is recognized, this with the intention of joining the identified points in order to overlap the part that is equal in both images. And finally in section *C* in bottom left, the transformation model estimation is done by aligning both images according to its correspondence, while in right bottom is the image resampling and transformation to have the complete image registration and a good stitching.

3 Homography

Homography is any projective transformation that determines a correspondence between two flat geometrical figures, in such a way that given points in an image, respectively correspond to points given in the other image. The homography can be used to perform different image transformations:

- *Symmetry*, this concept is associated to all the transformations, translations, rotations and reflections, and it is said that two objects are symmetrical to each other in what concerns a given group of operations if one is obtained from another by some operations (and vice versa) (Fig. 5).
- *Translation*. The translations can be seen as direct movements without orientation changes, therefore, they maintain the shape and size of the figures or objects moved. A translation can be described as a rigid motion: the others rigid motions are rotations, reflections and glide reflections represented in image (Fig. 6).
- *Homology and its particular affinity case* is a transformation resulting from a projection from a certain point in which each point of a flat figure corresponds, respectively, a point of its homologous.

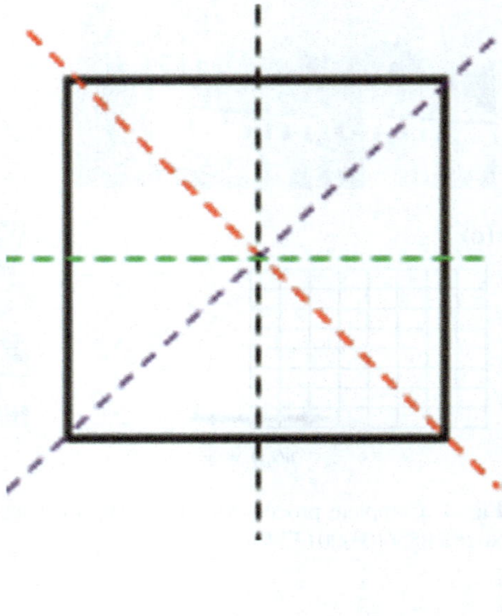

Fig. 5 Representation of symmetry

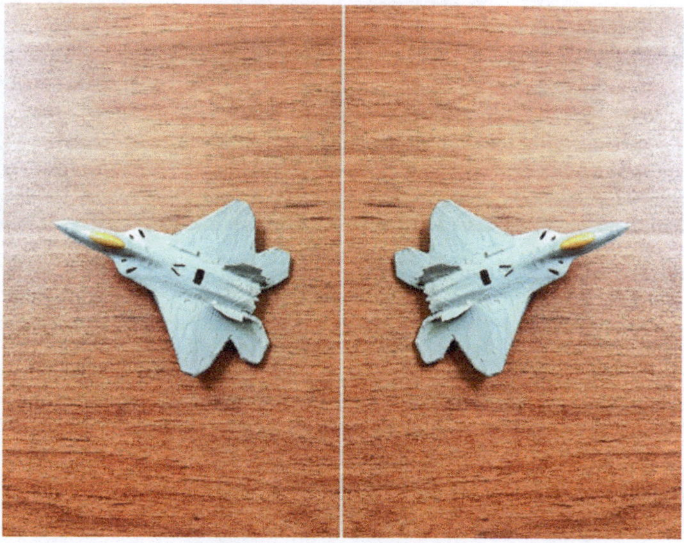

Fig. 6 Representation of geometry translation

The image of Fig. 8 is an image belonging to a frame of a video in which there is a tunnel where some animals are recorded while they have neuronal injuries and their evolution with different treatments, clearly the image is not aligned but this is not perceptible at the time of recording, so that image has to be corrected in order to obtain the correct statistics about the results. Figure 8 has four control points (Fig. 7) that will serve as a reference for the homography calculation.

The last image transformation (homology and its particular affinity case) will be used in this chapter to calculated the projective—mapping between any two projection planes with the same center of projection and it is represented as 3 × 3 matrix in homogenous coordinates (Fig. 9).

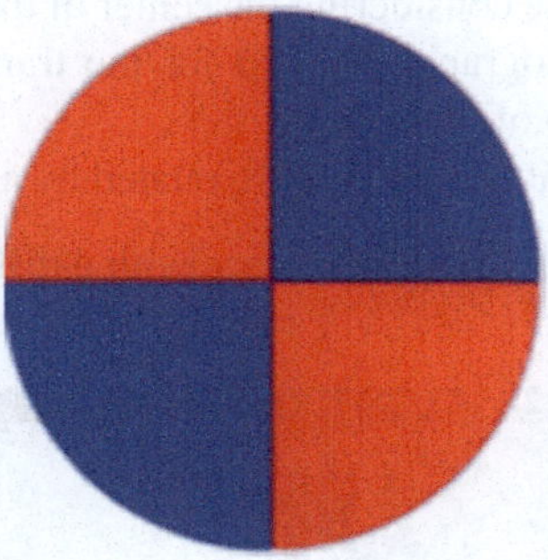

Fig. 7 Control point

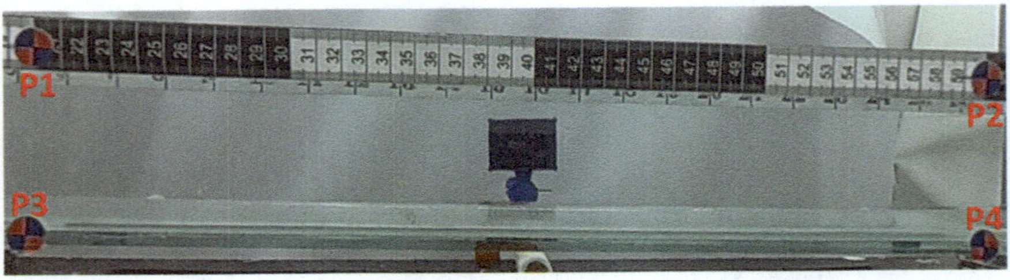

Fig. 8 Image with non-aligned tunnel

$$H = \begin{bmatrix} M_{11} & M_{12} M_{31} \\ M_{21} M_{22} M_{32} \\ M_{31} & M_{32} M_{33} \end{bmatrix}$$

Fig. 9 Homography matrix

$$\underset{\mathbf{P'}}{\begin{bmatrix} wx' \\ wy' \\ w' \end{bmatrix}} = \underset{\mathbf{H'}}{\begin{bmatrix} M_{11} & M_{12} M_{31} \\ M_{21} M_{22} M_{32} \\ M_{31} & M_{32} \, M_{33} \end{bmatrix}} \underset{\mathbf{P}}{\begin{bmatrix} x \\ y \\ 1 \end{bmatrix}}$$

Fig. 10 Applying the homography matrix

To apply a homography H first compute $p' = Hp$ (regular matrix multiply) and convert p' from homogeneous to image coordinates (divide by w) see Fig. 10. To observe the results in an image by applying homography matrix see Figs. 11 and 12 [7].

In Fig. 11 you can see the application of the homography to transform an image and correct it from another perception, the application of the homography and the clipping of the image is done considering the center of the four control points shown above. In Fig. 12 the homography is used for the transformation of two images coordinates and the overlap of these [8].

It must be considered that two images are related by a homography if and only if:

- Both images are pointing to the same place with a different angle.

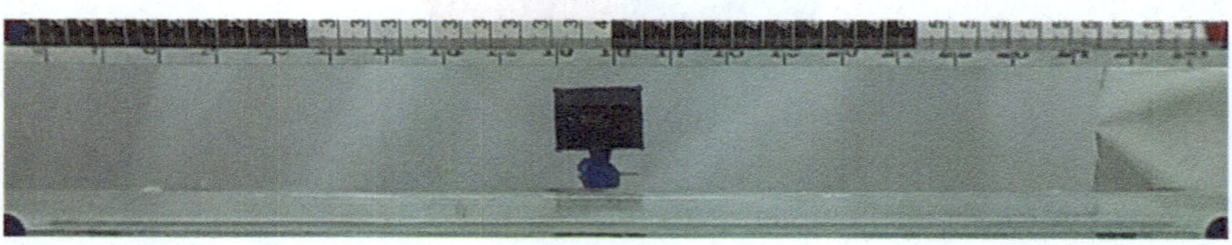

Fig. 11 Image with perspective correction applied by the homography matrix

Fig. 12 Use of the homography to transform the coordinates of two images and overlap them

- Both images are taken from the same camera but from a different angle.
- Camera is rotated about its center of projection without any translation or Camera is rotated about its center of projection without any translation.
- Note that the homography relationship is independent of the scene structure.
- Relationship holds regardless of what is seen in the images.
- It does not depend on what the cameras are looking at.

There are many methods to estimate the homography matrix, the most used is RANSAC. This method is used to estimate the homography matrix, there are multiple correspondences but the correct homography matrix is the one that has the maximum number of inliers, this method is the most fulfilled since it is a robust method and is opposed to the traditional of smoothing techniques. It uses as small an initial data as feasible and enlarges the consensus set.

RANSAC is a statistical method that shows the data until the convergence. Thus, an optimal solution cannot be obtained if the proportion of atypical values is unknown. This is because in the current time a solution must be determined, but there is another RANSAC-F algorithm of n competitive solutions where ordered lists are kept. It does not find a solution at every time but it determines if a solution is optimal with an average of the best n in the list [9].

In any case there are other ways of estimating the homography matrix, in this case an evolutionary algorithm will be used.

4 Evolutionary Algorithms

An evolutionary algorithm (EA) is a subset of evolutionary computation, a generic population-based metaheuristic optimization algorithm, with emphasis on extensions and analysis bio-inspired genetic algorithms that are the result of interdisciplinary research field with a relationship to biology, these algorithms are based on models of organic evolution, nature is the source of inspiration and they model the collective learning process within a population of individuals, each of which represents not only a search point in the space of potential solutions to a given problem, but also may be a temporal container of current knowledge about the "laws" of the environment [10].

The evolutionary computation methods have been consolidated as an alternative solution to many optimization problems with practical implications in many areas. They are also considered generic optimization tools that can solve very complex problems that have too large a search space, since they have the capacity to reduce the effective size of the search space through the use of effective strategies. In comparison with the heuristic methods, the EA allows solving the problems in a faster and more robust way and its techniques are easier to design and implement [11–13].

The starting population is initialized by an algorithm-dependent method, and involves towards successively better regions of the search space by means of (more or less) randomized process of recombination, mutation and selection. The environment delivers a quality information (fitness value) for new search points, and the selection process favors those individuals of higher quality to reproduce more often than worse individuals. The recombination mechanism allows for mixing of parental information while passing it to their descendants, and mutation introduces innovation into the population [14].

Therefore, the implementation can be seen as:

- Generate the initial population of individuals randomly (starting population).
- Evaluate the individual fitness of new individuals.
- Breed new individuals through crossover and mutation operations to give birth to offspring.

For better understanding see Fig. 13.

In summary, an evolutionary algorithm mathematically looks for a minimum or a maximum of a given function, according to the algorithm that is used, it follows a series of steps where it looks for a global optimum, this means that the algorithm must be prepared so as not to get stuck in some optimal local (Fig. 14).

In order to better understand the functioning of an evolutionary algorithm, an example will be given where an objective function is established and where the evolutionary algorithm works to find the global optimum of the established function, the development of the evolutionary algorithm is in the programming environment with MATLAB, the implementation of this evolutionary algorithm and more evolutionary algorithms can be seen more in detail in Erik et al. [11–13].

The evolutionary algorithm used is known as temper. All algorithms are named after the natural process on which they are based, if based on the behavior of some

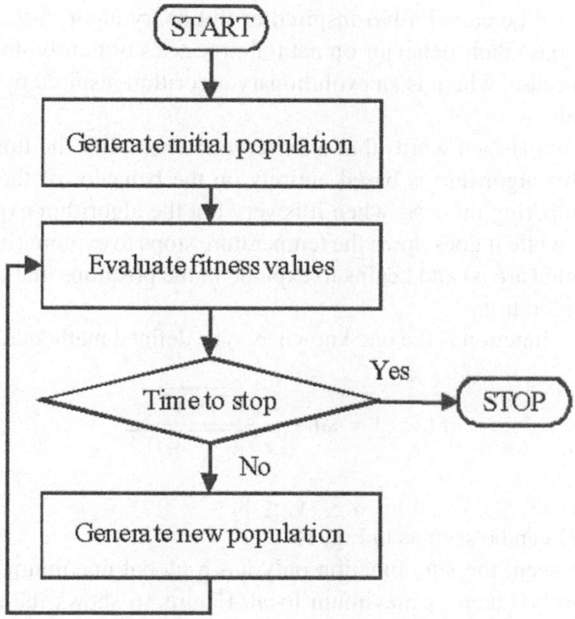

Fig. 13 Flow diagram of an evolutionary algorithm. Taken from Kachitvichyanukul [14]

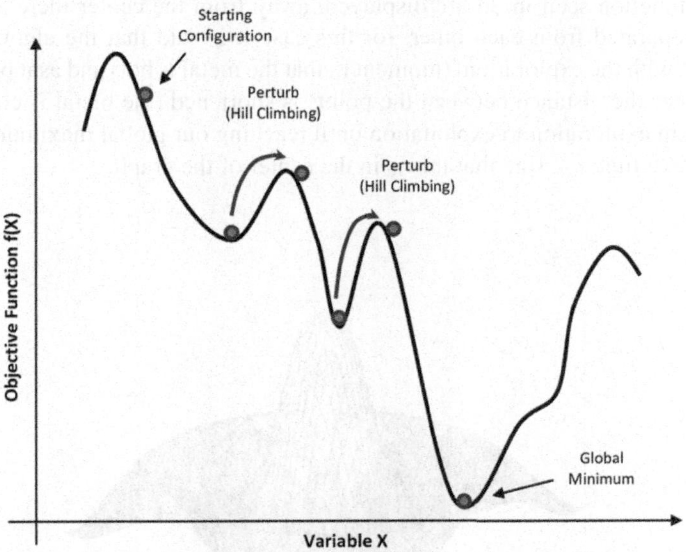

Fig. 14 Evolutionary algorithm avoiding local minimums (https://medium.com/@duoduoyunnini/introduction-implementation-and-comparison-of-four-randomized-optimization-algorithms-fc4d96f9feea)

living being it can be called a bio-inspired evolutionary algorithm. There are also algorithms that base their behavior on natural processes of non-biological phenomena, such as this case, which is an evolutionary algorithm inspired by the process of tempering metals.

With very summarized words that only serve to exemplify the functioning of an evolutionary, this algorithm is based entirely on the behavior of the particles of a metal in the tempering process, when it is very hot the algorithm explores possible solutions, then, while it goes down the temperature stops to explore (it begins to look for in more limited areas) and begins to explode in the positions that considered like better in the exploration.

The objective function is the one known as sync defined mathematically as Eq. 2:

$$f(x, y) = \sin \frac{\sqrt[2]{x^2 + y^2}}{\sqrt[2]{x^2 + y^2 + 0.1}} \tag{2}$$

Considering $-8 \leq x \leq 8 \quad -8 \leq y \leq 8$.

Plotted in 3D can be seen as in Fig. 15.

As it can be seen, the sinc function only has a global maximum but it also has an area that can be taken as maximum local. Figure 16 shows the values that the evolutionary algorithm took at the time of its execution.

In the performance of the evolutionary algorithm temple, a series of points on the sync function seen in 2d are displayed, away from the center there are points that are separated from each other, for this case it is said that the algorithm was beginning with the exploration, (moment in that the metal is hot) and as it progresses to the center the distance between the points is shortened (the metal is cooling) so it goes from exploration to exploitation until reaching our global maximum that for our objective function sinc that this is in the center of the graph.

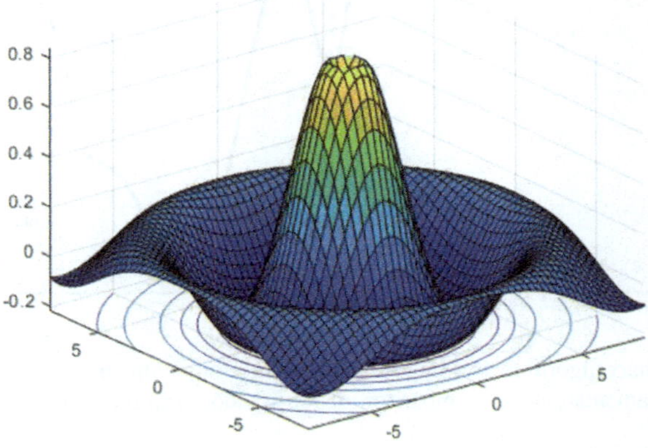

Fig. 15 3D sync function

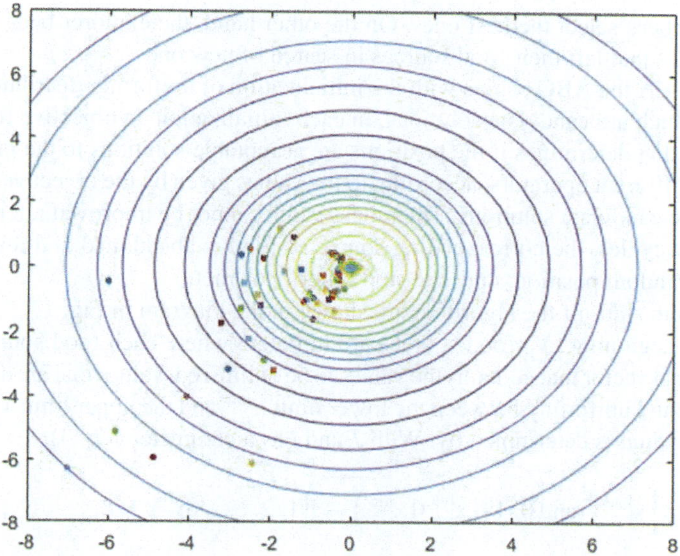

Fig. 16 Performance of the evolutionary algorithm temple

5 Artificial Bee Colony ABC

The evolutionary, artificial bee colony algorithm is a metaheuristic method in the field of artificial intelligence used to solve optimization problems. Based on the collective behavior of a swarm of bees, these anthophiles have demonstrated the ability to solve complex problems without centralized control [11–13].

This algorithm, like other evolutionary algorithms, is composed of a series of standard steps, mainly generating a population of agents, modified in three different phases through its operators. Dervis Karaboga proposed it in 2005 moment since it has been implemented in a variety of real-world applications demonstrating its good capabilities [15].

This method is constituted by three main parts: positions of the food sources, amount of nectar and different types of bees. Some numerical comparisons have shown that the ABC algorithm is competitive in comparison with other population algorithms, with the advantage that it requires the use of smaller numbers of control parameters [16, 17].

The ABC algorithm is classified into three groups of bees, each of these groups have their operators that serve to find a new candidate position of a food source, the groups of bees are: worker bees, observer bees and scout bees.

Half of the colony is composed of worker bees and the other half of scout bees, the source of food is equal to the number of worker bees in the hive, they are looking for food around the food source stored in their memory, while this information is passed to the observing bees. From the food sources found by the worker bees, the observing bees select the best ones. On the other hand, the explorer bees are a few worker bees that left their food sources in search of new ones.

As any EA the ABC begins with the initialization of uniformly distributed populations, which are candidate solutions, in each initialization an objective function is evaluated that determines if the positions are acceptable solutions to the problem.

Three different operators, according to the values given by the objective function, modify the candidate solutions. The fitness value cannot be improved after a certain number of cycles, the corresponding source of food is abandoned and reinitialized in a new random position, until the stop criterion is met.

The main steps of the algorithm are shown in the diagram in Fig. 17.

At the beginning, N_p food sources are initiated, where each food source is a *d*-dimensional vector that contains the values to be optimized, values that are distributed randomly and uniformly between the lower limit x_j^{low} and the upper limit x_j^{hig}. Thus, each individual is determined by: With *J* and *i* as a parameter (Eq. 3).

$$\mathrm{x_{j,i}}\left[\mathrm{x_j^{low}} \cdot \mathrm{rand}\, 0,\, 1\right]\left(\mathrm{x_j^{hig}}\, 0\, \mathrm{x_j^{low}}\right) - \mathrm{j}\left[1, 2, \ldots, \mathrm{d};\;\; \mathrm{i}\; 1, 2, \ldots, \mathrm{N_p};\right] \tag{3}$$

As mentioned, half of the bee colony is made up of worker bees and the other half by observing bees, for each food source there is a worker bee, every time a worker bee depletes its food source it becomes an explorer. Each worker bee generates a new source of food around its position in a way (Eq. 4):

$$v_{j,i} = x_{j,i} + \varnothing_{j,i}\left(x_{j,i} - x_{j,k}\right) \quad k \in \left\{1, 2, \ldots, N_p\right\}; \quad j \in \{1, 2, \ldots, d\} \tag{4}$$

where $x_{j,i}$ is a parameter *j* randomly selected from the individual *i* and *k* is any of the N_p food sources, such that the condition $i \neq k$ is satisfied. The adjustment factor $\varnothing_{j,i}$, is a random number between $[-1\ 1]$.

A candidate solution to the problem under consideration is each position of the food sources, also, the amount of nectar from the best sources of food, constitutes the quality of said solution according to its fitness value in the objective function. Thus, each time a new $V_{j,i}$ solution is generated, the fitness value must be evaluated for it.

In the stage of the observing bees, the process is emulated in which each observing bee must select one of the possible sources of food proposed, which depends completely on the fitness value, which is defined by the worker bees. The probability of selection for each food source is given by the following equation (Eq. 5):

$$prob = \frac{fit_i}{\sum_{i=1}^{N_p} fit_i} \tag{5}$$

where fit_i is the fitness value for the food source *i*, which is related to the objective function j_i. The probability that a food source will be selected increases with its fitness value.

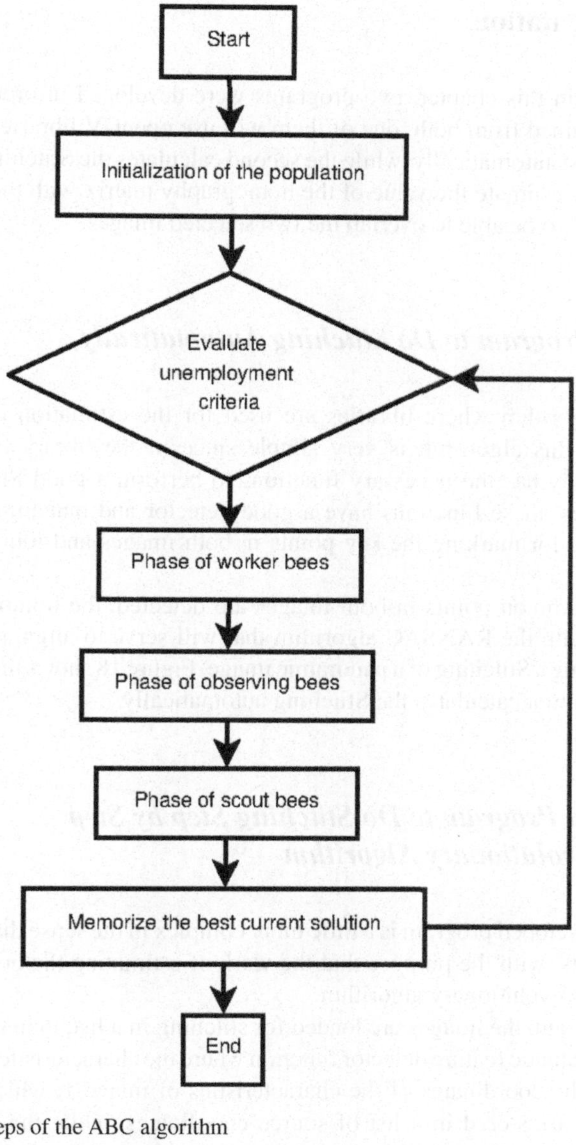

Fig. 17 Main steps of the ABC algorithm

To leave a food source the number of attempts must be equal to the "Limit", this parameter depends on the refinement of the obtained solutions and with this the quality of these food sources is determined. To know if the limit number of attempts is already reached, an *Ai* counter is added to each food source *i*, with this counter it is known if a food source cannot be improved after the limit of attempts, then the bee leaves this source of food.

6 Implementation

As mentioned in this chapter, two programs were developed in order to compare the results obtained from both, one of them will use openCV libraries that perform Stitching almost automatically, while the second calculates the Stitching step by step and in this way estimate the value of the homography matrix with the evolutionary algorithm ABC to be able to overlap the two selected images.

6.1 First Program to Do Stitching Automatically

In the first algorithm where libraries are used for the estimation of the homography matrix, this algorithm is very simple since in the library to perform the Stitching already has the necessary functions to perform a good Stitching of two or more images, these functions have a good detector and matching features that are responsible for marking the key points in both images and join those that are identical.

Once the common points in both images are detected, the homography matrix is calculated with the RANSAC algorithm that will serve to align and overlap the images obtaining a Stitching of a panoramic image. Figure 18 shows the flow diagram of the program that calculates the Stitching automatically.

6.2 Second Program to Do Stitching Step by Step with Evolutionary Algorithm

The second developed program is a little more complex in the sense that the Stitching is done by parts, with the purpose that the work of estimating the homography can be found in the evolutionary algorithm.

In this program, the images are loaded for stitching in a list, then they are passed through an automatic feature detector function where the characteristics of the images are detected, the coordinates of the characteristics of image 1, which we will call source image, are stored in a list of source coordinates, while the coordinates of the image, which we will call destination image, are stored in a list of destination coordinates.

The next step in this algorithm is to calculate the homography matrix with the evolutionary algorithm (ABC), in order to perform this step, the feature lists, source list and destination list are sent by parameters to the ABC algorithm.

The ABC algorithm will be responsible for estimating the homography matrix from the source and destination points, in order to achieve this, it must have an objective function that serves to calculate the matrix. The lists are made up of two coordinates for each characteristic so they are lists of tuples, thus, each list has the

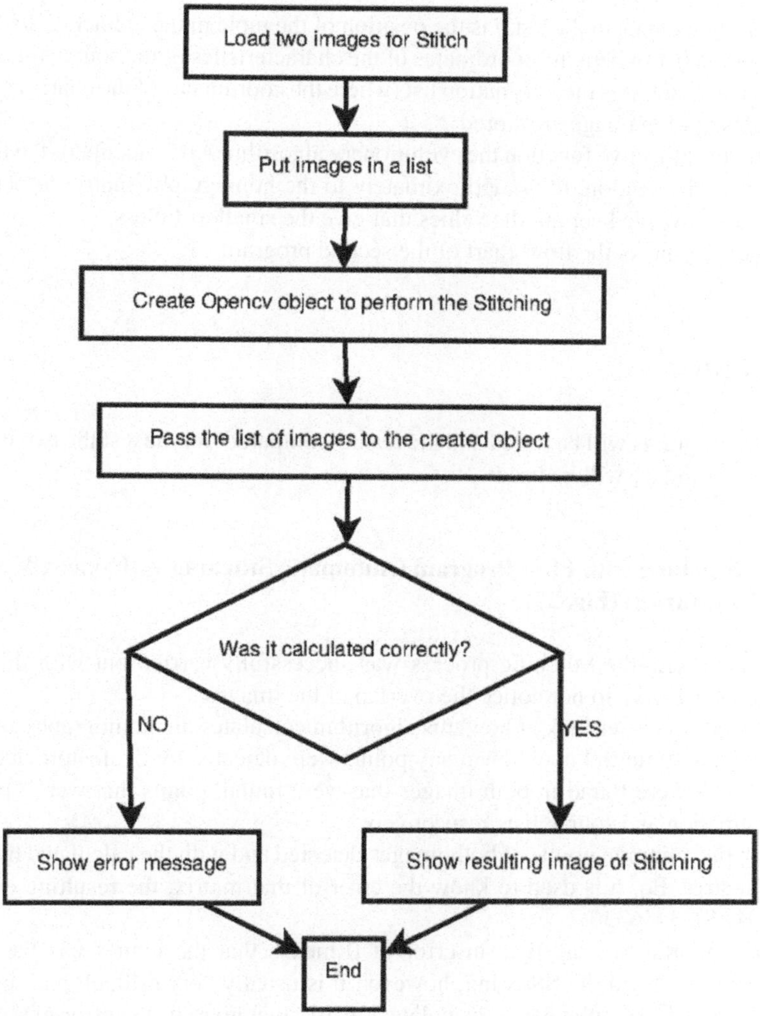

Fig. 18 Flow diagram of the first program

length of the number of characteristics of an image, in each element of this list the X and Y coordinates of that characteristic are stored. Mathematically, the objective function is defined as Eq. 6.

$$\sum_{i=0}^{i=ll} \sqrt[2]{\sum_{j=0}^{j=2} listSour[i][j] - listDest[i][j]} \qquad (6)$$

where ll is the length of the list (both lists have the same length since the number of characteristics of the source image is the same as that of the target image), i the number of elements in the list, *j* is the position of the tuple in the element *i*, *listSourc* is the source list (where the coordinates of the characteristics of the source image are stored) and *listDest* is the destination list (where the coordinates of the characteristics of the destination image are stored).

With this objective function the evolutionary algorithm ABC calculates the fitness value of each iteration, to go approximately to the homography matrix to perform the stitching, while keeping the values that give the smallest fitness.

Figure 19 shows the flow chart of the second program.

6.3 Tests

The two programs will be tested and the results compared, for the test the two images of Figs. 20 and 21 will be used.

6.3.1 Stitching with First Program (Automatic Stitching with openCV Libraries) (Fig. 22)

As you can see, the Stitching process was successfully carried out with this first program since you do not notice the overlap of the images.

Let's take a closer look at how this algorithm calculates the homography matrix, first it is important to know how many points were detected by the feature detector, 1280 points were found in both images that were found, points that were saved in the destination and source lists respectively.

With the common points of both images detected and with the calculated homography matrix, Eq. 6 is used to know the error of that matrix, the resulting error is 19.410474023436336.

It is important to say that an error of 0 means that the matrix has the exact numbers to perform the Stitching, however, it is usually very difficult and in some cases extremely complicated to calculate a matrix that gives us the error of 0 so that an error as close to 0 is what is sought.

6.3.2 Stitching with Second Program (Stitching Step by Step Using the Evolutionary Algorithm) (Fig. 23)

As you can see the small difference between the matrices of homography makes it to be noticed in the seam of the images, the target image is noted more up compared to the source image.

Like the previous algorithm, the objective of this program is to find the homogenous matrix of two images to be able to do a Stitching of them, in order to obtain a matrix that has an error of approximately 0 many tests were done, based on these tests the parameters of the evolutionary algorithm were adjusted.

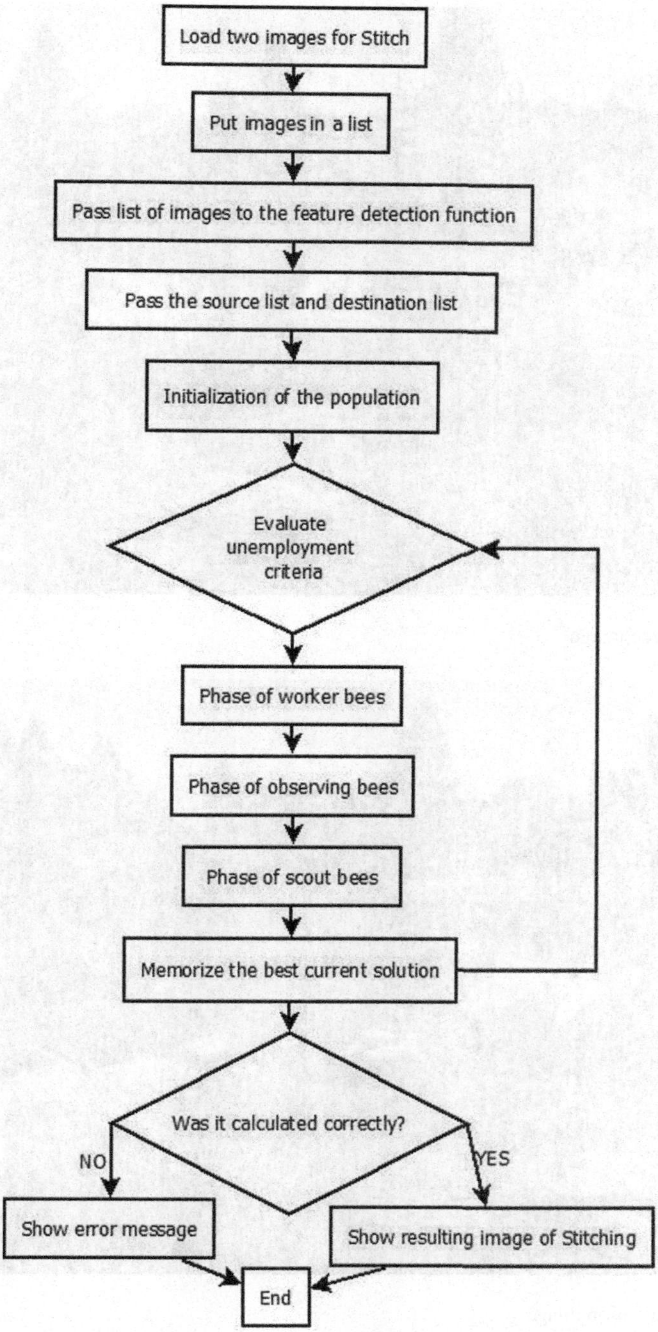

Fig. 19 Flow diagram of the second program

Fig. 20 Source image

Fig. 21 Destination image

Fig. 22 Image stitched with first program

Fig. 23 Image stitched with first program

Some of the most relevant results are presented below (Fig. 24) were calculated by the Hive algorithm (https://rwuilbercq.github.io/Hive/) open source algorithm, whose objective function was modified with Eq. 6 with which we got that the fitness of the algorithm is the error said above that corresponds to the matrix of homography.

As can be seen in Fig. 24, the number of bees in the algorithm plays an important factor, but it does not mean that the more bees the better the result, so it is conjectured that the number of bees is sufficient with the problem dimension number in our case, the homography matrix is made up of 9 digits, so our problem is 9-dimension.

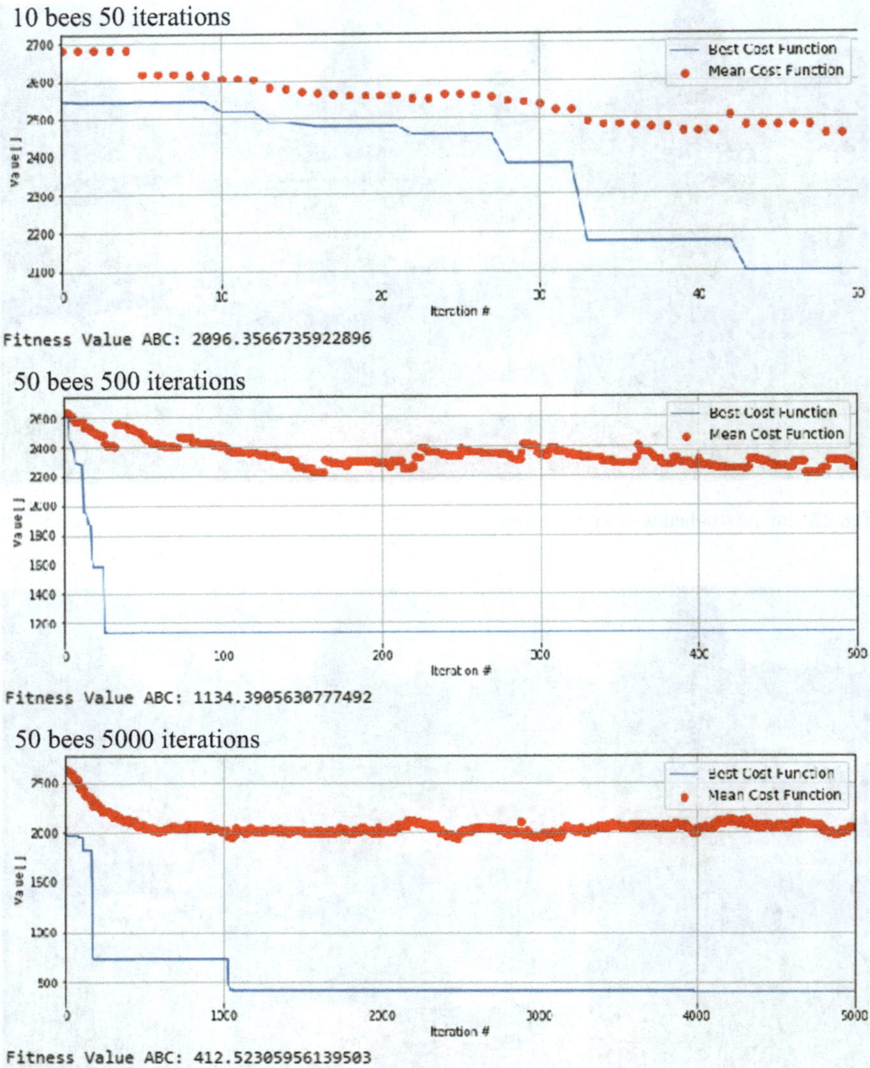

Fig. 24 Tests with the ABC algorithm with different parameters

7 Conclusions

There is no method that claims to have an error of 0 in the estimation of the matrix of homography, the methods tested here proved to be good, and although the matrix of homography of the two methods was similar, it is worth highlighting the time that the algorithm took in comparison to the method used by the openCV library, the

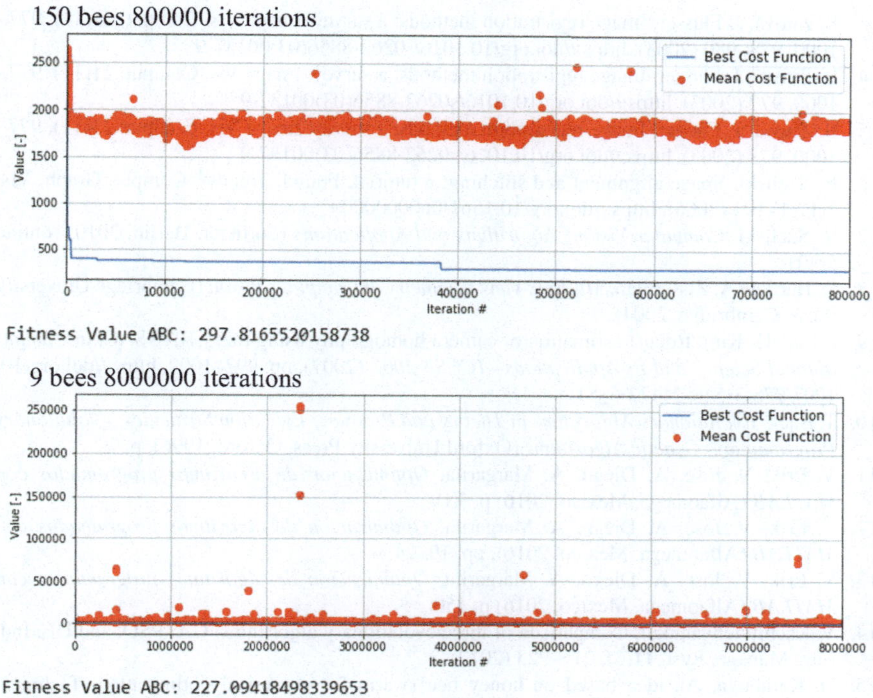

Fig. 24 (continued)

evolutionary algorithm took so much time, as it was iterated in the order of 20 million times each calculating a matrix of homography.

In the results obtained, a fitness of 19.000045345 was found, which is a mistake very similar to that found with openCV libraries but although the matrices are very similar and only a thousandths of difference in the elements of each matrix the results are visibly different, so it is concluded that despite the evolutionary algorithm found good results, it is not convenient to use it to calculate the homography matrix since although the experiment is repeated with the same parameters, it is not certain that it is as exact as in this case, it can be that with fewer iterations you find a better one as well as the opposite.

References

1. P. Brajendra, S. Sudeep, B. Usha, in *Innovations in Computational Intelligence: Best Selected Papers of the Third International Conference on REDSET 2016*. Studies in Computational Intelligence, vol 713 (2016), p. 212
2. S. Mann, R. Picard, Virtual bellows: constructing high-quality images from video, in *Proceedings of the IEEE First International Conference on Image Processing. IEEE International Conference*, 13–16 November 1994 (IEEE, Austin, Texas, 1994)

3. B. Zitová, J. Flusser, Image registration methods: a survey. Image Vis. Comput. **21**(11), 977–1000, 978–980 (2003). https://doi.org/10.1016/s0262-8856(03)00137-9
4. B. Zitová, J. Flusser, Image registration methods: a survey. Image Vis. Comput. **21**(11), 977–1000, 977 (2003). https://doi.org/10.1016/s0262-8856(03)00137-9
5. B. Zitová, J. Flusser, Image registration methods: a survey. Image Vis. Comput. **21**(11), 977–1000, 978 (2003). https://doi.org/10.1016/s0262-8856(03)00137-9
6. R. Szeliski, Image alignment and stitching: a tutorial. Found. Trends® Comput. Graph. Vis. **2**(1), 1–104 (2006). https://doi.org/10.1561/0600000009
7. R. Szeliski, *Computer Vision: Algorithms and Applications* (Springer, Berlin, 2010) (online draft)
8. R. Hartley, A. Zisserman, *Multiple View Geometry in Computer Vision* (Cambridge University Press, Cambridge, 2004)
9. J. Lee, G. Kim, Robust estimation of camera homography using fuzzy RANSAC, in *Computational Science and Its Applications—ICCSA 2007* (2007), pp. 992–1002. https://doi.org/10.1007/978-3-540-74472-6_81
10. T. Back, *Evolutionary Algorithms in Theory and Practice: Evolution Strategies, Evolutionary Programming, Genetic Algorithms* (Oxford University Press, Oxford, 1996), p. 7
11. V. Erik, V. Jose, A. Diego, A. Margarita, *Optimizacion de algoritmos programados con MATLAB* (Alfaomega, Mexico, 2016) p. XIV
12. V. Erik, V. Jose, A. Diego, A. Margarita, *Optimizacion de algoritmos programados con MATLAB* (Alfaomega, Mexico, 2016), pp. 19–24
13. V. Erik, V. Jose, A. Diego, A. Margarita, *Optimizacion de algoritmos programados con MATLAB* (Alfaomega, Mexico, 2016) p. 150.
14. V. Kachitvichyanukul, Comparison of three evolutionary algorithms: GA, PSO, and DE. Ind. Eng. Manage. Syst. **11**(3), 215–223 (2012)
15. D. Karaboga, An idea based on honey bee swarm for numerical optimization. Technical Report—TR06, Erciyes University, Kayseri, Turkey (2005)
16. D. Karaboga, B. Basturk, A powerful and efficient algorithm for numerical function optimization: artificial bee colony (ABC) algorithm. J. Global Optim. **39**, 171–459 (2007)
17. D. Karaboga, B. Basturk, A comparative study of artificial bee colony algorithm. Appl. Math. Comput. **214**, 108–132 (2009)
18. Department of Radiology—Medical Physics, University Medical Center Freiburg DCE-MRI-image-registration_en.png. https://www.uniklinik-freiburg.de/mr-en/research-groups/postprocessing
19. Complete Process to Stitch Images (Image registration methods) RegistrationMethod.png. https://doi.org/10.1016/S0262-8856(03)00137-9
20. Evolutionary Algorithm Avoiding Local Minimums (Simulated-annealing-optimization-of-a-one-dimensional-objective-function.png). https://medium.com/@duoduoyunnini/introduction-implementation-and-comparison-of-four-randomized-optimization-algorithms-fc4d96f9feea

Active Contour Model in Deep Learning Era: A Revise and Review

T. Hoang Ngan Le, Khoa Luu, Chi Nhan Duong, Kha Gia Quach, Thanh Dat Truong, Kyle Sadler and Marios Savvides

Abstract Active Contour (AC)-based segmentation has been widely used to solve many image processing problems, specially image segmentation. While these AC-based methods offer object shape constraints, they typically look for strong edges or statistical modeling for successful segmentation. Clearly, AC-based approaches lack a way to work with labeled images in a supervised machine learning framework. Furthermore, they are unsupervised approaches and strongly depend on many parameters which are chosen by empirical results. Recently, Deep Learning (DL) has become the go-to method for solving many problems in various areas. Over the past decade, DL has achieved remarkable success in various artificial intelligence research areas. DL is supervised methods and requires large volume ground-truth. This paper first provides a fundamental of both Active Contour techniques and Deep

---

T. Hoang Ngan Le (✉) · K. Luu · K. Sadler
Department of Computer Science and Computer Engineering, University of Arkansas, Fayetteville, AR 72701, USA
e-mail: thile@uark.edu

K. Luu
e-mail: khoaluu@uark.edu

K. Sadler
e-mail: krs028@uark.edu

C. N. Duong · K. G. Quach
Department of Computer Science and Software Engineering, Concordia University, Montreal, Canada
e-mail: dcnhan1909@gmail.com

K. G. Quach
e-mail: khagia1001@gmail.com

T. D. Truong
Department of Computer Science, University of Science, Ho Chi Minh, Vietnam
e-mail: thanhdattrg@gmail.com

M. Savvides
Department of Electrical and Computer Engineering, Carnegie Mellon University, 5000 Forbes Ave, Pittsburgh, PA 15213, USA
e-mail: msavvides@andrew.cmu.edu

D. Oliva and S. Hinojosa (eds.), *Applications of Hybrid Metaheuristic Algorithms for Image Processing*, Studies in Computational Intelligence 890,
https://doi.org/10.1007/978-3-030-40977-7_11

Learning framework. We then present the state-of-the-art approaches of Active Contour techniques incorporating in Deep Learning framework.

Keywords Level set · Active contour · Deep learning

1 Introduction

Among numerous segmentation methods developed in last few decades, Active Contour (AC), or Deformable Models, based on variational models and partial differential equations (PDEs), can be considered as one of the most widely used approaches in medical image segmentation. Among many AC-based approaches in the last few decades for image segmentation, variational LS methods [1–12] have obtained promising performance under some constraints, e.g. resolution, illumination, shape, noise, occlusions, etc.

The key idea behind the Active Contour (AC) for image segmentation is to start with an initial guess boundary represented in a form of closed curves i.e. contours C. The curve is then iteratively modified by applying shrink or expansion operations and moved by image-driven forces under given constraints to more accurately detect the object boundaries to the boundaries of the desired objects. The entire process is called contour evolution, denoted as $\frac{\partial C}{\partial t}$ where C is the object boundary (contour). The existing active contour models are generally categorized into two groups, depend on the kind of information used: *edge-based models* [13–16], *region-based models* [17–27]. The first category, *edge-based models*, utilizes image gradient as an additional constraint to stop the contours on the boundaries of the desired objects. For instance, geodesic active contour (GAC) model [14] constructs an gradient stop function to attract the AC to the object boundaries. Later, Li et al. [15] proposes an novel AC model which is able to eliminate the expensive re-initialization procedure by penalizing the deviation of the level set (LS) function. In general, these kinds of models have the ability to handle only images with well-defined edge information.

However, the accuracy in the LS methods dramatically drops when dealing with images collected in the wild. Meanwhile, the recent advanced deep learning (DL) approaches [28–40] have achieved the state-of-the-art performance on various computer vision problems. The goal of the presented paper is to review approaches based on AC framework where the AC contour evolution is combined with deep learning (DL) in different mechanisms such as (i) AC is utilized as a post-processing process while DL plays a role as feature extraction; (ii) AC acts as a segmentation procedure in DL framework; (iii) AC is utilized as a regularization term in a loss function in DL framework; (iv) DL is employed to lean and estimate LS parameters.

2 Active Contour Techniques

2.1 Active Contour—Background

Take the AC proposed by Kass et al. [13] as an example. In this approach, a contour parameterized by arc length s as $C(s) = \{(x(s), y(s)) : 0 \leq s \leq 1\}$. An energy function $E(C)$ can be defined on the contour such as:

$$E(C) = \int_0^1 E_{int} + E_{ext} \tag{2.1}$$

where E_{int} and E_{ext} are the internal energy and external energy functions, respectively. The internal energy function determines the regularity, i.e. smooth shape, of the contour.

$$E_{int}(C(s)) = \alpha|C'(s)|^2 + \beta|C''(s)|^2 \tag{2.2}$$

Here α controls the tension of the contour, and β controls the rigidity of the contour while $C^{'}(s)$ makes the spline act like a membrane (like "elasticity") and $C^{''}(s)$ makes it act like a thin-plate (like "rigidity"). The external energy term determines the criteria of contour evolution depending on the image $\mathbf{I}(x, y)$, and can be defined as in Eq. 2.3.

$$E_{image} = w_{line}E_{line} + w_{edge}E_{edge} + w_{term}E_{term} \tag{2.3}$$

$$E_{line} = \mathbf{I}(x, y) \tag{2.4}$$

$$E_{edge} = -|\bigtriangledown \mathbf{I}(x, y)|^2 \tag{2.5}$$

$$E_{term} = \frac{(\phi_x^2\phi_{yy} - 2\phi_x\phi_y\phi_{xy} + \phi_y^2\phi_{xx})}{|\bigtriangledown \phi|^3} \tag{2.6}$$

The first term is given in Eq. 2.4 and it depends on the sign of w_{line} which guides the snake towards the lightest or darkest nearby contour. The second term defined in Eq. 2.5 attracts the snake to large intensity gradients. The third term E_{term} attracts the snake toward termination of line segments and corners. E_{term} is defined in Eq. 2.6 using curvature of level lines.

The snake provide an accurate location of the edges only if the initial contour is given sufficiently near the desired edges. Moreover, snake cannot detect more than one boundary simultaneously because the snakes maintain the same topology during the evolution stage. However, they are sensitive to image noise and weak boundaries. To overcome those problems, the second category, *region-based models*, uses the statistical information inside and outside AC to guide the curve evolution. Clearly, global region-based models have several advantages over edge-based models such as less sensitivity to image noise and a higher capability to detect weak boundaries (even without boundaries) because they do not use the image gradient. Furthermore, global region-based models are robust to initial contours which means the initial contour can start anywhere in the image. One of the most successful region-based models is the piecewise constant model [17] with the assumption that each image region is statistically homogeneous.

2.2 Implicit Active Contour—Level Set

LS based or implicit AC models have provided more flexibility and convenience for the implementation of AC, thus, they have been used in a variety of image processing and computer vision tasks. The basic idea of the implicit AC is to represent the initial curve C implicitly within a higher dimensional function, called the level set function $\phi(x, y) : \Omega \rightarrow R$, such as:

$$C = \{(x, y) : \phi(x, y) = 0\}, \forall(x, y) \in \Omega \tag{2.7}$$

where Ω denotes the entire image plane. Figure 1 (left) shows the evolution of level set function $\phi(x, y)$, and Fig. 1 (right) shows the propagation of the corresponding contours C.

The evolution of the contour is equivalent to the evolution of the level set function, i.e. $\frac{\partial C}{\partial t} = \frac{\partial \phi(x,y)}{\partial t}$. One of the advantages of using the zero level set is that a contour can be defined as the border between a positive area and a negative area, so the contours can be identified by signed distance function as follows:

$$\phi(x) = \begin{cases} d(x, C)) & \text{if } x \text{ is inside } C \\ 0 & \text{if } x \text{ is on } C \\ -d(x, C)) & \text{if } x \text{ is outside } C \end{cases} \tag{2.8}$$

where $d(x, C)$ denotes the distance from an arbitrary position to the curve.

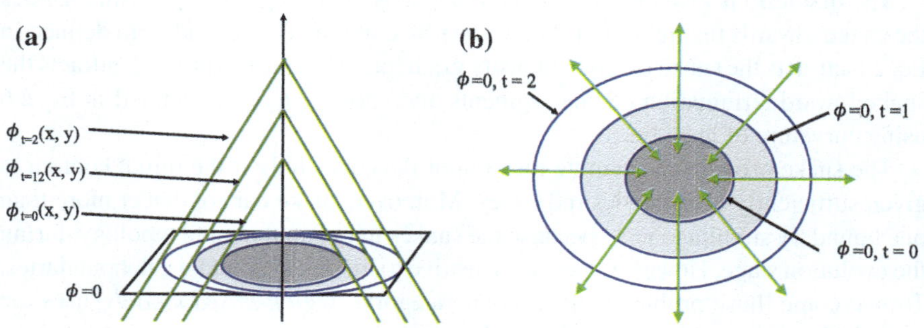

Fig. 1 Level set evolution and the corresponding contour propagation: **a** topological view of level set $\phi(x, y)$ evolution, **b** the changes on the zero level set $C = \{(x, y) : \phi(x, y) = 0\}$

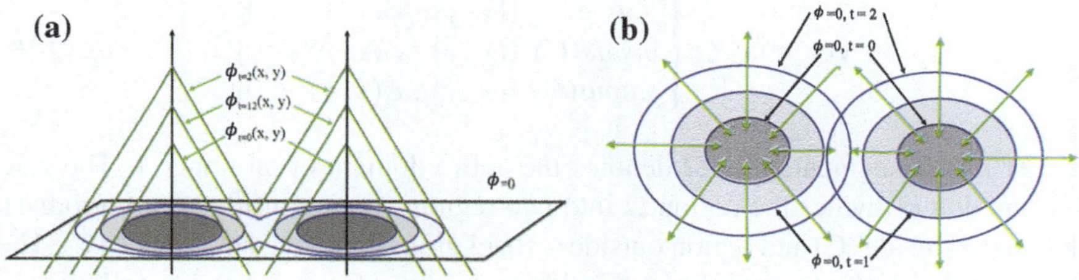

Fig. 2 Topology of level set function changes in the evolution and the propagation of corresponding contours: **a** the topological view of level set $\phi(x, y)$ evolution, **b** the changes on the zero level set $C : \{\phi(x, y) = 0\}$

The LS evolution can be written in the form as follows:

$$\frac{\partial \phi}{\partial t} + F\,|\nabla \phi| = 0 \tag{2.9}$$

where F is a speed function. In some particular cases, F is defined as mean curvature, $F = div\left(\frac{\nabla \phi}{||\nabla \phi||}\right)$.

An outstanding characteristic of LS methods is that contours can split or merge as the topology of the level set function changes. Therefore, LS methods can detect more than one boundary simultaneously, and multiple initial contours can be placed as shown in Fig. 2.

The computation is performed on the same dimension as the image plane Ω, therefore, the computational cost of LS methods is high and the the convergence speed is quite slow.

The process of evolving of the curve C (contour evolution), denoted as $\frac{\partial C}{\partial t}$ and illustrate in Fig. 3. In this example, iterative stages of active contour evolution for the segmentation of objects (cells in this case) in images is shown in red. The underlying image are the level-set function is typically defined over the entire 2D image domain.

Under the scenarios of image segmentation, we consider an image in 2D space, Ω. The LS is to find the boundary C of an open set $\omega \in \Omega$, which is defined as: $C = \partial \omega$. In LS framework, the boundary C can be represented by the zero LS ϕ as follows:

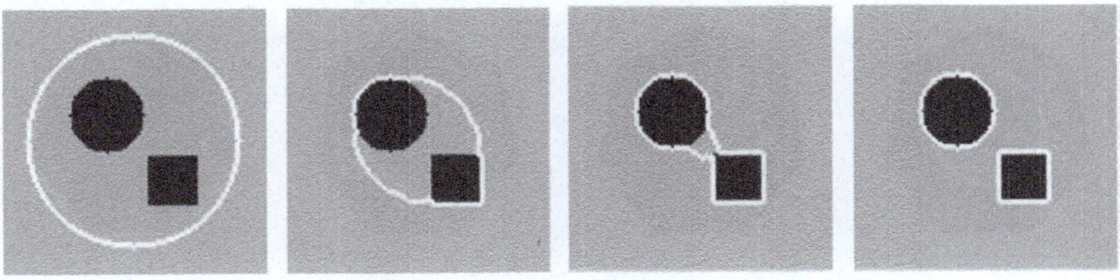

Fig. 3 Example of active contour evolution for image segmentation

$$\forall(x,y) \in \Omega \begin{cases} C = & \{(x,y) : \phi(x,y) = 0\} \\ inside(C) & \{(x,y) : \phi(x,y) > 0\} \\ output(C) & \{(x,y) : \phi(x,y) < 0\} \end{cases} \quad (2.10)$$

For image segmentation, Ω denotes the entire domain of an image **I**. The zero LS function ϕ divides the region Ω into two regions: region inside ω (foreground), denoted as inside (C) and region outside ω (background) denoted as outside (C). The length of the contour C and the area inside the contour C are defined as follows:

$$\begin{aligned} Length(C) &= \int_{\Omega} |\nabla H(\phi(x,y))| dxdy = \int_{\Omega} \delta(\phi(x,y)) |\nabla \phi(x,y)| dxdy \\ Area(C) &= \int_{\Omega} H(\phi(x,y)) dxdy \end{aligned} \quad (2.11)$$

where, $H_{\epsilon}(\cdot)$ is a Heaviside function.

2.3 LS-Based Image Segmentation

Typically, the LS-based image segmentation methods start with an initial level set ϕ_0 and an given image **I**. The LS updating process is performed via gradient descent by minimizing an energy function which defined based on the difference of image features, such as color and texture, between foreground and background. The fitting term in LS model is defined by the inside contour energy (E_1) and outside contour energy (E_2).

$$E = E_1 + E_2 = \int_{inside\,C} (I_{(x,y)} - c_1)^2 dxdy + \int_{outside\,C} (I_{(x,y)} - c_2)^2 dxdy$$

where c_1 and c_2 are the average intensity inside and outside the contour C, respectively. Figure 4 gives an example of all possible cases of the curve. It is easy to see that the energy E is minimized when the contour is right on the object boundary.

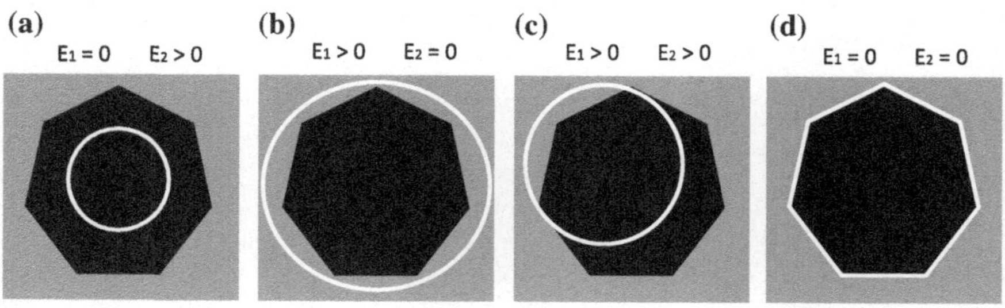

Fig. 4 An illustration of energy inside the contour C (E_1) and outside and the contour C (E_2)

Most region-based AC models consist of two components: regularity and energy minimization. The first part is to determine the smooth shape of contours whereas the second part searches for uniformity of a desired feature within a subset.

One of the most popular region based AC models is proposed by Chan-Vese (CV) [17]. In this model the boundaries are not defined by gradients and the curve evolution is based on the general Mumford-Shah (MS) [19] formulation of image segmentation as shown in Eq. 2.12.

$$E = \int_{\Omega} |\mathbf{I} - u|^2 dxdy + \int_{\Omega/C} |\nabla u|^2 dxdy + \nu Length(C) \tag{2.12}$$

CV's model is an alternative form of MS's model which restricts the solution to piecewise constant intensities and it has successfully segmented an image into two regions, each having a distinct mean of pixel intensity by minimizing the following energy functional.

$$\begin{aligned} E(c_1, c_2, \phi) &= \mu \text{Area}(\omega_1) + \nu \text{Length}(C) \\ &+ \lambda_1 \int_{\omega_1} |\mathbf{I}(x, y) - c_1|^2 dxdy + \lambda_2 \int_{\omega_2} |\mathbf{I}(x, y) - c_2|^2 dxdy \end{aligned} \tag{2.13}$$

where c_1 and c_2 are two constants. The parameters μ, ν, λ_1, λ_2 are positive parameters and usually fixing $\lambda_1 = \lambda_2 = 1$ and $\mu = 0$. Thus, we can ignore the first term in Eq. 2.13. Thus the energy functional is rewritten as follows:

$$\begin{aligned} E(c_1, c_2, \phi) &= \mu \int_{\Omega} H(\phi(x, y)) + \nu \delta(\phi(x, y)) |\nabla \phi(x, y)| dxdy \\ &+ \lambda_1 \int_{\omega_1} |\mathbf{I}(x, y) - c_1|^2 dxdy + \lambda_2 \int_{\omega_2} |\mathbf{I}(x, y) - c_2|^2 dxdy \end{aligned} \tag{2.14}$$

For numerical approximations, the δ function needs a regularizing term for smoothing. In most cases, the Heaviside function H and Dirac delta function δ are defined as in (2.15) and (2.16), respectively.

$$H_\epsilon(x) = \frac{1}{2}\left(1 + \frac{2}{\pi} \arctan\left(\frac{x}{\epsilon}\right)\right) \tag{2.15}$$

$$\delta_\epsilon(x) = H'(x) = \frac{1}{\pi} \frac{\epsilon}{\epsilon^2 + x^2} \tag{2.16}$$

As $\epsilon \to 0$, $\delta_\epsilon \to \delta$, and $H_\epsilon \to H$. Using Heaviside function H, the Eq. 2.14 becomes Eq. 2.17.

$$
\begin{aligned}
E(c_1, c_2, \phi) = \mu H(\phi(x, y)) + \nu\delta(\phi(x, y))|\nabla\phi(x, y)|dxdy \\
+ \lambda_1 \int_\Omega |\mathbf{I}(x, y) - c_1|^2 H(\phi(x, y))dxdy \\
+ \lambda_2 \int_\Omega |\mathbf{I}(x, y) - c_2|^2 (1 - H(\phi(x, y)))dxdy
\end{aligned}
\tag{2.17}
$$

In the implementation, they choose $\epsilon = 1$. For fixed c_1 and c_2, gradient descent equation with respect to ϕ is:

$$
\frac{\partial\phi(x, y)}{\partial t} = \delta_\epsilon(\phi(x, y)[\nu\kappa(\phi(x, y) - \mu - \lambda_1((\mathbf{I}(x, y) - c_1)^2 + \lambda_2((\mathbf{I}(x, y) - c_2)^2]
\tag{2.18}
$$

where δ_ϵ is a regularized form of Dirac delta function and c_1, c_2 are the mean of inside the contour ω_{in} and the mean of the outside of the contour ω_{out}, respectively. The curvature κ is given by:

$$
\kappa(\phi(x, y)) = -div\left(\frac{\Delta\phi}{|\Delta\phi|}\right) = -\frac{\phi_{xx}\phi_y^2 - 2\phi_x\phi_y\phi_{xy} + \phi_{yy}\phi_x^2}{(\phi_x^2 + \phi_y^2)^{1.5}}
\tag{2.19}
$$

where $\partial_x\varphi_t, \partial_y\varphi_t$ and $\partial_{xx}\varphi_t, \partial_{yy}\varphi_t$ are the first and second derivatives of φ_t with respect to x and y directions. For fixed ϕ, gradient descent equation with respect to c_1 and c_2 are:

$$
\begin{aligned}
c_1 = \frac{\sum_{x,y} \mathbf{I}(x, y)H(\phi(x, y))}{\sum_{x,y} H(\phi(x, y))} \\
c_2 = \frac{\sum_{x,y} \mathbf{I}(x, y)(1 - H(\phi(x, y)))}{\sum_{x,y}(1 - H(\phi(x, y)))}
\end{aligned}
\tag{2.20}
$$

herein, we use notation φ_t to indicate φ at the iteration tth in order to distinguish the φ at different iterations. Under this formulation, the curve evolution is shown as a time series process which helps to give better visualization of reformulating LS. From this point, we redefine the curve updating in a time series form for the LS function φ_t as in Eq. (2.21).

$$
\varphi_{t+1} = \varphi_t + \eta\frac{\partial\varphi_t}{\partial t}
\tag{2.21}
$$

The LS at time $t + 1$ depends on the previous LS at time t and the curve evolution $\frac{\partial\varphi_t}{\partial t}$ with a learning rate η.

2.4 State-of-the-Art AC Methods

Li et al. [21] solved the problem of segmenting images with intensity inhomogeneity by using a local binary fitting energy. By minimizing the unbiased pixel-wise average misclassification probability, Wu et al. [41] formulated an active contour to segment an image without any prior information about the intensity distribution of regions. By realizing curve evolution via simple operations between two linked lists, Shi and Karl [42] achieved a fast level set algorithm for real-time tracking. Also, they incorporated the smoothness regularization with the use of a Gaussian filtering process and proposed the two-cycle fast (TCF) algorithm to speed up the level set evolution.

To overcome the limitation of classic LS being binary-phase segmentation, Samson et al. [1] associated a LS function with each image region, and evolves these functions in a coupled manner. Later, Brox and Weickert [2] performed hierarchical segmentation by iteratively splitting previously obtained regions using the CLS. To deal with reinitialization, DRLSE [4] is proposed with a new variational level set formulation in which the regularity of the LS function is intrinsically maintained during the LS evolution. Different from the aforementioned methods that work on global, Li et al. [5] focused on intensity inhomogeneity which often occurs in real-world images. In their approach, they derived a local intensity clustering property and defined a local clustering criterion function in a neighborhood of each point. Lucas [6] suggested using a single LS function to perform the LS evolution for multi-region segmentation. It requires managing multiple auxiliary LS functions when evolving the contour, so that no gaps/overlaps are created. Bae and Tai [3] proposed to divide an image into multiple regions by a single, piecewise constant LS function, obtained using either augmented Lagrangian optimization, or graph-cuts. Later, to deal with local minima, LSE [10] uses some form of local processing to tackle intra-region inhomogeneity, which makes such methods susceptible to local minima. Recently, Dubrovina et al. [43] have developed a multi-region segmentation with single LS function. However, Dubrobina et al.'s approach was developed for contour detection and needs good initialization, namely, it requires specify more initial regions than it is expected to be in the final segmentation. Furthermore, the algorithm requires that initial contours cover the image densely, specifically the initial contour has to pass through all different regions. In addition to multi region segmentation problem, optimization [8, 12, 44], and shape prior [45] have also been considered.

Generally, the LS model minimizes a certain energy function via gradient descent [46], making the segmentation results prone to getting stuck in local minima. To conquer this problem, Chan et al. [47] restated the traditional Mumford-Shah image segmentation model [19] as a convex minimization problem to obtain the global minimum . The above methods have obtained promising performance in segmenting high quality images. However, when attempts are made to segment images with heavy noise, this leads to poor segmentation results. Existing methods assume that pixels in each region are independent when calculating the energy function. This underlying assumption makes the contour motion sensitive to noise. In addition, the implementation of level set methods is complex and time consuming, which limits their application to large scale image databases. To maintain numerical stability, the numerical scheme used in level set methods, such as the upwind scheme or finite difference scheme, must satisfy the Courant-Friedrichs-Lewy (CFL) condition [48], which limits the length of the time step in each iteration and wastes time.

Some limitations of AC-based approaches:

- They are unsupervised approaches and therefore require no learning properties from the training data. Thus, they have difficulty in dealing with noise and occlusions.
- While these AC methods offer object shape constraints, they typically look for strong edges or statistical modeling for successful segmentation. These techniques lack a way to work with labeled images in a supervised machine learning framework.
- There are many parameters which are chosen by empirical results.
- They are build off of gradient descent to implement the non-convex energy minimization and can get stuck in undesired local minima and thereby lead to erroneous segmentation.
- Most of the level set based approaches are not able to robustly segment images in the wild.
- They often give unpredictable segmentation results due to unsupervised behaviors.
- The accuracy of segmenting results strongly depends on the number of iterations which is usually set as a big number.

3 Deep Learning

Recently, deep learning is gaining attention as a technique for realizing artificial intelligence. This technique is known to be useful for recognition or categorization of images, speech recognition, and natural language processing. An artificial neural network is utilized as one of the methods of conventional machine learning techniques, and a deeply stacked artificial neural network is utilized in deep learning. Several types of deep neural networks exist, such as convolutional neural networks (CNNs) and recurrent neural networks (RNNs).

3.1 Multi-Layer Neural Network

Deep learning models, in simple words, are large and deep artificial neural networks. Let us consider the simplest possible neural network which is called "**neuron**" as illustrated in Fig. 5a. A computational model of a single neuron is called a perceptron which consists of one or more inputs, a processor, and a single output.

In this example, the neuron is a computational unit that takes $\mathbf{x} = [x_1, x_2, x_3]$ as input, the intercept term $+1$ as bias $\mathbf{b}$, and the output $\mathbf{o}$. The gold of this simple network is to learn a function $f : \mathrm{R}^{\mathrm{N}} \rightarrow \mathrm{R}^{\mathrm{M}}$ where N is the number of dimensions for input $\mathbf{x}$ and M is the number of dimensions for output which is computed as $\mathbf{o} = f(\mathbf{W}, \mathbf{x})$. Mathematically, the output $\mathbf{o}$ of a one output neuron is defined as:

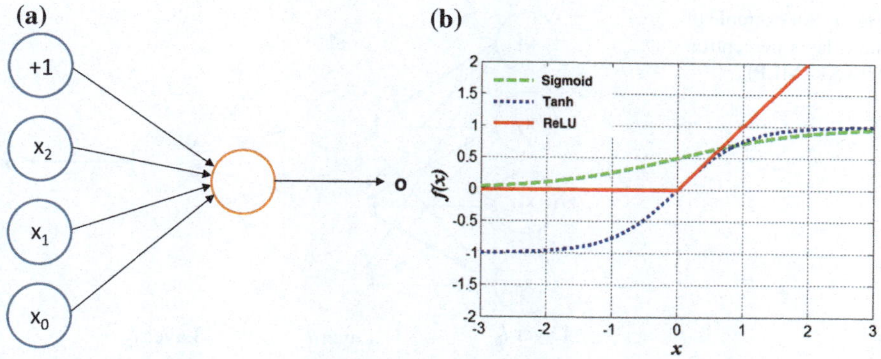

Fig. 5 **a** An example of one neuron which takes input $\mathbf{x} = [x_1, x_2, x_3]$, the intercept term $+1$ as bias, and the output **o**; **b** Plot of different activation functions, i.e. *Sigmoid*, *Tanh* and rectified linear (*ReLU*) functions

$$\mathbf{o} = f(\mathbf{x}, \theta) = \sigma\left(\sum_{i=1}^{N} w_i x_i + b\right) = \sigma(\mathbf{W}^T \mathbf{x} + b) \tag{3.1}$$

In this equation, σ is the point-wise non-linear activation function. The common non-linear activation function for hidden units are chosen as a hyperbolic tangent (*Tanh*) or logistic sigmoid as shown in Eq. 3.4. A different activation function, the rectified linear (*ReLU*) function, has been proved to be better in practice for deep neural networks. This activation function is different from *Sigmoid* and (*Tanh*) because it is not bounded or continuously differentiable. Furthermore, when the network goes very deep, ReLU activations are popular as they reduce the likelihood of the gradient to vanish. The rectified linear activation (*ReLU*) function is given by Eq. 3.4. These functions are used because they are mathematically convenient and are close to linear near origin while saturating rather quickly when getting away from the origin. This allows neural networks to model well both strongly and mildly nonlinear mappings. Figure 5 is the plot of *Sigmoid*, *Tanh* and rectified linear (*ReLU*) functions.

$$Sigmoid(\mathbf{x}) = \frac{1}{1 + exp^{-x}} \tag{3.2}$$

$$Tanh(\mathbf{x}) = \frac{exp^{2x-1}}{exp^{2x+1}} \tag{3.3}$$

$$ReLU(\mathbf{x}) = max(0, x) \tag{3.4}$$

Notably, the system becomes linear with matrix multiplications if removing the activation function. The *Tanh* activation function is actually a rescaled version of the *sigmoid*, and its output range is $[-1, 1]$ instead of $[0, 1]$. The rectified linear function is piece-wise linear and saturates at exactly 0 whenever the input is less than 0.

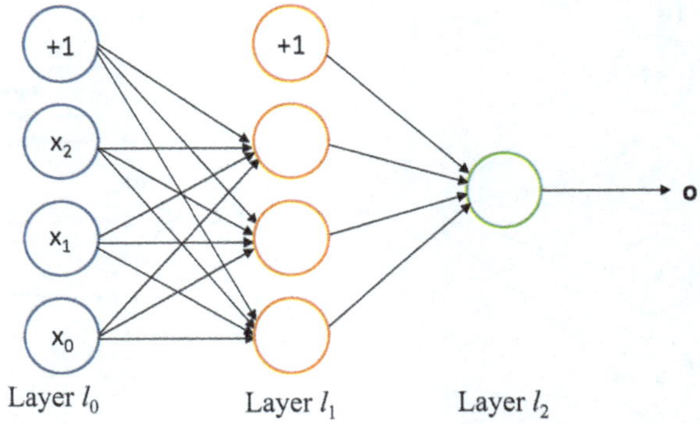

Fig. 6 An example of multi-layer perceptron network (MLP)

A neural network is composed of many simple "neurons," so that the output of a neuron can be the input to another. An special case of a neural networks is also called multi-layer perceptron network (MLP) and illustrated in Fig. 6.

A typical neural network is composed of one input layer, one output layer and many hidden layers. Each layer may contains many units. In this network, $\mathbf{x}$ is the input layer, $\mathbf{o}$ is the output layer. The middle layer is called hidden layer. In Fig. 6, the neural network contains 3 units of input layers, 3 units of hidden layer, and 1 unit of output layer.

In general, we consider a neural network with L hidden layers of units, one layer of input units and one layer of output units. The number of input units is N, output units M, and units in hidden layer l is N^l. The weight of the jth unit in layer l and the ith unit in layer $l+1$ is denoted by w_{ij}^l. The activation of the ith unit in layer l is $\mathbf{h}_i^l$. The input and output of the network are denoted as $\mathbf{x}(n)$, $\mathbf{o}(n)$, respectively, where n denotes training instance, not time.

3.2 Convolutional Neural Networks

Convolutional Neural Networks (CNNs) [49, 50] are a special case of fully connected multi-layer perceptrons that implement weight sharing for processing data that has a known, grid-like topology (e.g. images). CNNs use the spatial correlation of the signal to constrain the architecture in a more sensible way. Their architecture, somewhat inspired by the biological visual system, possesses two key properties that make them extremely useful for image applications: spatially shared weights and spatial pooling. These kind of networks learn features that are shift-invariant, i.e., filters that are useful across the entire image (due to the fact that image statistics are stationary). The pooling layers are responsible for reducing the sensitivity of the output to slight input shift and distortions. Since 2012, one of the most notable results in Deep Learning is the use of convolutional neural networks to obtain a remarkable improvement in object recognition for ImageNet classification challenge.

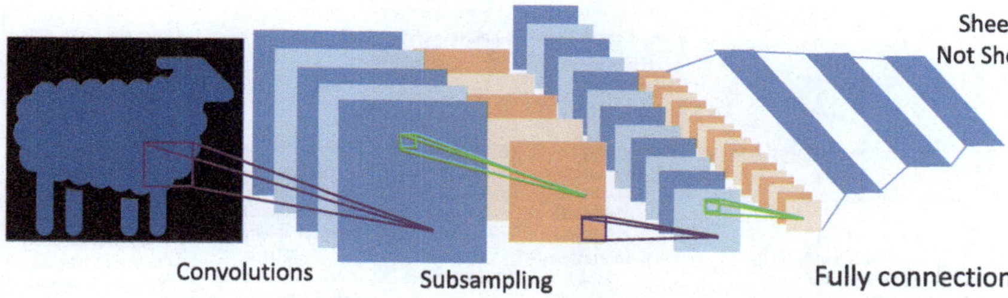

Fig. 7 Architecture of a typical convolutional network for image classification containing three basic layers: convolution layer, pooling layer and fully connected layer

A typical convolutional network is composed of multiple stages, as shown in Fig. 7. The output of each stage is made of a set of 2D arrays called feature maps. Each feature map is the outcome of one convolutional (and an optional pooling) filter applied over the full image. A point-wise non-linear activation function is applied after each convolution. In its more general form, a convolutional network can be written as:

$$\begin{aligned} \mathbf{h}^0 &= \mathbf{x} \\ \mathbf{h}^l &= pool^l(\sigma_l(\mathbf{w}^l\mathbf{h}^{l-1} + \mathbf{b}^l)), \forall l \in 1, 2, ...L \\ \mathbf{o} = \mathbf{h}^L &= f(\mathbf{x}, \theta) \end{aligned} \tag{3.5}$$

where $\mathbf{w}^l$, $\mathbf{b}^l$ are trainable parameters as in MLPs at layer l. $\mathbf{x} \in \mathrm{R}^{c\times h\times w}$ is vectorized from an input image with c is color channels, h is the image height and w is the image width. $\mathbf{o} \in \mathrm{R}^{n\times h'\times w'}$ is vectorized from an array of dimension $h' \times w'$ of output vector (of dimension n). $pool^l$ is a (optional) pooling function at layer l.

The main difference between MLPs and CNNs lies in the parameter matrices $\mathbf{w}_l$. In MLPs, the matrices can take any general form, while in CNNs these matrices are constraints to be Toeplitz matrices. That is, the columns are circularly shifted versions of the signal of various shifts for each columns in order to implement spatial correlation operation using matrix algebra. Moreover, these matrices are very sparse since the kernel is usually much smaller than the input image. Therefore, each hidden unit array $\mathbf{h}_l$ can be expressed as a discrete-time convolution between kernels from $\mathbf{w}_l$ and the previous hidden unit $\mathbf{h}_{l-1}$ (transformed through a point-wise non-linearity and possibly pooled). There are numerous variants of CNNs architectures in the literature. However, their basic components are very similar which contains of convolutional layer, pooling layer, activation layer and fully connected layer.

CNNs have achieved state-of-the-art performance in various domains in computer vision. *Object detection* has been a long-standing and important problem in computer vision [51]. Object proposal based methods attract a lot of interests and are widely studied in the literature [52, 53]. R-CNN [53] uses Selective Search (SS) to extract around 2000 bottom-up region proposals that are likely to contain objects. Spatial pyramid pooling network (SPP net) [54], Fast-RCNN, Faster R-CNN [55] are improvements of R-CNNs. Based on object detection, many successfully semantic

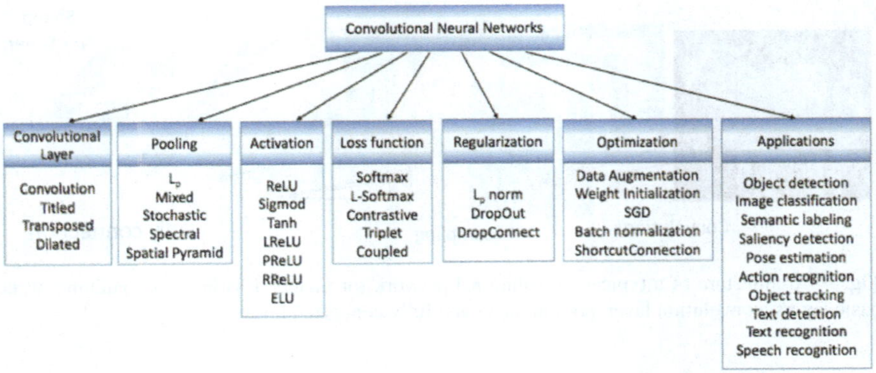

Fig. 8 Summary on the works and techniques related to implementing a CNN architecture

instance segmentation models have been proposed such as MNC [56], Mask R-CNN [57], DeepLab [37], CRLS [58], DRLS [59]. Scene understanding and parsing inherits all the merits of CNNs, specially fully convolutional (FCNs) [28], SegNet [36], PSPNet [39], CRRN [60], CNNs have been applied in *image classification* for a long time [61]. Compared to traditional methods, CNNs achieve better classification accuracy on large scale datasets [62]. With large number of classes, proposing a hierarchy of classifiers is a common strategy for image classification [63]. *Visual tracking* is an another application that turns the CNNs model from a detector into a tracker [64]. As an special case of image segmentation, *saliency detection* is another computer vision application that uses CNNs [65, 66]. In additional to the previous applications, *pose estimation* [67] is another interesting research that uses CNNs to estimate human-body pose. *Action recognition* in both still images and in videos are special case of recognition and are challenging problems. Gkioxari et al. [68] utilizes CNN-based representation of contextual information in which the most representative secondary region within a large number of object proposal regions together the contextual features are used to describe the primary region. CNNs-based action recognition in video sequences are reviewed in [69]. *Text detection and recognition* using CNNs is the next step of optical character recognition (OCR) [70], word spotting [71]. Going beyond still images and videos, *speech recognition, speech synthesis* is also an important research field that have been improved by applying CNNs [72, 73]. In short, CNNs have made breakthroughs in many computer vision areas i.e image, video, speech and text. The summary on the works related to CNNs is given in Fig. 8.

3.3 Recurrent Neural Networks

3.3.1 Vanilla Recurrent Neural Networks

The Recurrent Neural Networks (RNNs) is an extremely powerful sequence model and was introduced in the early 1990s [74]. A typical RNNs contains three parts, namely, sequential input data ($\mathbf{x}_t$), hidden state ($\mathbf{h}_t$) and sequential output data ($\mathbf{o}_t$) as shown in Fig. 9.

RNNs make use of sequential information and perform the same task for every element of a sequence where the output is dependent on the previous computations. The activation of the hidden states at timestep t is computed as a function f of the current input symbol $\mathbf{x_t}$ and the previous hidden states $\mathbf{h_{t-1}}$. The output at time t is calculated as a function g of the current hidden state $\mathbf{h_t}$ as follows:

$$\begin{aligned} \mathbf{h}_t &= f(\mathbf{U}\mathbf{x}_t + \mathbf{W}\mathbf{h}_{t-1}) \\ \mathbf{o}_t &= g(\mathbf{V}\mathbf{h}_t) \end{aligned} \tag{3.6}$$

where $\mathbf{U}$ is the input-to-hidden weight matrix, $\mathbf{W}$ is the state-to-state recurrent weight matrix, $\mathbf{V}$ is the hidden-to-output weight matrix. f is usually a logistic sigmoid function or a hyperbolic tangent function and g is defined as a softmax function.

Most work on RNNs has made use of the method of backpropagation through time (BPTT) [75] to train the parameter set ($\mathbf{U}$, $\mathbf{V}$, $\mathbf{W}$) and propagate error backward through time. In classic backpropagation, the error or loss function is defined as:

$$E(\mathbf{o}, \mathbf{y}) = \sum_t ||\mathbf{o}_t - \mathbf{y}_t||^2 \tag{3.7}$$

where $\mathbf{o}_t$ is prediction and $\mathbf{y}_t$ is labeled groundtruth.

For a specific weight $\mathbf{W}$, the update rule for gradient descent is defined as $\mathbf{W}^{new} = \mathbf{W} - \gamma\frac{\partial E}{\partial \mathbf{W}}$, where γ is the learning rate. In RNNs model, the gradients of the error with respect to our parameters $\mathbf{U}$, $\mathbf{V}$ and $\mathbf{W}$ are learned using Stochastic Gradient Descent (SGD) and chain rule of differentiation.

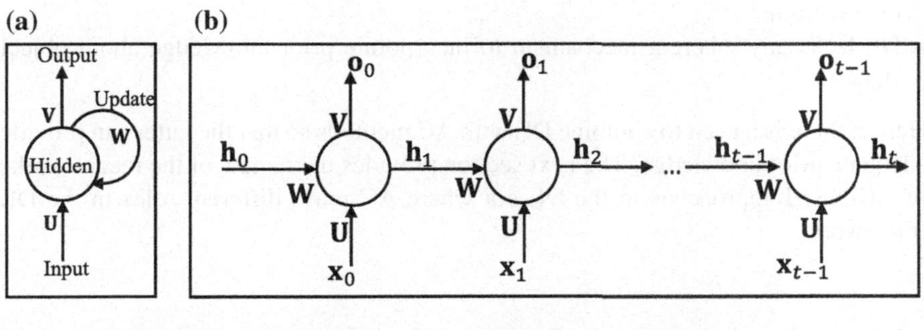

Fig. 9 An RNNs and the unfolding in time of the computation involved in its forward computation

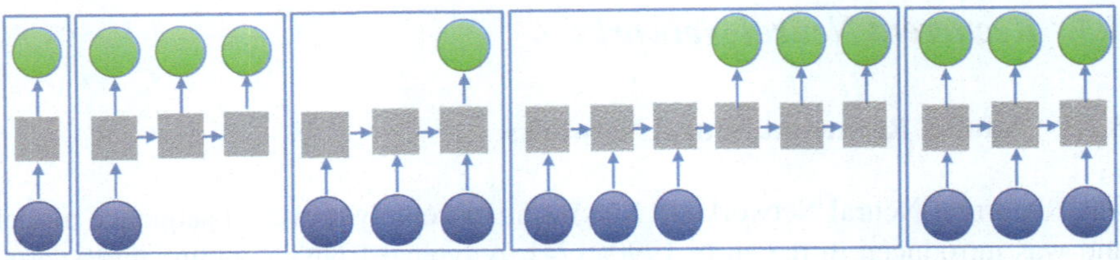

Fig. 10 From left to right: fixed-sized input to fixed-sized output (e.g. image classification); fixed-sized input (e.g. image) and sequence output (e.g. a set of words in image captioning); sequence input (set of words in a sentence) and sequence output (e.g machine translation); Synced sequence input and sequence output (e.g label every frame in a video)

In practice, there are different ways to design a RNNs architecture. In a simple case, RNNs accepts a fixed-sized vector as input (e.g. an image) and produce a fixed-sized vector as output (e.g. probabilities of different classes) as given in the first model in Fig. 10. More flexible, RNNs can have various architecture as shown in Fig. 10.

The difficulty of training an RNNs to capture long-term dependencies has been studied in [76]. To address the issue of learning long-term dependencies, Hochreiter and Schmidhuber [77] proposed Long Short-Term Memory (LSTM), which is able to maintain a separate memory cell inside it that updates and exposes its content only when deemed necessary. Recently, a Gated Recurrent Unit (GRU) was proposed by [78] to make each recurrent unit adaptively capture dependencies of different time scales. Like the LSTM unit, the GRU has gating units that modulate the flow of information inside the unit, but without having separate memory cells.

Several variants of RNNs have been later introduced and successfully applied to wide variety of tasks, such as natural language processing [9, 79], speech recognition [80, 81], machine Translation [82, 83], question answering [84], image captioning: [85, 86], and many more.

Limitations of DL

- DL is trained in a supervised manner, which requires large number of high quality ground-truth.
- DL relies on the volume of labeled images to implicitly learn about object shapes or constraints.
- DL lacks any inherent mechanism to incorporate prior knowledge about object shapes.

Hence, there is a need to combine DL with AC methods so that the latter can provide adequate prior knowledge. The next section provides the review of the recent works of AC-based approaches in the DL era where AC plays different roles in the DL framework.

4 Active Contour in Deep Learning

In order to inherit the merits of both deep learning and active contour as well as overcome the aforementioned limitations, numerous approaches have been proposed. Active contour has been utilized at different stage with different purposes such as (1) active contour works as a post-processing step after deep neural network; (2) the energy minimization in active contour is used as a loss function inside a deep framework; (3) active contour replaces the fully connected layer for segmentation task (4) deep learning is used to learn parameters in active contour. The following section reviews all the state-of-the-art works of active contour in the deep learning era.

4.1 Active Contour as Post-processing

The earliest works on employing neural networks as an optimization framework in order to evolve the contour were proposed by Villarino et al. [87] or Rekeczky et al. [88]. The employed networks, however, are neither deep, nor are used for training features of certain object categories. Later, Chen et al. [89] proposed a shape-driven approach using deep Boltzmann machine (DBM) to learn the hierarchical architecture of shape priors for object segmentation. Their approach is based on observation that shape is represented implicitly by signed distance functions (SDF) in LS method. However, SDF for shape representation are not closed under linear operations, e.g., the mean shape and linear combinations of training shapes are typically no longer SDF. In the proposed network, they make use of DBM for modeling object shapes by first employing DBM to extract the hierarchical architecture of shapes in the training set. It is then introduced into the variational framework as a shape prior term in an energetic form. By minimizing the proposed objective functional, this model is able to constrain an evolutional shape to follow global shape consistency while preserving its ability to capture local deformations. Sharing similar idea which first learns the shape prior using deep structured inference using a deep belief network (DBN), Ngo and Carneiro [90] propose a methodology for lung segmentation using a hybrid method based on a combination of distance regularized level set DRLS [21] and deep structured inference. Such combination aims at exploring the advantages of both approaches, which are the efficient optimization and the prior shape and appearance terms from DRLS, and the robust statistical segmentation models produced by deep learning methods. Continue making use of DBM to learn prior shape which can satisfy the global deformation and local deformation, Wu et al. [91] combined DBM and LS method to extract face region. By using of LS method and the local Gaussian distribution fitting energy, the image energy term is represented by local mean and local variance of image and the prior shape energy is introduced to construct the final energy term of image segmentation model. Images is segmented under different prior shapes, intensity inhomogeneity and partial occlusion and has the advantages in terms of computational accuracy and efficiency. To obtain better contour using trained CNN feature, Rupprecht et al. [92] relied on an explicit contour representation and sample small patches at each point on the evolving contour. In this approach, they train a CNN which predicts a vector feature for the respective point on the evolving contour towards the closest point on the actual boundary of the object of interest. These predictions form a vector field which is then used for evolving the contour. This work is strongly based on the initial seeds (point), patch center, where CNN is applied onto. Employing similar idea of using CNN for feature extraction and LS for segmentation, Bupphawat et al. [93] proposed a network framework for super- resolution land cover mapping on remote sensing images. In this network, CNN is used to find the probabilities that a subpixel belonging to a land cover class, and the LS M is employed to fine tune the boundaries among land cover classes. Not only on image segmentation, active contour also is utilized to improve the performance of saliency detection. Deep learning has been applied to saliency detection, but it may output maps with blurred saliency and inaccurate boundaries. To tackle such an issue, Hu et al. [94] proposes a deep LS network which drives the network to learn a LS function for salient objects so it can output more accurate boundaries and compact saliency. In this proposed method, they also extend a superpixel-based guided filter to be a layer in the network in order to propagate saliency information among pixels and recover full resolution saliency map. That means, different from previous work on saliency detection which trained binary groundtruth directly, Hu et al.'s method was trained to learn a level set function. Active contour is not only used as post-processing of a CNN networks, it also used to refine the segmenting results in combining with Deep Belief Network (DBN). As proposed in [95], distance regularised level sets [4] is employed to takes the estimated ROI from DBN. The proposed method makes used of an advantage of using DBN models that needs of smaller training sets. This merit is appropriate for many medical applications where available training sets is not big. In [95], DBN is used as a detector providing a ROI where the initial contour is obtained by Otsu's thresholding. Based on the observation that in absence of a regularizing prior and by disregarding domain specific image features, most DL approach fail to address the intricate details of the object boundary, Singhal et al. [96] proposed a hybrid variational curve propagation model which embeds a deep learned endometrium probability map in the segmentation energy functional. The model is initialized by Unet and the segmentation cost function consists of an image based external energy for curve propagation computed via specially designed endometrium plateness function, which selectively enhances the object from the background clutter. Variational LS methods have been successfully applied into medical imaging. Therefore, combining CNN and LS methods has been considered by many researcher to solve medical imaging. Cho et al. [97] proposed CNN-TD based segmentation approach. In this approach, a CNN-based segmentation scheme is first employed to obtain a feature map which then segmented by a topological derivative (TD)-based scheme. In the processing and analysis of retinal fundus images, the optic disc (OD), as the main anatomical structure of ocular fundus, its shape, border, size and pathological depression are very important auxiliary parameters for the diagnosis of fundus diseases. In order to automatically localizing and segmenting OD, Faster R-CNN [55] and LS method are employed in [98]. Faster R-CNN model is used to locate the OD via a bounding box. The main blood vessels in the bounding box are removed by Hessian matrix if necessary. Finally, a shape constrained level set algorithm is used to segment the boundary of the OD. Working on segmenting cardiac MR, Duan et al. [99] proposed deep nested level set (DNLS) to obtain the simultaneous probability maps over region and edge locations in CMR images using a FCN with 2 losses: softmax cross-entropy and class-balanced sigmoid cross- entropy. These probability maps can then be incorporated in a single nested LS optimisation framework to achieve multi-region segmentation with high efficiency. In DNLS, an initialisation of the level set function can be readily derived from the learned feature thus DNLS is fully automated. Table 1 summarizes all approaches which utilize DL as either extracting feature maps (probability maps) or learning prior and make use of AC methods (LS methods) as post processing for object segmentation.

4.2 Active Contour Is Used Within DL as End-to-End Framework

Different than using the LS model as a post-processing tool, Min et al. [117], integrated LS into the training phase to fine-tune the FCN in order to incorporate smoothing and prior information to achieve an accurate segmentation. The proposed method allows the use of unlabeled data during training in a semi-supervised setting. Far apart from the previous category where LS is utilized as post-processing and makes use of generative models hat is not trained by the joint model, Min, et al.'s method propose an integrated FCN-LS model that iteratively refines the FCN using a level set module. In order to integrates priors and constraints into the segmentation process [118], proposed Deep Structured Active Con- tours (DSAC) to combine the expressiveness of deep CNNs with the versatility of AC model. This network leverages the original AC model formulation by learning high-level features and prior parameterizations, including the balloon term of AC model. The balloon term is then converted to the energy formulation. The optimization of the AC model is casted as a structured prediction problem and find optimal features and parameters using a Structured Support Vector Machine. DSAC employs a CNN to learn the energy function that would allow an AC model to generate polygons close to a set of ground truth instances. DSAC is trained in an end-to-end framework and applied to learn geometric properties of buildings. The state-of-the-art end-to-end framework employing LS and DL-based object detector has been utilized in CRLS by Le et al.'s work [58] where the authors reformulate LS evolution process under a RNN network. Level set method is hardly applied in multi-class images, thus they utilize an object detection network to obtain single object images. The entire system is designed under cascade structure with three stages of localizing object detection, estimating the mask and classing object. The proposed CRLS is trained in end-to-end framework and has successfully solved the

Table 1 Summary of approaches which utilize DL as either extracting feature maps (probability maps) or learning prior shape and make use of AC methods (LS methods) as post processing for segmentation task. Deep Boltzmann Machine (DBM), Deep Belief Network (DBN), Multilayer perceptron (MLP)

	Net.	App.	Technique	Datasets
[89]	DBM [100]	Object	• DBM is to model object shapes	MPEG-7 [101]
			• The shape prior is used in an energetic form	Person [102]
[90]	DBN [103]	Lung	• Combine distance regularized LS and deep structured inference	JSRT [104]
[91]	DBM [100]	Face	• DBM is to learn prior shape	MSRC [105]
			• Variational LS and local Gaussian distribution are to segment	
[92]	AlexNet [106]	Medical object	• DL is to learn patch-based representation	STACOM [107]
			• AC framework is to segment	PASCAL [108]
[93]	MLP	Remote sensing	• CNN is to find the probabilities	QUICKBIRD [109]
			• LS is to fine tune boundaries	
[94]	VGG [110]	Saliency	• DL is for feature map extraction	PASCAL [108]
			• Use superpixel-based guided filter	SED2 [111]
			• LS function for salient objects	THUR [112] etc.
[95]	DBN [103]	Left ventricle	• DBN is to model variations	MICCAI 09 [113]
			• Otsu's thresholding is for initialization training	
[96]	Unet [114]	Endometrium	• Unet is to learn endometrium probability map	Internal
			• Probability map is used as a soft shape prior in variational functional	
[97]	VGG [110]	Prostate	• CNN-based is learn feature map TD is to refine	Protate [115]

(continued)

Table 1 (continued)

	Net.	App.	Technique	Datasets
[98]	Faster RCNN +ZF [55]	Fundus	• Faster R-CNN model is to locate a bounding box	MESSIDOR [116]
			• Hessian matrix is to remove main blood vessels	
			• Shape constrained LS is to segment	
[99]	Unet [114]	Cardiac MR	• DL is to estimate simultaneous probability maps over region and edge location	Internal
			• Probability maps is incorporated in LS optimisation	

semantic instance segmentation by taking advantages of both CNN and LS methods. Ordinary Differential Equations (ODEs) and Partial Differential Equations (PDEs) are an important parts when working with AC. There has been a number of attempts to solve ODEs and PDEs using artificial intelligence methods. In [119], the authors tested feed forward neural networks on several ODEs and PDEs. According to their test, neural network gives even better results than finite element method, on both accuracy and efficiency. In [120], NN and stochastic simulation were combined to solve viscoelastic flow problem. That means, it uses a "universal approximator" based on neural network methodology in combination with the kinetic theory of polymeric liquid in which the stress is computed from the molecular configuration rather than from closed form constitutive equations. In Baymani et al. [121], tried to solve Navier-Stokes equations in an analytical function form. The network contains two parts. The first part directly satisfies the boundary conditions and therefore, contains no adjustable parameters whereas the second part is constructed such that the governing equation is satisfied inside the solution domain, while the boundary conditions remain untouched. The NN is applied to the second part to adjust parameters such that minimizing the loss. Dealing with realtime fluid simulation, Ladicky et al. [122], used regress forest to perform Smoothed Particle Hydrodynamics (SPH) simulation. In the proposed method, Ladicky et al., formulates physics-based fluid simulation as a regression problem, estimating the acceleration of every particle for each frame. In order to give the method strong generalization properties to reliably predict positions and velocities of particles in a large time step setting on yet unseen test videos, the is designed by modelling individual forces and constraints from the Navier-Stokes equations. To compute steady flow field for cars, Guo et al. [123], proposed an CNN framework to deal with the limitation of computational fluid dynamics (CFD) solvers which is usually a computationally expensive, memory demanding and time consuming iterative process. Their method is general and flexible approximation model for real-time prediction of non-uniform steady laminar flow in a 2D or 3D domain. From experimental results, they also show that CNN can estimate the velocity field two orders of magnitude faster than a GPU-accelerated CFD solver and four orders of magnitude faster than a CPU-based CFD solver at a cost of a low error rate. Later, based on an observations that most CNN model only requires constant execution time, while numerical solving often relies on iterative solving until the output converges, thus using CNN is much higher efficiency. Table 2 summarizes the approaches which utilize LS as a segmentation component in and end-to-end DL frameworks.

Table 2 Summary of approaches which embeds LS within DL framework where LS plays the roles of segmentation task. Fully convolutional networks (FCN), Computational fluid dynamics (CFD), Multilayer perceptron (MLP)

	Net.	App.	Technique	Datasets
[117]	FCN [28]	Liver CT	• Iteratively refines the FCN using a LS module	Liver [124]
		Left ventricle MRI		Left ventricle [125]
[118]	Hyper-column [126]	Building segment	• CNN to learn energy function that allows an ACM to generate polygons	Toronto city [127]
[58]	VGG [110]	Instance segment	• Reform LS evolution as GRU	Pascal [108]
			• LS is embedded into end-to-end framework to differentiate object	COCO [128]
[121]	MLP	Solving Navier–Stokes equations	• CNN is used as a solution of the Navier–Stokes equations	–
[123]	CNN based CFD	Steady flow approximation	• Formulate fluid simulation as a regression problem to achieve 2–4 orders of magnitude	2D car 3D geometry
[129]	FCN [28]	Brain tumor segmentation	• Refine feature map by LS	BRATS13
			• Perform the nets under recurrent mechanism	BRATS15
				BRATS17

4.3 Active Contour in the Loss Function

CNN has widely used in semantic segmentation but it is limited in obtaining small objects with fine boundary information to refine spatial details of segmentation results. In order to refine spatial details of segmentation results, Kim et al. [130] proposed LS loss which utilizes spatial correlation in ground truth whereas most of the semantic segmentation frameworks use the cross-entropy loss. To address the issue of multiple classes, they separate the ground truth into the binary images of each class and each binary image consists of background and regions belonging to a class. In such design, LS functions is converted into class probability maps and the energy is calculated for each class. The network is trained to minimize the weighted sum of the LS loss and the cross-entropy loss. Continue address the issue of multi-phase image segmentation, Kim et al. [131] proposed a multi phase LS loss which acts as regularization function to enhance supervised semantic segmentation algorithms. In this proposed network, the multiphase LS function evolution is learned using a neural network. Different from other previous works which train the model in a supervised manner, B. Kim's network unrolls LS evolution using a CNN and it is able to work under semi-supervised or unsupervised or fully supervised setting. Thanks to the LS loss that depends on pixel statistics, the proposed network by B. Kim et al. does not require the weak-labeled supervision for unlabeled data, but still uses these unlabeled images as elements of the training data. Table 3 summarizes the approaches where LS is utilized as either a loss function or a regularization term within loss function in an end-to-end DL framework. Based on the observation that models based on deep learning have improved results but are restricted to pixel-wise setting of the segmentation map, Chen et al. [132] tackle this limitation by making use of area inside as well as outside the region of interest as well as the size of boundaries during learning. Their loss function incorporates area and size information and integrates this into a dense deep learning model. Taking the form of a distance metric on the space of contours not regions, Kervadec et al. [133], proposed boundary loss which mitigates the difficulties of regional losses in the context of highly unbalanced segmentation problems. In their approach, the boundary loss is inspired by discrete (graph-based) optimization techniques for computing gradient flows of curve evolution. Following an integral approach for computing boundary variations, they express a non-symmetric L_2 distance on the space of contour as a regional integral, which avoids completely local differential computations involving contour points. Their boundary loss is able to be combined with standard regional losses due to boundary loss expressed as the sum of linear functions of the regional softmax probability outputs of the network.

Table 3 Summary of approaches where LS is utilized as either a loss function or a regularization term in the loss function in an end-to-end DL framework

	Net.	App.	Technique	Datasets
[130]	FCN [28]	Semantic segment	• LS loss is to refine spatial details of segmentation results	COCO [128]
	Deeplab [37]		• LS functions is converted into class Prob. maps	
[131]	Semi-supervised CNN	• Object segment	• Replace Euler-Lagrangian by CNN	Pascal [108]
		• Medical segment	• Unrolling LS evolution using a CNN so that it can be used under semi/un/full-supervised setting	LiTS [134] etc.
[133]	UNet [114]	Medical segment	• Use contour information	ISLES [135]
			• Based on graph-based optimization	WMH [136] etc.
[132]	UNet [114]	CMR segment	• The loss is based on area inside, outside, size of boundaries	ACDC [137]

References

1. C. Samson, L. Blanc-Feraud, G. Aubert, J. Zerubia, A level set model for image classification. Int. J. Comput. Vision (IJCV) **40**(3), 187–197 (2000)
2. T. Brox, J. Weickert, Level set segmentation with multiple regions. IEEE Trans. Image Process. **15**(10), 32133218 (2006)
3. E. Bae, X.-C. Tai, Graph cut optimization for the piecewise constant level set method applied to multiphase image segmentation, in *2nd International Conference on Scale Space and Variational Methods in Computer Vision*, pp. 1–13 (2009)
4. C. Li, C. Xu, C. Gui, M.D. Fox, Distance regularized level set evolution and its application to image segmentation. IEEE Trans. Image Process. **19**(12), 3243–3254 (2010)
5. C. Li, R. Huang, Z. Ding, C. Gatenby, D.N. Metaxas, J.C. Gore, A level set method for image segmentation in the presence of intensity inhomogeneities with application to MRI. IEEE Trans. Image Process. **20**(7), 2007–2016 (2011)
6. B. Lucas, M. Kazhdan, R. Taylor, Multi-object spring level sets (muscle), in *15th International Conference on Medical Image Computing Computed-Assisted Intervention*, pp. 495–503 (2012)
7. T.H.N. Le, K. Luu, M. Savvides, Sparcles: dynamic l1 sparse classifiers with level sets for robust beard/moustache detection and segmentation. IEEE Trans. Image Process. **22**(8), 3097–3107 (2013)
8. Q. Huang, X. Bai, Y. Li, L. Jin, X. Li, Optimized graph-based segmentation for ultrasound images. Neurocomputing **129**, 216–224 (2014)

9. J. Li, M. Luong, D. Jurafsky, A hierarchical neural autoencoder for paragraphs and documents. arXiv preprint arXiv:1506.01057 (2015)
10. S. Mukherjee, S. Acton, Region based segmentation in presence of intensity inhomogeneity using legendre polynomials. IEEE Signal Process. Lett. **22**(3), 298–302 (2015)
11. T.H.N. Le, M. Savvides, A novel shape constrained feature-based active contour (SC-FAC) model for lips/mouth segmentationin the wild. Pattern Recogn. **54**, 23–33 (2016)
12. J. Shen, Y. Du, X. Li, Interactive segmentation using constrained laplacian optimization. IEEE Trans. Circuits Syst. Video Tech. **24**(7), 1088–1100 (2014)
13. M. Kass, A. Witkin, D. Terzopoulos, Snakes: active contour models. Int. J. Comput. Vision (IJCV) **1**(4), 321–331 (1988)
14. V. Caselles, R. Kimmel, G. Sapiro, Geodesic active contours. Int. J. Comput. Vision (IJCV) **22**, 61–79 (1997)
15. C. Li, C. Xu, C. Gui, M. Fox, Level set evolution without reinitialization: a new variational formulation, in *CVPR*, pp. 430–436 (2005)
16. N. Paragios, R. Deriche, Geodesic active regions and level set methods for supervised texture segmentation. Int. J. Comput. Vision (IJCV) **46**, 223–247 (2002)
17. T.F. Chan, L.A. Vese, Active contours without edges. IEEE Trans. Image Process. **10**, 266–277 (2001)
18. J. Lie, M. Lysaker, X. Tai, A binary level set model and some application to Mumford Shah image segmentation. IEEE Trans. Image Process. pp. 1171–1181 (2010)
19. D. Mumford, J. Shah, Optimal approximation by piecewise smooth functions and associated variational problems. Commun. Pure Appl. Math. **42**(5), 577–685 (1989)
20. L.A. Vese, T.F. Chan, A multiphase level set framework for image segmentation using the Mumford and Shah model. Int. J. Comput. Vision (IJCV) **50** (2002)
21. C. Li, C. Kao, J. Gore, Z. Ding, Implicit active contours driven by local binary fitting energy, in *CVPR*, pp. 1–7 (2007)
22. C. Li, C.-Y. Kao, J.C. Gore, Z. Ding, Minimization of region-scalable fitting energy for image segmentation. IEEE Trans. Image Process. (2008)
23. K. Zhang, H. Song, L. Zhang, Active contours driven by local image fitting energy. Pattern Recogn. **43**(4), 1199–1206 (2010)
24. L. Wang, C. Pan, Robust level set image segmentation via a local correntropy-based k-means clustering. Pattern Recogn. **47**(5), 1917–1925 (2014)
25. Y. Han, W. Wang, X. Feng, A new fast multiphase image segmentation algorithm based on non-convex regularizer. Pattern Recogn. **45**(1), 363–372 (2012)
26. S. Liu, Y. Peng, A local region-based chan-vese model for image segmentation. Pattern Recogn. **45**(7), 2769–2779 (2012)
27. Y. Wang, S. Xiang, C. Pan, L. Wang, G. Meng, Level set evolution with locally linear classification for image segmentation. Pattern Recogn. **46**(6), 1734–1746 (2013)
28. J. Long, E. Shelhamer, T. Darrell, Fully convolutional networks for semantic segmentation, in *CVPR* (2015)
29. S. Zheng, S. Jayasumana, B. Romera-Paredes, V. Vineet, Z. Su, D. Du, C. Huang, P. Torr, Conditional random fields as recurrent neural networks, in *ICCV* (2015)
30. T.N. Le, C. Zhu, Y. Zheng, K. Luu, M. Savvides, Robust hand detection in vehicles, in *ICPR*, pp. 573–578 (2016)
31. T.N. Le, Y. Zheng, C. Zhu, K. Luu, M. Savvides, Multiple scale FasterRCNN approach to drivers cell-phone usage and hands on steering wheel detection, in *CVPRW*, pp. 46–53 (June 2016)
32. K. Luu, C.C. Zhu, C. Bhagavatula, T.N. Le, M. Savvide, A deep learning approach to joint face detection and segmentation, in *Advances in Face Detection and Facial Image Analysis*, ed. by M. Kawulok, M. Celebi, B. Smolka (Springer, Cham, 2016)
33. Y. Zheng, C. Zhu, K. Luu, C. Bhagavatula, T.N. Le, M. Savvides, Towards a deep learning framework for unconstrained face detection, in *IEEE 8th International Conference on Biometrics: Theory, Applications and Systems (BTAS)*, pp. 1–8 (2016)

34. T.N. Le, K. Luu, C. Zhu, M. Savvides, Semi self-training beard/moustache detection and segmentation simultaneously. Image Vision Comput. **58**, 214–223 (2017)
35. T.N. Le, C. Zhu, Y. Zheng, K. Luu, M. Savvides, Deepsafedrive: a grammar-aware driver parsing approach to driver behavioral situational awareness (DB-SAW). Pattern Recogn. **66**, 229–238 (2017)
36. V. Badrinarayanan, A. Kendall, R. Cipolla, Segnet: a deep convolutional encoder-decoder architecture for image segmentation. IEEE Trans. Pattern Anal. Mach. Intell. **39**(12), 2481–2495 (2017)
37. L.-C. Chen, G. Papandreou, I. Kokkinos, K. Murphy, A.L. Yuille, Deeplab: semantic image segmentation with deep convolutional nets, atrous convolution, and fully connected CRFs. Comput. Vision Pattern Recogn. **40**(4), 834–848 (2018)
38. P.O. Pinheiro, R. Collobert, P. Dollár, Learning to segment object candidates, in *NIPS* (2015)
39. H. Zhao, J. Shi, X. Qi, X. Wang, J. Jia, Pyramid scene parsing network, in *CVPR*, pp. 2881–2890 (2017)
40. Y. Li, H. Qi, J. Dai, X. Ji, Y. Wei, Fully convolutional instance-aware semantic segmentation, in *CVPR*, pp. 2359–2367 (2017)
41. H. Wu, V.V. Appia, A.J. Yezzi, Numerical conditioning problems and solutions for nonparametric i.i.d. statistical active contours. IEEE Trans. Software Eng. **35**(6), 1298–1311 (2013)
42. Y. Shi, W.C. Karl, Real-time tracking using level sets **2**, 34–41 (2005)
43. A. Dubrovina, G. Rosman, R. Kimmel, Multi-region active contours with a single level set function. IEEE Trans. Pattern Anal. Mach. Intell. **37**, 1585–1601 (2015)
44. K. Zhang, Q. Liu, H. Song, X. Li, A variational approach to simultaneous image segmentation and bias correction. IEEE Trans. Cybernetics **45**(8), 1426–1437 (2015)
45. T.H.N. Le, M. Savvides, A novel shape constrained feature-based active contour model for lips/mouth segmentation in the wild. Pattern Recogn. **54**, 23–33 (2016)
46. H. Zhou, X. Li, G. Schaefer, M.E. Celebi, P.C. Miller, Comput. Vision Image Underst. **117**(9), 1004–1016 (2013)
47. T.F. Chan, S. Esedoglu, M. Nikolova, Algorithms for finding global minimizers of image segmentation and denoising models. SIAM J. Appl. Math. (Technical Report) (2006)
48. J. Weickert, B.M.T.H. Romeny, M.A. Viergever, Efficient and reliable schemes for nonlinear diffusion filtering. IEEE Trans. Image Process. **7**, 398–410 (1998)
49. Y. LeCun, D. Touresky, G. Hinton, T. Sejnowski, A theoretical framework for back-propagation, in *Proceedings of the 1988 Connectionist Models Summer School* (Morgan Kaufmann, CMU, Pittsburgh, PA, 1988), pp. 21–28
50. Y. LeCun, L. Bottou, G.B. Orr, K.-R. Müller, Efficient backprop, in *Neural networks: Tricks of the Trade*, pp. 9–50 (Springer, Berlin, 1998)
51. D.T. Nguyen, W. Li, P.O. Ogunbona, Human detection from images and videos: a survey. Pattern Recogn. **51**, 148–175 (2016)
52. J.R. Uijlings, K.E. Van De Sande, T. Gevers, A.W. Smeulders, Selective search for object recognition. Int. J. Comput. Vision (IJCV) **104**(2), 154–171 (2013)
53. R. Girshick, J. Donahue, T. Darrell, J. Malik, Rich feature hierarchies for accurate object detection and semantic segmentation, in *CVPR*, pp. 580–587 (2014)
54. K. He, X. Zhang, S. Ren, J. Sun, Spatial pyramid pooling in deep convolutional networks for visual recognition. IEEE Trans. Pattern Anal. Mach. Intell. **37**(9), 1904–1916 (2015)
55. S. Ren, K. He, R.B. Girshick, J. Sun, Faster R-CNN: towards real-time object detection with region proposal networks. arXiv preprint arXiv:1506.01497 (2015)
56. J. Dai, K. He, J. Sun, Instance-fully convolutional instance-aware semantic segmentation via multi-task network cascades, in *CVPR* (2016)
57. K. He, X. Zhang, S. Ren, J. Sun, Identity mappings in deep residual networks, in *ECCV* (2016)
58. T.H.N. Le, K.G. Quach, K. Luu, C.N. Duong, M. Savvides, Reformulating level sets as deep recurrent neural network approach to semantic segmentation. Comput. Vision Pattern Recogn. **27**(5), 2393–2407 (2018)

59. T.H.N. Le, R. Gummadi, M. Savvides, Deep recurrent level set for segmenting brain tumors, in *MICCAI*, pp. 646–653 (2018)
60. T.H.N. Le, C.N. Duong, L. Han, K. Luu, K.G. Quach, M. Savvides, Deep contextual recurrent residual networks for scene labeling. Pattern Recogn. **80**, 32–41 (2018)
61. M. Egmont-Petersen, D. de Ridder, H. Handels, Image processing with neural networks: a review. Pattern Recogn. **35**(10), 2279–2301 (2002)
62. O. Russakovsky, J. Deng, H. Su, J. Krause, S. Satheesh, S. Ma, Z. Huang, A. Karpathy, A. Khosla, M. Bernstein et al., Imagenet large scale visual recognition challenge. Int. J. Comput. Vision (IJCV) **115**(3), 211–252 (2015)
63. A.-M. Tousch, S. Herbin, J.-Y. Audibert, Semantic hierarchies for image annotation: a survey. Pattern Recogn. **45**(1), 333–345 (2012)
64. J. Fan, W. Xu, Y. Wu, Y. Gong, Human tracking using convolutional neural networks. IEEE Trans. Neural Networks **21**(10), 1610–1623 (2010)
65. L. Wang, H. Lu, X. Ruan, M.-H. Yang, Deep networks for saliency detection via local estimation and global search, in *CVPR* (IEEE, New York, 2015), pp. 3183–3192
66. G. Li, Y. Yu, Visual saliency based on multiscale deep features. arXiv preprint arXiv:1503.08663 (2015)
67. M. Patacchiola, A. Cangelosi, Head pose estimation in the wild using convolutional neural networks and adaptive gradient methods. Pattern Recogn. **71**, 132–143 (2017)
68. G. Gkioxari, R. Girshick, J. Malik, Contextual action recognition with R * CNN, in *ICCVn*, pp. 1080–1088 (2015)
69. J. Zhang, W. Li, P.O. Ogunbona, P. Wang, C. Tang, Rgb-d-based action recognition datasets: a survey. Pattern Recogn. **60**, 86–105 (2016)
70. H. Xu, F. Su, Robust seed localization and growing with deep convolutional features for scene text detection, in *Proceedings of the 5th ACM on International Conference on Multimedia Retrieval* (ACM, New York, 2015), pp. 387–394
71. M. Jaderberg, A. Vedaldi, A. Zisserman, Deep features for text spotting, in *ECCV* (Springer, Berlin, 2014), pp. 512–528
72. D. Yu, W. Xiong, J. Droppo, A. Stolcke, G. Ye, J. Li, G. Zweig, Deep convolutional neural networks with layer-wise context expansion and attention, in *Interspeech*, pp. 17–21 (2016)
73. L.-H. Chen, T. Raitio, C. Valentini-Botinhao, J. Yamagishi, Z.-H. Ling, DNN-based stochastic postfilter for hmm-based speech synthesis, in *Interspeech*, pp. 1954–1958 (2014)
74. M.I. Jordan, Artificial neural networks, in *Attractor Dynamics and Parallelism in a Connectionist Sequential Machine*, pp. 112–127 (1990)
75. D.E. Rumelhart, G.E. Hinton, R.J. Williams, Neurocomputing: foundations of research, in *Learning Representations by Back-propagating Errors* (MIT Press, Cambridge, MA, 1988), pp. 696–699
76. Y. Bengio, P. Simard, P. Frasconi, Learning long-term dependencies with gradient descent is difficult. IEEE Trans. Neural Networks **5**(2), 157–166 (1994)
77. S. Hochreiter, J. Schmidhuber, Long short-term memory. Neural Comput. **9**(8), 1735–1780 (1997)
78. K. Cho, B. van Merrienboer, Ç. Gülçehre, F. Bougares, H. Schwenk, Y. Bengio, Learning phrase representations using RNN encoder-decoder for statistical machine translation. arXiv preprint arXiv:1406.1078 (2014)
79. T. Mikolov, S. Kombrink, L. Burget, J. Cernocký, S. Khudanpur, Extensions of recurrent neural network language model, in *ICASSP*, pp. 5528–5531 (2011)
80. A. Graves, A. Mohamed, G.E. Hinton, Speech recognition with deep recurrent neural networks. arXiv preprint arXiv:1303.5778 (2013)
81. J. Chorowski, D. Bahdanau, D. Serdyuk, K. Cho, Y. Bengio, Attention based models for speech recognition. arXiv preprint arXiv:1506.07503 (2015)
82. N. Kalchbrenner, P. Blunsom, Recurrent continuous translation models (Association for Computational Linguistics, October 2013)
83. T. Luong, I. Sutskever, Q.V. Le, O. Vinyals, W. Zaremba, Addressing the rare word problem in neural machine translation. arXiv preprint arXiv:1410.8206 (2014)

84. F. Hill, A. Bordes, S. Chopra, J. Weston, The goldilocks principle: reading children's books with explicit memory representations. arXiv preprint arXiv:1511.02301 (2015)
85. J. Mao, W. Xu, Y. Yang, J. Wang, A.L. Yuille, Deep captioning with multimodal recurrent neural networks (M-RNN). arXiv preprint arXiv:1412.6632 (2014)
86. J. Donahue, L.A. Hendricks, S. Guadarrama, M. Rohrbach, S. Venugopalan, K. Saenko, T. Darrell, Long-term recurrent convolutional networks for visual recognition and description. arXiv preprint arXiv:1411.4389 (2014)
87. D. Vilariño, V. Brea, D. Cabello, X. Pardo, Discrete-time CNN for image segmentation by active contours. Pattern Recogn. Lett. **19**, 721–734 (1998)
88. T. Kozek, D.L. Vilariño, An active contour algorithm for continuous time cellular neural networks. J. VLSI Signal Process. Syst. Signal Image Video Tech. **23**, 403–414 (1999)
89. F. Chen, H. Yu, R. Hu, X. Zeng, Deep learning shape priors for object segmentation, in *CVPR*, pp. 1870–1877 (2013)
90. T.A. Ngo, G. Carneiro, Lung segmentation in chest radiographs using distance regularized level set and deep-structured learning and inference, in *ICIP*, pp. 2140–2143 (2015)
91. X. Wu, J. Zhao, H. Wang, Face segmentation based on level set and deep learning prior shape, in *2017 10th CISP-BMEI*, pp. 1–5 (2017)
92. C. Rupprecht, E. Huaroc, M. Baust, N. Navab, Deep active contours. arXiv preprint arXiv:1607.05074 (2016)
93. W. Bupphawat, T. Kasetkasem, I. Kumazawa, P. Rakwatin, T. Chanwimaluang, Super-resolution land cover mapping based on deep learning and level set method, in *2017 14th International Conference on ECTI-CON*, pp. 557–560 (2017)
94. P. Hu, B. Shuai, J. Liu, G. Wang, Deep level sets for salient object detection, in *The CVPR* (July 2017)
95. T. Ngo, Z. Lu, G. Carneiro, Combining deep learning and level set for the automated segmentation of the left ventricle of the heart from cardiac cine magnetic resonance. Med. Image Anal. **35**, 05 (2016)
96. N. Singhal, S. Mukherjee, C. Perrey, Automated assessment of endometrium from transvaginal ultrasound using deep learned snake, in *ISBI*, pp. 283–286 (2017)
97. C. Cho, Y.H. Lee, S. Lee, Prostate detection and segmentation based on convolutional neural network and topological derivative, in *ICIP*, pp. 3071–3074 (2017)
98. D. Zhang, W. Zhu, H. Zhao, F. Shi, X. Chen, Automatic localization and segmentation of optical disk based on faster R-CNN and level set in fundus image, in *Medical Imaging 2018: Image Processing*, Houston, TX, USA, 10–15 February 2018, p. 105741U (2018)
99. J. Duan, J. Schlemper, W. Bai, T.J.W. Dawes, G. Bello, G. Doumou, A.M.S.M. de Marvao, D.P. O'Regan, D. Rueckert, Deep nested level sets: fully automated segmentation of cardiac MR images in patients with pulmonary hypertension. arXiv preprint arXiv:1807.10760 (2018)
100. R. Salakhutdinov, G. Hinton, Deep Boltzmann machines, in *Proceedings of the Twelfth International Conference on Artificial Intelligence and Statistics*, pp. 448–455 (2009)
101. D. Cremers, F.R. Schmidt, F. Barthel, Shape priors in variational image segmentation: convexity, lipschitz continuity and globally optimal solutions, in *CVPR* (June 2008)
102. L.J. Latecki, R. Lakamper, T. Eckhardt, Shape descriptors for non-rigid shapes with a single closed contour, in *CVPR*, vol. 1, pp. 424–429 (2000)
103. G.E. Hinton, Deep belief networks. Scholarpedia **4**(5), 5947 (2009)
104. J. Shiraishi, S. Katsuragawa, J. Ikezoe, T. Matsumoto, T. Kobayashi, K.-I. Komatsu, M. Matsui, H. Fujita, Y. Kodera, K. Doi, Development active contour model in deep learning era: a revise and review 29 of a digital image database for chest radiographs with and without a lung nodule: receiver operating characteristic analysis of radiologists' detection of pulmonary nodules. Am. J. Roentgenol. (AJR) **174**, 71–74 (2000)
105. M. Everingham, L. Van Gool, C.K.I. Williams, J. Winn, A. Zisserman, The Pascal visual object classes (VOC) challenge. Int. J. Comput. Vision (IJCV) **88**, 303–338 (2010)
106. A. Krizhevsky, I. Sutskever, G.E. Hinton, Imagenet classification with deep convolutional neural networks, in *NIPS*, pp. 1097–1105 (2012)

107. A collaborative resource to build consensus for automated left ventricular segmentation of cardiac MR images. Med. Image Anal. **18**(1), pp. 50–62 (2014)
108. M. Everingham, S.M. Eslami, L. Gool, C.K. Williams, J. Winn, A. Zisserman, The Pascal visual object classes challenge: a retrospective. Int. J. Comput. Vision **111**, 98–136 (2015)
109. J. Jeong, C. Yang, T. Kim, Geo-positioning accuracy using multiplesatellite images: IKONOS, QuickBird, and KOMPSAT-2 stereo images. Remote Sens. **7**(4), 4549–4564 (2015)
110. K. Simonyan, A. Zisserman, Very deep convolutional networks for largescale image recognition. arXiv preprint arXiv:1409.1556 (2014)
111. S. Alpert, M. Galun, A. Brandt, R. Basri, Image segmentation by probabilistic bottom-up aggregation and cue integration. IEEE Trans. Pattern Anal. Mach. Intell. **34** (2012)
112. M. Cheng, N.J. Mitra, X. Huang, S. Hu, Salientshape: group saliency in image collections. Visual Comput. **30**(4), 443–453 (2014)
113. P. Radau, Y. Lu, K. Connelly, G. Paul, A. Dick, G. Wright, Evaluation framework for algorithms segmenting short axis cardiac MRI (2009)
114. O. Ronneberger, P. Fischer, T. Brox, U-net: convolutional networks for biomedical image segmentation, in*MICCAI*, vol. 9351 of LNCS, pp. 234–241 (2015)
115. G. Litjens et al., Evaluation of prostate segmentation algorithms for MRI: the promise12 challenge. Med. Image Anal. **18**, 359–373 (2014)
116. E. Decencière, X. Zhang et al., **33**, 231 (2014)
117. M. Tang, S. Valipour, Z.V. Zhang, D. Cobzas, M. Jägersand, A deep level set method for image segmentation. arXiv preprint arXiv:1705.06260 (2017)
118. D. Marcos, D. Tuia, B. Kellenberger, L. Zhang, M. Bai, R. Liao, R. Urtasun, Learning deep structured active contours end-to-end, in *IJCV* (IEEE Computer Society, 2018), pp. 8877–8885
119. I.E. Lagaris, A. Likas, D.I. Fotiadis, Artificial neural networks for solving ordinary and partial differential equations. Trans. Neural Networks **9**, 987–1000 (1998)
120. C. Tran, T. Tran-Cong, Computation of viscoelastic ow using neural networks and stochastic simulation. Korea-Australia Rheol. J. **14**, 161–174 (2002)
121. M. Baymani, S. Effati, H. Niazmand, A. Kerayechian, Artificial neural network method for solving the Navier-Stokes equations. Neural Comput. Appl. **26**, 765–773 (2015)
122. L. Ladický, S. Jeong, B. Solenthaler, M. Pollefeys, M. Gross, Data-driven UID simulations using regression forests. ACM Trans. Graph. **34** (2015)
123. X. Guo, W. Li, F. Iorio, Convolutional neural networks for steady ow approximation, in *SIGKDD*, KDD '16 (2016)
124. B.V. Ginneken, T. Heimann, M. Styner, M.: 3d segmentation in the clinic: a grand challenge, in *MICCAI Workshop on 3D Segmentation in the Clinic: A Grand Challenge* (2007)
125. A. Suinesiaputra, B.R. Cowan et al., Left ventricular segmentation challenge from cardiac MRI: a collation study, in *Statistical Atlases and Computational Models of the Heart*. Imaging and Modelling Challenges, Springer, Berlin, Heidelberg, pp. 88–97 (2012)
126. B. Hariharan, P. Arbelaez, R. Girshick, J. Malik, Hyper-columns for object segmentation and fine-grained localization, in *CVPR* (2015)
127. S. Wang, M. Bai, G. Máttyus, H. Chu, W. Luo, B. Yang, J. Liang, J. Cheverie, S. Fidler, R. Urtasun, Torontocity: seeing the world with a million eyes. arXiv preprint arXiv:1612.00423 (2016)
128. T. Lin, M. Maire, S.J. Belongie, L.D. Bourdev, R.B. Girshick, J. Hays, P. Perona, D. Ramanan, P. Dollár, C.L. Zitnick, Microsoft COCO: common objects in context. arXiv preprint arXiv:1405.0312 (2014)
129. T.H.N. Le, R. Gummadi, M. Savvides, Deep recurrent level set for segmenting brain tumors, in *Medical Image Computing and Computer Assisted Intervention – MICCAI 2018*, ed. by A.F. Frangi, J.A. Schnabel, C. Davatzikos, C. Alberola-López, G. Fichtinger (2018)
130. Y. Kim, S. Kim, T. Kim, C. Kim, CNN-based semantic segmentation using level set loss, in *WACV*, pp. 1752–1760 (2019)
131. B. Kim, J.C. Ye, Multiphase level-set loss for semi-supervised and unsupervised segmentation with deep learning. arXiv preprint arXiv:1904.02872 (2019)

132. X. Chen, B.M. Williams, S.R. Vallabhaneni, G. Czanner, R. Williams, Y. Zheng, Learning active contour models for medical image segmentation, in *Proceedings of the IEEE Conference on Computer Vision and Pattern Recognition*, pp. 11,632–11,640 (2019)
133. H. Kervadec, J. Bouchtiba, C. Desrosiers, E. Granger, J. Dolz, I. Ben Ayed, Boundary loss for highly unbalanced segmentation, in *Proceedings of The 2nd International Conference on Medical Imaging with Deep Learning*, ed. by M.J. Cardoso, A. Feragen, B. Glocker, E. Konukoglu, I. Oguz, G. Unal, T. Vercauteren, vol. 102 of Proceedings of Machine Learning Research, (London, UK, 08–10 Jul 2019), pp. 285-296
134. P. Bilic aet al., The liver tumor segmentation benchmark (liTS). arXiv preprint arXiv:1901.04056 (2019)
135. ISLES, Ischemic stroke lesion. http://www.isles-challenge.org
136. WMH, White matter hyperintensities. http://wmh.isi.uu.nl
137. ACDC, ACDC 2017 challenge. https://acdc.creatis.insa-lyon.fr/

Linear Regression Techniques for Car Accident Prediction

Miguel Islas Toski, Karla Avila-Cardenas and Jorge Gálvez

Abstract This chapter explains the basics of simple linear regression. Showing different approaches to solve a prediction problem of this type. First, we explain the theory, then, it is solved by the algebraic method of least squares, checking the procedure and results through MATLAB. Finally, a basic example of the field of evolutionary computing is shown, using several evolutionary techniques such as PSO, DE, ABC, CS and the classical method of descending gradient. Which optimize the function to find the best coefficients for an estimated straight line. This is applied to a set of fatal traffic accident data in the U.S. states.

Keywords Simple linear regression · Evolutionary computation techniques · Predictive analytics · Optimization

M. Islas Toski (✉) · K. Avila-Cardenas · J. Gálvez
Departamento de Electrónica, Universidad de Guadalajara, CUCEI, Guadalajara, Jalisco, México
e-mail: miguel.islas@alumnos.udg.mx

K. Avila-Cardenas
e-mail: karla.acardenas@alumnos.udg.mx

J. Gálvez
e-mail: Jorge.galvez@cutonala.udg.mx

D. Oliva and S. Hinojosa (eds.), *Applications of Hybrid Metaheuristic Algorithms for Image Processing*, Studies in Computational Intelligence 890,
https://doi.org/10.1007/978-3-030-40977-7_12

1 Introduction

In recent years, predictive analytics has increased a lot, due to the prediction of future events. To mention some areas of them, we could list the next ones: finance, biology, security, medicine, automotive, aerospace, engineering, manufacturing, and areas of big data and machine learning.

But what is predictive analytics? And how does predictive analytics work? Above-board, predictive analytics is a process that uses data to estimate next behaviors or predictions of future events, this is based on data analysis, stats, and machine learning techniques. And its operation is shown in (Fig. 1).

For a better explanation predictive analytics are divided into the following 4 sections:

1. **Data**: Input sources included in a file, preferably in CSV format.
2. **Pre-process data**: Data cleaning. Then subsequently combine the different sources of data.
3. **Development of a predictive model**: Use mathematical and computational methods to predict an outcome depending on the inputs (Data).
4. **Integrate analytics with systems**: Once the predictive model was developed is possible to create software programs or devices.

One example could be a time-series regression model for predicting the response in a timeline of fatal traffic accidents in the U.S. based on a linear regression model.

A fundamental component in the presentation of concepts in this text is the tutorial support that represents the demonstration within a robust and consolidated simulation environment such as Matlab. This environment is recognized worldwide as a simulation and design platform that has permeated all areas of knowledge, and whose operational structure supports the development of functions and algorithms in a modular way based on its basic functions.

On the other hand, Matlab also provides an environment for graphic presentations and particularly mathematical and simulation tools. All these characteristics include a broad collaboration of a community of users.

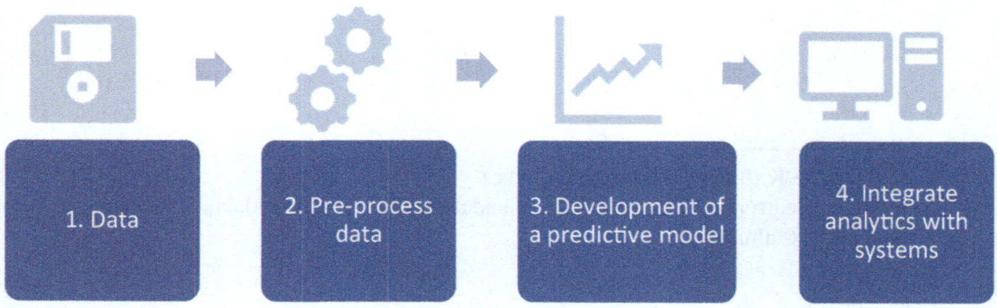

Fig. 1 Predictive analytics workflow

2 Linear Regression

Linear regression is a statistical modeling method, which is used to describe a continuous response variable in the function mode of one or more predictive variables.

It is commonly used to predict the behavior of systems or analyze experimental data in different branches.

Said the above, we could mention and classify different models of linear regression:

- **Simple**: Models with one predictor
- **Multiple**: Models with two or more predictors
- **Multivariate**: Models with more than one response variables.

The creation of a linear model is based on different linear regression techniques [1]. With the purpose to describe the relationship between a dependent variable $\boldsymbol{y}$ (response) as a function of one or more independent variables $\boldsymbol{X}_i$ (Predictors).

The linear regression model is represented by the Eq. (1):

$$y = \beta_0 + \sum \beta_i X_i + \epsilon_i \tag{1}$$

where $\boldsymbol{\beta}$ are the parameter estimates that need to be calculated and ϵ represents the error term. We can see a similarity between the equation of the line (2).

$$y = mx + b \tag{2}$$

where $\boldsymbol{m}$ represents the slope or regression coefficient and $\boldsymbol{b}$ is the $y\text{-}intercept$.

First of all, to build the linear regression prediction model we need to know the trend and behavior of our data.

2.1 Scatter Diagram

A **Scatter diagram** will provide us with the correct visualization of our data on a Cartesian plane. From the data in Table 1. We build the scatter diagram (Fig. 2).

2.2 Estimated Straight Line

From the scatter diagram (Fig. 2) we can conclude that there is a positive linear relationship. The next thing now is to find an **Estimated straight line** (6) that fits the best cross with the data. That means, we could have different solutions, consequently, to know which is the optimum we must calculate **the Error** (3) or the distance that

Table 1 Dataset of fatal traffic accidents in U.S. states. Where x is the population of the state and y represent the fatal traffic accidents per state

n	x	y
1	493,782	164
2	572,059	43
3	608,827	98
4	626,932	101
5	642,200	100
6	754,844	197
7	783,600	134
8	902,195	229
9	1,048,319	83
10	1,211,537	142
11	1,235,786	171
12	1,274,923	194
13	1,293,953	260
14	1,711,263	254
15	1,808,344	411
16	1,819,046	521
17	1,998,257	395
18	2,233,169	296
19	2,673,400	704
20	2,688,418	461
21	2,844,658	900
22	2,926,324	390
23	3,405,565	291
24	3,421,399	456
25	3,450,654	774
26	4,012,012	1046
27	4,041,769	964
28	4,301,261	665
29	4,447,100	1154
30	4,468,976	904
31	4,919,479	567
32	5,130,632	1150
33	5,296,486	643
34	5,363,675	792
35	5,595,211	1130
36	5,689,283	1288
37	5,894,121	563

(continued)

Table 1 (continued)

n	x	y
38	6,080,485	947
39	6,349,097	476
40	7,078,515	925
41	8,049,313	1557
42	8,186,453	1634
43	8,414,350	731
44	9,938,444	1159
45	11,353,140	1286
46	12,281,054	1490
47	12,419,293	1356
48	15,982,378	3244
49	18,976,457	1493
50	20,851,820	3583
51	33,871,648	4120

Fig. 2 Scatter diagram

exists between the points and the proposed line as shown in (Fig. 3). A complex engineering problem tends to produce a multimodal error surface [2].

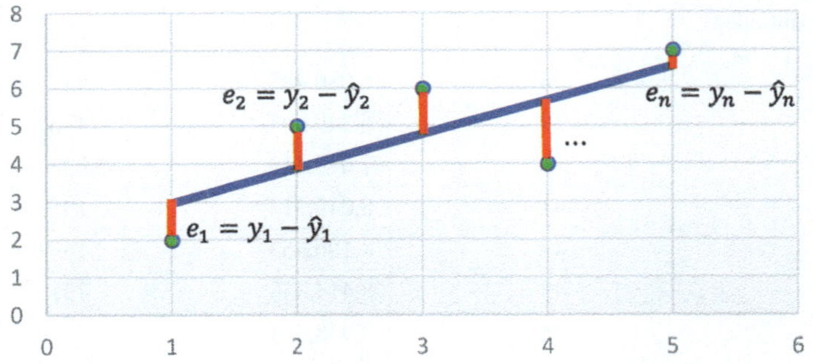

Fig. 3 Error in linear regression

2.3 *Mean Squared Error*

By calculating **the Squared error** (4), instead of the simple error, we make sure that the error is always positive. In this way, we know that the perfect error is 0.

$$e = y_i - \hat{y}_i \tag{3}$$

$$e^2 = (y_i - \hat{y}_i)^2 \tag{4}$$

Otherwise, if we did not square the error, sometimes the error would be positive and sometimes negative. Another possibility would be to use the absolute value, instead of squaring it.

However, if we use the absolute value, we will obtain a non-derivable function. And as we will see later, having a derivable function makes it possible to use optimization algorithms. For example, the gradient descent.

Now that we know how to calculate the error at each point, we can calculate what the average error is. To do this, we add all the errors and divide them by the total number of points. If we call M the total number of points, we have the formula of **the Mean Square Error** (5).

$$MSE = \frac{1}{M}\sum_{i=1}^{M}(y_i - \hat{y}_i)^2 \tag{5}$$

2.4 *Least Square Method*

Using **the Least square method,** we will find an **Estimated straight line** with the best fit for a set of data. It is necessary to mention that there are other methods that can be consulted in [1, 3].

$$\hat{y} = \boldsymbol{\beta}_0 + \boldsymbol{\beta}_1 \boldsymbol{x} \tag{6}$$

$$\boldsymbol{\beta}_0 = \bar{\boldsymbol{y}} - \boldsymbol{\beta}_1 \bar{\boldsymbol{x}} \tag{7}$$

$$\boldsymbol{\beta}_1 = \frac{n \sum_{i=1}^{n} xy - \sum_{i=1}^{n} x \cdot \sum_{i=1}^{n} y}{n \sum_{i=1}^{n} x^2 - \left(\sum_{i=1}^{n} x\right)^2} \tag{8}$$

Conceptually with this process, we are penalizing with greater intensity those points that are farthest from our line and with less intensity to those that are closer. To know (6) we will start solving (8) **(slope)** and (7) **(y-*intercept*)**. We could see the necessary values in Table 2. Which we get by helping us from Table 1.

Substituting in (8) previously knowing the total samples $\mathbf{n = 51}$:

$$\beta_1 = \frac{n \sum_{i=1}^{n} xy - \sum_{i=1}^{n} x \cdot \sum_{i=1}^{n} y}{n \sum_{i=1}^{n} x^2 - \left(\sum_{i=1}^{n} x\right)^2}$$

$$\beta_1 = \frac{51\left(4.739535590460000 \times 10^{11}\right) - (281421906)(42636)}{51\left(3.452668323818234 \times 10^{15}\right) - (281421906)^2}$$

$$\beta_1 = \mathbf{1.256394273876983} \times \mathbf{10^{-4}}$$

Now substituting in (7):

$$\beta_0 = \bar{y} - \beta_1 \bar{x}$$

$$\beta_0 = 836 - (1.256394273876983 \times 10^{-4}) \cdot (5.518076588235294 \times 10^{6})$$

$$\beta_0 = \mathbf{1.427120171726536 \times 10^2}$$

Finally substituting in (6) we obtain $\hat{y}$:

$$\hat{y} = \beta_0 + \beta_1 x$$

$$\hat{y} = \mathbf{1.427120171726536 \times 10^2} + \left(\mathbf{1.256394273876983 \times 10^{-4}}\right) \cdot \boldsymbol{x}$$

Table 2 Necessary values to get (6)

$\sum x$	281,421,906	$\sum x^2$	$3.452668323818234 \times 10^{15}$
$\sum y$	42,636	$\sum xy$	$4.739535590460000 \times 10^{11}$
$\bar{y}$	836	$\bar{x}$	$5.518076588235294 \times 10^{6}$

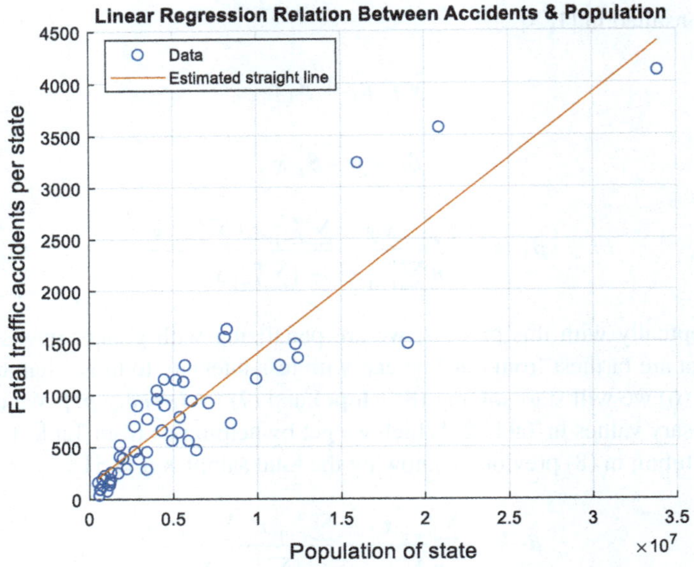

Fig. 4 Linear regression relation between accidents and population

Later we can plot our estimated straight line in the scatter diagram and observe its trajectory on (Fig. 4).

3 Developing a Predictive Model Using MATLAB

The above procedure can be solved using MATLAB explaining the procedure of [4]. Forming a system of linear equations represented in a matrix form.

$$\boldsymbol{Y} = \boldsymbol{X\beta} \tag{9}$$

where we have:

$$\begin{bmatrix} y_1 \\ y_2 \\ \vdots \\ y_n \end{bmatrix} = \begin{bmatrix} 1 & x_1 \\ 1 & x_2 \\ \vdots & \vdots \\ 1 & x_n \end{bmatrix} \begin{bmatrix} \beta_0 \\ \beta_1 \end{bmatrix}$$

An advantage that MATLAB offers is that we can find in an easy way $\boldsymbol{\beta}$ using the *mldivide* operator as $\boldsymbol{\beta} = \boldsymbol{X} \backslash \boldsymbol{Y}$. The \ operator performs at least-squares regression. The first three lines on Matlab code are just to load the respective data that are going to use it for the simple linear regression example.

```
load accidents
x = hwydata(:,14); %Population of states
y = hwydata(:,4); %Accidents per state
format long
b1 = x\y
```

b1 is the slope or regression coefficient. As a result, we obtain a linear relation:

$$y = \boldsymbol{\beta}_1 \boldsymbol{x} = (\mathbf{1.37271673556487} \times \mathbf{10^{-4}}) \cdot (\boldsymbol{x})$$

We continue plotting $\boldsymbol{y} = \boldsymbol{\beta}_1 \boldsymbol{x}$ on a scatter diagram (Fig. 5).

```
yCalc1 = b1*x;
scatter(x,y)
hold on
plot(x,yCalc1)
xlabel('Population of state')
ylabel('Fatal traffic accidents per state')
title('Linear Regression Relation Between Accidents &
Population')
grid on
```

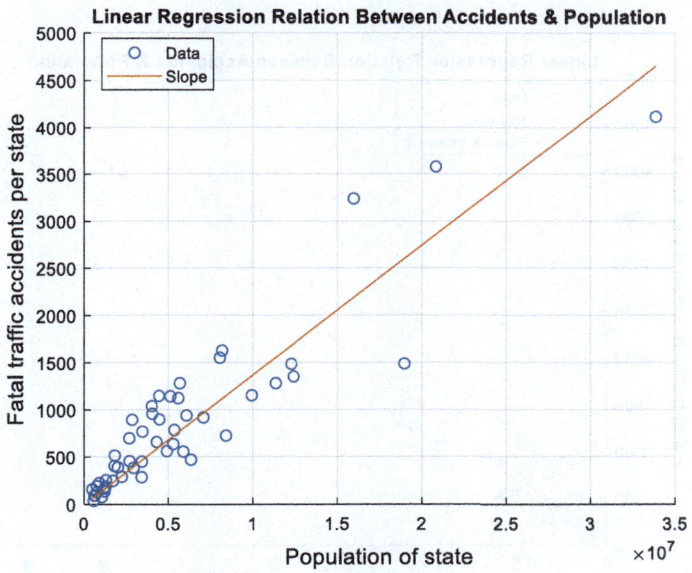

Fig. 5 Linear regression relation just with the slope

But this represents just one estimated straight line with a slope. As previously seen, this is one of the possible solutions but is not the best solution. We would improve the result if we add a *y-intercept* β_0 in the model as $y = \beta_0 + \beta_1 x$. We will calculate β_0 by padding x with a column of ones and using the same operator as we use previously \.

```
X = [ones(length(x),1) x];
b = X\y
```

Once we have the slope and *y-intercept* the relation is represented as:

$$y = \boldsymbol{\beta}_0 + \boldsymbol{\beta}_1 \boldsymbol{x}$$

$$y = \mathbf{1.427120171726536} \times \mathbf{10^2} + (\mathbf{1.256394273876983} \times \mathbf{10^{-4}}) \cdot \boldsymbol{x}$$

And we can visualize and compare the two results on (Fig. 6).

```
yCalc2 = X*b;
plot(x,yCalc2,'--')
legend('Data','Slope','Slope&Intercept','Location','best');
```

Analyzing (Fig. 6) we can see that the two fits look similar. To know which is the best we will calculate the **coefficient of determination** that is the square of

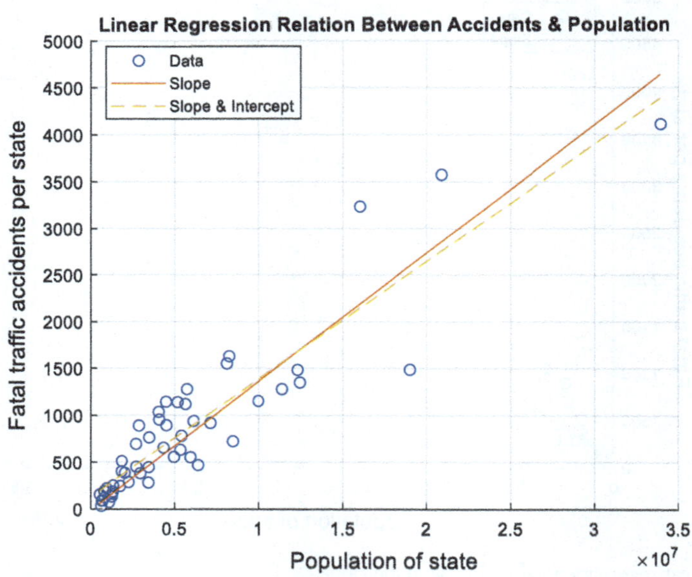

Fig. 6 Comparison of linear regression relation models

the **coefficient of correlation**. The coefficient of determination is one measure of how well a model can predict the data and falls between 0 and 1, being better when approaching 1. Where $\widehat{y}$ represents the calculated values of y and $\overline{y}$ is the mean of y, the coefficient of determination is represented by R^2 and defined as:

$$R^2 = 1 - \frac{\sum_{i=1}^{n} \left(y_i - \widehat{y}_i\right)^2}{\sum_{i=1}^{n} \left(y_i - \bar{y}_i\right)^2} \tag{10}$$

Solving (10) on MATLAB we will get the coefficient of determination to know the best fit.

```
Rsq1 = 1 - sum((y - yCalc1).^2)/sum((y - mean(y)).^2)
Rsq2 = 1 - sum((y - yCalc2).^2)/sum((y - mean(y)).^2)
```

For the relation with just the slope

$$y = \beta_1 x = (1.37271673556487 \times 10^{-4}) \cdot (x)$$

We get

$$\boxed{R_1^2 = 0.822235650485566}$$

And for the relation with slope and *y-intercept*

$$y = 1.427120171726536 \times 10^2 + (1.256394273876983 \times 10^{-4}) \cdot x$$

We get

$$\boxed{R_2^2 = 0.838210531103428}$$

And we could conclude that R_2^2 is the best fit.

4 Evolutionary Techniques

As previously seen to know which estimated straight line is the best, the quadratic error was reduced by the least-squares method. There is various algebraic methods and algorithms [5] that can be used to solve the linear regression problem, but now we will talk about the optimization methods by some popular evolutionary techniques.

Evolutionary algorithms have been applied to many research fields such as image processing [6, 7], control systems [8, 9], power systems, and others.

Optimization has become an important part of all disciplines. One reason for this inclusion is the motivation to produce quality products or services at competitive prices. In general, optimization represents the process of finding the "best solution" to a problem among a very large set of possible solutions [10].

The cost function to be optimized will be the MSE given by the (11).

$$MSE = \frac{1}{M}\sum_{i=1}^{M}(y_i - \hat{y}_i)^2 \tag{11}$$

4.1 Gradient Descent

Classical methods of optimization are based on the use of the gradient of a function $f(\boldsymbol{x})$ for the generation of new solutions. The gradient descendant (Fig. 7) is one of the techniques used for the minimization of objective-multidimensional functions. It is frequently used for the optimization of nonlinear functions due to its ease of programming, however, it is of slow convergence.

Having an initial point $\boldsymbol{x}^0$, the decision vector is modified iteratively until finding the optimal solution $\boldsymbol{x}^*$. This modification is expressed as:

$$\boldsymbol{x}^{k+1} = \boldsymbol{x}^k - \alpha g(f(\boldsymbol{x})) \tag{12}$$

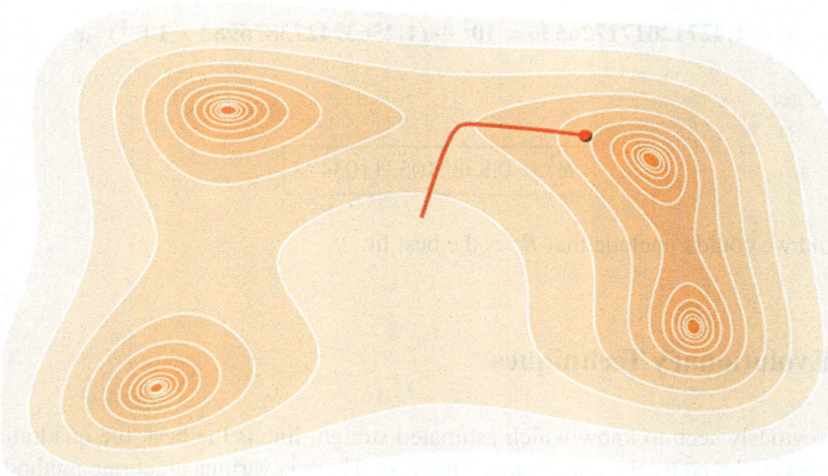

Fig. 7 Gradient descendant on a multimodal function

where $\boldsymbol{x}^k$ is the current point, $\boldsymbol{\alpha}$ is the size of the search step and $\boldsymbol{g}(\boldsymbol{f}(\boldsymbol{x}))$ is the gradient of the function.

The gradient $\boldsymbol{g}$ of a function $\boldsymbol{f}(\boldsymbol{x})$ at point $\boldsymbol{x}$ expresses the direction in which the function $\boldsymbol{f}(\boldsymbol{x})$ presents its maximum growth. Otherwise, in a minimization problem, the descent direction can be obtained by multiplying by -1 to $\boldsymbol{g}$. Ensuring that $\boldsymbol{f}(\boldsymbol{x}^{k+1}) < \boldsymbol{f}(\boldsymbol{x}^k)$ said, in other words, the new solution generated will be better than the previous one.

The gradient of a multidimensional function $\boldsymbol{f}(\boldsymbol{x})(\boldsymbol{x} = (\boldsymbol{x}1, ..., \boldsymbol{x}\boldsymbol{d}) \in \mathbb{R}^d)$ represents the way in which the function varies with respect to its $\boldsymbol{d}$ dimensions. In this way the gradient $\boldsymbol{g}_{x_1}$ expresses how it varies $\boldsymbol{f}(\boldsymbol{x})$ with respect to $\boldsymbol{x}1$. The gradient is defined as:

$$\boldsymbol{g}\boldsymbol{x}_1 = \frac{\partial \boldsymbol{f}(\boldsymbol{x}_1)}{\partial \boldsymbol{x}_1} \tag{13}$$

To calculate numerically the gradient $\boldsymbol{g}_{x_1}$ the following procedure is performed [11]:

1. A new decision vector $\widetilde{\boldsymbol{x}_i}$ is generated. This vector is the same in all the decision variables that $\boldsymbol{x}$ except in $\boldsymbol{x}_i$. This value will be replaced by $\boldsymbol{x}_i + \boldsymbol{h}$ where $\boldsymbol{h}$ is a very small value. Under these conditions, the new vector is defined as:

$$\widetilde{\boldsymbol{x}_i} = (\boldsymbol{x}_1, \boldsymbol{x}_2, \cdots, \boldsymbol{x}_i + \boldsymbol{h}, \cdots, \boldsymbol{x}_d) \tag{14}$$

2. Calculate the gradient $\boldsymbol{g}_{x_i}$ using the following model:

$$\boldsymbol{g}_{xi} \approx \frac{\boldsymbol{f}\left(\widetilde{\boldsymbol{x}_i}\right) - \boldsymbol{f}(\boldsymbol{x})}{\boldsymbol{h}} \tag{15}$$

Below is shown an example of the gradient descendant algorithm for simple linear regression on Matlab.

The first part of a Matlab code is to close all the windows and clear all the variables previously stored on the workspace memory using the commands *close*, *clear* and *clc*.

```
close all;
clear all;
```

The next section is for settings, we are going to set the training data that we are going to use for this example.

```
3  % Training data
4  load accidents
5  x = hwydata(:,14); %Population of states
6  y = hwydata(:,4); %Accidents per state
```

Then the next part of the code is for the settings of the initial parameters. For this example, we need to propose a straight line which will be adjusted after the gradient descendant process.

Notice that we propose a straight line with a slope of 120 (*theta0*) and a *y-intercept* (*theta1*) in 0. As we have previously seen, we know that a close value for the slope is 142, so the gradient descendant is going to fix this error iteration by iteration until it got the best value. Alpha indicates the learning rate or training rate. And also we set a maximum of iterations, variable *m* indicates the quantity of data, to help us with the mean in the function.

```
7  % Initial parameters
8  theta0 = 120;
9  theta1 = 0;
10 alpha = 0.000000000000003;
11 iter_max = 100;
12 m = numel(x);
```

Below is the section to make the graph, to create a window, use the command *figure(1)*, and for the scatter diagram use the command *scatter(x,y)*, and to hold the graph, write *hold on*. As the reader can see on line 17 there is the first straight line, and on line 19 is the command to graph it in color green. From line 20 to line 22 are the commands just for the labels of the respective axes. Using the command *grid on* we establish a grid on the graph for better viewing.

```
13 figure(1);
14 scatter(x,y);
15 %plot(x, y, 'bo', 'MarkerFaceColor', 'r',
   'MarkerSize', 10);
16 hold on;
17 h = theta0 + theta1*x;
18 %h = theta1*x;
19 plot(x, h, 'g');
20 xlabel('Population of state')
21 ylabel('Fatal traffic accidents per state')
22 title('Linear Regression Relation Between Accidents
   & Population')
23 grid on
```

The next part shows the cost function, for this case we are using the least square method. And we are going to start from the iteration 1 to 100, we could see the iterations on a new window *figure(2);* graphing the iterations with respect to the cost function *J* and holding the previous ones using *hold on*.

```
24 J = (1/(2*m))*sum((h-y).^2); %%cost function
25 iter = 1;
26 figure(2);
27 plot(iter, J, '*');
28 hold on
```

Now the explanation of the iterative process of the gradient descendant is explained from line 30 to line 44, as the reader can see on line 31 and 32, we are getting the partial derivatives of the function and those results are using on the lines 33 and 34 substituting respectively to be evaluated on line 35 iteration by iteration. Finally on line 37 we have the graph for the new straight line proposed. Furthermore, on line 41, the graph of iterations is shown respectively with the cost function *J*.

```
29 % Iterative Process
30 while(iter < iter_max)
31     g_theta0 = (1/m) * sum(h - y);
32     g_theta1 = (1/m) * sum((h - y).*x);
33     theta0 = theta0 - alpha * g_theta0;
34     theta1 = theta1 - alpha * g_theta1;
35     h = theta0 + theta1*x;
36     %h = theta1*x;
37     figure(1);
38     plot(x, h, 'g');
39     J = (1/(2*m))*sum((h-y).^2);
40     iter = iter + 1;
41     figure(2);
42     plot(iter, J, '*');
43     pause(0.1);
44 end
```

4.2 *Particle Swarm Optimization*

There are some problems that do not allow the use of traditional methods of optimization as the descendant gradient. Because in addition to performing complex operations, they can get stuck in local solutions [12].

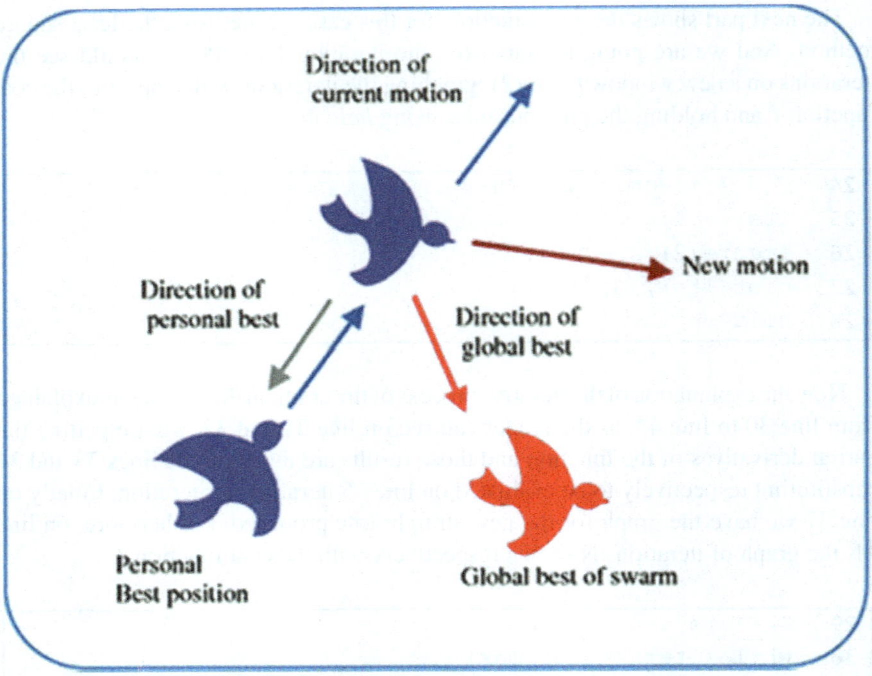

Fig. 8 Schematic representation of updating the velocity of a particle [14]

The PSO algorithm (Fig. 8) through the distribution of a population of particles in the search space and its own operators avoids falling into local optima and delivering unwanted solutions. PSO is based and inspired by the behavior of flocks of birds or schools of fish. Was proposed by Kennedy and Eberhart [13].

In summary, we have a population of particles, which operate on a search space. Each iteration generates new positions for the particles, which are obtained using a speed that is calculated considering the best global position and the best current position of each particle.

To determine the quality of the population an objective function is required, where individuals are evaluated each time, they take a new position. For the selection of members of the population should know if it is a problem of maximization or minimization. The information is shared among the elements of the population, this allows to approach the global solution of the problem with greater or lesser speed.

The initial population of particles is generated randomly, distributing the possible solutions in a search space bounded by the upper and lower limits previously defined along with other parameters such as the size and restrictions that may exist. Equation (13) describes the optimization of the particles in the PSO algorithm.

$$x_k^{i,t} = l_k + rand(u_k - l_k), x_k^{i,t} \in x^t \tag{16}$$

where $x_k^{i,t}$ is the i-th particle of the population $\boldsymbol{x}^t$, $\boldsymbol{i}$ is the index that refers to the number of particles and has the maximum size of the population ($\boldsymbol{i} = 1, 2, \ldots, \boldsymbol{n}$). $\boldsymbol{K}$ is the dimension of the problem, $\boldsymbol{l}_k$ and $\boldsymbol{u}_k$ are the lower and upper limits of the search space, the number of iterations is $\boldsymbol{t}$ and $\boldsymbol{rand}$ is a random number in the range from zero to one.

Before obtaining the new position that the particles will have in the search space, **the speed** of each one must be calculated.

To carry out this, we will need the previous value of the speed, if it is the first iteration, this value will be equal to zero. In addition, the best global and local values of each particle are needed.

The calculation of the speed is presented in (14).

$$\boldsymbol{v}^{t+1} = \boldsymbol{v}^t + \boldsymbol{rand}1 \times \left(\boldsymbol{P} + \boldsymbol{X}^t\right) + \boldsymbol{rand}2 \times (\boldsymbol{G} - \boldsymbol{X}) \tag{17}$$

where $\boldsymbol{v}^{t+1}$ is the value of the speed that is calculated for the iteration $\boldsymbol{t} + 1$, $\boldsymbol{v}^t$ is the velocity in the previous iteration $\boldsymbol{t}$. $\boldsymbol{x}^t$ is the vector that contains the positions of each particle, $\boldsymbol{P}$ contains the best current positions associated with the velocity of each particle, while $\boldsymbol{G}$ is the best current particle globally. $\boldsymbol{rand}1$ and $\boldsymbol{rand}2$ are random numbers with values between zero to one this with the purpose that the particles choose any trajectory if the direction of the local and global optimum is followed. After calculating the speed, the particles move to new positions in the current iteration. This is known as **the movement of particles** and behaviors in (15).

$$\boldsymbol{x}^{t+1} = \boldsymbol{x}^t + \boldsymbol{v}^{t+1} \tag{18}$$

where $\boldsymbol{x}^{t+1}$ is the vector of the new positions obtained in the iteration $\boldsymbol{t} + 1$, $\boldsymbol{x}^t$ represents the previous positions of the particles and $\boldsymbol{v}^{t+1}$ is the velocity vector that was obtained using the (14). We recommend [10] and [15] to address the issue in depth.

On (Fig. 9) we can see the flow chart of the PSO algorithm, notice that the first square of every algorithm is the initialization and also the initial setting of the position of the population, the values of the speed. The next thing is the evaluation of the population on the objective function to extract the best current particle of the iteration, then, is necessary to generate new speeds and also new positions, one of the most important things is to be sure if all the particles have been evaluated, if not, it is necessary to repeat the process, on the other hand, the best particle of the population will be updated until the detection criteria rule is reached.

4.3 Differential Evolution

The differential evolution algorithm (Fig. 10) was proposed by Storn and Price in 1995 [16]. It works with a population of solutions and searches stochastically through the search space. It works with the general framework of evolutionary algorithms

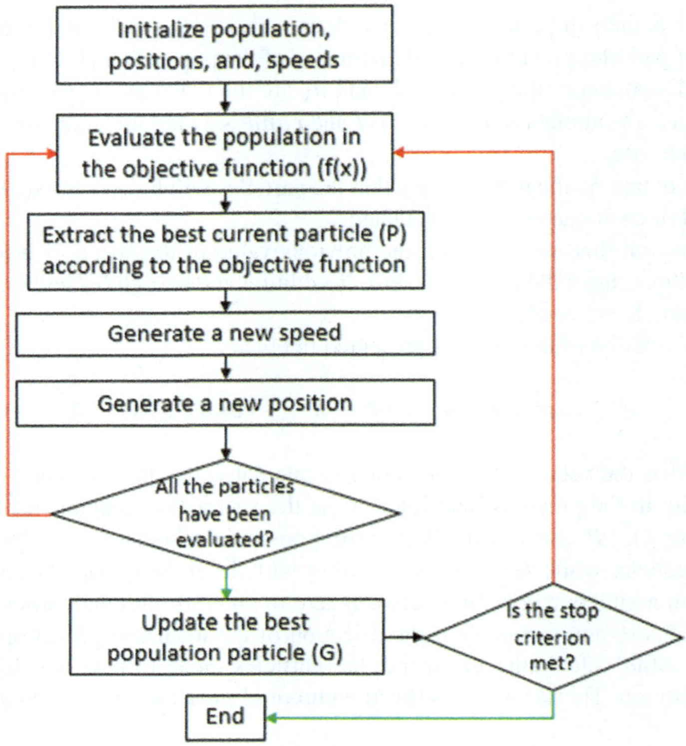

Fig. 9 Flow chart of the PSO algorithm

Fig. 10 Differential evolution algorithm on an unimodal function

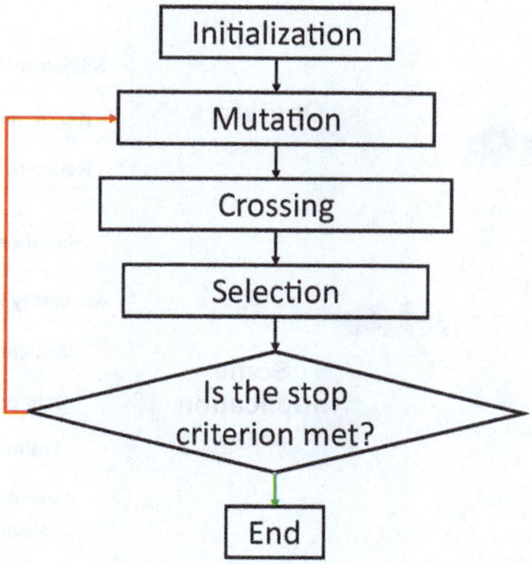

Fig. 11 Flow chart of the differential evolution algorithm

and uses many of its concepts such as multipoint searching, use of recombination, and selection operators.

It starts to explore the search space by sampling at multiple, randomly chosen initial points. Thereafter, the algorithm guides the population toward the vicinity of the global optimum through repeated cycles of reproduction and selection.

The differential evolution algorithm works with the following components: **initialization**, **mutation**, **crossing,** and **selection**. These components are mentioned in the flow chart of (Fig. 11).

We recommend addressing these concepts in [10] and [15] where each of these stages is thoroughly described and how it works.

4.4 Artificial Bee Colony

Swarm intelligence is a metaheuristic method in the field of artificial intelligence that is used to solve optimization problems. It is based on the collective behavior of some social insects, flocks of birds and schools of fish.

The Artificial Bee Colony algorithm (Fig. 12) is composed similar to an evolutionary algorithm. It is mainly based on generating a population of agents which forms the bee colony and is modified into three groups by corresponding operators. It was proposed by Dervis Karaboga in 2005 [17].

This algorithm is inspired by the intelligent search behavior of bee swarms. The method consists of three essential parts: positions of the food sources, amount of nectar and different kinds of bees. There are three types of bees, workers, observers, and explorers, each type performs a specific task collaboratively with the purpose

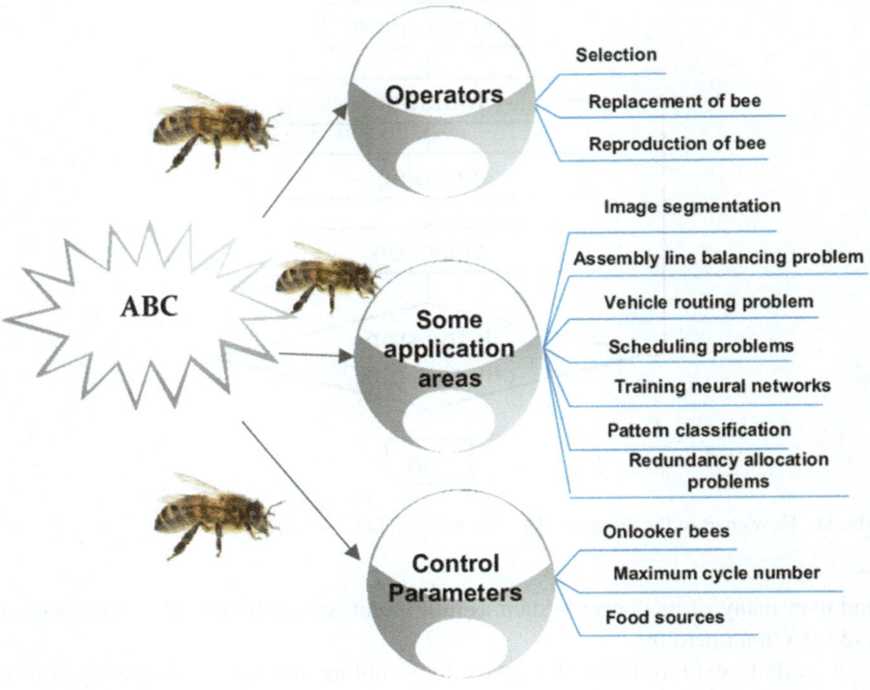

Fig. 12 ABC algorithm [18]

of finding a new candidate position of a food source. The observer and worker bees carry out the exploitation process in the search space and the scouting bees control the exploration process.

Half of the colony is composed of worker bees and in turn, they are the same number of food sources, the other half includes the observers. The worker bees look for food around the food source stored in their memory, while they pass this information to the observing bees, which are in charge of selecting the best sources of food from those that were found by the worker bees. The scout bees are transferred from a few worker bees, who previously left their food sources in search of new sources.

This algorithm begins with the random production of an initial population, which will be the candidate solutions. Then a cost function is evaluated which determines if the population positions are acceptable solutions to the problem.

The candidate solutions are modified by the three different operators of the algorithm, in a manner consistent with the values given by the cost function. Once the fitness value cannot be improved after a certain number of cycles, the corresponding source of food is abandoned, and it is reinitialized in a new random position, this until the stop criterion is met. We can see the flow chart of the algorithm (Fig. 13). We recommend [10] to address the issue in depth.

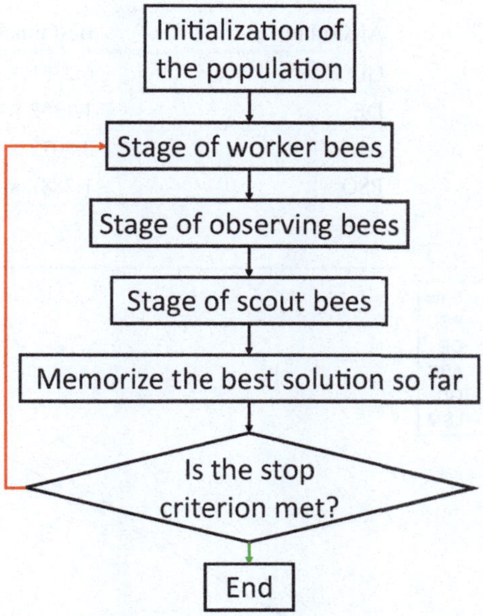

Fig. 13 Flow chart of the artificial bee colony algorithm

5 Results

To evaluate the performance, we conducted numerical experiments and compared the solution search performance of conventional GD, DE, ABC, and PSO versus the algebraic LSM (Least square method).

The following parameters were fixed for all tests: number of iterations (1500), number of particles (50). To ensure the fairness of the comparison tests, the tests were conducted such that each function was evaluated 200 times. The results of the experiment are shown in Tables 3 and 4 and (Fig. 14).

As we can infer and observe from the tables and the graph all the lines look similar between them, but the real difference is given by the best slope and also the best *y-intercept*. Notice that the gradient descendant for this problem shows a poor performance on its *y-intercept*, on the graph the best straight line of the gradient

Table 3 Experimental results

Algorithm	Best R^2	Best *y-intercept*	Best slope
LSM	0.838210531103428	142.7120	0.0001256
GD	0.822235650485832	$11.890546 \times 10^{-12}$	0.0001372
DE	0.838210414039373	142.4206	0.0001256
ABC	0.838210531103428	142.7120	0.0001256
PSO	0.798233589601288	149.2380	0.0001454

Table 4 Fitness function

Algorithm	Best fitness
GD	6.2351×10^4
DE	1.1352×10^5
ABC	8.8072×10^{-6}
PSO	1.2887×10^5

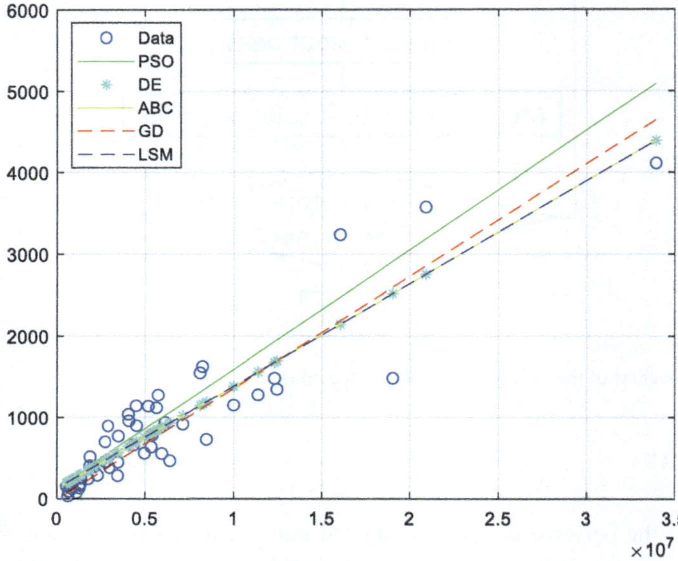

Fig. 14 Comparison of experimental results

descendant is represented by the color red and looks like if it has best position than the PSO line represented by green. And of course, it is clearly better than the PSO, we can conclude this because de R^2 of the gradient, descendant is close to the number 1. Regardless of whether the PSO algorithm made an adjustment in its interception. Finally the yellow and blue lines have the best R^2, the yellow one is the ABC line and the blue one was not from an algorithm, was the LSM. Also, there is another line represented by * in color cyan, this one is for the DE result. We can conclude that it is a good result but not the best. For a tiny difference on its R^2.

6 Conclusions

Simple linear regression has become one of the most widely statical tools for analyzing data. It is important to mention that we can use the same theory for multiple linear regression problems but whit little changes, another thing to consider is, that there are other popular methods to find the best line. As we saw in the chapter this analysis is a simple method for investigating functional relationships among two variables, accidents, and, states, this relationship was expressed in the form of a model connecting the response (dependent variable) and one predictor variable.

On the other hand, evolutionary algorithms are a big field of techniques to find solutions on statical functions and there is a branch for dynamic functions, but the concept is the same for both. The majority of the evolutionary algorithms set a population with different features and variables for movement to have a good performance on its exploration and exploitation techniques and it is interesting the way that the algorithms work. Some of them were inspired by social behaviors and others in biological and physic phenomenons.

We can conclude that there is no better method than all, every algorithm or every mathematical method have advantages and disadvantages between them, so the application is the main factor to decide which one to use.

References

1. S. Chatterjee, A.S. Hadi, *Regression Analysis by Example*, 5th edn. (2012)
2. E.V. Cuevas Jimenez, O. Avalos, J. Gálvez, Parameter estimation for chsaotic fractional systems by using the locust search algorithm. Comput y Sist **21** (2), 369–380 (2017)
3. D.C. Montgomery, E.A. Peck, G.G. Vining, V. González Pozo, Introducción al análisis de regresión lineal. Patria Cultural (2002)
4. MATLAB, Linear Regression—MATLAB. [Online]. Available: https://la.mathworks.com/help/matlab/data_analysis/linear-regression.html?lang=en. Accessed: 22 Apr 2019
5. H. Späth, Mathematical Algorithms for Linear Regression (1987)
6. D. Oliva, S. Hinojosa, E.V. Cuevas Jimenez, G. Pajares, O. Avalos, J. Gálvez, Cross entropy based thresholding for magnetic resonance brain images using crow search algorithm. Expert Syst. Appl. **79**, 164–180, Aug (2017)
7. O. Avalos et al., A comparative study of evolutionary computation techniques for solar cells parameter estimation. Comput. y Sist. **23**(1), 231 (2019)
8. P. Fleming, R. Purshouse, Evolutionary algorithms in control systems engineering: a survey. Control Eng. Pract. **10**(11), 1223–1241 (2002)
9. J.B. Pollack, H. Lipson, S. Ficici, P. Funes, G. Hornby, R.A. Watson, *Evolutionary techniques in physical robotics* (Springer, Berlin, Heidelberg, 2000), pp. 175–186
10. E.V. Cuevas Jimenez, J.V. Osuna Enciso, D.A. Oliva Navarro, M.A. Diaz Cortez, *Optimizacion. Algoritmos programados con MATLAB*, 1st edn. (ALFAOMEGA Grupo Editor, Guadalajara, 2016)
11. M. John, F. Kurtis, *Métodos numericos con MATLAB*. Prentice Hall (2000)
12. R. Reinhardt, A. Hoffmann, T. Gerlach, *Nichtlineare optimierung* (Springer, 2013)
13. J. Kennedy, R. Eberhart, Particle swarm optimization, in *Proceedings of the IEEE International Conference on Neural Networks* (1995)
14. A. Ahmadi, F. Karray, M.S. Kamel, Flocking based approach for data clustering. Nat. Comput. **9**, 767–791 (2010). https://doi.org/10.1007/s11047-009-9173-5
15. H. Iba, *Evolutionary Approach to Machine Learning and Deep Neural Networks: Neuro-evolution and Gene Regulatory Networks* (2018)
16. R. Storn, K. Price, Differential Evolution-A Simple and Efficient Heuristic for Global Optimization Over Continuous Spaces (Kluwer Academic Publishers, 1997)

17. K. Dervis, An Idea Based on Honey Bee Swarm for Numerical Optimization (Kayseri/Türkiye, 2005)
18. C. Rajan, K. Geetha, Investigation on bio-inspired population based metaheuristic algorithms for optimization problems in ad hoc networks. World Acad. Sci. Eng.Technol. **9**, 111–118 (2015)

Salp Swarm Algorithm: A Comprehensive Review

Essam H. Houssein, Ibrahim E. Mohamed and Yaser M. Wazery

Abstract Swarm Intelligence (SI) is referred to the social conduct emerging within decentralized and self-organization of swarms. These swarms are summarized as the well-known examples such as bird groups, fish schools, and the most social in insects species for instance bees, termites, and ants. Among those, Salp Swarm Algorithm (SSA), that has been successfully utilized and held in different fields of optimization, engineering practice, and real-world problems, so far. This review carries out a extensive study for the present status of publications, advances, applications, variants with SSA including its modifications, population topology, hybridization, extensions, theoretical analysis, and parallel implementation in order to show its potential to show its potential to overcome many practical optimization issues. Further, this review will be greatly useful for the researchers and algorithm developers analyzing at Swarm Intelligence, especially SSA to use this simple and yet very efficient approach for several tough optimization issues.

Keywords Swarm intelligence · Salp swarm algorithm · Nature-inspired algorithm · Engineering optimization problems · Multi-objective problems

1 Introduction

Artificial Intelligence (AI) is that the smartness displayed by machineries. It can be described as "the study and design of intelligent agents" [1], in which associate intelligent agents represent systems that identify their setting and take actions to exploit their goals. Now general methodologies of AI contain conventional statistical approaches [2], conventional symbolic AI, and Computational Intelligence (CI) [3]. It's a set of nature-inspired computing paradigms to capture information and make

E. H. Houssein (✉) · Y. M. Wazery
Faculty of Computers and Information, Minia University, Minia, Egypt
e-mail: essam.halim@mu.edu.eg

I. E. Mohamed
Faculty of Computers and Information, Luxor University, Luxor, Egypt

D. Oliva and S. Hinojosa (eds.), *Applications of Hybrid Metaheuristic Algorithms for Image Processing*, Studies in Computational Intelligence 890,
https://doi.org/10.1007/978-3-030-40977-7_13

sense of it to which conventional methodologies are inefficient or useless includes Evolutionary Computation (EC), Artificial Neural Network and fuzzy logic [4].

Swarm Intelligence (SI) is highly adapted to a group of mobile agents that are directly or indirectly communicated with each other, and collectively solve a set of basic problems that cannot be solved if the agents are operating independently. SI is in a pioneer tributary in computer science, a complex discipline that expresses a set of nature-inspired mathematical models that inspire from the collective behavior of natural or artificial decentralized and self-organized systems cum habits of organisms like plants, animals, fish, birds, ants and other elements in our ecosystem that employ the intuitive intelligence of the entire swarm/herd to solve some complex problems for a single agent [5–7]. The characteristics of swarm-based techniques are:

- They are population-based.
- Their searches are done using multiple agents.
- The agents that frame the population are typically homogenized.
- The collective behaviors of the system arise from each individual interaction with the other and with their environment.
- The agents are always moving randomly in a haphazard way.
- The agents' actions, principally movements are responsive to the environment.
- There is no centralized control, leaders' performances are solely the standard for their emergence in each iteration [8].

A collection of animals such as fish, fowls and insects acts on local information within a swarm, follow simple rules, and there is no centralized control. Due to perception in the neighborhood, these swarms (individual agents) are acting stochastically and without supervising. Among-st these individuals and the environment, the networks of interactions contain the intelligence of the swarm [9, 10]. Recently, methods of SI are employed in engineering, machine learning [11], image processing, data mining [12], MRI liver image [13], demanding robustness and flexibility [14].

A class of optimization algorithms called meta-heuristics has earned more popularity over precise methodologies to solve optimization problem due to the clarity and robustness of outcomes generated while being adopted in a wide range of different areas like engineering, business and even humanities. Which use a group of standards or theoretical constructs, the search process could be carried out via several agents that basically form a system of adapting solutions through different iterations. These iterations hold before some predetermined criterion reaches the solution. The final step is called optimal or near optimal solution [15–18].

In 2017, Mirjalili et al. [19] suggested a recent meta-heuristic,SSA, deeply influenced by the swarming behavior of deep-sea Salps. SSA aims to develop a new optimizer based on populations by attempting to mimic Salps ' swarming conduct in the natural environment [20]. The SSA is displayed in parallel and serial modes as proficient, adaptable, straightforward and user-friendly. The main goal of this article is to review the SSA, a relatively recent natural computing paradigm, Focused on the natural selection process mechanics, and connections with EC. Salp has been widely applied in complex optimization domains, and is currently an ongoing area of research, offering a competition to the current formulated EC techniques that may be employed in many of the same domains.

The remainder of the review is divided as follows; Sect. 2 first briefly touches the inspiration and mathematical model proposed for the SSA. Diverse methods of SSA are Shown and explained in Sect. 3. Section 4 summarizes several challenging real problems of SSA in terms of applications. Section 5 brings the discussion of SSA behavior, And routes for this algorithm's more improvements are covered. Lastly, Sect. 6 concludes the work properly and indicates multiple future directions for research.

2 Salp Swarm Algorithm (SSA)

SSA is an SI algorithm devised for continuous problem optimization. Compared to some existing algorithmic techniques, this algorithm seems to have a better or equivalent performance [21]. SSA is a multivariate algorithm in which optimization operation can be started with random initial population of solutions, and then improving these solutions over the time in two stages; Exploring and exploiting. In first stage, the promising regions are discovered by exploring the search space, while in exploitation, better solutions than the existing ones are hoped to be found by searching the neighborhood of specific solutions.

The wide popularity of nature inspired algorithms is because of their ability to find a better solution to real-world issues, their ability to avoid the local optimum, and their simplicity as well as their flexibility [22]. This is because they are inspired by the intelligent behaviors of swallows, such as human, swarms, physics, etc. There are a lot of research on SSA. The materials were gathered in this publication as a review article using the SSA as a key word via two phases. First, credit-worthy publishers like Elsevier's Journals, Springer's journals, and IEEE Xplore; in addition to various journals decided to search using academic search engine discussed the articles published on SSA.

Figure 1 displays the count of articles written on SSA by multiple academics between 2017 and 2019 as well as their variations based on various article data sets such as Elsevier, Springer, IEEE and more.

2.1 *Inspiration Analysis*

Salps fit into the gelatinous Salpidae family which are barrel-shaped and zoo plankton to form large swarms. They move slowly forwards through the sea as each zooid rhythmically contracts. This flow, hyped by muscle action, concurrently provides chemosensory details, food, exchange of respiratory gas, removal of solid and dissolved waste, sperm dispersal and propulsion by jet. To move forward, The body

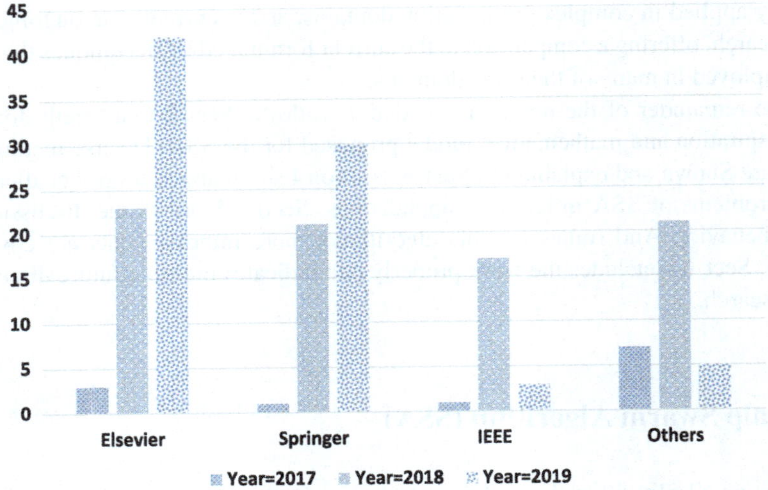

Fig. 1 Trends of SSA application from year 2017–2019

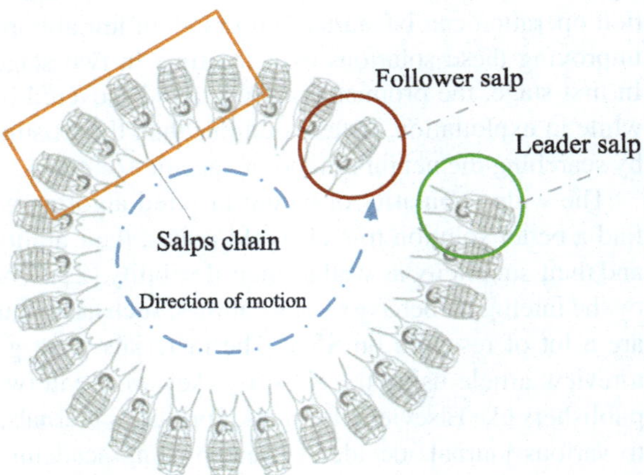

Fig. 2 Demonstration of Salp's series

pumps water as propulsion [23]. SSA is the first technique to imitative the behavior of Salps in nature. The Salps are marine organisms living in oceans and seas. They resemble Jellyfish in their tissues and movement towards food sources [23]. Groups (swarms) called Salp chains represent Salps; a leader and a set of followers are contained in each Salp chain. The leader Salp attacks directly the target (feeding Source), whereas all followers start moving direct or indirect to the rest of the Salps (and leader). Figure 2 Illustrates Salp chain. Each individual grows in size. However, while swimming and feeding, they remain attached together. This peculiar behavior is still obscure, but these attempts are believed to achieve better movement using foraging and rapid coordinated changes [24].

2.2 *Mathematical Model for Salp Chains*

The swarming behaviors [25] and population of Salp [26] are little in the literature to be mathematically modeled. Furthermore, to solve optimization problems, swarms of various animals (such as bees and ants) were commonly designed and utilized as mathematical model, while it is rare to find mathematical pattern physical processes (salp swarms) to solve various optimization issues. Through the next sub section, the stander model of Salp chains in the review is proposed [19] to solve different problems with optimizing process. Mathematically, the Salp chains are divided into two groups by random division of the population (Salps): leader and followers. The first Salp in the series of salps is called the leader, whereas the remaining of Salps are regarded followers. Through the given name of both types of these Salps, the leader directs swarms and the remaining of thesis series follow each other (and leader either explicitly or implicitly).

Given M is a counter for variables in a particular problem, Like the others SI-based methodologies, the Salps' location is denoted in a M-dimensional search space. Therefore, the population of Salps X is composed of N swarms with M dimension. It could therefore be identified by a $N \times M$ matrix, as outlined in the equation below:

$$X_i = \begin{bmatrix} x_1^1 & x_2^1 & \dots & x_M^1 \\ x_1^2 & x_2^2 & \dots & x_M^2 \\ \vdots & \vdots & \ddots & \vdots \\ x_1^N & x_2^N & \dots & x_M^N \end{bmatrix} \tag{1}$$

A feeding source termed F is also thought to be the target of the swarm in the search space. The position of the leader is updated by the following equation:

$$X_j^1 = \begin{cases} T_j + c_1((ub_j - lb_j)c_2 + lb_j) & c_3 \geq 0 \\ T_j - c_1((ub_j - lb_j)c_2 + lb_j) & c_3 < 0 \end{cases} \tag{2}$$

where X_j^1 and T_j denote the positions of leaders and feeding source in the jth dimension, respectively. The ub_j and lb_j indicates the upper (superior) and lower (inferior) bounds of jth dimension. c_2 and c_3 are two random floats from the closed interval [0, 1]. Actuality, they guide the next location in jth dimension toward the $+\infty$ or $-\infty$ besides determining the step size. Equation 2 indicates that the leader only updates its location with respect to the feeding source. The coefficient c_1, the most effective parameter in SSA, gradually decreases over the course of iterations to balance exploration and exploitation, and defined as follows:

$$c_1 = 2e^{-(\frac{4l}{L})^2} \tag{3}$$

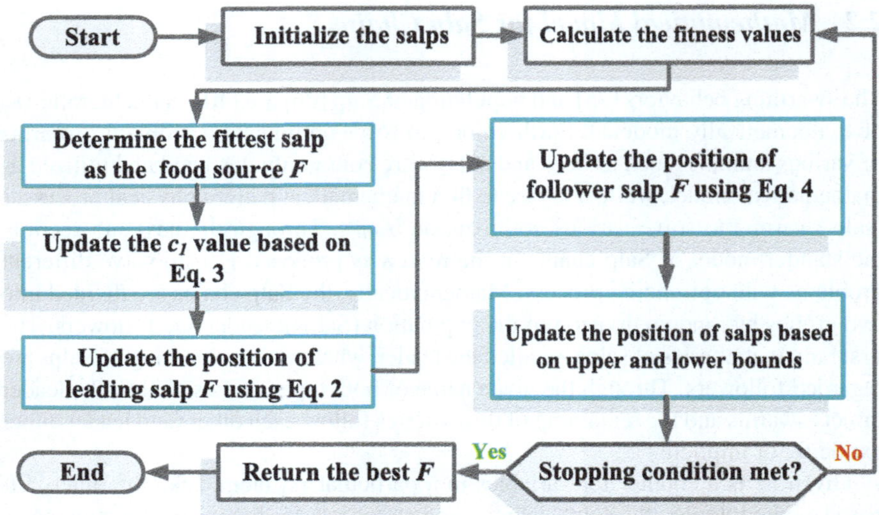

Fig. 3 The flowchart of SSA

where l and L represent the current iteration and maximum number of iterations, respectively. To update the position of the followers, the next equations is used (Newton's motion law):

$$X_j^i = \frac{X_j^i + X_j^{i-1}}{2} \tag{4}$$

where $i \geq 2$ and X_j^i is the location of the ith follower at the jth dimension. In SSA, the followers move toward the leader, whereas the leader moves toward the feeding source. During the process, the feeding source location can be changed and consequently the leader will move towards the new feeding source location. The flowchart of SSA is demonstrated in Fig. 3.

3 Various Methods of SSA

There are many techniques of SSA algorithms in the literature. To categorize them, a certain classification scheme is required. Figure 4 displays different SSA approaches.

3.1 Hybridization

Merge two or more algorithms to take the advantages of different algorithms while reducing their constraints. Without hybridization, certain techniques have failed to

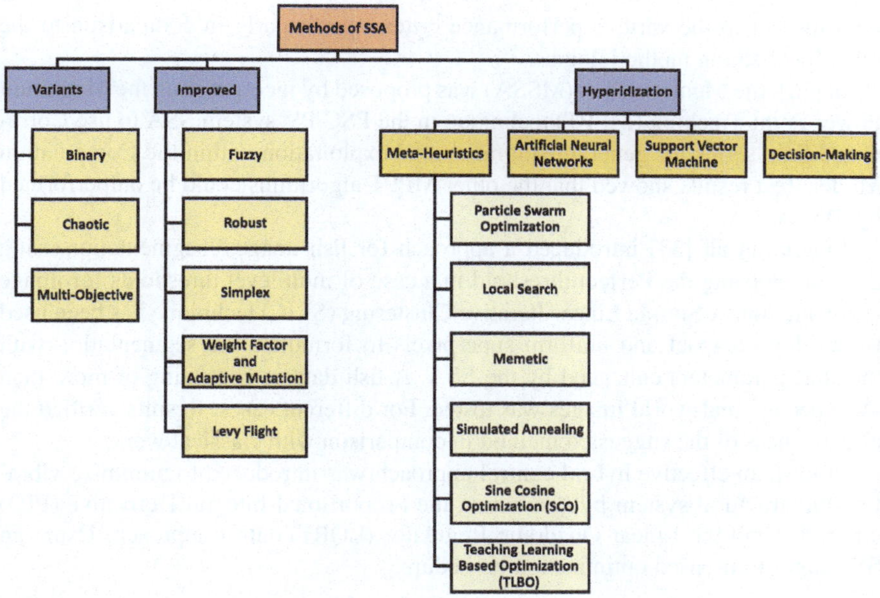

Fig. 4 Categorization of various methods of SSA

deliver better performance to solve a particular problem and improve its results. By hybridizing algorithms, we can fully enhance the algorithm's exploration and exploitation [27, 28].

3.1.1 Meta-Heuristic Algorithms

To enhance SSA's capacity for exploration and exploitation, Ibrahim et al. [29] have implemented the Hybrid SSA-PSO Algorithm (SSAPSO), a new population-based meta-heuristic hybrid algorithm. The implemented algorithm makes use of the PSO strategy's attributes to enhance the SSA's feasibility in finding solutions. The convergence percentage is therefore risen. The accuracy of SSAPSO is greater than the accuracy of the other algorithms, depending on the classification.

Liu et al. [30] suggested a novel method combining the SSA with a local search strategy to figure out the non-linear problem of Time-Difference-Of-Arrival (TDOA) passive position. Experimental test has demonstrated that, compared to PSO and Enhanced PSO, SSA has significantly increased location accuracy, fewer control parameters, and more robust performance, the developed method can converge rapidly and stably to the passive TDOA target location. In [31], Kanoosh et al. addressed the node location issue as an optimization issue by using the SSA-based node location algorithm to solve it. Different demographic nodes and anchor nodes have been used in various Wireless Sensor Networks (WSNs) installations to enforce and validate the proposed algorithm. The proposed algorithm has proved superior with respect to the various performance criteria particularly in comparison to the other localization methodologies.

In [32], the Memetic SSA (MSSA) was proposed by incorporating the Maximum Power Point Tracking (MPPT) technique in the PSC PV system SSA to use Correspond Salp series for general exploration and exploitation within the Computation Model. Test results showed that the other MPPT algorithms could be outperformed by MSSA.

Ibrahim et al. [33] introduced a approach for fish images' segmentation using SSA in choosing the Perfect threshold in a case of multilevel thresholds for image segmentation. A Simple Linear Iterative Clustering (SLIC) technique has been used to develop compact and uniform superpixels to formulate the segmentation with the start parameters enhanced by the SSA. A fish dataset consisting of more than 400 species' real-world images was tested. For different cases, Results verified the effectiveness of the suggested method in comparison with classical work.

In [34], an effective hybrid control approach was introduced to minimize vibration on structural system by combining the Proportional Integral Derivative (PID) controller and the Linear Quadratic Regulator (LQR) control approach. Using an SSA algorithm-based optimization procedure.

3.1.2 Artificial Neural Networks

The Artificial Neural Networks (ANNs) are responsible for learn phase through practice [35]. Neural Components are widely connected with various modules such as prediction, modeling and control of the system. In [36], in order to perform pattern classification, an SSA-based effective method was presented to adjust the parameters of NN connections weights.

In [37], SSA is utilized to train Feed-Forward Neural Networks (FNNs). The planned strategy applies to various conventional classification and regression datasets. Results also appreciate the better and balanced exploitation-exploration properties of the SSA optimization algorithm, making the algorithm favorable for neural network training.

3.1.3 Support Vector Machine

Support Vector Machine (SVM) [38] is a administered algorithms for machine learning. Those machines are built to support Categorization as well as degeneration tasks by analysing information records and discovering specific patterns.

In [39], to predict carbon dioxide emissions (CO_2) using the principles of SVM, a Least Squares Support Vector Machine (LS-SVM) model has been Suggested by Zhao et al. Compared to the other selected models, the proposed model has shown more predicting performance. The statistical results reported SSA-LSSVM model's great superiority and ability to enhance the accuracy and reliability of CO_2 emissions predicting.

3.1.4 Decision-Making

Game Theory (GT), which is a principle for life form and informatics of intelligent decision-making, applies to a wide range of behavioral relationships [40]. The coalition-based GT model has been used in [41] to formalize the cost scheme model. Usually, electricity is charged per unit using Demand Response (DR); dynamic or time-based price models. These price rates shift the user load and provide the incentives. By using the proposed model, the per-unit price is determined by both energy consumption and extra generation costs. The extra cost of generation is distributed among each user using the shapely value according to the electricity load profile.

A Comparison of the Meta-heuristic and other methods in publication papers has been made. Ratio of SSA hybrid publication papers with meta-heuristic, SVM and ANNs is shown in Fig. 5. Based on the numerical analysis shown in Fig. 5, there is more ratio of meta-heuristic algorithms.

Figure 6 displays the Diagram of the number of SSA publication according to published year. As we can see in the Diagram, The highest possible number of articles published in 2019 is clearly part of the area of meta-heuristic methodologies. Highest possible SSA hybridization is associated with meta-heuristic methodologies.

3.2 *Improved SSA*

This part is about enhancing SSA by using Chaotic, Robust, Simplex, and Weight Factor and Adaptive Mutation.

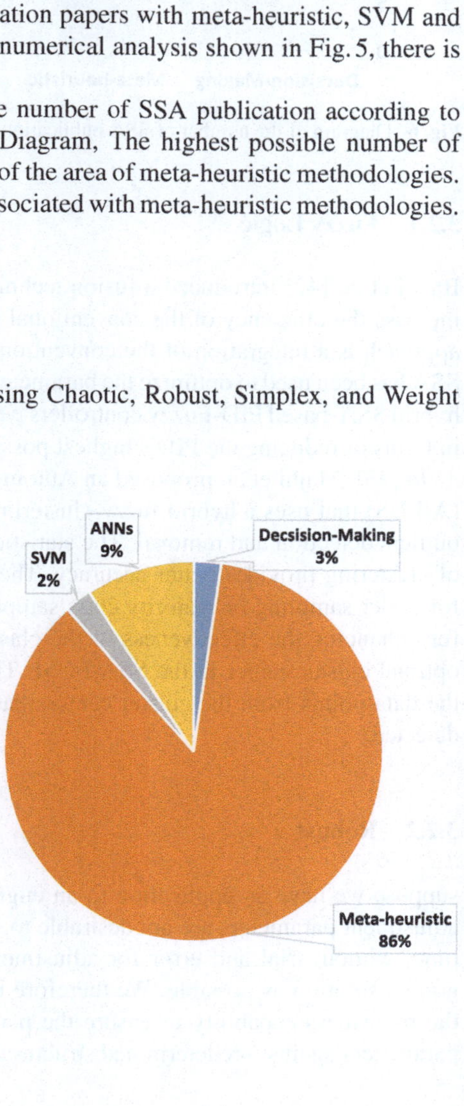

Fig. 5 Ratio of SSA hybrid publication with meta-heuristic, SVM and ANNs

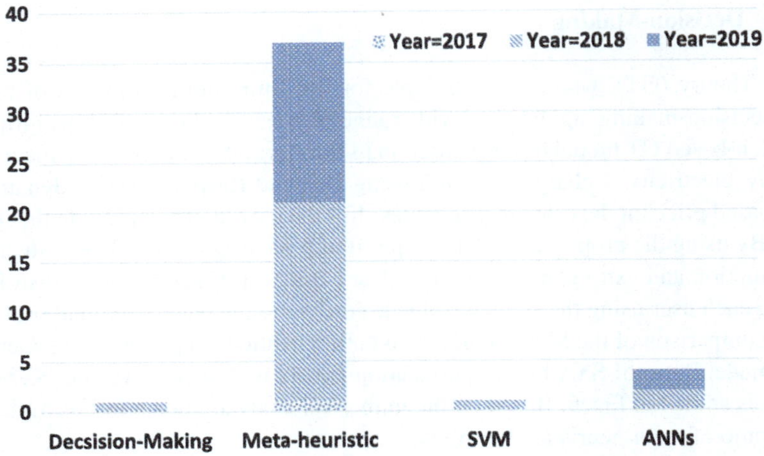

Fig. 6 Diagram of the number of SSA publication based on published year

3.2.1 Fuzzy Logic

Baygi et al. [42] introduced a fusion technique known as PID-Fuzzy controllers to increase the efficiency of the conventional PID and fuzzy controller. The proposed approach is a integration of the conventional PID with the fuzzy logic controllers. SSA has been used to optimize the parameters of the PID. This study indicate that the hybrid SSA-based PID-Fuzzy controllers perform greater solitary than other controls in terms of reducing the PID's highest possible displacement and maximum speed.

In [43], Majhi et al. proposed an Automobile Insurance Fraud Detection System (AIFDS) that uses a hybrid fuzzy clustering technique using SSA (SSA-FCM) for outliers detection and removal. The statistical tests revealed that the used approach of clustering provides better accuracy. The proposed fuzzy clustering was applied for under-sampling of majority class samples of the automobile insurance data set for enhancing the effectiveness of the classifiers. The SSA helps in obtaining the optimal cluster center in the SSA-FCM. The SSA-FCM calculates the distance of the data-points from the cluster centers based on which the suspicious classes are detected.

3.2.2 Robust

suppose we have an application to an engineering optimization problem, multiple adjustment parameters are not desirable to solve a problem. In a limited calculation time, without trial and error for adjustment parameters, it is necessary to get as good a solution as possible. We therefore believe that meta-heuristics should have the robustness capability to ensure the performance of searching for pre-adjusted parameters against predetermined structural variation of problems to be solved.

Tolba et al. [44] developed a novel methodology rely on SSA to find and optimize the size of Renewable Distribution Generator (RDGs) and Shunt Capacitor Banks (SCBs) on Radial Distributed Network (RDNs). Approach with various objectives, functions and different constraint conditions to enhance the voltage level, properly reduce energy loss and annual operating costs.

Fathy et al. adopted SSA in [45] to efficiently manage fuel cell, chargers and super-capacitors in the fusion generating systems. The hybrid system was combined with a DC-DC Adapter to provide the highly fluctuated load. The purpose of the designated technique was to reduce the total hydrogen consumption via enlarging the energy supplied from both the super-capacitors and the batteries. The outcomes of the proposed SSA have been compared with other traditional meta-heuristic approaches. The comparison concerned the quantity of hydrogen usage and the strategy's performance. The suggested approach was observed identical to other methods with a consumption of hydrogen and better efficiency.

3.2.3 Simplex

The simplex methods are strategies of stochastic variants, which maximizes population diversity and improves the algorithm's local search capability. This approach helps to accomplish a better trade off between the swarm algorithm's exploration and exploitation capabilities and makes swarm algorithm more robust and faster. The simplex method [46] has the strong ability to avoid local optimum and enhance the ability of searching the global optimum.

In contrast with some traditional algorithms, the SSA has demonstrated better performance, but there are still certain drawbacks such as spend too long time in research phase, and need to enhance the ability of convergence speed and calculation accuracy. To overcome the above problems, a Simplex Method-based SSA (SMSSA) was Suggested in [47] to enhance the precision of the convergence of basic SSA. Experimental results showed that SMSSA achieves not only faster convergence speed, but also better solutions compared with other algorithms. By combining the advantages of both techniques, SMSSA can get a balance between exploitation and exploration to deal with classical engineering problems.

3.2.4 Weight Factor and Adaptive Mutation

A weighting factor is usually used for calculating a weighted mean, to give less (or more) importance to group members. For balancing between global exploration and local exploitation, dynamic weight factor is Included in Updating the formulation of the Place of population. In addition, during the process of evolution, an adaptive mutation strategy is introduced to avoid premature convergence and stagnation of evolution.

Wu et al. [48] offered an enhanced SSA, by introducing two enhancement methods depend on Weight Factor and Adaptive Mutation (WASSA) to enhance the convergence accuracy, accelerate the convergence rate and avoid early convergence.

To increase the improvement of the convergence ratio, reliableness and accuracy, Hegazy et al. [49] tried to improve the architecture of indigenous SSA by adding a new control parameter and inertia weight to provide the current best solution. The novel approach called enhanced SSA was tested in FS task. To evaluate the efficiency of resulting algorithm where it works better than other algorithms.

To solve hybrid TDOA and angle of arrival (TDOA-AOA) nonlinear equation problems, Chen et al. [50] proposed an improved SSA (ISSA). By modifying the leadership fraction and designing a new convergence factor equation, The ISSA can better maintain the capacity for exploration and exploitation than the indigenous SSA. Experimental result showed that ISSA has better exploration capabilities and the algorithm's threshold effect is improved.

In [51], for solving FS problem, KHAMEES et al. developed two new FS approaches. The first strategy implemented the indigenous SSA to identify the ideal features group based on wrapper mode to fix FS problems in information mining assignments. The second strategy was hybrid of SSA with mutation operator (SSAMUT), it proved obviously notable efficiency.

3.2.5 Levy Flight

Levy flight is a specific random walk category that distributes the step size according to the tails of heavy power law. Sometimes, the large step help an approach performs a global search. It'd be helpful to use the Levy flight trajectory [52, 53] to gain a better trade-off between exploring and exploiting the algorithm and based on the local optimum avoidance, it has plus points.

The image can be adaptively segmented according to the pixels rely on the Grayscale Co-occurrence Matrices (GLCM), but as the number of threshold's increases, the segmentation effect becomes worse. For solving this kind of problems, The modern Diagonal Class Entropy (DCE) was used as the fitness function of the GLCM algorithm to enhance multi-threshold GLCM image segmentation approach to enhance the efficiency of SSA according to a Levy flight trajectory-based SSA (LSSA) [54]. Different evaluation techniques is used to determine the value of segmented images through algorithm experiments of the Berkeley images, the satellite images, and color natural images such as displacement error, global consistency error, boundary, feature similarity, peak signal-to-noise ratio, variation of information, and probability rand index. The experimental results of segmentation and robustness affirmed the sublimity of GLCM-LSSA algorithm. Therefore, due to good image segmentation ability, complex image segmentation tasks can be better handled by the GLCM-LSSA algorithm.

From the statistical results contained in Fig. 7, it can be inferred that the chaotic algorithms excel in the ratio more than SSA enhancement publications

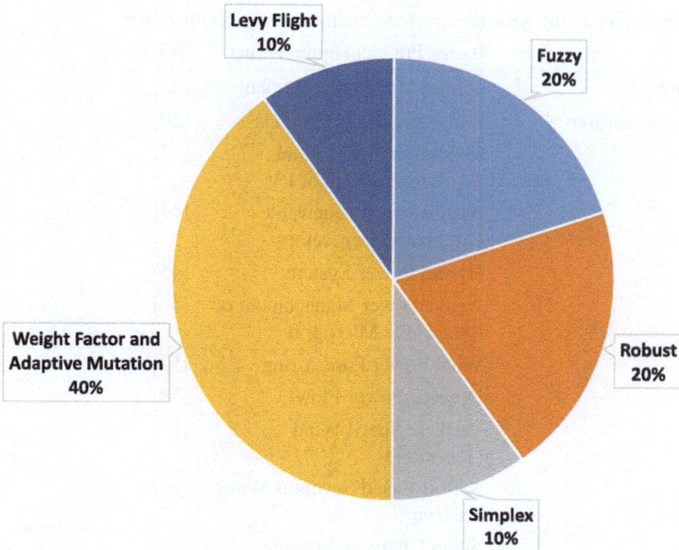

Fig. 7 Percentage of SSA enhancement publications

3.3 *Variants of SSA*

SSA and its variants have been applied for solving many optimization and classification problems, as well as several industrial fields. The taxonomy of the developed SSA algorithm applications is illustrated in Table 1.

3.3.1 Binary

Different types of optimization problem cannot be solved by meta-heuristics. Binary optimization problem has diverse decision factors that are the components of interval [0, 1]. In term, 0 or 1 represents a digital meaning of each decision variable of the binary problem. Problems such as Features Selection (FS) [88], unit commitment problem [89, 90], and Knapsack Problems (KP) [91, 92].

Using a modified Arctan transformation, Rizk-Allah et al. [93] suggested a novel binary approach of the SSA called BSSA with the objective of transforming the continuous space to binary space. To enhance the exploration and exploitation capabilities, multiplicity and mobility are two advantages with regard to the modification transfer function.

Table 1 A summary of the SSA applications in engineering optimization

Class	Issues/Practical applications	References
Cloud Computing	Virtual Machine Placement	[55]
Electrical Engineering	Photovoltaic Cell Models	[20]
	Maximum Power Point Tracking (MPPT) of PV	[32]
	Allocation of Renewable Distributed Generators	[44]
	Hybrid Power System	[45]
	Active Power Management of Renewable Microgrid	[56]
	Wind Power Forecasting	[57]
	Optimal Power Flow	[58]
	Variable Speed Wind Generators	[59]
	Airfoil Based Savonius Wind Turbine	[60]
	Smart Charging Scheme	[61]
	Wind Energy Forecasting	[62]
	High Power Amplifier (PA)	[63]
	Power Loss Reduction	[64]
	Power System Stabilizer	[65]
	Simultaneous Optimal Distributed Generators (DGs)	[66]
	Polymer Exchange Membrane Fuel Cells (PEMFCs)	[67]
	Economic Dispatch and Emission Problem	[68]
Civil Engineering	Control Design of Structural System	[34]
	Control Design of Structural System against Earthquake	[42]
	Mechanically Stabilized Earth Wall (MSEW)	[69]
Classification	Extreme Learning Machines (ELMs)	[21]
	Multi-Objective Feature Selection	[70]
	Training Neural Networks for Pattern Classification	[36]
	Feed-Forward Neural Network (FNN)	[37]
	Feature Selection for Data Classification	[71]
	Feature Selection	[72, 73]

(continued)

Table 1 (continued)

Class	Issues/Practical applications	References
Clustering	Automobile Insurance Fraud Detection	[43]
Image Processing	Fish Image Segmentation	[33]
	Multilevel Color Image Segmentation	[54]
Mechanical Engineering	Spring Design Problem	[47]
Chemistry	Predicting Chemical Compound Activities	[74]
Control Engineering	Load Frequency Control	[75]
	Automatic Generation Control (AGC)	[76]
	Multistage PDF Plus (1+PI) Controller in AGC	[77]
Robotics	Mother-son Robotic System	[78]
Networks	Node Localization in Wireless Sensor Networks	[31]
	SDN Multi-Controller Placement Problem	[79]
Task Scheduling	Scheduling of User Appliances	[80]
	Job Shop Scheduling Problems	[81]
	Energy Efficient Scheduling	[82]
Other Problems	Time Difference of Arrival (TDOA)	[30]
	Optimal Design of CMOS Differential Amplifier	[83]
	Forecasting Energy-Related CO_2 Emissions	[39]
	Demand Side Management	[41]
	Global Optimization	[84]
	Graph Coloring Problem	[85]
	Soil Water Retention Curve Model	[86]
	Doha Reverse Osmosis Desalination Plant	[87]

3.3.2 Chaotic

SSA can approximate optimum solution with high convergence, but SSA is not yet beneficial in searching the optimum solution which effect the algorithm performance. Therefore, to decrease this impact and to enhance its potential and effectiveness, Chaotic SSA (CSSA) was proposed by Sayed et al. [84] by merging between SSA and chaos algorithm. Chaos a novel numerical approach has recently been used to improve the execution of meta-heuristic approaches. Chaos is described as simulation for self-motivated conduct of non-linear system [94]. Population of meta-heuristic methods have the same advantages include scalable approach, simplicity, and reduce computation time. However, these Approaches have two intrinsic weaknesses; low convergence rate and recession in local optima [95].

To solve the graph coloring problem, Meraihi et al. [85] proposed a new Chaotic Binary SSA (CBSSA). First, the Binary SSA (BSSA) was gained from the stander SSA where the S-Shaped transfer function is used. Second, a common chaotic map, called logistic map was used. Using the well-known DIMACS benchmark instances, the performance of the proposed approach asserted in comparison with various relevant colouring methods. The experimental results verified the performance and strength of the proposed CBSSA approach compared to aforesaid algorithms.

In [71], five chaotic maps have been utilized for Diagnosing and designing different feature selection methods based on SSA for data classification. Ateya et al. [79] introduced the latency and cost aware Software-Defined Networking (SDN) controller placement problem. To minimize the deployment cost and the latency, to achieve the optimum amount of controllers and also the ideal allocation of switches to controllers, a CSSA method was constructed. By introducing chaotic maps, optimizer efficiency was enhanced and local optima was prevented. The method has been evaluated for different real functionality from the topology of the zoo. The effect of variation of different network parameters on the performance was checked. In terms of reliability and execution time, simulation outcomes proved that the introduced algorithm outperforms a GT-based approach in addition to meta-heuristic methodologies.

To get over the potential shortcomings of native SSA, Zhang et al. [96] improved an SSA-based optimizer. The designed variant was defined as Chaos-induced and Mutation-driven SSA (CMSSA) that simultaneously combines two approaches. First, to boost the exploitation of the algorithm, the basic SSA was used to introduce a chaotic exploitative mechanism with "shrinking" mode. Then, to get full benefit from the reliable diversification of Cauchy mutation and the strong intensification of Gaussian mutation, a combined mutation scheme was adapted. The statistical tests on representative benchmark showed the effectiveness of the proposed method in solving optimization and engineering design problems by alleviating the precocious convergence of SSA.

3.3.3 Multi-Objective Problems (MOPS)

This part of our publication denotes to present the basics of MOPS [97, 98]. The target of MOPS is assumed to minimize or maximize incompatible numerous objectives function [99, 100]. On the contrary, to improve individual target problems, including multiple objective and MOPS objectives contradict each other. A difficult task of MOPS to find a optimal solution to optimize objectives of each function concurrently. Therefore, balancing should be done between all objectives of each function to achieve an optimum solution collection.

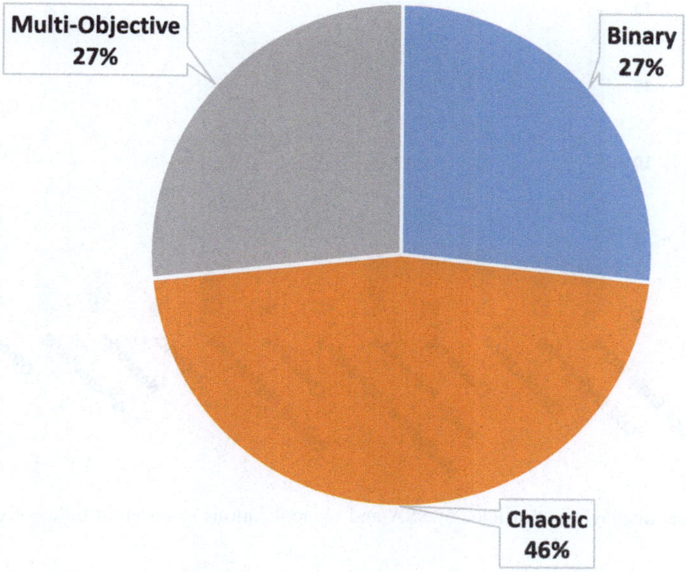

Fig. 8 Percentage of SSA variants papers

To find out a appropriate solution for virtual machine locating problem, Alresheedi et al. [55] combined the SSA and sine cosine algorithm(SCA) with the means of improving MOPS techniques (MOSSASCA). The main purposes of the proposed MOSSASCA are to minimizing quality of services infringements, to reduce power consumption, and maximizing average time before a agent shutdown in addition to minimizing conflict between the three objectives. In SCA, to increase convergence speed and to avoid getting trapped on a local optimum solution, a local search technique is followed to increase the performance of SSA. Various virtual and physical machines were in a set of experiments to evaluate the performance of the combined Algorithm. Well-known MOP methods were compared with the results of MOSSASCA. Results indicates that a balance between achieving the three objectives. Based on the statistical result shown in Fig. 8, the chaotic Algorithm and its variants have a higher proportion on SSA's publications by about 46% while Binary algorithms and Multi-objective algorithms come in second place with percentage about 27% from available publications in different journals.

4 Applications of SSA in Optimization Problems

SSA can be helpful Algorithm in problem solving and it intensely used in solve optimization problems in engineering and other several fields. The goals of any optimization algorithm for any problem in real world are defining suitable factors

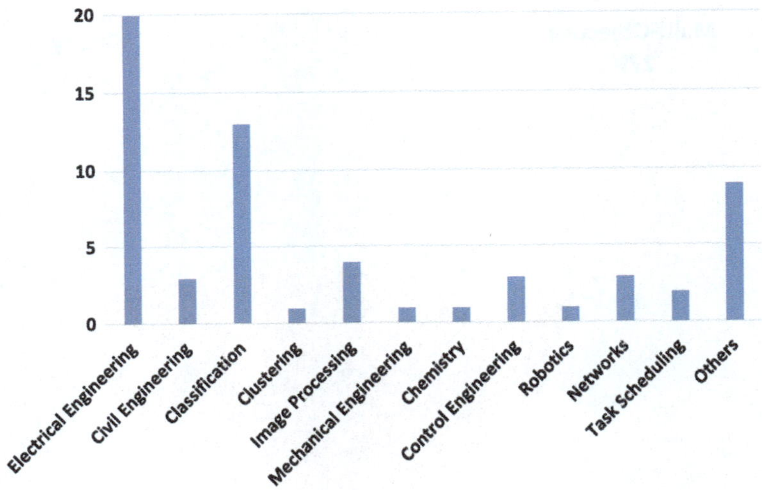

Fig. 9 Chart displaying of articles of SSA and its applications in different fields of engineering problems

and maximize objective function while dealing with engineering Problem. Here, a brief review is carried out by classifying the application fields. Several engineering optimization and fields of the SSA are classified into the next categories: Electrical systems, Civil Engineering, Machine learning, Images Processing, Mechanical Systems, Control Systems, robotics, Network Management, manufacturing Engineering, and other several applications. Table 1 provides a overview of the applications of SSA in optimization problems. Each row of the Table 1 enumerates the functionalists provided by the SSA and its variants and issues or piratical application.

Figure 9 shows statistics of articles of SSA and its application in several fields of optimization problems. Overall, many academics have made a lot of improvements and modifications previously. In short, to achieve the expected results, distinct targets or objectives of the literature review demand different methods. It is not possible to increase the solution of any problem to just one technique. The solution's advancement will therefore be expanded continuously.

5 Discussion

The method of acquisition optimum solutions of a problem or mathematical pattern by increasing or decreasing the value of objective function is known as global numerical optimization. The improvement of effective optimization practices is becoming further important, serious, and urgent than before because optimization tasks reused in several real-world issues are complex to develop.

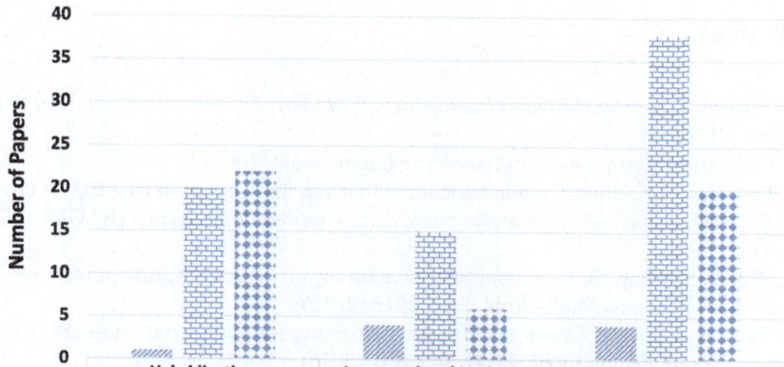

	Hybridization	Improved and Variants	Applications
Year=2017	1	4	4
Year=2018	20	15	38
Year=2019	22	6	20

Fig. 10 The total number published for the SSA publication in various ways such as hybridization and enhanced and adaptation methods plus applications based on publishing time

From the review, trend patterns have taken place between 2017 and 2019. Figure 10 displays several types of hybridisation, upgraded, modifications, and applications as far as the absolute number of publication reviews from every year. As showed in the figure, 2018 and 2019 years demonstrate a high use of the SSA compared to 2017 year. This shows that the use of SSA in the area of optimization has increased significantly in the last two years.

6 Conclusion

We reviewed one of the latest successful meta-heuristic algorithms called Salp Swarm Algorithm (SSA) in this chapter, which is influenced by the swarming behavior of salps in waters. Moreover, we provided a brief summarizing for the core of SSA As well as its applications. Results from different studies have been investigated. We included those studies from various publications as summarized in Table 1. We examined SSA's effectiveness of SSA in different fields such as engineering, classification, optimization, industrial and other hard issues. The paper performers a wide-ranging review of the SSA in various terms such as classification, optimization, feature selection and intelligent decision-making. We investigated a compare between original SSA and its variants in term of performance for exploration and exploitation phase. The future studies can enhance the original exploration and exploitation operators of this algorithm. They also can utilize the proposed SSA and its variants for other applications.

References

1. S. Russell, P. Norvig, *Artificial Intelligence: A Modern Approach Prentice-Hall* (Englewood cliffs, NJ, 1995)
2. B.L. Agarwal, *Basic Statistics* (New Age International, 2006)
3. K.E. Voges, N.K. Pope, Computational intelligence applications in business: A cross-section of the field, in *Business Applications and Computational Intelligence* (Igi Global, 2006), pp. 1–18
4. Y. Zhang, S. Wang, G. Ji, A comprehensive survey on particle swarm optimization algorithm and its applications. Math. Probl. Eng. **2015** (2015)
5. V. Pandiri, A. Singh, Swarm intelligence approaches for multidepot salesmen problems with load balancing. Appl. Intell. **44**(4), 849–861 (2016)
6. A.A. Ewees, M.A. Elaziz, E.H. Houssein, Improved grasshopper optimization algorithm using opposition-based learning. Expert. Syst. Appl. **112**, 156–172 (2018)
7. A.G. Hussien, E.H. Houssein, A.E. Hassanien, A binary whale optimization algorithm with hyperbolic tangent fitness function for feature selection, in *2017 Eighth International Conference on Intelligent Computing and Information Systems (ICICIS)* (IEEE, 2017), pp. 166–172
8. R.S. Parpinelli, H.S. Lopes, New inspirations in swarm intelligence: a survey. Int. J. Bio-Inspired Comput. **3**(1), 1–16 (2011)
9. A. Hamad, E.H. Houssein, A.E. Hassanien, A.A. Fahmy, Hybrid grasshopper optimization algorithm and support vector machines for automatic seizure detection in eeg signals, in *International Conference on Advanced Machine Learning Technologies and Applications* (Springer, 2018), pp. 82–91
10. M.M. Ahmed, E.H. Houssein, A.E. Hassanien, A. Taha, E. Hassanien, Maximizing lifetime of wireless sensor networks based on whale optimization algorithm, in *International Conference on Advanced Intelligent Systems and Informatics* (Springer, 2017), pp. 724–733
11. A. Hamad, E.H. Houssein, A.E. Hassanien, A.A. Fahmy, A hybrid eeg signals classification approach based on grey wolf optimizer enhanced svms for epileptic detection, in *International Conference on Advanced Intelligent Systems and Informatics* (Springer, 2017), pp. 108–117
12. A.E. Hassanien, M. Kilany, E.H. Houssein, H. AlQaheri, Intelligent human emotion recognition based on elephant herding optimization tuned support vector regression. Biomed. Signal Process. Control. **45**, 182–191 (2018)
13. S. Said, A. Mostafa, E.H. Houssein, A.E. Hassanien, H. Hefny, Moth-flame optimization based segmentation for mri liver images, in *International Conference on Advanced Intelligent Systems and Informatics* (Springer, 2017), pp. 320–330
14. D. Karaboga, B. Gorkemli, C. Ozturk, N. Karaboga, A comprehensive survey: artificial bee colony (abc) algorithm and applications. Artif. Intell. Rev. **42**(1), 21–57 (2014)
15. A.G. Hussien, A.E. Hassanien, E.H. Houssein, S. Bhattacharyya, M. Amin, S-shaped binary whale optimization algorithm for feature selection, in *Recent Trends in Signal and Image Processing* (Springer, 2019), pp. 79–87
16. A.A. Ismaeel, I.A. Elshaarawy, E.H. Houssein, F.H. Ismail, A.E. Hassanien, Enhanced elephant herding optimization for global optimization. IEEE Access **7**, 34738–34752 (2019)
17. M.M. Ahmed, E.H. Houssein, A.E. Hassanien, A. Taha, E. Hassanien, Maximizing lifetime of large-scale wireless sensor networks using multi-objective whale optimization algorithm. Telecommun. Syst. 1–17 (2019)
18. E.H. Houssein, A. Hamad, A.E. Hassanien, A.A. Fahmy, Epileptic detection based on whale optimization enhanced support vector machine. J. Inf. Optim. Sci. **40**(3), 699–723 (2019)
19. S. Mirjalili, A.H. Gandomi, S.Z. Mirjalili, S. Saremi, H. Faris, S.M. Mirjalili, Salp swarm algorithm: A bio-inspired optimizer for engineering design problems. Adv. Eng. Softw. **114**, 163–191 (2017)
20. R. Abbassi, A. Abbassi, A.A. Heidari, S. Mirjalili, An efficient salp swarm-inspired algorithm for parameters identification of photovoltaic cell models. Energy Convers. Manag. **179**, 362–372 (2019)

21. H. Faris, S. Mirjalili, I. Aljarah, M. Mafarja, A.A. Heidari, Salp swarm algorithm: Theory, literature review, and application in extreme learning machines, in *Nature-Inspired Optimizers* (Springer, 2020), pp. 185–199
22. M. Mafarja, D. Eleyan, S. Abdullah, S. Mirjalili, S-shaped vs. v-shaped transfer functions for ant lion optimization algorithm in feature selection problem, in *Proceedings of the International Conference on Future Networks and Distributed Systems* (ACM, 2017), p. 21
23. L.P. Madin, Aspects of jet propulsion in salps. Can. J. Zool. **68**(4), 765–777 (1990)
24. P. Anderson, Q. Bone, Communication between individuals in salp chains. ii. physiology. Proc. R. Soc. London. Ser. B. Biol. Sci. **210**(1181), 559–574 (1980)
25. V. Andersen, P. Nival, A model of the population dynamics of salps in coastal waters of the ligurian sea. J. Plankton Res. **8**(6), 1091–1110 (1986)
26. N. Henschke, J.A. Smith, J.D. Everett, I.M. Suthers, Population drivers of a thalia democratica swarm: insights from population modelling. J. Plankton Res. **37**(5), 1074–1087 (2015)
27. R. Šenkeřík, I. Zelinka, M. Pluhacek, A. Viktorin, J. Janostik, Z. K. Oplatkova, Randomization and complex networks for meta-heuristic algorithms, in *Evolutionary Algorithms, Swarm Dynamics and Complex Networks* (Springer, 2018), pp. 177–194
28. I. Fister, D. Strnad, X.-S. Yang, Adaptation and hybridization in nature-inspired algorithms, in *Adaptation and Hybridization in Computational Intelligence* (Springer, 2015), pp. 3–50
29. R.A. Ibrahim, A.A. Ewees, D. Oliva, M.A. Elaziz, S. Lu, Improved salp swarm algorithm based on particle swarm optimization for feature selection. J. Ambient. Intell. Hum. Ized Comput. 1–15 (2018)
30. X. Liu, H. Xu, Application on target localization based on salp swarm algorithm, in *37th Chinese Control Conference (CCC)*. (IEEE, 2018), pp. 4542–4545
31. H.M. Kanoosh, E.H. Houssein, M.M. Selim, Salp swarm algorithm for node localization in wireless sensor networks. J. Comput. Netw. Commun. **2019** (2019)
32. B. Yang, L. Zhong, X. Zhang, H. Shu, T. Yu, H. Li, L. Jiang, L. Sun, Novel bio-inspired memetic salp swarm algorithm and application to mppt for pv systems considering partial shading condition. J. Clean. Prod. **215**, 1203–1222 (2019)
33. A. Ibrahim, A. Ahmed, S. Hussein, A.E. Hassanien, Fish image segmentation using salp swarm algorithm, in *International Conference on Advanced Machine Learning Technologies and Applications* (Springer, 2018), pp. 42–51
34. S.M.H. Baygi, A. Karsaz, A hybrid optimal pid-lqr control of structural system: A case study of salp swarm optimization, in *2018 3rd Conference on Swarm Intelligence and Evolutionary Computation (CSIEC)* (IEEE, 2018), pp. 1–6
35. G. Villarrubia, J.F. De Paz, P. Chamoso, F. De la Prieta, Artificial neural networks used in optimization problems. Neurocomputing **272**, 10–16 (2018)
36. A.A. Abusnaina, S.Ahmad, R.Jarrar, M.Mafarja, Training neural networks using salp swarm algorithm for pattern classification, in *Proceedings of the 2nd International Conference on Future Networks and Distributed Systems* (ACM, 2018), p. 17
37. D. Bairathi, D. Gopalani, Salp swarm algorithm (ssa) for training feed-forward neural networks, in *Soft Computing for Problem Solving* (Springer, 2019), pp. 521–534
38. B. Ghaddar, J. Naoum-Sawaya, High dimensional data classification and feature selection using support vector machines. Eur. J. Oper. Res. **265**(3), 993–1004 (2018)
39. H. Zhao, G. Huang, N. Yan, Forecasting energy-related co2 emissions employing a novel ssa-lssvm model: Considering structural factors in china. Energies **11**(4), 781 (2018)
40. R.B. Myerson, *Game Theory* (Harvard University Press, 2013)
41. A. Khalid, Z.A. Khan, N. Javaid, Game theory based electric price tariff and salp swarm algorithm for demand side management, in *Fifth HCT Information Technology Trends (ITT)*. (IEEE, 2018), pp. 99–103
42. S.M.H. Baygi, A. Karsaz, A. Elahi, A hybrid optimal pid-fuzzy control design for seismic exited structural system against earthquake: A salp swarm algorithm, in *6th Iranian Joint Congress on Fuzzy and Intelligent Systems (CFIS)*. (IEEE, 2018), pp. 220–225
43. S.K. Majhi, S. Bhatachharya, R. Pradhan, S. Biswal, Fuzzy clustering using salp swarm algorithm for automobile insurance fraud detection. J. Intell. Fuzzy Syst. **36**(3), 2333–2344 (2019)

44. M. Tolba, H. Rezk, A. Diab, M. Al-Dhaifallah, A novel robust methodology based salp swarm algorithm for allocation and capacity of renewable distributed generators on distribution grids. Energies **11**(10), 2556 (2018)
45. A. Fathy, H. Rezk, A.M. Nassef, Robust hydrogen-consumption-minimization strategy based salp swarm algorithm for energy management of fuel cell/supercapacitor/batteries in highly fluctuated load condition. Renew. Energy **139**, 147–160 (2019)
46. X.-S. Yang, *Engineering Optimization: An Introduction with Metaheuristic Applications* (Wiley, 2010)
47. D. Wang, Y. Zhou, S. Jiang, X. Liu, A simplex method-based salp swarm algorithm for numerical and engineering optimization, in *International Conference on Intelligent Information Processing* (Springer, 2018), pp. 150–159
48. J. Wu, R. Nan, L. Chen, Improved salp swarm algorithm based on weight factor and adaptive mutation. J. Exp. Theor. Artif. Intell. 1–23 (2019)
49. A.E. Hegazy, M. Makhlouf, G.S. El-Tawel, Improved salp swarm algorithm for feature selection. J. King Saud Univ.-Comput. Inf. Sci. (2018)
50. T. Chen, M. Wang, X. Huang, Q. Xie, Tdoa-aoa localization based on improved salp swarm algorithm, in *2018 14th IEEE International Conference on Signal Processing (ICSP)* (IEEE, 2018), pp. 108–112
51. M. KHAMEES, A.Y. ALBAKR, K. SHAKER, A new approach for features selection based on binary slap swarm algorithm. J. Theor. Appl. Inf. Technol. **96**(7) (2018)
52. X.-S. Yang, S. Deb, Cuckoo search via lévy flights, in *World Congress on Nature and Biologically Inspired Computing (NaBIC)*. (IEEE, 2009), pp. 210–214
53. A.F. Kamaruzaman, A.M. Zain, S.M. Yusuf, A. Udin, Levy flight algorithm for optimization problems-a literature review, in *Applied Mechanics and Materials*, vol. 421. (Trans Tech Publ, 2013), pp. 496–501
54. Z. Xing, H. Jia, Multilevel color image segmentation based on glcm and improved salp swarm algorithm. IEEE Access (2019)
55. S.S. Alresheedi, S. Lu, M.A. Elaziz, A.A. Ewees, Improved multiobjective salp swarm optimization for virtual machine placement in cloud computing. Hum.-Centric Comput. Inf. Sci. **9**(1), 15 (2019)
56. A.K. Barik, D.C. Das, Active power management of isolated renewable microgrid generating power from rooftop solar arrays, sewage waters and solid urban wastes of a smart city using salp swarm algorithm, in *Technologies for Smart-City Energy Security and Power (ICSESP)*. (IEEE, 2018), pp. 1–6
57. P. Jiang, R. Li, H. Li, Multi-objective algorithm for the design of prediction intervals for wind power forecasting model. Appl. Math. Model. **67**, 101–122 (2019)
58. A.A. El-Fergany, H.M. Hasanien, Salp swarm optimizer to solve optimal power flow comprising voltage stability analysis. Neural Comput. Appl. 1–17 (2019)
59. M.H. Qais, H.M. Hasanien, S. Alghuwainem, Enhanced salp swarm algorithm: Application to variable speed wind generators. Eng. Appl. Artif. Intell. **80**, 82–96 (2019)
60. M. Masdari, M. Tahani, M.H. Naderi, N. Babayan, Optimization of airfoil based savonius wind turbine using coupled discrete vortex method and salp swarm algorithm. J. Clean. Prod. **222**, 47–56 (2019)
61. K. Kasturi, M.R. Nayak, Assessment of techno-economic benefits for smart charging scheme of electric vehicles in residential distribution system. Turk. J. Electr. Eng. Comput. Sci. **27**(2), 685–696 (2019)
62. W. Yang, J. Wang, H. Lu, T. Niu, P. Du, Hybrid wind energy forecasting and analysis system based on divide and conquer scheme: a case study in china. J. Clean. Prod. (2019)
63. M. Malhotra, A.S. Sappal, Ssa optimized digital pre-distorter for compensating non-linear distortion in high power amplifier. Telecommun. Syst. pp. 1–10 (2019)
64. D. Yodphet, A. Onlam, A. Siritaratiwat, P. Khunkitti, Electrical distribution system reconfiguration for power loss reduction by salp swarm algorithm. Int. J. Smart Grid Clean Energy
65. S. Ekinci, B. Hekimoglu, Parameter optimization of power system stabilizer via salp swarm algorithm, in *2018 5th International Conference on Electrical and Electronic Engineering (ICEEE)* (IEEE, 2018), pp. 143–147

66. M.S. Asasi, M. Ahanch, Y.T. Holari, Optimal allocation of distributed generations and shunt capacitors using salp swarm algorithm, in *Iranian Conference on Electrical Engineering (ICEE)* (IEEE, 2018), pp. 1166–1172
67. A.A. El-Fergany, Extracting optimal parameters of pem fuel cells using salp swarm optimizer. Renew. Energy **119**, 641–648 (2018)
68. B. Mallikarjuna, Y. S. Reddy, R. Kiranmayi, Salp swarm algorithm to combined economic and emission dispatch problems. Int. J. Eng. Technol. **7**(3.29), 311–315 (2018)
69. A.B. Sereshki , A. Derakhshani, Optimizing the mechanical stabilization of earth walls with metal strips: Applications of swarm algorithms. Arab. J. Sci. Eng. 1–14 (2018)
70. M. Khamees, A. Albakry, K. Shaker, Multi-objective feature selection: Hybrid of salp swarm and simulated annealing approach, in *International Conference on New Trends in Information and Communications Technology Applications* (Springer, 2018), pp. 129–142
71. A.E. Hegazy, M. Makhlouf, G.S. El-Tawel, Feature selection using chaotic salp swarm algorithm for data classification. Arab. J. Sci. Eng. 1–16 (2018)
72. S. Ahmed, M. Mafarja, H. Faris, I. Aljarah, Feature selection using salp swarm algorithm with chaos, in *Proceedings of the 2nd International Conference on Intelligent Systems, Metaheuristics and Swarm Intelligence* (ACM, 2018), pp. 65–69
73. I. Aljarah, M. Mafarja, A.A. Heidari, H. Faris, Y. Zhang, S. Mirjalili, Asynchronous accelerating multi-leader salp chains for feature selection. Appl. Soft Comput. **71**, 964–979 (2018)
74. A.G. Hussien, A.E. Hassanien, E.H. Houssein, Swarming behaviour of salps algorithm for predicting chemical compound activities, in *2017 Eighth International Conference on Intelligent Computing and Information Systems (ICICIS)* (IEEE, 2017), pp. 315–320
75. P.C. Sahu, S. Mishra, R.C. Prusty, S. Panda, Improved-salp swarm optimized type-ii fuzzy controller in load frequency control of multi area islanded ac microgrid. Sustain. Energy, Grids Netw. **16**, 380–392 (2018)
76. T.K. Mohapatra, B.K. Sahu, Design and implementation of ssa based fractional order pid controller for automatic generation control of a multi-area, multi-source interconnected power system, in *Technologies for Smart-City Energy Security and Power (ICSESP)* (IEEE, 2018), pp. 1–6
77. P.C. Sahu, R.C. Prusty, S. Panda, Salp swarm optimized multistage pdf plus (1+ pi) controller in agc of multi source based nonlinear power system, in *International Conference on Soft Computing Systems* (Springer, 2018), pp. 789–800
78. S. Guo, S. Sun, J. Guo, Design of a sma-based salps-inspired underwater microrobot for a mother-son robotic system, in *2017 IEEE International Conference on Mechatronics and Automation (ICMA)* (IEEE, 2017), pp. 1314–1319
79. A.A. Ateya, A. Muthanna, A. Vybornova, A.D. Algarni, A. Abuarqoub, Y. Koucheryavy, A. Koucheryavy, Chaotic salp swarm algorithm for sdn multi-controller networks, Eng. Sci. Technol. Int. J. (2019)
80. H.M. Faisal, N. Javaid, U. Qasim, S. Habib, Z. Iqbal, H. Mubarak, An efficient scheduling of user appliances using multi objective optimization in smart grid, in *Workshops of the International Conference on Advanced Information Networking and Applications* (Springer, 2019), pp. 371–384
81. Z.-X. Sun, R. Hu, B. Qian, B. Liu, G.-L. Che, Salp swarm algorithm based on blocks on critical path for reentrant job shop scheduling problems, in *International Conference on Intelligent Computing* (Springer, 2018), pp. 638–648
82. S. Khan, Z.A. Khan, N. Javaid, S.M. Shuja, M. Abdullah, A. Chand, Energy efficient scheduling of smart home, in *Workshops of the International Conference on Advanced Information Networking and Applications* (Springer, 2019), pp. 67–79
83. S. Asaithambi, M. Rajappa, Swarm intelligence-based approach for optimal design of cmos differential amplifier and comparator circuit using a hybrid salp swarm algorithm. Rev. Sci. Instrum. **89**(5), 054702 (2018)
84. G.I. Sayed, G. Khoriba, M.H. Haggag, A novel chaotic salp swarm algorithm for global optimization and feature selection. Appl. Intell. **48**(10), 3462–3481 (2018)

85. Y. Meraihi, A. Ramdane-Cherif, M. Mahseur, D. Achelia, A chaotic binary salp swarm algorithm for solving the graph coloring problem, in *International Symposium on Modelling and Implementation of Complex Systems*(Springer, 2018), pp. 106–118
86. J. Zhang, Z. Wang, X. Luo, Parameter estimation for soil water retention curve using the salp swarm algorithm. Water **10**(6), 815 (2018)
87. N. Patnana, S. Pattnaik, V. Singh, Salp swarm optimization based pid controller tuning for doha reverse osmosis desalination plant. Int. J. Pure Appl. Math. **119**(12), 12707–12720 (2018)
88. H. Faris, M.M. Mafarja, A.A. Heidari, I. Aljarah, A.-Z. Ala'M, S. Mirjalili, H. Fujita, An efficient binary salp swarm algorithm with crossover scheme for feature selection problems. Knowl.-Based Syst. **154**, 43–67 (2018)
89. L.K. Panwar, S. Reddy, A. Verma, B.K. Panigrahi, R. Kumar, Binary grey wolf optimizer for large scale unit commitment problem. Swarm Evol. Comput. **38**, 251–266 (2018)
90. Y.-K. Wu, H.-Y. Chang, S.M. Chang, Analysis and comparison for the unit commitment problem in a large-scale power system by using three meta-heuristic algorithms. Energy Procedia **141**, 423–427 (2017)
91. Y. He, X. Wang, Group theory-based optimization algorithm for solving knapsack problems. Knowl.-Based Syst. (2018)
92. E. Ulker, V. Tongur, Migrating birds optimization (mbo) algorithm to solve knapsack problem. Procedia Comput. Sci. **111**, 71–76 (2017)
93. R.M. Rizk-Allah, A.E. Hassanien, M. Elhoseny, M. Gunasekaran, A new binary salp swarm algorithm: development and application for optimization tasks. Neural Comput. Appl. 1–23 (2018)
94. L. dos Santos Coelho, V.C. Mariani, Use of chaotic sequences in a biologically inspired algorithm for engineering design optimization. Expert. Syst. Appl. **34**(3), 1905–1913 (2008)
95. K.-L. Du, M. Swamy, Particle swarm optimization, in *Search and Optimization by Metaheuristics* (Springer, 2016), pp. 153–173
96. Q. Zhang, H. Chen, A.A. Heidari, X. Zhao, Y. Xu, P. Wang, Y. Li, C. Li, Chaos-induced and mutation-driven schemes boosting salp chains-inspired optimizers. IEEE Access **7** 31243–31261 (2019)
97. S.Z. Mirjalili, S. Mirjalili, S. Saremi, H. Faris, I. Aljarah, Grasshopper optimization algorithm for multi-objective optimization problems. Appl. Intell. **48**(4), 805–820 (2018)
98. A. Tharwat, E.H. Houssein, M.M. Ahmed, A.E. Hassanien, T. Gabel, Mogoa algorithm for constrained and unconstrained multi-objective optimization problems. Appl. Intell. 1–16 (2017)
99. A. Zhou, B.-Y. Qu, H. Li, S.-Z. Zhao, P.N. Suganthan, Q. Zhang, Multiobjective evolutionary algorithms: A survey of the state of the art. Swarm Evol. Comput. **1**(1), 32–49 (2011)
100. B. Qu, Y. Zhu, Y. Jiao, M. Wu, P.N. Suganthan, J. Liang, A survey on multi-objective evolutionary algorithms for the solution of the environmental/economic dispatch problems. Swarm Evol. Comput. **38**, 1–11 (2018)

Health Applications

Segmentation of Magnetic Resonance Brain Images Through the Self-Adaptive Differential Evolution Algorithm and the Minimum Cross-Entropy Criterion

Itzel Aranguren, Arturo Valdivia and Marco A. Pérez

Abstract The segmentation is regarded as a vital step in preprocessing techniques for image analysis. Automatic segmentation of brain magnetic resonance images has been extensively investigated since with a precise segmentation can be identified and diagnosed several brain diseases. Thresholding is an important simple but efficient technique of image segmentation. Various strategies have been submitted to find optimal thresholds. Amongst those methods, the minimum cross-entropy (MCE) has been broadly implemented due to its simpleness. Although MCE is quite effective in bilevel thresholding, the computational cost increases exponentially the higher the number of thresholds (*th*) to find. This article introduces a new approach called MCE-SADE for multilevel thresholding using the Self-Adaptive Differential Evolution (SADE) algorithm. SADE is a robust metaheuristic algorithm (MA) that resolve general problems efficiently since, through evolution, the parameters and the proper learning strategy are continuously adjusted pursuant to prior knowledge. The optimum *th* values are found minimizing cross-entropy through SADE algorithm. The proposed method is tested in two groups of reference images; the primary group is formed of standard test images, while the following group consists of brain magnetic resonance images. In turn, MCE-SADE is compared with two metaheuristic algorithms, Grey Wolf Optimizer (GWO) and Competitive Imperialist Algorithm (ICA). From the experimental results, it is observed that MCE-SADE results improve in terms of consistency and quality in contrast to GWO and ICA based methods.

Keywords Magnetic resonance image · Metaheuristic algorithms · Minimum cross-entropy · Multilevel thresholding · Segmentation · Self-Adaptive Differential Evolution

I. Aranguren (✉) · A. Valdivia · M. A. Pérez
División de Electrónica y Computación, Universidad de Guadalajara, CUCEI, Guadalajara, Jalisco, México
e-mail: niasandiu.aranguren@alumno.udg.mx

A. Valdivia
e-mail: arturo.valdivia@academicos.udg.mx

M. A. Pérez
e-mail: marco.perez@cucei.udg.mx

D. Oliva and S. Hinojosa (eds.), *Applications of Hybrid Metaheuristic Algorithms for Image Processing*, Studies in Computational Intelligence 890,
https://doi.org/10.1007/978-3-030-40977-7_14

1 Introduction

Medical imaging has become essential for medical diagnosis, treatment evaluation, and surgical planning. Different modalities are used to acquire medical images as Positron Emission Tomography (PET), Ultrasonography (US), Magnetic Resonance (MR) and Computed Tomography (CT). MR is a non-invasive type system that provides high spatial resolution and detailed information of anatomical structures. Nevertheless, the analysis of magnetic resonance (MR) images is complex since they are affected due to artifacts as the intensity non-uniformity (INU), the voluntary or involuntary movement of the patient and the partial volume effect (PVE) [1]. Segmentation is a vital step in preprocessing techniques for medical image analysis [2–4].

Segmentation befalls through the division into non-overlapped consistent areas of an image that share certain attributes such as texture, shape or intensity value of gray. The segmentation of magnetic resonance (MR) brain images has been widely researched since with a correct segmentation of brain can be identified brain illnesses such as multiple sclerosis (MS), schizophrenia, Alzheimer's disease (AD) and dementia [3]. Frequently, MR brain images are analyzed based on the experience and the visual capacity of the expert; notwithstanding, is a time-wasting complex task limited by human vision that is not able to distinguish most of the gray levels found in an MR image [3]. Therefore, computer-aided techniques are necessary to automatically analyze and segment MR images. In the area of MR brain images segmentation have been submitted diverse methodologies, see for example [4–6]. They can be classified into thresholding techniques, region growing methods, clustering approaches, and model-based techniques [7].

Thresholding is a simple but efficient image segmentation technique, which aims to distribute into different sets the pixels of an image by setting a distinct threshold (*th*) intensity values. Thresholding techniques are classified according to the number of *th* levels as bilevel and multilevel thresholding (BTH and MTH, respectively). The BTH divides the image pixels by a single *th* value into two classes, whereas in MTH multiple *th* values separate the image into several significant classes. Numerous research studies have been proposed to determine the optimum *th* value, for example, the works of Otsu [8], Kittler and Illingworth [9], Pal [10] and Shanbhag [11] to list some. Sezgin and Sankur [12] distributed the thresholding techniques according to the data they are exploiting in six categories. Between these categories, the methods founded on entropy have caught the attentiveness of researchers, because they have demonstrated to be more efficient in image segmentation [13].

Entropy establishes an index of statistic diversity on the gray intensity levels present in an image. Several entropy-based methods have been proposed, such as Shannon entropy [14], Kapur entropy [15], minimum cross-entropy [16], Tsallis Entropy [17], Renyi entropy [18], among others. In the literature, the minimum cross-entropy (MCE) formulated by Li and Lee [16] has been broadly employed for image segmentation. This method elects the optimum *th* value that minimizes the cross-entropy among the initial image and the processed image. Initially, the MCE, like other entropy-based methods, was produced for a BTH and later expanded to MTH. Whereas that bi-level thresholding searches for the optimal threshold value quickly and efficiently, the multi-level thresholding exhaustively calculates the values of the thresholds, which exponentially increases the computational cost. To improve the search for optimal thresholds and reduce the computational cost of conventional thresholding methods, researchers have implemented metaheuristic algorithms (MA) to settle the MTH problem [19].

Metaheuristic algorithms (MA) provide high-quality solutions to different optimization problems. The MTH is a problem of optimization whose fitness function is a thresholding technique. Over the years, various MA had been applied to multilevel thresholding problem such as Genetic Algorithm (GA) [20, 21], Particle Swarm Optimization (PSO) [22, 23], Differential Evolution (DE) [24–26], Artificial Bee Colony (ABC) [27, 28] and Harmony Search (HS) [29, 30] to name a few; all the works cited above have been tested on general purpose images normally utilized in the image treatment literature. Significant contributions have also been made to the segmentation of brain MR images using MA, as the method proposed by Kaur, Saini, and Gupta that segments brain tumors in MR images using the PSO algorithm with a two-dimensional minimum cross entropy [31]. Oliva et al. present a MR brain image segmentation methodology founded on the Crow Search Algorithm (CSA) with MCE as fitness function [32]. Sathya and Kayalvizhi apply an upgraded Bacterial Foraging Optimization (BFO) algorithm for MTH of MR brain images [33]. Ali et al. propose to segment MR medical images using a hybrid DE algorithm aided with a Gaussian curve fitting [34].

Although the cited approaches are able to obtain acceptable results, they present a deficiency of balance rate among their exploration and exploitation traits. The ability of a metaheuristic algorithm to reach global optima depends on the adequate balance rate of exploration and exploitation. In general, the procedure of the aforementioned algorithms performs a modification the position of each member of the population in the current generation, then in the next generation, the members of the population are attracted towards the best member solution known so-far, leaving aside the exploration. Then most of the algorithms perform a fast movement towards the vicinity of the best candidate found, provoking an early convergence [35]. In turn, the MA are not designed to solve all the problems, they are planned to solve certain problems in a significantly better way than others [36]. Each year several new algorithms are proposed, which are evaluated according to a generic comparison methodology that involves numerical bench problems whose results are statistically analyzed to define which algorithm surpasses others [37]. However, few studies evaluate the performance of MA according to an application, so it is important to continue developing application context assessments in metaheuristic algorithms.

The several outcomes published by multiple authors to this particular optimization problem motivated the implementation of a novel and effective MTH image segmentation technique founded on the Self-Adaptive Differential Evolution (SADE) metaheuristic algorithm. From the literature, it is noted that the advantages of the SADE algorithm for solving the multilevel threshold image segmentation problem have not been investigated. The submitted approach called MCE-SADE selects optimal threshold values using as fitness function the minimum cross-entropy (MCE). Two sets of reference images were used to verify the quality of the segmentation of our method. The first group is formed from standard test images broadly employed in the field of image processing. The following group consists of magnetic resonance brain images. To examine the effectiveness and robustness of MCE-SADE, its outcomes are compared with those provided through two metaheuristic algorithms, Grey Wolf Optimizer (GWO) [38] and Competitive Imperialist Algorithm (ICA) [39]. The experiments are validated in an objective, subjective and statistical comparisons. The principal work contribution is the employment of the Self-Adaptive Differential Evolution (SADE) metaheuristic algorithm for MTH image segmentation through applying the minimum cross-entropy criterion. Experimental outcomes evidence that MCE-SADE obtains better results in respect of consistency and quality, in comparison to GWO and ICA based methods, it is also demonstrated that the technique can be implemented to segment both standard test images and complex MR brain images.

The remnant of this work is orderly from the following form. Section 2 examines the entropy-based techniques and details the minimum cross-entropy. Section 3 introduces the self-adaptive optimization algorithm. Section 4 describes the submitted multilevel thresholding technique. All experimental results of MCE-SADE and its discussion are exhibited in Sect. 5. The statistical analysis for both reference images groups are presented in Sect. 6. Lastly, in Sect. 7 the conclusions of the work are summed.

2 Entropy-Based Thresholding Segmentation

A description of uncertainty related to a particular data set is measured by the entropy. In the literature, there are several entropy-based thresholding methods. These are classified into three categories according to [12] as entropic thresholding (ET), fuzzy entropic thresholding (FET) and cross-entropic thresholding (CET). The ET estimates two distinct signal sources on the image, forefront and backdrop. In the FET, the fuzzy memberships evidence how mightily a gray intensity value relates to backdrop or forefront. Lastly, the CET formulates the optimum *th* value as the minimizing of a data theoretic distance.

2.1 Minimum Cross-Entropy

The cross-entropy proposed by Kullback [40] establishes a data theoretic distance D amongst two probability functions. In their work of minimum cross-entropy (MCE), Li and Lee [16] modify Kullback's method to select the optimum *th* value that minimizes the cross-entropy among initial image and processed image. Suppose $R = \{r_1, r_2, \ldots, r_N\}$ be the probability function of reference image and $S = \{s_1, s_2, \ldots, s_N\}$ the probability function of segmented image, then the D among the two probability functions $D(R, S)$ is determined by:

$$D(R,S) = \sum_{i=1}^{N} r_i \log\left(\frac{r_i}{s_i}\right) \tag{1}$$

where the set r_i and s_i proceeds from a selfsame position in the reference image and segmented image respectively. Let I be the reference image and $h^r(i) = 1, 2, \ldots, L$ be its histogram with L denoting the total number of values of gray intensities present in the image. Thus, the processed image defined through I_{th} employing th as the threshold value that partition the image into two different areas (forefront and backdrop) is computed as:

$$I_{th}(x,y) \equiv \begin{cases} \mu(1, th), & I(x,y) < th \\ \mu(th, L+1), & I(x,y) \geq th \end{cases} \tag{2}$$

where,

$$\mu(a,b) = \sum_{i=a}^{b-1} ih^r(i) \Big/ \sum_{i=a}^{b-1} h^r(i) \tag{3}$$

therefore, the cross-entropy for a BTH is estimated by:

$$D(th) = \sum_{i=1}^{th-1} ih^r(i) \log\left(\frac{i}{\mu(1,th)}\right) + \sum_{i=th}^{L} ih^r(i) \log\left(\frac{i}{\mu(th, L+1)}\right) \tag{4}$$

Based on Eq. (4), the MCE algorithm seeks the optimum th values t^* through the minimization of cross-entropy, as follows:

$$t^* = \arg\min_t (D(th)) \tag{5}$$

2.2 *Multilevel Thresholding*

This MCE bilevel method its used when the image contains only two classes and thus required only one threshold. As the image complexity increases, MCE can be expanded into an MTH approach. Suppose $\boldsymbol{Th} = [th_1, th_2, \ldots, th_n]$ is a vector that contains n thresholds values, where $th_1 < th_2 < \cdots < th_n$, then the multilevel cross-entropy among the image of reference and the thresholded image is described by:

$$D(\boldsymbol{Th}) = \sum_{i=1}^{th_1-1} ih^r(i)\log\left(\frac{i}{\mu(1, th_1)}\right) + \sum_{i=th_1}^{th_2-1} ih^r(i)\log\left(\frac{i}{\mu(th_1, th_2)}\right) + \cdots + \sum_{i=th_n}^{L} ih^r(i)\log\left(\frac{i}{\mu(th_n, L+1)}\right) \quad (6)$$

Multilevel MCE determine the optimal thresholds $\boldsymbol{t}^* = [t_1^*, t_2^*, \ldots, t_n^*]$ through the minimization of cross-entropy into the objective function $\boldsymbol{t}^* = \arg\min_t(D(\boldsymbol{Th}))$, therefore Eq. (6) is expressed as:

$$\begin{aligned} D(\boldsymbol{Th}) = &\sum_{i=1}^{L} ih^r(i)\log(i) - \sum_{i=1}^{th_1-1} ih^r(i)\log(\mu(1, th_1)) \\ &- \sum_{i=th_1}^{th_2-1} ih^r(i)\log(\mu(th_1, th_2)) \\ &- \cdots - \sum_{i=th_{k-1}}^{th_k-1} ih^r(i)\log(\mu(th_{k-1}, th_k)) \\ &- \cdots - \sum_{i=th_n}^{L} ih^r(i)\log(\mu(th_n, L+1)), \quad k = 1, 2, \ldots, n-1 \end{aligned} \quad (7)$$

For the case of bilevel thresholding the MCE method result efficacious, nonetheless, when the problem its extended to multilevel thresholding, the computation time increases considerably. Therefore, to enhance the computation speed and precision in image segmentation, this paper incorporated SADE algorithm to choose the optimum *th* values through minimization of the fitness function given in Eq. (7).

3 Self-Adaptive Differential Evolution

The differential evolution (DE) is a robust and minimalist vector-based optimization method proposed by Storn and Price in the following reference [41]. DE has a stable convergence property in the continuous search space in problems of global optimization. This optimization method has been successfully applied in distinct fields, for example [42–46].

Nevertheless, the control parameters CR and ρ, the mutation and crossover strategies (MCS) involved in DE are extremely reliant on the problem in question. It is well known that a huge amount of time to tune the parameters and found adequate MCS.

The development of a Self-Adaptive Differential Evolution (SADE) algorithm that was proposed by Qin and Suganthan in 2009 [47] motivated to solve the parameters and MCS dilemma. This algorithm is able to solve problems more efficiently due that during the evolution, the suitable MCS and parameters are gradually self-adapted conformable to the learning experience.

3.1 Differential Evolution

Assume that the search space of DE is D-dimensional. The standard version of DE works as ensues:

Initialization stage: The total population is formed by NP vectors of dimension D, each vector is defined as follows $X_i^G = \left[X_{i,1}^G, X_{i,2}^G, \ldots X_{i,D}^G\right]$ where every vector (also called "individual") form a population defined as $P^G = \left[X_1^G, X_2^G, \ldots 1X_{NP}^G\right]$. Through a uniform distribution, all the individuals in the population are initialized as follows:

$$X_{i,j}^{(0)} = X_{i,j} + rand_{i,j}(0, 1) \times \left(X_{i,j}^U - X_{i,j}^L\right) \tag{8}$$

$$\{X_{i,j}^{(0)} | X_{i,j}^L \leq X_{i,j}^0 \leq X_{i,j}^U, i = 1, 2, \ldots, NP; j = 1, 2, \ldots, D\} \tag{9}$$

where, X_i^L, X_i^U are the lower bound and upper bound of variable X_i, $rand\,(0, 1)$ is a uniform distribution with mean 0 and standard deviation 1.

Mutation stage: The mutation operation is where a mutant vector, so-called "target vector" is produced $V_{i,j}^{(G+1)}$, starting of each individual vector $X_{i,j}^{(G)}$, some useful and famous mutation strategies are listed as ensue:

$$\text{(1) "DE/rand/1/bin": } V_{ij}^{(G+1)} = X_{r1,j}^{(G)} + \rho\left(X_{r2,j}^{(G)} + X_{r3,j}^{(G)}\right) \tag{10}$$

$$\begin{aligned} \text{(2) "DE/rand/2/bin": } V_{ij}^{(G+1)} =& X_{r1,j}^{(G)} + \rho\left(X_{r2,j}^{(G)} + X_{r3,j}^{(G)}\right) \\ &+ \rho\left(X_{r4,j}^{(G)} + X_{r5,j}^{(G)}\right) \end{aligned} \tag{11}$$

$$\text{(3) "DE/best/1": } V_{ij}^{(G+1)} = X_{best,j}^{(G)} + \rho\left(X_{r2,j}^{(G)} + X_{r3,j}^{(G)}\right) \tag{12}$$

$$\begin{aligned} \text{(4) "DE/rand to best/1": } V_{ij}^{(G+1)} =& X_{r1,j}^{(G)} + \lambda\left(X_{best,j}^{(G)} + X_{r1,j}^{(G)}\right) \\ &+ \rho\left(X_{r2,j}^{(G)} + X_{r3,j}^{(G)}\right) \end{aligned} \tag{13}$$

$$\begin{aligned} \text{(5) "DE/rand to best/2/bin": } V_{ij}^{(G+1)} =& X_{i,j}^{(G)} + \rho\left(X_{best,j}^{(G)} + X_{i,j}^{(G)}\right) \\ &+ \rho\left(X_{r1,j}^{(G)} + X_{r2,j}^{(G)}\right) + \rho\left(X_{r3,j}^{(G)} + X_{r4,j}^{(G)}\right) \end{aligned} \tag{14}$$

$$\text{(6) ``DE/current to rand /1'':}\ V_{ij}^{(G+1)} = X_{i,j}^{(G)} + \lambda\left(X_{r1,j}^{(G)} + X_{i,j}^{(G)}\right) + \rho\left(X_{r2,j}^{(G)} + X_{r3,j}^{(G)}\right) \quad (15)$$

where, ρ and λ are real-valued mutation factors ($\rho > 0, \lambda > 0$), regard the following r_1, r_2, r_3 and r_4 satisfy $r_1, r_2, r_3, r_4 \in \{1, 2, \ldots, NP\}$ and $i \neq r_1 \neq r_2 \neq r_3 \neq r_4$. The feasibility of solutions must be guarantee, due to this all individual out of boundaries must be generated randomly into the search space.

Crossover stage: The Crossover operator is employed to increment the diversity of the perturbed parameter vectors, so called "trial vector". This operation ensures that at least one component in $X_{i,j}^{(G+1)}$ is generated from $X_{i,j}^{(G)}$.

$$U_{i,j}^{(G+1)} = \begin{cases} V_{ij}^{(G+1)} & \textit{if } rand_{i,j}(0,1) \leq CR \textit{ or } j = \text{int}\left(D \times rand_{i,j}(0,1)\right) \\ X_{i,j}^{(G)} & \textit{otherwise}, j = 1, 2, \ldots, D \end{cases} \quad (16)$$

where, the crossover probability is as ensues $CR \in [0, 1]$.

Selection stage: Selection method is a tourney between the target vector and trial vector, the one with better fitness value is the one who enters to the next generation.

$$X_{i,j}^{(G+1)} = \begin{cases} V_{ij}^{(G+1)} & \textit{if}\left(f\left(U_{i,j}^{(G+1)}\right) \leq f\left(X_{i,j}^{(G)}\right)\right) \\ X_{i,j}^{(G)} & \textit{otherwise} \end{cases} \quad (17)$$

Thus, the individuals in the new generation are better than the previous generation. The algorithmic description of Differential Evolution is shown in Algorithm 1.

Algorithm 1 Algorithmic description of DE

```
Initialize the population with random values X_{i,j}, with NP individuals based on Eq. (8)
Set the weight ρ ∈ [0,2] and Crossover Probability Cr = ∈ [0,1]
WHILE (Stop criterion)
        FOR i=1 to NP
                Produce a mutated Vector V_i^G = {v_{i,1}^G, v_{i,2}^G, ..., v_{i,D}^G } for each target
                vector X_i^G according to Eq. (10)
                Produce a trial Vector U_i^G = {u_{i,1}^G, u_{i,2}^G, ..., u_{i,D}^G } for each target
                vector X_i^G according (Eq. 16)
                Apply the selection process between U_i^G and X_i^G with Eq. (17)
        END
Augment the generation count G = G + 1
END
```

3.2 SADE Algorithm

The differential evolution (DE) employs three vital control parameters: crossover rate (*CR*), mutation rate factor (ρ and λ) and the size of the population (*NP*). The adequate MCS and parameter improved the performance of DE although how to select the best parameters and MCS is a complex assignment itself. It was observed that during distinct stages of the evolutive process, distinct MCS linked with modifying parameters value in consecutive generations, produce a robust algorithm in terms of exploration and exploitation capabilities.

3.2.1 Trial Vector Generation Thorough the Adaptation of MCS

SADE employs a collection of efficient and diverse: MCS and control parameters. Through the evolutionary process, regarding their respective target vector in the population in the present generation, one MCS strategy will be selected from the mutant and crossover strategy collection (MCSC) according to the probability adapted by generating individuals who aptly pass to next generation in last learning period window (LPW). The probability is adapted taking in count the MCS in the previous generations of generating promising individuals, this increase the probability of being selected in the current generation to produce a mutant vector and then new trial vector. During evolution, these probabilities adapt continuously.

With regard to all the strategies probabilities, these are initialized as $1/K$ where K represents the quantification of strategies available in the MCSC. Each target vector selects a GS in its mutation stage, according to the probability and employing the stochastic universal selection method [48]. All the newly generated trial vectors are evaluated after the generation G, Now is proceeded to record in ns_k^G all the trial vectors produced employing the *kth* strategy that aptly enters the into next generation. On the other hand, all the trial vectors generated by the *kth* strategy discarded by the selection stage are recorded in nf_k^G. Then ns_k^G sand nf_k^G are accumulated within a specific generation number called "Learning Period Window" (LPW) as shown in Fig. 1. After the first LPW generations, the probability of select different MCS is subsequently updated according to Eq. (18) in the las LPW generations.

$$p_k^G = \frac{S_k^G}{\sum_{k=1}^{K} S_k^G} \tag{18}$$

where,

$$S_k^G = \frac{\sum_{g=G-LPW}^{G-1} ns_k^G}{\sum_{g=G-LPW}^{G-1} ns_k^G + \sum_{g=G-LPW}^{G-1} nf_k^G} + \epsilon, (k = 1, 2, \ldots, K; G > LPW) \tag{19}$$

The S_k^G value represents the degree of achievement of employing the kth strategy to produce trial vectors that aptly enrolling into the next generation in the prior LPW

Fig. 1 Success memory and failure memory (SFM)

Success memory

Index	Strategy 1	Strategy2	...	Strategy K
1	ns_1^{G-LP}	ns_2^{G-LP}	...	ns_k^{G-LP}
2	ns_1^{G-LP+1}	ns_2^{G-LP+1}	...	ns_k^{G-LP+1}
⋮	⋮	⋮	⋮	⋮
LP	ns_1^{G-1}	ns_2^{G-1}	...	ns_k^{G-1}

Failure memory

Index	Strategy 1	Strategy 2	...	Strategy K
1	nf_1^{G-LP}	nf_2^{G-LP}	...	nf_k^{G-LP}
2	nf_1^{G-LP+1}	nf_2^{G-LP+1}	...	nf_k^{G-LP+1}
⋮	⋮	⋮	⋮	⋮
LP	nf_1^{G-1}	nf_2^{G-1}	...	nf_k^{G-1}

generations. The ϵ is a fixed value that is used to avoid null success degrees. The probability of *kth* strategy of been selected depends on the correspondent S_k^G value. Then larger achievement degree, therefore, larger is the probability of being selected to produce trial vectors.

3.2.2 Parameter Adaptation of the Mutation and Crossover Stages

In the standard DE, the ensure control parameters *CR* and ρ must be tuned in relation to the problem in question. The control of convergence speed is controlled by ρ its value is assigned employing a normal distribution with a standard deviation of 0.3 and a mean value 0.5. The parameter *CR*, control the part of the exploration process of the search, a wrong choice decays the general performance. In this approach is suggest a continuously adjusting the range of values of *CR* employed to produce trial vectors aptly entering the next generation. The algorithmic description of SADE is presented in Algorithm 2.

The *CR* value is adapted with respect to each MCS in the collection. Regard the kth strategy in the MCSC, in the reminder that the value Crm_k is set to 0.5 initially. Through a normal distribution defined with the following parameters w $\mathcal{N}(Crm_k, 0.1)$ a group of values is produced and assigned to *CR* now is proceeding to generate the target vectors of the *kth* strategy employing the values of *CR*. It is necessary establishing the CR_{Memory} is the way to adapt the value *CR*. The CR_{Memory} record those *CR* values that produce trial vectors that entering aptly to the next generation in prior LPW with regard to the kth MCS. During the first LPW generations employing a normal distribution, the follows parameter $\mathcal{N}(Crm_k, 0.1)$ the *CR* values are generated. Concluding the first LPW generations, is taking in count the median of the values stored in CR_{Memory} per strategy, now the value of Crm_k is updated at each generation taking in count only the values of CR_{Memory} in the last LPW generations.

Algorithm 2 Algorithmic description of SADE

Step 1 Initialization
Produce a population with randomly values $X_{i,j}$ with NP individuals according to Eq. (8).
Initialize the value of $CR = (Crm_k)$ and the probability p_k^G, $k = 1, \ldots, K$;
K is the number of available of MCS. Learning
period window LPW=50. Generation counter G=0, Set the weight $and\ \rho \in [0,2]$.
Step 2 Evaluate the population
Step 3 WHILE (Stop criterion)
Step 3.1 calculate Strategy probability p_k^G and update the Success and Failure memory (SFM)
IF G> LPW
FOR k=1 to K
Update the p_k^G by Eq. (18)
Remove ns_k^{G-LP} and ns_k^{G-LP} out of the SFM respectively.
END
END
Step 3.2 Assign trial vector generation strategy and parameters to each target vector $X_{i,j}^G$
Employ universal sampling to select one strategy k for each $X_{i,j}^G$
// Assign control parameter ρ
FOR i=1 to NP ρ_i =Nornrnd (0.5,0.3) **END**
// Assign control parameter CR
IF G>= LP
FOR k= 1 to K CRm_k = median (CR_{Memory}) **END**
END
FOR k=1 to K
FOR i=1 to NP $CR_{k,i}$ =Nornrnd (CRm_k,0.3) **END**
END
Step 3.3 Generate a new population where each trial vector $U_{k,i}^G$ is generated according to associated trial vector generalization strategy k and parameters ρ_i and $CR_{k,i}$ in the step 3.2
Step 3.4 Randomly reinitialize the trial vector $U_{k,i}^G$ within the search space if any variable is outside of its boundaries.
Step 3.5 Selection:
FOR i=1 to NP
Evaluated the trial vector $U_{k,i}^G$
IF $f(U_{k,i}^G) \leq f(X_i^G)$
$X_i^{G+1} = U_{k,i}^G$, $f(X_i^{G+1}) = f(U_{k,i}^G)$, $ns_k^G = ns_k^G + 1$
Store into CR_{Memory}
IF $f(U_{k,i}^G) \leq f(X_{Best}^G)$
$X_{Best}^G = U_{k,i}^G$, $f(X_i^{G+1}) = f(U_{k,i}^G)$,
END
ELSE
$nf_k^G = nf_k^G + 1$
END
END
Store ns_k^G and nf_k^G into the SFM respectively.
Step 3.6 Increase the generation count $G = G + 1$
END

---

4 Minimum Cross-Entropy by SADE Algorithm

A new methodology called MCE-SADE for image thresholding is presented in this work. The minimum cross-entropy (MCE) thresholding is a procedure where a pixel map information is separated into a specific number of information classes through the determination of threshold values, where the optimization problem of thresholding increase of complexity as the number of thresholds increments, the MCE thresholding procedure face an optimization problem with a high modality subject to the search space due to the histogram irregularities of the pixel map.

The SADE is an optimization algorithm that is employed to minimize the MCE among the total classes or levels. SADE perform the ensue actions, the whole population of SADE encodes a set of feasible threshold solutions of the search space. The MCE is employed as fitness function to specify the quality of each individual. The operators, MCS and rules of SADE where the values of fitness function of each individual in the population will guide the evolutive process and produce new promising individuals while the segmentation quality improves.

The problem related to multilevel thresholding can be approach as an optimization problem where a fitness function regards the criterion of the minimum cross-entropy stated as ensue:

$$\arg\min_{\boldsymbol{Th}} D(\boldsymbol{Th}) = \left\{\boldsymbol{Th} \in \mathbb{R}^n | 0 \leq th_j \leq 255, j = 1, 2, \ldots, n\right\} \tag{20}$$

The MCE function (Eq. 7) is represented by $D(\boldsymbol{Th})$ and $\boldsymbol{Th}$ corresponds to the arg min for n number of thresholds that is the constrained to the 0–255 different values related a grayscale pixmap. Through the employment of SADE algorithm found the threshold values $\boldsymbol{Th}$ that solves Eq. (7). The proposed MCE-SADE takes advantages of convenient self-adaptive strategy and parameters proposed by Qin et al. It is well known that most metaheuristic algorithms (MA) invest time tuning the parameters to solve a specific problem, while SADE only requires a parameter called learn generations. The SADE algorithm which adapts the control parameter and MCS learning from their previous generation producing more suitable individuals along of the evolutive process. MCE-SADE benefits from the definition of MCE by Li and Lee [16] which present an approach more exact and with faster performance than a traditional approach such as Otsu's among class variance [8] and fuzzy entropy [49].

4.1 Solution Representation

Each individual in the population is a threshold $\boldsymbol{Th}$ form up by number decision variable. Hence, the population is represented as follows:

$$P^G = [\boldsymbol{Th}_1, \boldsymbol{Th}_2, \ldots \boldsymbol{Th}_{NP}], \boldsymbol{Th}_i = [th_1, th_2, \ldots, th_n] \tag{21}$$

where, G is related to current generation, NP is the size of the population and n is the number of thresholds applied to the image.

4.2 SADE Implementation

For the implementation of the MCE-SADE method the following steps are carried out:

Step 1 Read the pixmap I.
Step 2 Calculate histogram h^r of I.
Step 3 Initialize SADE's parameters λ, ρ, CR and p_k^G.
Step 4 Initialize a population P^G of NP random individuals with n dimensions.
Step 5 Evaluate each member of population P^G employing the cost function (D), Eq. (7).
Step 6 Calculate Strategy probability p_k^G and update (SFM).
Step 7 Assign a specific MCS and parameters to each vector X_i^G.
Step 8 Produce a novel population where each trial vector $U_{k,i}^G$is produced by the MCS assigned and parameters $\rho_i$and $CR_{k,i}$.
Step 9 Randomly reinitialize any trial vector $U_{k,i}^G$ outside its boundaries.
Step 10 Perform the selection stage over the $U_{k,i}^G$ and X_i^G vectors. Update ns_k^G and nf_k^G according to the case.
Step 11 If the termination criteria are not accomplished, then entry into step 5 again.
Step 12 Produce the resultant pixmap I_{th} employing X_{best}^G which is the best individual across all generations.

4.3 Multilevel Thresholding

After SADE the concludes iterative process, the X_{best}^G that represents the best solution founded by minimization the MCE to segment the original pixmap. The following expression defines generic rules to perform a segmentation of two thresholds:

$$I_{th}(x, y) = \begin{cases} I(x, y) \; if & I(x, y) \leq th_1 \\ th_1 \quad if & th_1 \leq I(x, y) < th_2 \\ I(x, y) \; if & I(x, y) > th_2 \end{cases} \tag{22}$$

where, the term $I(x, y)$ is related to pixel position x, y in the pixmap, to this position has a corresponding gray level value and the terms th_1, th_2 are t threshold values. For this work is extended to n levels due to the multilevel approach.

$$I_{th}(x,y) = \begin{cases} I(x,y)\ if & I(x,y) \le th_1 \\ th_{j-1}\ \ if\ th_{j-1} \le I(x,y) < th_j, & j = 2,3,\ldots,n-1 \\ I(x,y)\ if & I(x,y) > th_n \end{cases} \tag{23}$$

5 Experimental Results

This section conducted several experiments in order to appraise the aptness of the purposed method with regard to stability and solution quality. Two groups of reference images are employed for the experiments; the first group consists of nine standard test images with the dimension of 512 × 512 pixels each, while the second group is integrated of eight magnetic resonance axial cut brain images. The purpose of the first reference group is to estimate the exploration skills from SADE when is applied to a high-dimensional space. Most associated literature quest for optimum *th* values in an of up to 5-dimensions space, while in this work, the quest of the optimum *th* values is performed in a 16-dimensions space since the superiority of an algorithm against other approaches becomes remarkable in terms of threshold selection quality as threshold levels augment. The levels of *th* to search in this first group are $L_{Th} = 2, 4, 8, 16$.

The second group of reference images appraises the effectiveness of the MCE-SADE when tested on MR brain images. As mentioned, this group consists only in magnetic resonance axial cut brain images, therefore these images show similar structures, which allows us to set an estimated quantity of *th* levels. In this second group, the thresholds levels to search are $L_{Th} = 2, 3, 4, 5$. The brain MR images employed for this work can be found in the BrainWeb databank (https://brainweb.bic.mni.mcgill.ca/brainweb). For the first group as well as for the second, 35 experiments were carried out for each image to avoid any discrepancy in the results.

To appraise the results of this approach, the MCE-SADE is likened against two metaheuristic algorithms (MA), Grey Wolf Optimizer (GWO) [38] and Imperialist Competitive Algorithm (ICA) [39], both MA apply MCE as a fitness function. The total number of search agents in the population is defined to 50 regarding a max_number of iterations set to 2500 for each experiment performed in SADE, GWO, and ICA. The following criterion is regarded for each run of the experiment, if during the 10% of the max_number of iterations of the evolutive process the best fitness value so-far is not enhanced, the algorithm stops. Parametric settings from each algorithm are presented in Table 1. All experiments were performed with MATLAB R2016a at a 3.41 GHz Intel Core i7 CPU with 12 GB of RAM.

In both groups of reference images, segmentation quality is evaluated through three image processing metrics the Peak Signal to Noise Ratio (PSNR), the Structural Similarity Index (SSIM) and the Feature Similarity Index (FSIM). The PSNR [52] is calculated straight through the gray intensity values of the image. This metric is employed as a quality measure among the initial image and the processed image. The

Table 1 Parametric settings for SADE, GWO, and ICA

SADE [47]	GWO [50]	ICA [51]
Population N: 50 Learn generations: 50 Type of strategies employs: (a) DE/rand/1/bin (b) DE/rand-to-best/2/bin (c) DE/rand/2/bin (d) DE/current-to-rand/1	Population N: 50	Population N: 50 Number of initial empires: 5 Selection pressure: 1 Assimilation coefficient: 2 Revolution probability: 0.05 Revolution rate: 0.1 Colonies mean cost coefficient: 0.2

upper the value of PSNR, the better segmentation quality of an image. The Eq. (24) presents the formula in decibel (dB) to calculate the PSNR.

$$PSNR = 20\log_{10}\left(\frac{255}{RMSE}\right) \tag{24}$$

where the RMSE (Root Mean Square Error) is calculated as indicated in Eq. (25).

$$RMSE = \sqrt{\frac{\sum_{i=1}^{M}\sum_{j=1}^{N}(I_r(i,j) - I_s(i,j))}{MN}} \tag{25}$$

where, the total number of rows and columns are represented by M and N respectively, the reference image is I_r and I_s is the segmented image. Structural similarity index (SSIM) [53] is founded on the deterioration of structural data. It compares the present structures among the initial and the processed image. The Eq. (26) defines the SSIM as:

$$SSIM(I_r, I_s) = \frac{(2\mu_{I_r}\mu_{I_s} + C_1)(2\sigma_{I_rI_s} + C_2)}{(\mu_{I_r}^2 + \mu_{I_s}^2 + C_1)(\sigma_{I_r}^2 + \sigma_{I_s}^2 + C_2)} \tag{26}$$

where, the average intensity value from the initial and the processed image are represented by μ_{I_r} and μ_{I_s} correspondingly, σ_{I_r} and σ_{I_s} depict the standard deviation of each image, $\sigma_{I_rI_s}$ denotes a local unit correlation factor among I_r and I_s. Lastly, the constants C_1 and C_2 ($C_1 = C_2 = 0.065$) dodge the unstableness if $\mu_{I_r}^2 + \mu_{I_s}^2$ is quite near to zero. A higher SSIM value indicates better segmentation performance.

The FSIM [54] compute the feature similitude among the initial and the processed image (Eq. 27), is formulated as:

$$FSIM = \frac{\sum_{X\in\Omega} S_L(X)PC_m(X)}{\sum_{X\in\Omega} PC_m(X)} \tag{27}$$

where,

$$S_L(X) = S_{PC}(X)S_G(X) \tag{28}$$

$$S_{PC}(X) = \frac{2PC_1(X)PC_2(X) + T_1}{PC_1^2(X) + PC_2^2(X) + T_1} \tag{29}$$

$$S_G(X) = \frac{2G_1(X)G_2(X) + T_2}{G_1^2(X) + G_2^2(X) + T_2} \tag{30}$$

G depicts the gradient magnitude (GM) from a given image. This is calculated through the following equation:

$$G = \sqrt{G_x^2 + G_y^2} \tag{31}$$

The Eq. (32) computed the phase congruence PC.

$$PC(X) = \frac{E(X)}{\left(\varepsilon + \sum_n A_n(X)\right)} \tag{32}$$

where, the magnitude of the respond vector is designated thru $E(X)$ and this is found at an X position on the n scale. The local amplitude is represented through $A_n(X)$. Last, ε is a small positive constant. An upper FSIM value indicates a better quality in the segmentation process. The Standard Deviation (STD) [55] is another metric applied to measure the performance of the MCE-SADE, which evaluates the algorithm stability and consistency. The algorithm becomes more inconsistent the higher the value of the standard deviation. The STD is defined in Eq. (33) as:

$$STD = \sqrt{\sum_{i=1}^{N_{It}} \left(\frac{S_i - \mu}{R}\right)} \tag{33}$$

where, the best fitness value acquired at the *ith* iteration is indicated by S_i, μ is the mean, R is the total number of executions which is equal to 35.

5.1 Standard Test Images

As stated previously, the first group of reference images contains nine standard test images in TIFF format of 512 × 512 pixels each. The chosen images are broadly employed in image processing, such as Cameraman, Lena, Peppers, Sailboat, among others. For each image 35 experiments are carried out, that evaluate 4 different thresholds levels $L_{Th} = 2, 4, 8, 16$. In this subsection, are exposed and analyzed the results of the experiments, when the MCE is implemented to SADE, GWO and ICA.

Table 2 presents the optimal threshold values acquired when applying the MCE to SADE, GWO and ICA, with four threshold levels $L_{Th} = 2, 4, 8, 16$ for the eight standard test images. At low threshold levels $L_{Th} = 2, 4$ can be perceived that for most of

Table 2 Optimum threshold values acquired through SADE, GWO and ICA based at MCE

Image	L_{Th}	SADE	GWO	ICA
Blonde	2	38 122	40 124	37 122
	4	31 96 132 165	31 96 132 165	30 96 133 165
	8	11 49 77 95 116 138 158 177	11 50 77 99 122 144 162 179	11 48 74 92 113 135 155 176
	16	5 23 46 65 77 87 97 110 121 132 141 149 159 168 178 188	5 23 46 65 77 87 97 110 121 132 141 149 159 168 178 188	4 31 54 68 75 83 90 99 106 108 116 125 137 151 166 182
Cameraman	2	50 137	50 137	50 137
	4	29 77 126 158	29 77 125 158	29 77 122 156
	8	13 25 51 84 115 137 158 176	13 24 46 76 104 128 151 172	24 52 84 112 134 156 172 201
	16	11 15 23 38 53 70 87 103 115 125 137 148 158 168 179 210	12 21 41 59 73 77 80 87 97 107 117 129 141 157 173 200	12 21 41 59 73 77 80 87 97 107 117 129 141 157 173 200
Couple	2	73 134	74 133	72 133
	4	38 83 124 161	38 83 124 162	38 82 124 162
	8	22 42 57 73 90 115 141 177	23 47 72 98 120 138 157 186	21 44 67 93 115 133 154 185
	16	10 19 32 46 58 71 85 99 110 121 131 142 154 168 185 206	11 23 36 51 63 76 89 101 112 121 132 141 152 165 181 205	9 23 39 53 64 74 82 87 93 99 108 120 133 146 166 191
House	2	81 159	81 159	81 159
	4	53 88 132 184	53 87 131 182	54 87 130 181
	8	40 60 82 104 126 158 190 220	40 59 78 96 109 128 156 190	40 59 77 95 108 125 152 188
	16	26 40 53 64 76 88 97 105 111 121 137 152 168 183 202 219	26 42 58 69 81 95 105 113 122 134 151 168 184 199 216 234	27 41 53 62 70 80 88 97 102 107 113 121 134 151 170 195
Jet	2	97 162	97 162	94 161
	4	64 107 150 193	64 107 150 191	64 105 150 192
	8	54 84 106 128 153 179 198 211	51 81 101 122 147 173 194 209	53 83 105 126 152 177 196 209
	16	38 56 72 84 95 106 117 129 143 158 171 183 194 202 209 216	34 53 69 80 90 96 103 109 116 127 139 155 173 190 203 213	26 39 54 69 81 95 108 121 136 152 167 180 191 200 207 215
Lena	2	83 142	83 142	83 140
	4	71 109 141 177	71 108 141 177	71 108 140 177
	8	52 69 90 111 130 147 166 191	53 71 91 109 126 143 162 188	47 61 77 97 114 138 159 186

(continued)

Table 2 (continued)

Image	L_{Th}	SADE	GWO	ICA
	16	43 52 61 71 81 92 102 112 122 132 141 151 161 173 188 204	38 43 50 58 68 83 95 108 119 131 140 150 161 175 189 204	38 46 55 65 75 85 95 104 113 122 130 141 153 166 184 204
Peppers	2	54 127	54 127	54 127
	4	37 77 119 165	37 77 119 165	36 74 115 162
	8	22 43 67 87 109 143 159 182	21 43 66 88 105 128 154 180	25 47 67 87 105 124 150 176
	16	12 22 34 49 64 77 88 101 115 127 138 150 162 174 189 204	13 20 34 49 64 77 87 96 107 120 133 149 165 181 196 251	14 31 46 57 61 67 73 81 86 94 103 115 131 148 166 189
Pirate	2	73 128	73 128	73 127
	4	58 89 122 157	58 89 122 157	58 90 122 158
	8	43 53 68 85 103 124 147 176	46 58 76 94 114 136 157 184	47 64 81 103 122 142 160 190
	16	29 44 52 61 70 80 91 99 110 121 132 143 155 168 179 195	30 39 42 46 52 60 69 75 82 92 102 114 129 144 161 187	25 44 49 55 64 75 85 97 109 123 136 148 160 173 192 256
Sailboat	2	74 142	74 142	74 142
	4	57 91 143 195	57 91 143 195	56 89 142 194
	8	40 54 68 89 118 148 174 202	38 52 68 92 121 151 177 203	38 53 68 87 112 140 172 203
	16	30 39 45 54 64 75 90 104 122 140 156 170 181 191 203 218	24 34 42 52 61 71 83 96 111 129 146 163 177 189 202 216	26 40 51 64 79 92 105 107 112 120 130 144 160 175 192 209

the images, the convergence tendency from elected algorithms (SADE, GWO, and ICA) is towards the same threshold values. At high threshold levels, the obtained results are quite different among the algorithms owing to the high dimensions thresholding problem is a complex task since it has a wider search space. As the amount of *th* increments, the advantage of an algorithm against other approaches becomes remarkable in terms of threshold selection quality.

Since the MA are stochastic, the achieved results in each execution might not be the same. Hence, the STD and the mean are used to assess the consistency and stableness from selected algorithms (SADE, GWO, and ICA) when the minimum cross-entropy Eq. (7) is applied as the objective function. Each algorithm is executed 35 times per image to find the values of the mean and STD, the average values are recorded in Table 3. The bold values indicate the best outcomes.

The algorithm is more stable if the STD value is lower. In turn, if the mean value is low the algorithm is more consistent since it is a minimization problem. The outcomes within Table 3 reveal that SADE performance along with MCE as objective function is more stable and consistent than the compared techniques GWO and ICA.

Table 3 Mean and standard deviation for standard test images obtained by SADE, GWO and ICA based on MCE

Image	L_{Th}	SADE		GWO		ICA	
		Mean	STD	Mean	STD	Mean	STD
Blonde	2	**1.4242**	**0.0008**	1.4976	0.0051	1.5132	0.0282
	4	**0.5384**	**0.0007**	0.5768	0.0014	0.6282	0.0067
	8	**0.1068**	**0.0014**	0.1131	0.0101	0.1638	0.0139
	16	**0.0335**	**0.0028**	0.0350	0.0034	0.0971	0.0178
Cameraman	2	**1.4306**	**0.0020**	1.4956	0.0022	1.5267	0.0130
	4	**0.4357**	**0.0003**	0.4376	0.0015	0.5973	0.0140
	8	**0.2218**	**0.0026**	0.2274	0.0031	0.5486	0.0496
	16	**0.1311**	**0.0017**	0.1331	0.0025	0.2183	0.0335
Couple	2	**1.8949**	**0.0004**	1.9113	0.0008	1.9643	0.0018
	4	**0.4032**	**0.0001**	0.4045	0.0004	0.4262	0.0005
	8	**0.1388**	**0.0014**	0.1400	0.0076	0.2513	0.0154
	16	**0.0442**	**0.0022**	0.0475	0.0044	0.1681	0.0264
House	2	**0.9896**	**0.0007**	0.9961	0.0009	1.2589	0.0035
	4	**0.4128**	**0.0002**	0.4156	0.0009	0.9899	0.0252
	8	**0.1453**	**0.0026**	0.1511	0.0086	0.3900	0.0138
	16	**0.0461**	**0.0018**	0.0489	0.0041	0.2637	0.0216
Jet	2	**0.9957**	**0.0004**	0.9983	0.0006	1.4610	0.0017
	4	**0.4082**	**0.0006**	0.4093	0.0007	0.7021	0.0018
	8	**0.1314**	**0.0021**	0.1343	0.0034	0.2235	0.0087
	16	**0.0398**	**0.0022**	0.0417	0.0035	0.1123	0.0266
Lena	2	**0.9911**	**0.0006**	0.9955	0.0012	1.6689	0.0024
	4	**0.3395**	**0.0004**	0.3406	0.0009	0.9375	0.0051
	8	**0.1137**	**0.0023**	0.1189	0.0025	0.4944	0.0032
	16	**0.0376**	**0.0021**	0.0382	0.0044	0.0673	0.0095
Peppers	2	**0.9874**	**0.0007**	0.9965	0.0018	1.0821	0.0030
	4	**0.3697**	**0.0002**	0.3734	0.0005	1.0189	0.0015
	8	**0.1254**	**0.0032**	0.1346	0.0177	0.3101	0.0243
	16	**0.0360**	**0.0026**	0.0381	0.0041	0.5123	0.0400
Pirate	2	**0.9848**	**0.0008**	0.9959	0.0016	1.3536	0.0043
	4	**0.3595**	**0.0003**	0.3631	0.0011	0.8106	0.0025
	8	**0.1087**	**0.0033**	0.1137	0.0110	0.1294	0.0120
	16	**0.0295**	**0.0032**	0.0335	0.0036	0.0690	0.0236
Sailboat	2	**0.9830**	**0.0004**	0.9929	0.0016	1.4610	0.0038
	4	**0.4320**	**0.0003**	0.4365	0.0008	0.8673	0.0018
	8	**0.1388**	**0.0018**	0.1390	0.0020	0.5558	0.0266
	16	**0.0439**	**0.0023**	0.0459	0.0034	0.1413	0.0036

Table 4 Comparison of PSNR results acquired through SADE, GWO, and ICA for MCE, Otsu's and Fuzzy Entropy method

Image	L_{Th}	Cross-entropy			Otsu's method			Fuzzy entropy		
		SADE	GWO	ICA	SADE	GWO	ICA	SADE	GWO	ICA
Blonde	2	**15.9642**	15.8003	15.4050	15.3811	15.1279	14.9581	14.2248	13.6361	13.0661
	4	21.0680	19.5540	19.5060	**21.4044**	20.3210	18.9921	18.8155	18.5042	18.1464
	8	**26.9135**	26.5987	26.5421	25.8667	25.1746	24.4581	24.4149	24.3018	23.3816
	16	**32.9041**	32.5921	30.8145	30.6215	28.3206	26.7416	26.5494	26.5071	26.4640
Cameraman	2	16.3755	16.2664	16.0684	**16.5778**	15.6829	15.3558	15.3682	14.1838	12.9750
	4	**21.6290**	21.5150	21.5031	20.1521	19.5633	19.3887	19.4414	19.0695	18.4392
	8	26.5487	26.3559	26.2142	**26.7021**	26.2806	26.2017	24.9133	24.1764	24.0479
	16	**31.7434**	31.6328	29.6971	30.6245	30.3134	30.0214	29.9846	29.1005	26.3556
Couple	2	**16.5101**	16.4114	16.3831	16.0526	15.8935	15.6880	14.0349	11.8053	11.7662
	4	**20.5587**	20.5400	20.4478	20.2837	20.0895	19.7679	19.2677	19.2530	18.9219
	8	**25.8696**	25.7751	25.6906	25.5367	24.9669	22.9648	22.6839	21.9013	21.7957
	16	31.2515	**31.3517**	29.3449	29.2753	28.9813	27.0745	27.1835	26.2161	24.7810
House	2	**15.5192**	15.2836	15.1140	15.0184	14.6283	14.1158	11.3978	11.0003	9.5649
	4	**20.6859**	20.3406	20.2221	19.9235	19.1152	18.8969	18.9081	17.5952	11.7406
	8	**25.0016**	24.8483	24.3546	24.1093	23.8912	23.2620	23.2503	23.1167	23.0171
	16	**32.7590**	31.8802	28.2597	28.2571	27.0394	26.9691	26.7870	26.2568	24.5073
Jet	2	**15.4365**	15.2622	15.0418	14.5379	14.0463	14.0339	16.4695	15.8893	15.6889
	4	**21.4357**	21.3154	21.2915	21.2742	20.9871	20.8347	20.1785	19.7352	19.6543
	8	26.8754	**26.8782**	26.8273	26.6272	25.8462	21.2958	21.9629	21.7573	20.3927
	16	**32.9380**	32.9254	30.8341	30.5637	30.4014	30.1754	28.9338	27.2271	26.2977

(continued)

Table 4 (continued)

Image	L_{Th}	Cross-entropy			Otsu's method			Fuzzy entropy		
		SADE	GWO	ICA	SADE	GWO	ICA	SADE	GWO	ICA
Lena	2	**15.6639**	15.6615	15.6599	15.3106	14.9080	14.6856	13.9333	11.9759	11.8929
	4	19.0236	18.8632	18.8610	**19.9207**	19.7538	18.2718	18.4054	18.0151	17.6004
	8	24.8956	24.3625	24.0586	**25.0826**	24.6118	19.9754	19.3434	19.3004	15.6240
	16	**31.4855**	30.9206	30.7625	29.6929	29.5067	29.1213	29.5388	29.2869	23.3387
Peppers	2	**15.9794**	15.9266	15.8446	15.8195	15.4466	15.2463	14.0140	11.8710	11.7371
	4	**20.5978**	20.5602	20.4911	20.2649	20.2000	20.0125	19.3131	18.1327	17.9329
	8	**25.7334**	25.6461	25.5473	24.7444	24.4299	23.8265	23.4218	23.3745	22.5741
	16	**31.6854**	31.5408	29.4629	29.2637	29.0713	28.4964	28.6151	28.4216	24.0493
Pirate	2	**16.2543**	16.2209	16.1852	15.7066	15.0854	14.9617	13.4012	12.4890	12.4540
	4	**19.5489**	19.5128	19.5079	18.9663	18.8762	18.7226	18.3940	18.1969	17.9734
	8	23.2140	22.7892	22.6506	**23.5957**	23.3085	20.5093	21.5197	20.9990	20.8253
	16	**31.2296**	29.9214	27.5656	24.7478	24.6091	23.9383	23.6682	22.8807	21.5943
Sailboat	2	**14.8556**	14.6492	14.4087	14.3428	14.3155	14.1636	14.5631	11.5734	11.4884
	4	**18.2065**	18.1877	18.0620	17.8864	17.5783	17.5166	17.4214	17.3427	16.7641
	8	**24.9954**	24.8028	24.4601	24.2171	24.1186	23.8303	23.5712	23.2938	19.5157
	16	**31.3345**	31.1462	29.5515	29.0477	29.0094	28.5220	28.2033	27.9521	19.4059

Table 4 is realized a comparison of the PSNR results acquired by SADE, GWO, and ICA when applying the MCE, also two reference models are established to contrast the results from the submitted approach. First reference model applies as fitness function Otsu's between-class-variance, the second reference model uses fuzzy entropy as an objective function, both models are applied and evaluated in the eight standard test images with the selected algorithms SADE, GWO, and ICA. Both Otsu method [8] and fuzzy entropy [49] are maximization problems, while minimum cross-entropy is a minimization problem, therefore it cannot straight compare to fitness values.

Hence, the PSNR evaluate the segmentation quality of the image; an upper PSNR value demonstrates a better segmentation quality process. The outcomes in Table 4 proof that when minimum cross-entropy is applied to the selected algorithms (SADE, GWO, and ICA) competitive PSNR values are obtained for most of the images in comparison to Otsu's method and fuzzy entropy. These results also prove that SADE based method provides better segmentation quality compared to GWO and ICA based methods. The bold values indicate the best outcomes.

As described at the beginning of Sect. 5, the SSIM and FSIM appraise the segmentation quality of the processed image in comparison with the initial image. Utilizing the Eq. (23) are produced the processed images; just as PSNR, a higher value denotes a proper image segmentation. Table 5 presents the results of SSIM and FSIM when the MCE is applied as an objective function to SADE, GWO, and ICA.

The reported values in Table 5 are the average of the 35 executions for each threshold and for each selected algorithm. From Table 5 it can be noted that the MCE-SADE acquires upper SSIM and FSIM values most often than the GWO and ICA based methods, therefore, SADE based method shows more accuracy and quality in the segmentation. The bold values indicate the best outcomes.

Table 6 exposes the visual results afterward implementing MCE-SADE to the group of standard test images. For illustrative purposes, only three images from the group were chosen to visually display the results. For each image, the first row exhibits the processed image at corresponding threshold level from left to right 2, 4, 8 and 16 respectively, the following row shows the image histogram along with the best threshold values found by MCE-SADE.

Figures 2 and 3, present the convergence curve of the fitness values from Lena and Jet images, respectively at 16 levels of thresholds.

5.2 MR Brain Images

As mentioned at the beginning of Sect. 5, the second group of benchmark images is integrated of eight magnetic resonance axial cut brain images.

These images are obtained of the z-planes from magnetic resonance with valor from 1, 2, 5, 10, 36, 72, 108 and 144 to the z-axle; these values allow us to obtain meaningful images of different brain slices. To obtain the optimal thresholds are carried out 35 experiments for each MR brain image, that evaluate 4 different thresholds

Table 5 Comparison of SSIM and FSIM results employing MCE for standard test images

Image	L_{Th}	SADE		GWO		ICA	
		SSIM	FSIM	SSIM	FSIM	SSIM	FSIM
Blonde	2	**0.5911**	**0.7502**	0.5903	0.7154	0.5293	0.7149
	4	**0.7226**	**0.8791**	0.7066	0.8728	0.7059	0.8727
	8	**0.8450**	**0.9583**	0.8427	0.9551	0.8399	0.9517
	16	**0.9351**	**0.9904**	0.9342	0.9894	0.8958	0.9688
Cameraman	2	**0.5895**	**0.7963**	0.5894	0.7953	0.5870	0.7942
	4	**0.6730**	**0.8849**	0.6593	0.8843	0.6204	0.8834
	8	**0.7666**	**0.9466**	0.7543	0.9458	0.7341	0.9244
	16	**0.8976**	**0.9758**	0.8885	0.9756	0.8584	0.9383
Couple	2	**0.5458**	**0.7578**	0.5443	0.7576	0.5441	0.7567
	4	**0.7138**	**0.8814**	0.7134	0.8812	0.7132	0.8810
	8	**0.8479**	**0.9562**	0.8461	0.9555	0.8439	0.9498
	16	0.9343	0.9866	**0.9358**	**0.9892**	0.8887	0.9600
House	2	**0.6754**	**0.7791**	0.6749	0.7786	0.6742	0.7777
	4	**0.8069**	**0.8628**	0.8066	0.8627	0.8055	0.8626
	8	**0.8846**	**0.9450**	0.8701	0.9449	0.8628	0.9275
	16	**0.9550**	**0.9858**	0.9520	0.9846	0.9472	0.9476
Jet	2	**0.7439**	**0.8020**	0.7433	0.8017	0.7431	0.8015
	4	**0.8056**	**0.8846**	0.8039	0.8843	0.8033	0.8836
	8	0.8602	0.9571	**0.8617**	**0.9581**	0.8589	0.9560
	16	**0.9418**	**0.9884**	0.9379	0.9872	0.8867	0.9663
Lena	2	**0.5601**	**0.7672**	0.5599	0.7669	0.5579	0.7662
	4	**0.6511**	**0.8551**	0.6509	0.8546	0.6507	0.8538
	8	**0.7845**	**0.9078**	0.7815	0.9070	0.7791	0.9054
	16	**0.9082**	**0.9738**	0.9079	0.9645	0.8923	0.9610
Peppers	2	**0.5554**	**0.7531**	0.5552	0.7529	0.5546	0.7527
	4	**0.6567**	**0.8497**	0.6556	0.8495	0.6549	0.8495
	8	**0.7965**	**0.9401**	0.7895	0.9359	0.7794	0.9173
	16	**0.9198**	**0.9815**	0.9176	0.9773	0.8453	0.9297
Pirate	2	**0.4923**	**0.7902**	0.4909	**0.7902**	0.4900	0.7894
	4	**0.6358**	**0.8833**	0.6356	**0.8833**	0.6341	0.8828
	8	**0.7579**	**0.9318**	0.7525	0.9305	0.7510	0.9214
	16	**0.8854**	**0.9523**	0.8775	0.9511	0.8419	0.9377
Sailboat	2	**0.5315**	**0.8318**	0.5309	0.8311	0.5290	0.8289
	4	**0.6354**	**0.8794**	0.6336	0.8791	0.6321	0.8783
	8	**0.8314**	**0.9462**	0.8291	0.9461	0.8269	0.9407
	16	**0.9348**	**0.9874**	0.9304	0.9843	0.9018	0.9653

Table 6 Results of applying MCE-SADE on the standard test images

	$L_{Th} = 2$	$L_{Th} = 4$	$L_{Th} = 8$	$L_{Th} = 16$
Cameraman				
	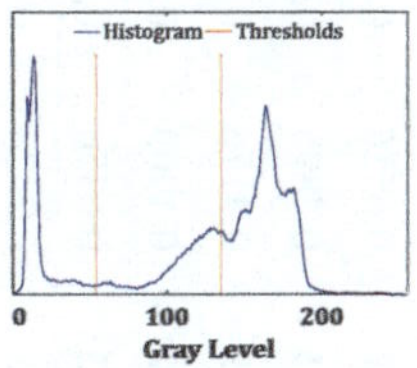	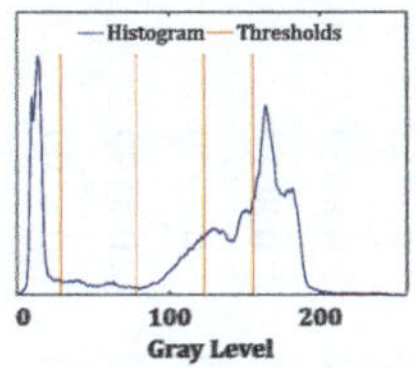	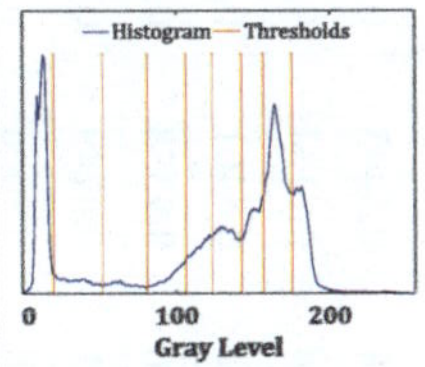	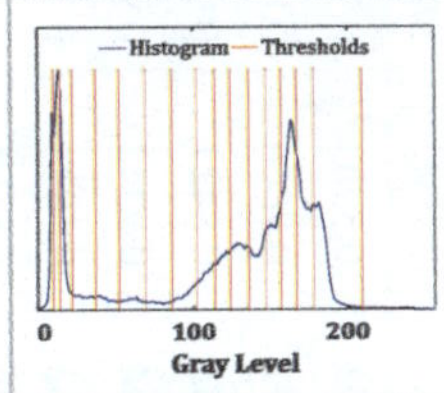
House				

(continued)

Table 6 (continued)

	$L_{Th} = 2$	$L_{Th} = 4$	$L_{Th} = 8$	$L_{Th} = 16$
	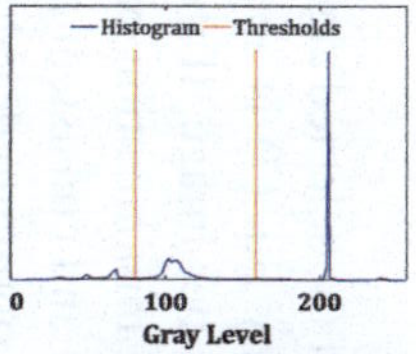	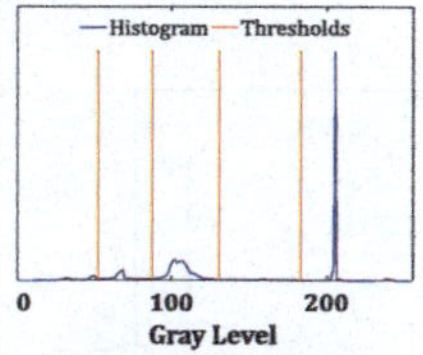	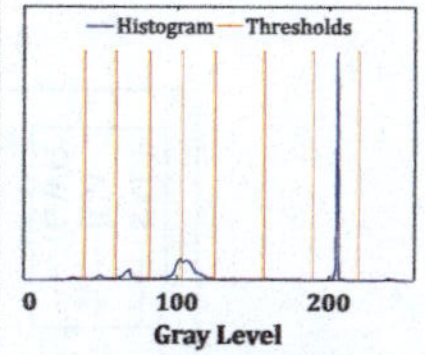	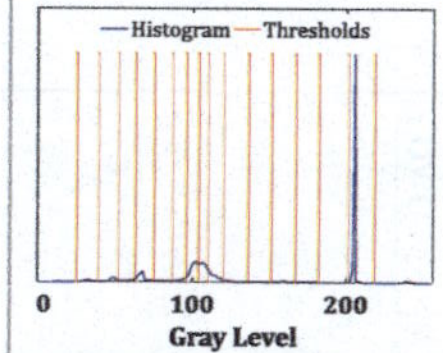
Jet				
	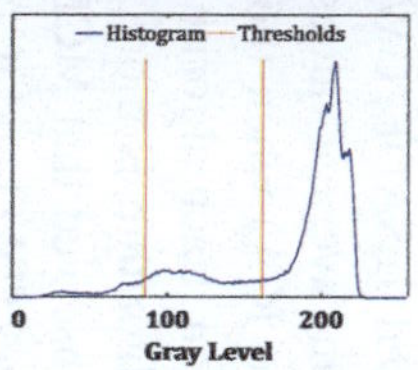	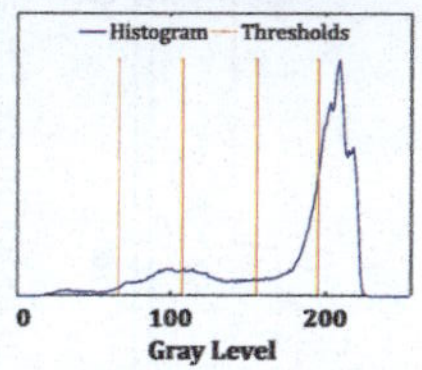	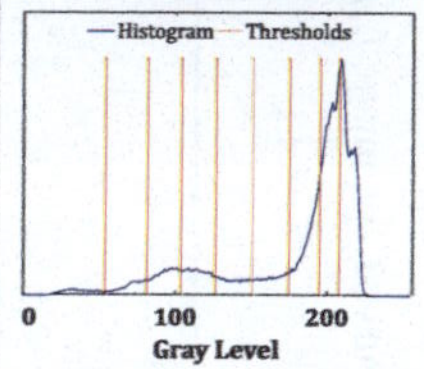	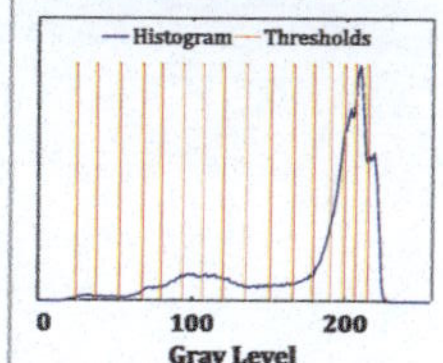

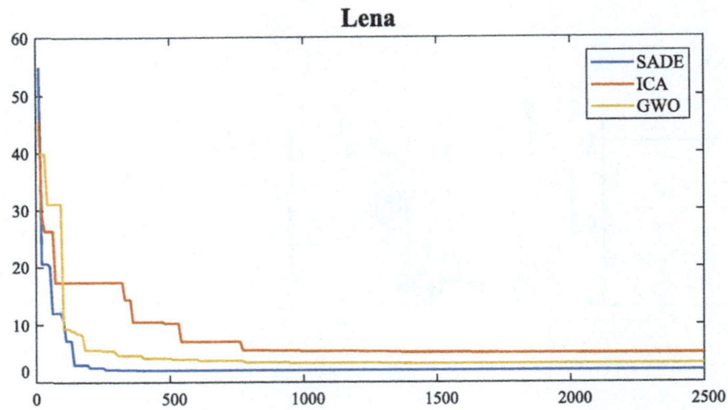

Fig. 2 Convergence curve Lena image $L_{Th} = 16$

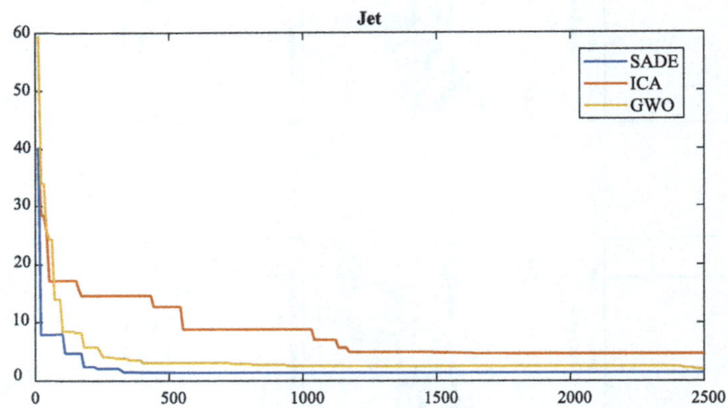

Fig. 3 Convergence curve Jet image $L_{Th} = 16$

levels $L_{Th} = 2, 3, 4, 5$. This subsection exhibits the experimental results when the MCE is implemented to SADE, GWO, and ICA.

Table 7 records the average values from STD and mean of the selected algorithms (SADE, GWO, and ICA) when the minimum cross-entropy Eq. (7) is applied as an objective function. Each algorithm is executed 35 times per image. If the mean value is low the algorithm is more consistent since it is a minimization problem. From Table 7 its observed that MCE-SADE is more stable and consistent than the compared techniques GWO and ICA since is able to obtain optimum *th* values. The bold values indicate the best outcomes.

To appraise the brain image segmentation quality is realized a PSNR, SSIM and FSIM values comparison obtained with SADE, GWO, and ICA when applying the

Table 7 Mean and standard deviation for MR brain images obtained by SADE, GWO and ICA based on MCE

Image	L_{Th}	SADE		GWO		ICA	
		Mean	STD	Mean	STD	Mean	STD
Z1	2	**3.2599**	**0.0092**	3.3965	0.0309	3.4774	0.0316
	3	**1.6975**	**0.0068**	1.6976	0.0105	1.7944	0.0150
	4	**0.9921**	**0.0047**	0.9924	0.0142	1.0276	0.0145
	5	0.7054	**0.0003**	**0.7046**	0.0138	0.7561	0.0113
Z2	2	**3.3009**	**0.0071**	3.3598	0.0219	3.4165	0.0223
	3	1.6694	**0.0060**	**1.6671**	0.0181	1.7732	0.0192
	4	0.9881	**0.0020**	**0.9877**	0.0107	1.0405	0.0113
	5	**0.7032**	**0.0002**	0.7054	0.0090	0.7260	0.0093
Z5	2	**3.4111**	**0.0310**	3.4256	0.0469	3.5066	0.0483
	3	**1.5924**	**0.0089**	1.5929	0.0206	1.6172	0.0211
	4	0.9991	**0.0047**	**0.9986**	0.0149	1.0300	0.0161
	5	**0.6918**	**0.0002**	0.6953	0.0142	0.7504	0.0144
Z10	2	**3.3429**	**0.0189**	3.3676	0.0461	3.4322	0.0470
	3	**1.5998**	**0.0049**	1.6002	0.0307	1.6227	0.0313
	4	**1.0006**	**0.0042**	1.0720	0.0279	1.1748	0.0306
	5	**0.6812**	**0.0003**	0.7039	0.0149	0.7166	0.0151
Z36	2	**1.9376**	**0.0205**	2.0589	0.0307	2.0847	0.0322
	3	**1.1413**	**0.0203**	1.1419	0.0280	1.1785	0.0283
	4	**0.6405**	**0.0043**	0.6513	0.0227	0.6842	0.0240
	5	**0.4811**	**0.0003**	0.4812	0.0131	0.5097	0.0135
Z72	2	**1.9049**	**0.0191**	2.0541	0.0289	2.1075	0.0295
	3	**1.1323**	**0.0116**	1.1330	0.0262	1.2235	0.0269
	4	**0.6267**	**0.0073**	0.6268	0.0220	0.6401	0.0238
	5	**0.4185**	**0.0003**	0.3803	0.0145	0.4993	0.0190
Z108	2	**1.9638**	0.0819	2.0553	**0.0514**	2.2180	**0.0514**
	3	1.1327	**0.0170**	**1.1325**	0.0248	1.1334	0.0268
	4	**0.5182**	**0.0161**	0.6276	0.0243	0.6388	0.0248
	5	0.4781	**0.0002**	**0.3913**	0.0136	0.5262	0.0183
Z144	2	**1.7320**	**0.0260**	1.8052	0.0394	1.8364	0.0419
	3	**1.1041**	**0.0083**	**1.1041**	0.0240	1.1721	0.0244
	4	**0.6736**	**0.0041**	0.6916	0.0125	0.7354	0.0133
	5	**0.3325**	**0.0002**	0.4128	0.0114	0.4226	0.0117

MCE, the average results are presented in Table 8. A higher PSNR, SSIM and FSIM value means a better segmentation quality process. Results from Table 8 show that MCE-SADE has more accuracy and brain image segmentation quality than the other compared methods. The bold values indicate the best outcomes. Figures 4 and 5, present the convergence curve of the fitness values from Z2 and Z36 axial cuts MR brain images, respectively at 5 levels of thresholds.

For illustrative purposes, only four images from the group were chosen to visually display the segmentation results in MR brain images, the results are recorded in Table 9. For each brain image, the first column presents initial MR brain image, while second, third and fourth columns present the segmented MR brain image when the MCE-SADE, MCE-GWO and MCE-ICA methods apply respectively and the $L_{Th} = 5$. All images are displayed in a colormap called jet instead of grayscale, to have better visibility of them. The cold colors denote low intensity values, the hot colors indicate high intensity values. From Table 9 its noticed that segmented images through the MCE-SADE method have sharper edges, this makes the representation of the brain image more precise.

6 Statistical Analysis

To statistically analyze the outcomes from the submitted method, a non-parametric statistical hypothesis known as Wilcoxon's rank test [56, 57] is utilized to assess the results variations among two associated samples. Comparison of the results is divided into two different statistical analyses. The first statistical analysis evaluates the results of the submitted approach when this is applied to standard test images and compares them with GWO and ICA based methods. The second analysis aims to assess the yield of the submitted method when implemented in MR brain images.

6.1 Statistical Analysis Standard Test Images

With aim of evaluating acquired results from the objective function of the purposed method, MCE-SADE is compared with GWO and ICA algorithms, both implement the minimum cross-entropy as an objective function. Each algorithm performs 35 experiments per standard test image with the elected threshold levels. In four tables are summarized the results for the standard test images, Table 2 presents the optimal threshold values found by each method, Table 3 exhibits the outcomes of mean and STD, Table 4 records PSNR values, Table 5 manifests SSIM and FSIM quality metrics results. Results in Table 3 expose that SADE is more stable and consistent than the compared techniques GWO and ICA. The acquired results from Tables 4 and 5, demonstrate that MCE-SADE is more accurate and have a better-quality segmentation of the standard test images than compared methods.

Table 8 PSNR, SSIM and FSIM values comparison achieved by SADE, GWO, and ICA applying MCE for MR brain images

Image	L_{Th}	SADE			GWO			ICA		
		PSNR	SSIM	FSIM	PSNR	SSIM	FSIM	PSNR	SSIM	FSIM
Z1	2	**14.3900**	**0.4887**	**0.5897**	14.3076	0.4876	0.5890	14.2168	0.4873	0.5887
	3	**17.8391**	**0.6605**	**0.7074**	17.8261	0.6581	0.7055	17.8136	0.6257	0.6785
	4	**19.8313**	**0.7221**	**0.7875**	19.7545	0.7205	0.7861	19.1298	0.6668	0.6733
	5	**21.8553**	**0.7772**	**0.8325**	19.7114	0.7769	0.8209	19.1498	0.6692	0.6975
Z2	2	**14.3595**	**0.4874**	**0.5882**	14.3489	0.4857	0.5872	14.2563	0.4840	0.5865
	3	**17.9259**	**0.6604**	**0.7087**	17.2593	0.6592	0.7085	17.1926	0.6315	0.6858
	4	**19.9051**	**0.7212**	**0.7894**	19.6172	0.7200	0.7881	19.5508	0.6713	0.6914
	5	**22.0159**	**0.7774**	**0.8335**	21.1737	0.7772	0.8262	20.9001	0.6877	0.7106
Z5	2	**14.5936**	**0.4881**	**0.5925**	14.5236	0.4869	0.5917	14.4609	0.4858	0.5909
	3	**18.1523**	**0.6642**	**0.7254**	18.1380	0.6640	0.7252	18.1054	0.6261	0.6945
	4	**19.9149**	**0.7151**	**0.7948**	19.8062	0.7135	0.7936	19.7947	0.6770	0.6988
	5	**21.9762**	**0.7752**	**0.8298**	21.2054	0.7747	0.8295	19.0112	0.6859	0.7181
Z10	2	**16.2249**	**0.5744**	**0.6441**	16.1420	0.5741	0.6438	16.1111	0.5736	0.6437
	3	**18.5690**	**0.6545**	**0.7423**	18.5535	0.6532	0.7415	18.4913	0.6169	0.7014
	4	**20.2326**	**0.6976**	**0.7934**	20.0725	0.6902	0.7916	20.0519	0.6897	0.7255
	5	**21.5348**	**0.7700**	**0.8395**	21.5189	**0.7700**	0.8393	19.2355	0.6976	0.7362
Z36	2	**15.9209**	**0.6169**	**0.6609**	15.9186	0.6167	0.6607	15.8369	0.6166	0.6606
	3	**18.3350**	**0.6997**	**0.7702**	18.2984	**0.6997**	0.7700	18.2820	0.6619	0.7121
	4	**20.0809**	**0.7393**	**0.8195**	20.0542	**0.7393**	0.8136	19.2836	0.7127	0.7420
	5	**21.3916**	**0.8083**	**0.8611**	21.3197	0.8058	0.8575	20.3066	0.7010	0.7281

(continued)

Table 8 (continued)

Image	L_{Th}	SADE			GWO			ICA		
		PSNR	SSIM	FSIM	PSNR	SSIM	FSIM	PSNR	SSIM	FSIM
Z72	2	**15.9694**	**0.6540**	**0.6929**	15.8250	0.6529	0.6916	15.6298	0.6523	0.6915
	3	**19.0492**	**0.7667**	**0.7987**	19.0261	0.7663	**0.7987**	19.0213	0.7141	0.7425
	4	**20.7019**	**0.8303**	**0.8465**	20.5702	0.8290	0.8458	20.3702	0.7383	0.7667
	5	**23.1588**	**0.8835**	**0.8934**	23.1190	0.8761	0.8862	21.2153	0.6936	0.7451
Z108	2	**16.2432**	**0.6702**	**0.6768**	16.1182	0.6696	0.6754	15.9924	0.6690	0.6750
	3	**19.7396**	**0.7913**	**0.7672**	19.7264	0.7902	0.7653	19.6795	0.7394	0.7201
	4	**21.4008**	**0.8588**	**0.8308**	21.2898	0.8585	0.8304	21.2174	0.7575	0.7436
	5	**23.2721**	**0.8974**	**0.8775**	23.2633	0.8956	0.8752	20.7299	0.6869	0.7140
Z144	2	**18.3950**	**0.7758**	**0.7521**	18.2765	0.7757	**0.7521**	18.2163	0.7752	0.7519
	3	**21.0007**	**0.8291**	**0.8232**	20.9649	0.8289	0.8229	20.9488	0.8219	0.7881
	4	**22.4532**	**0.8857**	**0.8587**	22.3827	0.8848	0.8580	22.2860	0.8404	0.7944
	5	**24.2066**	**0.9050**	**0.8943**	24.1552	0.9029	0.8898	22.1812	0.8431	0.7934

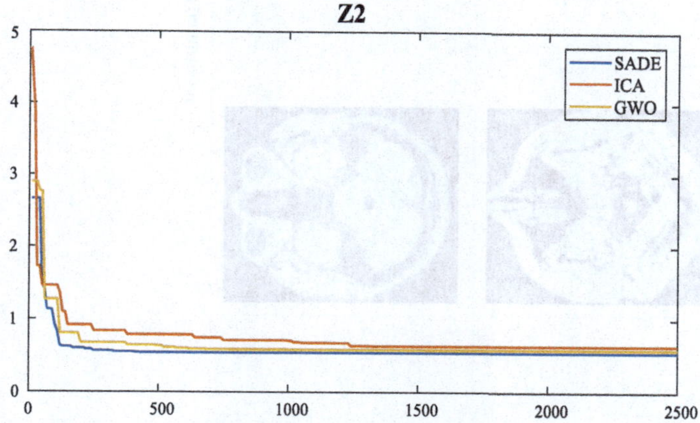

Fig. 4 Convergence curve Z2 MR brain image $L_{Th} = 5$

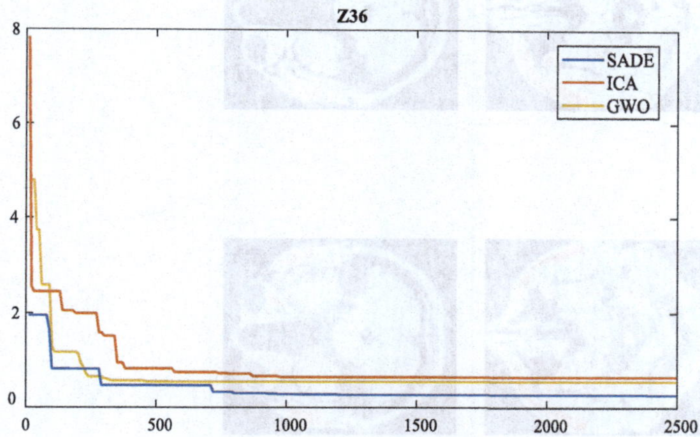

Fig. 5 Convergence curve Z36 MR brain image $L_{Th} = 5$

Wilcoxon's rank test statistical analysis is performed with 35 independent samples for each selected algorithm with a level of significance of 5% over fitness function outcomes for each threshold level. The resultant p-values are recorded in Table 10; these values are presented in two comparison groups SADE vs GWO and SADE vs ICA. The null hypothesis considers that there is none meaningful variation among the results of the tested algorithms. The alternate hypothesis regards that there is a meaningful variation between the results from both algorithms.

A p-value fewer than 0.05 imply that the null hypothesis can be discarded at a 5% significance level. Results in Table 10 show that there is a meaningful variation among the compared techniques since all the p-values are less than 0.05 and therefore null hypothesis can be discarded. Those outcomes also indicate that the values did

Table 9 Visual results of MR brain images obtained by SADE, GWO and ICA with $L_{Th} = 5$

	Original	SADE	GWO	ICA
Z1	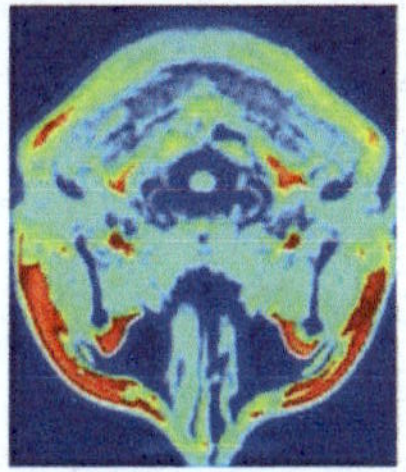	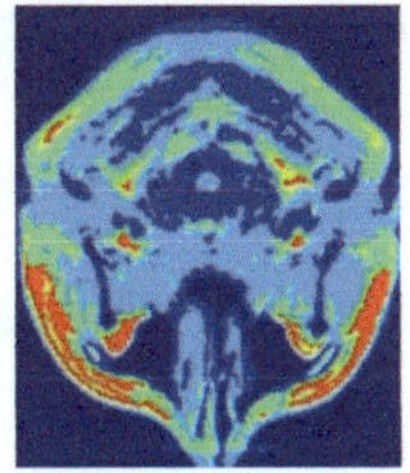	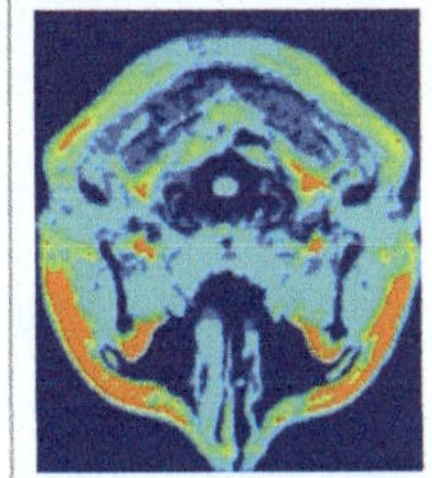	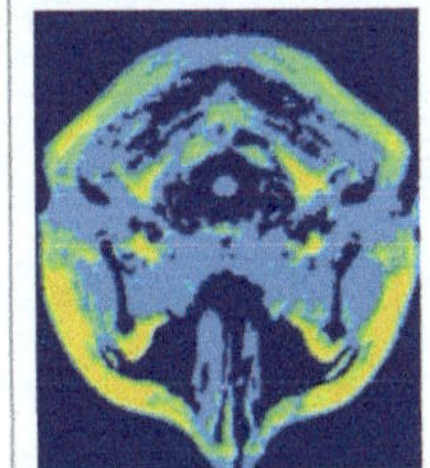
Z36	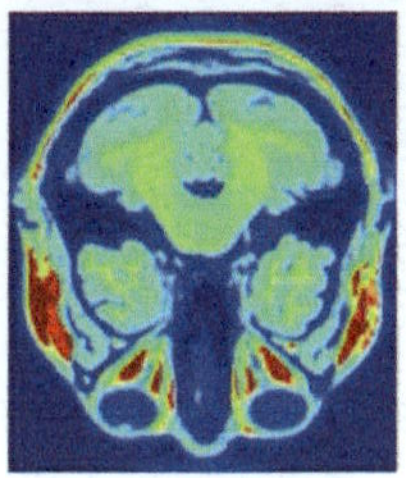	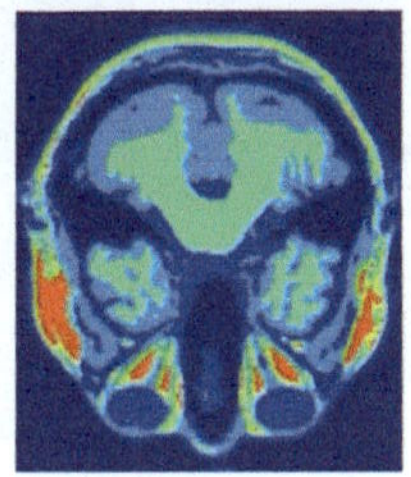	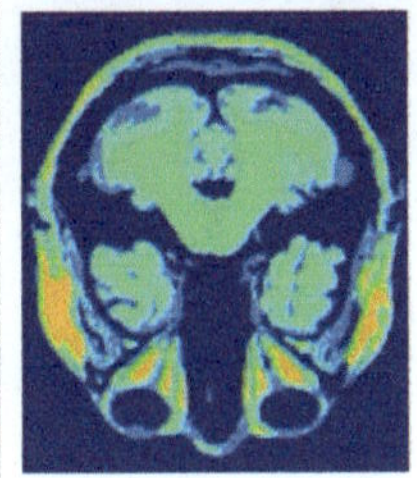	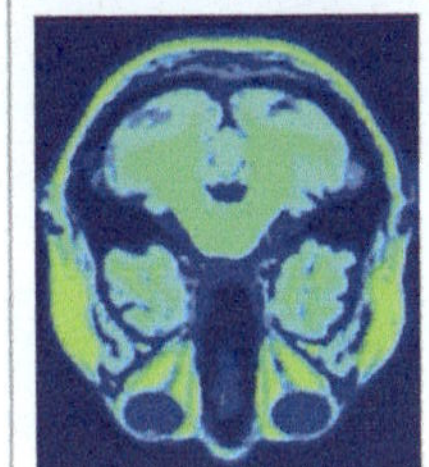

(continued)

Table 9 (continued)

	Original	SADE	GWO	ICA
Z72	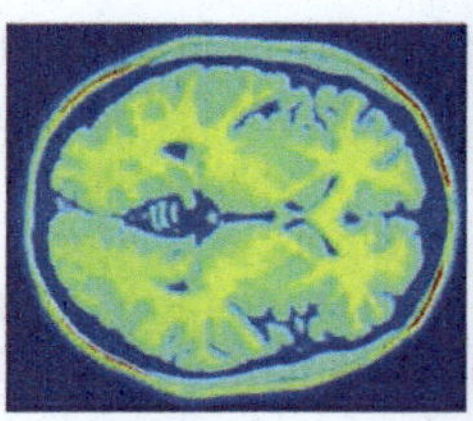	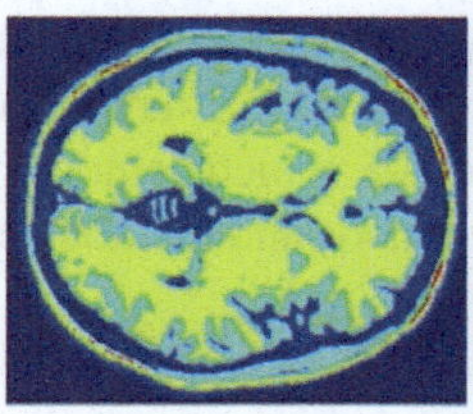	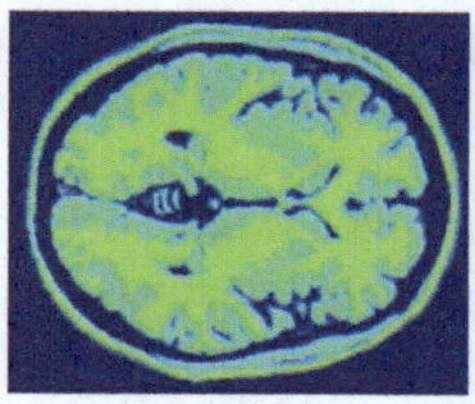	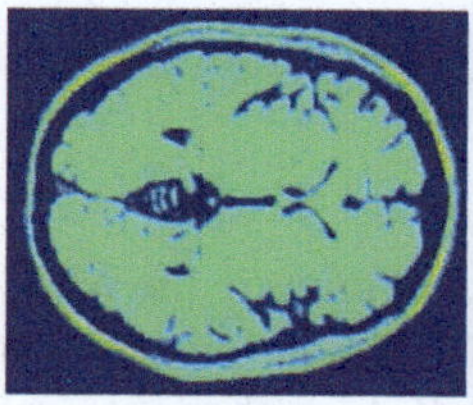
Z108	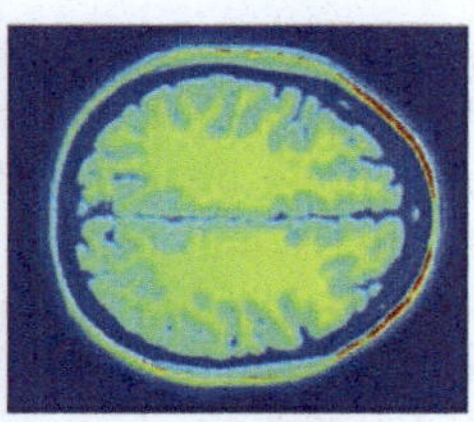	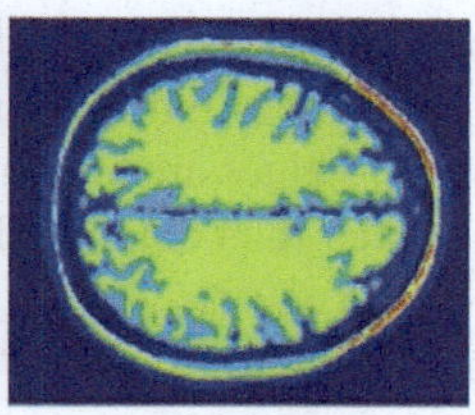	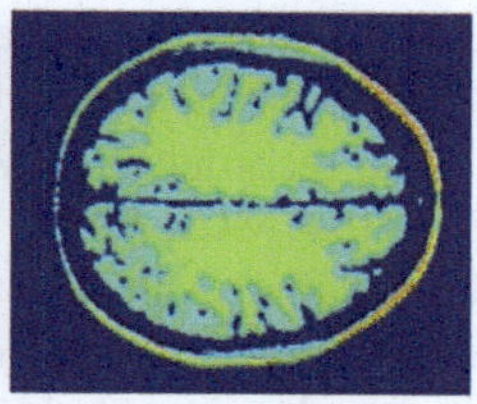	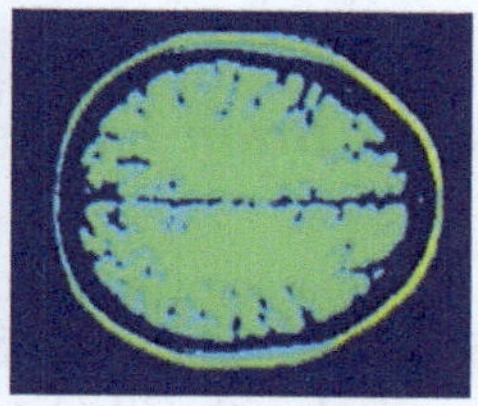

not happen by chance and that the SADE based method performs better than the GWO and ICA based methods.

6.2 Statistical Analysis MR Brain Images

The main objective of this statistical analysis is to determine if the MCE-SADE can provide a quality segmentation when applied to MR brain images. Just like in Sect. 6.1, MCE-SADE is compared with GWO and ICA methods, to assess the results of the MR brain images group. Each algorithm performs 35 experiments per MR brain image with the elected threshold levels. The results are recorded in three tables, Table 7 exhibits the outcomes of mean and STD, Table 8 shows the results of the PSNR, SSIM and FSIM quality metrics, Table 9 expose a visual contrast from the segmented MR brain images. In Table 7, the results from SADE are more consistent and stable than the STD and mean recorded for GWO and ICA. Outcomes for the quality metrics of Table 8 expose that MCE-SADE method has more accuracy and quality in the segmentation of brain images than the compared methods.

Wilcoxon's rank test statistical analysis is carried out at a 5% significance level with 35 independent samples for each selected algorithm (SADE, GWO and ICA) over the objective function values for each threshold level ($L_{Th} = 2, 3, 4, 5$). As in the previous section two hypothesis are formulated, the null hypothesis and the alternative hypothesis. A p-value fewer than 0.05 implies that the null hypothesis can be discarded at a level of significance of 5%. Table 11 presented the resultant p-values for the MR brain images set in two comparison groups SADE versus GWO and SADE versus ICA. The obtained p-values are less than 0.05, which demonstrate that SADE based method performs better than compared methods and that the outcomes did not befall through chance.

7 Conclusions

This study proposed a novel method founded on the Self-Adaptive Differential Evolution (SADE) algorithm for solving the multilevel thresholding (MTH) segmentation problem in Magnetic Resonance brain images (MRBI). Given that the MTH is a problem of optimization whose fitness function is a thresholding technique. The proposed approach called MCE-SADE searches the optimal threshold values through the Minimum Cross-Entropy (MCE) which is SADE algorithm fitness function. The MCE-SADE method was tested utilizing two groups of reference images; primary group consists of nine standard test images that are broadly employed in the literature of image processing such as Cameraman, Lena, Peppers, Sailboat, among others, while the following group of reference images is composed of MRBI extracted from Brainweb database. The purpose of the first reference group was to estimate the exploration skills from SADE when is applied to a high-dimensional space. While the

Table 10 Wilcoxon's test p-values from SADE versus GWO and SADE versus ICA on the standard test images

Image	L_{Th}	*p*-values	
		SADE versus GWO	SADE versus ICA
Blonde	2	2.40E−05	4.35E−07
	4	1.38E−06	4.24E−06
	8	9.13E−10	3.24E−11
	16	7.06E−10	5.49E−13
Cameraman	2	2.05E−06	8.46E−06
	4	1.12E−06	9.06E−07
	8	4.99E−08	1.39E−12
	16	1.52E−11	6.48E−13
Couple	2	2.74E−04	1.36E−05
	4	4.95E−06	1.99E−07
	8	7.48E−09	8.50E−10
	16	6.51E−10	5.78E−10
House	2	2.35E−06	4.61E−07
	4	5.17E−08	2.82E−07
	8	4.16E−10	6.52E−12
	16	1.01E−12	6.53E−13
Jet	2	2.18E−05	1.88E−05
	4	5.03E−06	1.53E−05
	8	6.13E−06	7.96E−06
	16	6.51E−13	2.07E−09
Lena	2	5.16E−05	5.42E−06
	4	7.93E−06	1.55E−07
	8	6.86E−08	7.16E−10
	16	6.00E−12	5.03E−13
Peppers	2	1.43E−05	8.41E−06
	4	3.85E−06	2.03E−07
	8	2.72E−10	2.45E−12
	16	6.52E−11	3.55E−12
Pirate	2	5.44E−05	7.80E−08
	4	3.40E−06	1.75E−07
	8	2.97E−07	1.33E−08
	16	4.39E−11	2.09E−13
Sailboat	2	1.13E−08	4.35E−04
	4	4.20E−08	6.59E−06
	8	1.21E−11	5.78E−07
	16	7.09E−13	2.35E−11

Table 11 Wilcoxon's test p-values from SADE versus GWO and SADE versus ICA on the MR brain images

Image	L_{Th}	*p*-values	
		SADE versus GWO	SADE versus ICA
Z1	2	1.34E−08	9.56E−07
	3	4.01E−10	3.47E−09
	4	6.02E−13	7.83E−12
	5	6.38E−13	4.38E−13
Z2	2	4.85E−10	7.78E−08
	3	1.45E−11	1.85E−07
	4	6.35E−13	8.29E−12
	5	6.42E−13	4.81E−08
Z5	2	4.18E−13	6.62E−05
	3	1.26E−13	2.97E−06
	4	5.96E−13	1.90E−13
	5	6.17E−13	4.28E−10
Z10	2	4.30E−13	4.07E−05
	3	6.30E−13	6.01E−07
	4	5.03E−13	7.69E−08
	5	1.57E−13	1.61E−10
Z36	2	1.66E−10	1.57E−08
	3	3.90E−12	5.90E−10
	4	5.58E−13	2.79E−12
	5	5.23E−13	5.98E−14
Z72	2	8.53E−08	9.06E−05
	3	1.99E−10	1.14E−08
	4	3.45E−11	1.37E−11
	5	1.64E−12	7.57E−12
Z108	2	1.67E−12	4.41E−05
	3	3.84E−12	1.48E−06
	4	5.96E−13	2.67E−08
	5	5.17E−13	7.76E−08
Z144	2	4.76E−11	1.51E−07
	3	1.42E−11	4.80E−07
	4	9.01E−13	6.18E−09
	5	3.15E−13	4.33E−15

objective of the second group of reference images was to evaluate the effectiveness of MCE-SADE when implemented in MRBI. To validate the above was compared the yield of MCE-SADE with two metaheuristics algorithms based methods, Grey Wolf Optimizer (GWO) and Imperialist Competitive Algorithm (ICA). Segmentation quality of the images was appraised employing three image processing metrics, PSNR, SSIM, and FSIM.

The experimental results from MCE-SADE approach demonstrate that the outcomes obtained are further consistent, stable and offer better solutions quality than GWO and ICA based methods. The statistical analysis proved that the values found by SADE did not happen by chance and therefore it can be asserted that the purposed approach has a better yield than GWO and ICA techniques. The values obtained in PSNR, SSIM and FSIM quality metrics suggest that MCE-SADE has a high segmentation quality of MRBI. In addition, segmented images by the proposed method show well-defined sections in comparison with GWO and ICA methods. As future work will aim our effort on developing a simpler and more efficient SADE algorithm focus only on image segmentation and computer vision problems regarding others multilevel thresholding methods such as Tsallis entropy and Renyi's entropy.

References

1. T. Budrys, V. Veikutis, S. Lukosevicius, et al., Artifacts in magnetic resonance imaging: how it can really affect diagnostic image quality and confuse clinical diagnosis? J. Vibroeng. **20**, 1202–1213 (2018). https://doi.org/10.21595/jve.2018.19756
2. S. Simu, S. Lal, A study about evolutionary and non-evolutionary segmentation techniques on hand radiographs for bone age assessment. Biomed. Signal Process. Control **33**, 220–235 (2017). https://doi.org/10.1016/J.BSPC.2016.11.016
3. S. González-Villà, A. Oliver, S. Valverde et al., A review on brain structures segmentation in magnetic resonance imaging. Artif. Intell. Med. **73**, 45–69 (2016). https://doi.org/10.1016/j.artmed.2016.09.001
4. T.X. Pham, P. Siarry, H. Oulhadj, Integrating fuzzy entropy clustering with an improved PSO for MRI brain image segmentation. Appl. Soft. Comput. J. **65**, 230–242 (2018). https://doi.org/10.1016/j.asoc.2018.01.003
5. Z. Yang, Y. Shufan, G. Li, D. Weifeng, Segmentation of MRI brain images with an improved harmony searching algorithm. Biomed. Res. Int. (2016). https://doi.org/10.1155/2016/4516376
6. P. Moeskops, M.A. Viergever, A.M. Mendrik et al., Automatic segmentation of MR brain images with a convolutional neural network. IEEE Trans. Med. Imaging **35**, 1252–1261 (2016). https://doi.org/10.1109/TMI.2016.2548501
7. R. Hiralal, H.P. Menon, *A survey of brain MRI image segmentation methods and the issues involved* (Springer, Cham, 2016), pp. 245–259
8. N. Otsu, A threshold selection method FROM gray-level histograms. IEEE Trans. Syst. Man. Cybern. **9**, 62–66 (1979). https://doi.org/10.1109/TSMC.1979.4310076
9. J. Kittler, J. Illingworth, Minimum error thresholding. Pattern Recognit. **19**, 41–47 (1986). https://doi.org/10.1016/0031-3203(86)90030-0
10. N.R. Pal, On minimum cross-entropy thresholding. Pattern Recognit. **29**, 575–580 (1996). https://doi.org/10.1016/0031-3203(95)00111-5
11. A.G. Shanbhag, Utilization of information measure as a means of image thresholding. CVGIP Graph. Model Image Process. **56**, 414–419 (1994). https://doi.org/10.1006/CGIP.1994.1037

12. M. Sezgin, B. Sankur, Survey over image thresholding techniques and quantitative performance evaluation. J. Electron. Imaging **13**, 146 (2004). https://doi.org/10.1117/1.1631315
13. C.-I. Chang, Y. Du, J. Wang et al., Survey and comparative analysis of entropy and relative entropy thresholding techniques. IEEE Proc.—Vis. Image Signal Process. **153**, 837 (2006). https://doi.org/10.1049/ip-vis:20050032
14. M.L. Menendez, Shannon's entropy in exponential families: statistical applications. Appl. Math. Lett. **13**, 37–42 (2000). https://doi.org/10.1016/S0893-9659(99)00142-1
15. J.N. Kapur, P.K. Sahoo, A.K.C. Wong, A new method for gray-level picture thresholding using the entropy of the histogram. Comput. Vis. Graph. Image Process. **29**, 273–285 (1985). https://doi.org/10.1016/0734-189X(85)90125-2
16. C.H. Li, C.K. Lee, Minimum cross entropy thresholding. Pattern Recognit. **26**, 617–625 (1993). https://doi.org/10.1016/0031-3203(93)90115-D
17. C. Tsallis, Computational applications of nonextensive statistical mechanics. J. Comput. Appl. Math. **227**, 51–58 (2009). https://doi.org/10.1016/J.CAM.2008.07.030
18. E. Beadle, J. Schroeder, B. Moran, S. Suvorova, An overview of Renyi Entropy and some potential applications, in *2008 42nd Asilomar Conference on Signals, Systems and Computers* (IEEE, 2008), pp. 1698–1704
19. V. Osuna-Enciso, E. Cuevas, H. Sossa, A comparison of nature inspired algorithms for multi-threshold image segmentation. Expert Syst. Appl. **40**, 1213–1219 (2013). https://doi.org/10.1016/J.ESWA.2012.08.017
20. J. Zhang, H. Li, Z. Tang, et al., (2014) An improved quantum-inspired genetic algorithm for image multilevel thresholding segmentation. Math. Probl. Eng. (2014). https://doi.org/10.1155/2014/295402
21. Y. Wang, Improved OTSU and adaptive genetic algorithm for infrared image segmentation, in *2018 Chinese Control and Decision Conference (CCDC)* (IEEE, 2018), pp. 5644–5648
22. Y. Li, S. Wang, J. Xiao, Image segmentation based on dynamic particle swarm optimization for crystal growth. Sensors **18**, 3878 (2018). https://doi.org/10.3390/s18113878
23. M.F. Di, S. Sessa, PSO image thresholding on images compressed via fuzzy transforms. Inf. Sci. (Ny) **506**, 308–324 (2020). https://doi.org/10.1016/J.INS.2019.07.088
24. H.V.H. Ayala, F.M. dos Santos, V.C. Mariani, L. dos Santos Coelho, Image thresholding segmentation based on a novel beta differential evolution approach. Expert Syst. Appl. **42**, 2136–2142 (2015). https://doi.org/10.1016/J.ESWA.2014.09.043
25. E. Cuevas, D.P.-C.M. Zaldivar, A novel multi-threshold segmentation approach based on differential evolution optimization. Expert Syst. Appl. **37**, 5265–5271 (2010). https://doi.org/10.1016/j.eswa.2010.01.013
26. U. Mlakar, B. Potočnik, J. Brest, A hybrid differential evolution for optimal multilevel image thresholding. Expert Syst. Appl. **65**, 221–232 (2016). https://doi.org/10.1016/j.eswa.2016.08.046
27. E. Cuevas, F. Sención, D. Zaldivar et al., A multi-threshold segmentation approach based on artificial bee colony optimization. Appl. Intell. **37**, 321–336 (2012). https://doi.org/10.1007/s10489-011-0330-z
28. S. Zhan, W. Jiang, S. Satoh, Multilevel thresholding color image segmentation using a modified artificial bee colony algorithm. IEICE Trans. Inf. Syst. **101**, 2064–2071 (2018). https://doi.org/10.1587/transinf.2017EDP7183
29. D. Oliva, E. Cuevas, G. Pajares, et al., Multilevel thresholding segmentation based on harmony search optimization. J. Appl. Math. (2013). https://doi.org/10.1155/2013/575414
30. V. Tuba, M. Beko, M. Tuba, *Color Image Segmentation by Multilevel Thresholding Based on Harmony Search Algorithm* (Springer, Cham, 2017), pp. 571–579
31. T. Kaur, B.S. Saini, S. Gupta, *Optimized Multi Threshold Brain Tumor Image Segmentation Using Two Dimensional Minimum Cross Entropy Based on Co-occurrence Matrix* (Springer, Cham, 2016), pp. 461–486
32. D. Oliva, S. Hinojosa, E. Cuevas, G. Pajares, O.G.J. Avalos, Cross entropy based thresholding for magnetic resonance brain images using Crow Search Algorithm. Expert Syst. Appl. **79**, 164–180 (2017). https://doi.org/10.1016/j.eswa.2017.02.042

33. P.D. Sathya, R. Kayalvizhi, Amended bacterial foraging algorithm for multilevel thresholding of magnetic resonance brain images. Meas. J. Int. Meas. Confed. **44**, 1828–1848 (2011). https://doi.org/10.1016/j.measurement.2011.09.005
34. M. Ali, P. Siarry, M. Pant, *Multi-level image thresholding based on hybrid differential evolution algorithm. Application on Medical Images* (Springer, Berlin, Heidelberg, 2017), pp. 23–36
35. G. Chen, Z. Yang Z (2009) Preserving and exploiting genetic diversity in evolutionary programming algorithms. IEEE Trans. Evol. Comput. **13**. https://doi.org/10.1109/TEVC.2008.2011742
36. D.H. Wolpert, W.G. Macready, No free lunch theorems for optimization. IEEE Trans. Evol. Comput. **1**, 67–82 (1997). https://doi.org/10.1109/4235.585893
37. D. Shilane, J. Martikainen, S. Dudoit, S.J. Ovaska, A general framework for statistical performance comparison of evolutionary computation algorithms. Inf. Sci. (Ny) **178**, 2870–2879 (2008). https://doi.org/10.1016/J.INS.2008.03.007
38. S. Mirjalili, S.M. Mirjalili, A. Lewis, Grey wolf optimizer. Adv. Eng. Softw. **69**, 46–61 (2014). https://doi.org/10.1016/j.advengsoft.2013.12.007
39. A.E. Gargari, C. Lucas, Imperialist competitive algorithm : an algorithm for optimization inspires by imperialistic competition. IEEE Congr. Evol. Comput., 4661–4667 (2007)
40. S. Kullback, *Information Theory and Statistics* (Dover Publications, 1968)
41. R. Storn, K. Price, Differential evolution—A simple and efficient heuristic for global optimization over continuous spaces. J. Glob. Optim. **11**, 341–359 (1997)
42. A.T. Buba, L.S. Lee, A differential evolution for simultaneous transit network design and frequency setting problem. Expert Syst. Appl. **106**, 277–289 (2018). https://doi.org/10.1016/J.ESWA.2018.04.011
43. B. Bošković, J. Brest, Protein folding optimization using differential evolution extended with local search and component reinitialization. Inf. Sci. (Ny) **454–455**, 178–199 (2018). https://doi.org/10.1016/J.INS.2018.04.072
44. D.M. Diab, K. El Hindi, Using differential evolution for improving distance measures of nominal values. Appl. Soft. Comput. **64**, 14–34 (2018). https://doi.org/10.1016/J.ASOC.2017.12.007
45. M.G. Villarreal-Cervantes, E. Mezura-Montes, J.Y. Guzmán-Gaspar, Differential evolution based adaptation for the direct current motor velocity control parameters. Math. Comput. Simul. **150**, 122–141 (2018). https://doi.org/10.1016/J.MATCOM.2018.03.007
46. S. Maggi, Estimating water retention characteristic parameters using differential evolution. Comput. Geotech. **86**, 163–172 (2017). https://doi.org/10.1016/J.COMPGEO.2016.12.025
47. A.K. Qin, V.L. Huang, P.N. Suganthan, Differential evolution Algorithm with strategy adaptation for global numerical optimization. IEEE Commun. Mag. **13**, 398–417 (2009). https://doi.org/10.1109/TEVC.2008.927706
48. J.E. Baker, Reducing bias and inefficiency in the selection algorithm, in *Proceedings of the Second International Conference on Genetic Algorithms*, 28–31 July 1987, Massachusetts Institute of Technology, MA (1987)
49. S. Sarkar, S. Paul, R. Burman et al., *A Fuzzy Entropy Based Multi-Level Image Thresholding Using Differential Evolution* (Springer, Cham, 2015), pp. 386–395
50. A.K.M. Khairuzzaman, S. Chaudhury, Multilevel thresholding using grey wolf optimizer for image segmentation. Expert Syst. Appl. **86**, 64–76 (2017). https://doi.org/10.1016/j.eswa.2017.04.029
51. M. Nejad, M. Fartash, Applying chaotic imperialist competitive algorithm for multi-level image thresholding based on Kapur's entropy. Adv. Sci. Technol. Res. J. **10**, 125–131 (2016). https://doi.org/10.12913/22998624/61940
52. B. Sankur, B. Sankur, K. Sayood, Statistical evaluation of image quality measures. J Electron. Imaging **11**, 206 (2002). https://doi.org/10.1117/1.1455011
53. Z. Wang, A.C. Bovik, H.R. Sheikh, E.P. Simoncelli, Image quality assessment: from error visibility to structural similarity. IEEE Trans. Image Process. **13**, 600–612 (2004). https://doi.org/10.1109/TIP.2003.819861

54. L. Zhang, L. Zhang, X. Mou, D. Zhang, FSIM: a feature similarity index for image quality assessment. IEEE Trans. Image Process. **20**, 2378–2386 (2011). https://doi.org/10.1109/TIP.2011.2109730
55. P. Ghamisi, M.S. Couceiro, J.A. Benediktsson, N.M.F. Ferreira, An efficient method for segmentation of images based on fractional calculus and natural selection. Expert Syst. Appl. **39**, 12407–12417 (2012). https://doi.org/10.1016/J.ESWA.2012.04.078
56. F. Wilcoxon, Individual comparisons by ranking methods. Biometrics Bull. **1**, 80–83 (1945)
57. S. García, D. Molina, M. Lozano, F. Herrera, A study on the use of non-parametric tests for analyzing the evolutionary algorithms' behaviour: a case study on the CEC'2005 Special Session on Real Parameter Optimization. J. Heuristics **15**, 617–644 (2009). https://doi.org/10.1007/s10732-008-9080-4

Automatic Detection of Malignant Masses in Digital Mammograms Based on a MCET-HHO Approach

Erick Rodríguez-Esparza, Laura A. Zanella-Calzada, Daniel Zaldivar and Carlos E. Galván-Tejada

Abstract Digital image processing techniques have become an important process within medical images. These techniques allow the improvement of the images in order to facilitate their interpretation for specialists. Within these are the segmentation methods, which help to divide the images by regions based on different approaches, in order to identify details that may be complex to distinguish initially. In this work, it is proposed the implementation of a multilevel threshold segmentation technique applied to mammography images, based on the Harris Hawks Optimization (HHO) algorithm, in order to identify regions of interest (ROIs) that contain malignant masses. The method of minimum cross entropy thresholding (MCET) is used to select the optimal threshold values for the segmentation. For the development of this work, four mammography images were used (all with presence of a malignant tumor), in their two views, craniocaudal (CC) and mediolateral oblique (MLO), obtained from the Digital Database for Screening Mammography (DDSM). Finally, the ROIs calculated were compared with the original ROIs of the database through a series of metrics, to evaluate the behavior of the algorithm. According to the results obtained, where it is shown that the agreement between the original ROIs and the calculated ROIs is significantly high, it is possible to conclude that the proposal of the MCET-HHO algorithm allows the automatic identification of ROIs containing malignant tumors in mammography images with significant accuracy.

E. Rodríguez-Esparza (✉) · D. Zaldivar
Universidad de Guadalajara, CUCEI, Blvd. Gral. Marcelino García Barragán 1421, Olímpica, 44430 Guadalajara, Jalisco, Mexico
e-mail: erick.rodrigueze@alumno.udg.mx

D. Zaldivar
e-mail: daniel.zaldivar@cucei.udg.mx

L. A. Zanella-Calzada (✉)
University of Lorraine, LORIA, Campus Scientifique, 615 Rue du Jardin-Botanique, 54506 Vandœuvre-lès-Nancy, Lorraine, France
e-mail: laura.zanella-calzada@univ-lorraine.fr

C. E. Galván-Tejada
Universidad Autónoma de Zacatecas "Francisco García Salinas", Jardín Juárez 147, Centro, 98000 Zacatecas, Zacatecas, Mexico
e-mail: ericgalvan@uaz.edu.mx

D. Oliva and S. Hinojosa (eds.), *Applications of Hybrid Metaheuristic Algorithms for Image Processing*, Studies in Computational Intelligence 890,
https://doi.org/10.1007/978-3-030-40977-7_15

Keywords Digital mammograms · Breast tumors detection · Multilevel threshold segmentation · Harris Hawks Optimization · Minimum cross entropy

1 Introduction

Digital image processing techniques have become an elementary process for the enhancement of medical images since they are usually affected by noise from different sources or phenomena that deteriorates the acquisition and measurement processes of images. Those artifacts that can be affecting medical images, such as the physiological system, are able to diminish the contrast and the visibility of significant details [41].

Image enhancement techniques are based on mathematical methods and they consist on producing another image, highlighting specific features of interest that may not be very evident in the original image. These techniques can be developed and applied complemented by an optimization method with reference to specific requirements and objective criteria [4]. In addition, although most enhancement techniques are applied in order to improve images for human observation use, such as the noise suppression, contrast enhancement and sharpening of the details, some techniques are used to derive images for use by a subsequent algorithm for computer processing, including edge enhancement and object segmentation [5].

Object or image segmentation plays a fundamental role in image processing [28] and it can be defined as the partitioning of digital images into multiple segments or sets of pixels. The main objective of this process is to simplify or change the representation of an image into a more meaningful and easier to analyze objects. Image segmentation is usually represented by the locations of objects and boundaries in images, assigning labels to each pixel, being the pixels with the same labels those that share specific features [25].

According to the literature, there are several image segmentation methods that have been presented such as clustering, graph cut, edge detection and thresholding [34]. However, none of them have reached to be universally applicable, since it is still a challenge to segment the object accurately when the images present noise, complex background and inhomogeneous intensity [28].

The thresholding approach is the most popular method for image segmentation since it is simple to implement with high accuracy and robustness against the other methods [17], and it works taking the information of the histogram from an image and defining the optimal threshold values (th) which separate adequately the distinct regions of pixels in the image being processed [16]. Image threshold methods can be classified into two different types based on the number of thresholds required for the image: bilevel and multilevel thresholding. In the bilevel method, the pixels of the objects of interest have an intensity value higher and are clearly distinguished from the background of an image by a single threshold value, while in the multilevel thresholding, several thresholds are used to separate the pixels in different regions that represent objects that are in the image [13, 35].

The thresholding problem can be summarized as the search of the optimal threshold values for an image, and to find them, there are two types of approaches: parametric and non-parametric. In parametric approach, it is necessary to estimates the parameters of the probability density functions and the combination of those classes will represent all the pixels in the image, the problem with this approach is that it is computationally expensive. However, the non-parametric approach employs a discriminative criteria (between-class variance, entropy and error rate) to separate the pixels into homogeneous regions, this criterion must be optimized to determine the optimal threshold values [11].

During the last years, multiple works have been presented in the literature in which the development of information theory provides opportunities to explore the use of several entropies to find efficient forms of segmenting images. Some examples of this entropies are the Kapur entropy [23], the Tsallis entropy [12], and the cross entropy [26]. The Minimum Cross Entropy Thresholding (MCET) is a widely algorithm used in the literature to minimizing the cross entropy between the original image and its segmented image to find optimal thresholding. Nevertheless, the searching of the best thresholds in multilevel thresholding is computationally expensive and the computing time for exhaustive search increases exponentially, especially when the number of thresholds is increased. In order to solve these problems, the optimization of metaheuristics algorithms (MA) inspired by nature has attracted attention in the field of threshold multilevel images [36]. There are many types of MA such as Particle Swarm Optimization (PSO) [24], Ant Colony Optimization (ACO) [15], Differential evolution (DE) [43], Firefly Algorithm (FA) [46], Social Spider Optimization (SSO) [9], Locus Swarm Optimization (LSO) [10] and Harris Haks Optimization (HHO) [19].

One application of these techniques is presented on mammograms, which are X-ray images of the breast, since even when induced noise does not affect in an important way the quality of the image, they present limited contrast because of the nature and the superimposition of the soft tissues of the breast, which are compressed during the imaging procedure. Then, the small differences between normal and abnormal tissues are affected by noise and artifacts, causing the implementation of digital image processing techniques for the improvement in the appearance and visual quality of the images, as well as the need to apply segmentation to highlight regions of interest (ROIs). This enhancement of the mammograms is very useful since it assists in their interpretation by medical specialists [37].

Therefore, many research has been focused on distinguish the breast tissue according to its histology for the mammograms examination based on these digital image processing, such as in the work of Shi et al. [42], which present a fully automated pipeline of mammogram image processing which estimates skin-air boundary based on a gradient weight map and, detects pectoral-breast boundary and calcifications inside the breast region using a texture filter. Nayak et al. [32] propose an automatic breast masses detection and extraction of breast density based on the Watershed algorithm, which is a technique for both contour analysis and region growing for its segmentation, providing an accuracy comparable to state-of-the-art techniques. Mughal et al. [31] present an hybrid methodology based on an adaptive hysteresis thresholding segmentation technique for localizing the breast masses in the curve stitching domain, where they remove the pectoral muscle and segment a breast region of interest in digital mammography images. Also, Min et al. [30] propose a simultaneous detection and segmentation technique for mammographic lesions based on a sifting architecture, where a novel region candidate selection approach using a multi-scale morphological sifting (MMS) method and cascade learning techniques are applied, orienting linear structuring elements to be used to sieve out the ROIs in mammograms, including stella patterns. In the work of Hmida et al. [22] is proposed an automatic breast mass segmentation based on a series of stages, beginning with the contour initialization applied to a given region of interest, then, the construction of fuzzy contours and estimation of fuzzy membership maps of different classes in the considered image and finally, the integration of these maps in the Chan-Vese method to obtain a model based on fuzzy-energy used for the delineation of the mass.

On the other hand, Valvano et al. [45], where a system based on convolutional neural networks (CNNs) for the detection and segmentation of microcalcifications clusters in the breast is presented, obtaining a significant accuracy that demonstrates the potentialities of deep learning for the support of radiologists. Al-antari et al. [2] also propose a system based on deep learning, implementing a full resolution convolutional network (FrCN) for the segmentation of breast masses, obtaining results that outperform the latest conventional deep learning methodologies. A segmentation method of a modified pulse coupled neural networks (MSPCNN) based on human visual system (HVS) for medical images is presented by Lian et al. [27], attempting to deduce the sub-intensity range of central neurons firing by introducing a neighboring matrix and calculating the intensity distribution range based on the MSPCNN, subsequently revealing the way of how sub-intense range parameters generate the input stimulus closer to HVS, improving the segmentation accuracy and reducing the complexity accuracy because of the parameter setting method.

Breast cancer is an approach that has been widely studied since is the most common cancer among women worldwide, being early detection crucial in lowering the mortality, where screening mammography is effective in the detection of breast cancer in its early stages [6]. However, even when screening mammography has shown to reduce mortality in women due to breast cancer, there are some aspects that do not allow to have the desired results in the diagnosis of breast cancer, being estimated that 70% of all breast cancer in digital mammography are due to misinterpretation while 30% are overlooked lesions [20]. Among the potential undesirable effects are including overdiagnosis, causing overtreatment, and false-positive mammography results, causing psychological distress and societal cost [29]. Another aspect that has become relevant in mammography screening is breast density, which radiologists determine according to the amount of radiopaque breast parenchyma that is visualized in mammography and which in turn can limit the screening accuracy because it is difficult to detect if the lesions overlap with dense tissue [38].

Computer-aided detection (CAD) has been developed to increase the sensitivity of mammographic examinations by marking suspicious regions on mammograms such as microcalcifications and masses, helping to achieve an effective medical analysis using the digital process of clinical treatment. One of the biggest challenges in CAD systems is the accurate localization or segmentation of breast mass regions due to the complicated structure of the overlapped tissues in the breast. This setup leads to erroneous location for finding precise ROI [31]. Therefore, different segmentation techniques have been implemented to improve the localization of breast masses.

In this chapter, a new method to automatically determine the region of interest (ROI) of mammograms is presented, the ROI locates the cancerous tumor. First a mammogram is processed, which consists of two stages, the first stage is used to eliminate the edges, the information on the type of scanner and the projection that was used to take the image, then the second stage is applied, which consists of eliminate noise through filters. The resulting image is map of the original image preserving the original intensity of the breast region free of artifacts. In the second part, this image of the breast of mammography is segmented using a new multilevel thresholding technique. This approach, called MCET-HHO, is composed of the criterion of Minimum Cross Entropy Thresholding (MCET) and the HHO algorithm. In the proposed method, the HHO algorithm is used to minimize the cross entropy among classes and thus, find the optimal threshold values for multilevel thresholding. Next, the mammogram is segmented using the bilevel threshold segmentation. The threshold used for this is the highest value of the optimal thresholds found by the MCET-HHO algorithm. Finally, the center of the ROI is the center of the mass of the image with that threshold.

The next sections of the chapter are as follows. Section 2 is described in detail the segmentation approach employed for this proposal. In Sect. 3 presents the overall description of the Harris Hawks Optimization algorithm. Then, in Sect. 4 presents the proposed methodology divided into five phases (data acquisition, preprocessing, MCET-HHO segmentation and selection of ROI, validation) with its sub-stages. The results and the discussions are provided in Sect. 5. In the end, conclusions of the proposed method are exposed in Sect. 6 and the future work in Sect. 7.

2 Image Segmentation

The problem of selecting multiple thresholds in images is presented using the minimization of cross entropy as a criterion for segmentation, which is a non-parametric approach. Image segmentation refers to the partition of an image into sets of regions that compose it, and the thresholding segmentation method is done by selecting optimal threshold values using the histogram information. The histogram can contain valleys and wide peaks with different heights, which makes difficult to find the optimal threshold levels, thus it is why the cross entropy threshold method is used to homogenize the histogram information between the original image and the segmented image.

2.1 Minimum Cross Entropy Method

The minimum cross entropy threshold algorithm (MCET) selects the optimal threshold value by minimizing the cross entropy between the original image and the threshold image. MCET algorithm seeks the minimum value because a lower value represents less uncertainty of the pixels and more homogeneity [21]. Let **I** be the original image and $h(i)$, $i = 1, 2, \ldots, L$ be the corresponding histogram with L being the number of gray levels. For bilevel thresholding, the thresholded image ($\mathbf{I_{th}}$) is established using the threshold value (th) by:

$$\mathbf{I_{th}}(x, y) = \begin{cases} \mu(1, th), & \text{if } \mathbf{I}(x, y) < th \\ \mu(th, L+1), & \text{if } \mathbf{I}(x, y) \geq th \end{cases} \tag{1}$$

where

$$\mu(a, b) = \sum_{i=a}^{b-1} ih(i) \Big/ \sum_{i=a}^{b-1} h(i) \tag{2}$$

The crossed entropy is calculated from the rewriting of the Eq. 1, since it generates a threshold image instead of an entropy value, to obtain an entropy value as an objective function, by:

$$f_{cross}(th) = \sum_{i=1}^{th-1} ih(i)log\left(\frac{i}{\mu(1, th)}\right) + \sum_{i=th}^{L} ih(i)log\left(\frac{i}{\mu(th, L+1)}\right) \tag{3}$$

Yin et al. [47] proposed a faster recursive programming technique to obtain the optimal threshold for digital images by reducing the computational cost by extending the Eq. 3 to a multilevel approach, calculated by:

$$\begin{aligned} f_{cross}(th) = & \sum_{i=1}^{L} ih(i)log(i) - \sum_{i=1}^{th-1} ih(i)log(\mu(1, th)) - \\ & \sum_{i=th}^{L} ih(i)log(\mu(th, L+1)) \end{aligned} \tag{4}$$

The multilevel approach of Eq. 4 is based on the vector $\mathbf{th} = [th_1, th_2 \ldots, th_{nt}]$, which contains nt different thresholds values, by:

$$f_{cross}(\mathbf{th}) = \sum_{i=1}^{L} ih(i)log(i) - \sum_{i=1}^{nt} H_i \tag{5}$$

where nt is the total number of thresholds and H_i is defined as:

$$H_1 = \sum_{i=1}^{th_1-1} ih(i)log(\mu(1, th_1))$$
$$H_k = \sum_{i=th_{k-1}}^{th_k-1} ih(i)log(\mu(th_{k-1}, th_k)), \quad 1 < k < nt \tag{6}$$
$$H_{nt} = \sum_{i=th_{nt}}^{L} ih(i)log(\mu(th_{nt}, L+1))$$

3 Harris Hawks Optimization

HHO is proposed by Heidari et al. [19] and it is an optimization technique without gradient, based on the population, which can be applied to any optimization problem subject to an adequate formulation. The principle of HHO is inspired in the behaviour of Harris' haws, which are listed among the most intelligent birds in nature, and their tactic to capture a prey, known as "seven kills" strategy. This tactic consists on several hawks attempting to attack cooperatively from different directions and converge simultaneously on a prey, which is usually a rabbit, that escapes from the outside. HHO is constituted by two phases, exploration and exploitation, as shown in Fig. 1.

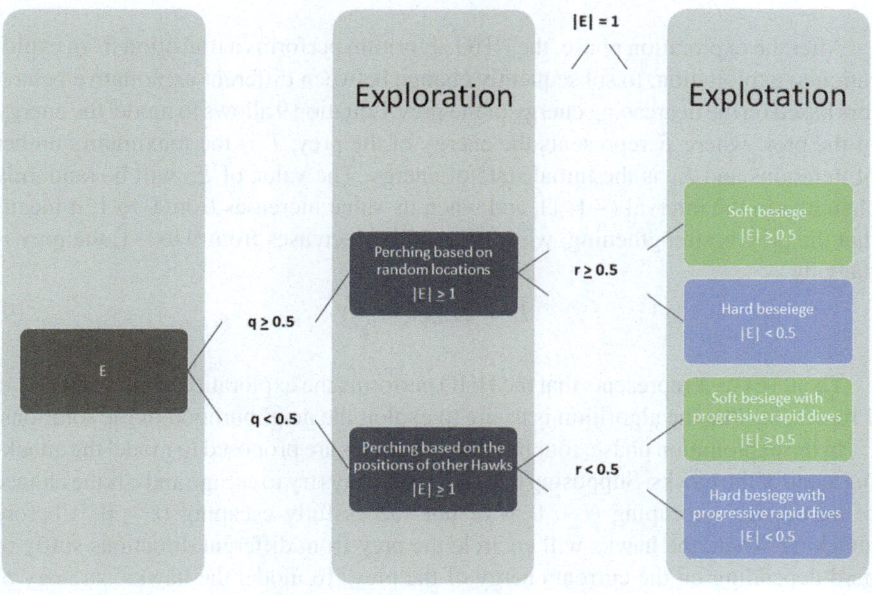

Fig. 1 Phase behavior of the HHO algorithm

The exploration phase is based in the tracking and detection of preys by Harris' hawks, where the hawks constitutes the candidate solutions, being the best candidate solution of each step considered as the intended prey or the possible optimum. For this stage, the hawks are randomly placed in some locations, waiting to detect a prey based on two strategies. If there is considered an equal chance q for each posed strategy, they are located according to the positions of other members and the prey, modeled by Eq. 7 where the condition is $q < 0.5$, and $X(t+1)$ is the position vector of haws in the next iteration t, $X_{rand}(t)$ is a randomly selected hawk from the current population, $X(t)$ is the current position vector of the haws r_1 and r_2, and q are random numbers inside (0, 1); or places on random locations inside the groups home range, modeled by Eq. 7 where the condition is $q \geq 0.5$, and $X_{prey}(t)$ is the position of the prey, r_3 and r_4 are hawks, LB and UB show the upper and lower bounds of variables and X_m is the average position of the current population.

$$X(t+1) = \begin{cases} X_{rand}(t) - r_1|X_{rand}(t) - 2r_2X(t)|, & q \geq 0.5 \\ (X_{prey}(t) - X_m(t)) - r_3(LB + r_4(UB - LB)), & q < 0.5 \end{cases} \tag{7}$$

To generate the random locations inside (LB, UB) a simple model is proposed, which is able to mimic the behaviors of hawks. The average position of hawks is attained by Eq. 8, where $X_i(t)$ indicates the location of each hawk in iteration t and N denotes the total number of hawks.

$$X_m(t) = \frac{1}{N}\sum_{i=1}^{N} X_i(t) \tag{8}$$

After the exploration phase, the HHO algorithm performs a transition from exploration to exploitation, to subsequently change between different exploitative behaviors based on the decreasing energy of the prey. Equation 9 allows to model the energy of the prey, where E represents the energy of the prey, T is the maximum number of iterations and E_o is the initial state of energy. The value of E_o will be randomly changing in the interval $(-1, 1)$, and when its value increases from 0 to 1, it means that the prey is strengthening, while if its value decreases from 0 to -1, the prey is flagging.

$$E = 2E_o(1 - \frac{t}{T}) \tag{9}$$

Then, $|E| \geq 1$ represents that the HHO performs the exploration phase, and $|E| < 1$ represents that the algorithm is trying to exploit the neighborhood of the solutions.

In the exploitation phase, four possible strategies are proposed to model the attacking stage of the hawks. Supposing that the prey always try to escape and r is the chance of successfully escaping ($r < 0.5$) or not successfully escaping ($r \geq 0.5$) before attacking. Then, the hawks will encircle the prey from different directions softly or hard depending on the current energy of the prey. To model the hawks' strategy of intensifying the besiege process to effortlessly catch the prey that has been loosing energy by trying to escape, and switching between soft and hard besiege processes, the $|E|$ is required, where $|E| \geq 0.5$ means soft besiege and $|E| < 0.5$ means hard besiege.

The first strategy consists on soft besiege, when $r \geq 0.5$ and $|E| \geq 0.5$, and the prey presents enough energy, trying to escape by taking randomly directions but finally being captured. During this process the hawks encircle it softly to make the prey more exhausted and then attacking. The model of this strategy is shown by Eqs. 10 and 11, where $\Delta X(t)$ is the difference between the position vector of the prey and the current location in iteration r, r_5 is a random number inside (0, 1) and $J = 2(1 - r_5)$ is the random movement of the prey trying to escape. The value of J changes randomly in each iteration to simulate the prey motions.

$$X(t-1) = \Delta X(t) - E|JX_{prey}(t) - X(t)| \tag{10}$$

$$\Delta X(t) = X_{prey}(t) - X(t) \tag{11}$$

The next strategy consists on hard besiege, when $r \geq 0.5$ and $|E| < 0.5$, and the prey is so exhausted, presenting low escaping energy. The hawks hardly encircle the intended prey to perform the final attack. To update the positions of the prey and the hawks, Eq. 12 is implemented.

$$X(t+1) = X_{prey} - E|\Delta X(t)| \tag{12}$$

In the strategy of soft besiege with progressive rapid dives, it is still $|E| \geq 0.5$ but $r < 5$, and the prey has enough energy to successfully escape and a soft besiege is constructed before the surprise attack. To perform a soft besiege, it is supposed that the hawks can decide their next move based on the rule of Eq. 13.

$$Y = X_{prey}(t) - E|JX_{prey}(t) - X(t)| \tag{13}$$

Then, the hawks compare the possible result of a movement according to a previous immersion to detect if it will be a good movement or not. If the movement does not present a good result, they need to perform irregular, abrupt and rapid movements when approaching the prey, based on Eq. 14, where D is the dimension of the problem, S is a random vector of size $1 \times D$ and LF is the levy flight function, which can be described by Eq. 15, where u are υ random values inside (0, 1), β is a default constant set to 1.5.

$$Z = Y + S \times LF(D) \tag{14}$$

$$LF(x) = 0.01 \times \frac{u \times \sigma}{|\upsilon|^{\frac{1}{\beta}}}, \sigma = \left(\frac{\Gamma(1+\beta) \times \sin(\frac{\pi\beta}{2})}{\Gamma(\frac{1+\beta}{2}) \times \beta \times 2(\frac{\beta-1}{2})} \right)^{\frac{1}{\beta}} \tag{15}$$

Therefore, the final stage for updating the positions of hawks is described by Eq. 16, where Y and Z are obtained by Eqs. 13 and 14.

$$X(t+1) = \begin{cases} Y, & F(Y) < F(X(t)) \\ Z, & F(Z) < F(X(t)) \end{cases} \tag{16}$$

The last strategy is when $E| < 0.5$ and $r < 0.5$, and the prey has not enough energy to escape and a hard besiege is applied before attacking the prey to be killed. In this situation the hawks are trying to decrease the distance of their average location with the escaping prey. Then, Eq. 17 is performed to apply the hard besiege condition, where Y and Z are obtained by Eqs. 18 and 19, and where $X_m(t)$ is obtained by Eq. 8.

$$X(t+1) = \begin{cases} Y, & F(Y) < F(X(t)) \\ Z, & F(Z) < F(X(t)) \end{cases} \tag{17}$$

$$Y = X_{prey}(t) - E|JX_{prey}(t) - X_m(t)| \tag{18}$$

$$Z = Y + S \times LF(D) \tag{19}$$

4 Automatic Detection of ROI Using MCET-HHO

The methodology proposed by this work is presented in Fig. 2. Initially, a preprocessing step is carried out to improve the contrast of the mammography images and to reduce the noise caused by artifacts. Then, a multilevel threshold segmentation based on the MCET-HHO algorithm is applied, followed by the selection of the ROI based on the optimal values found by the thresholding. These ROIs contain the malignant masses inside the breast. Finally, a validation stage is included to evaluate the performance of the methodology applied.

Each experiment consists of 35 independent executions of the MCET-HHO algorithm that use five gray levels in each of the images, and the average and standard deviation are reported. The images of the four patients used for the experimental step have the same format, png. All experiments were performed using Matlab 9.4 on an Intel Core i5 CPU at 2.7 GHz with 8 GB of RAM.

4.1 Datasets Description

The digital images used in this proposed method are breast images taken from patients collected from the Digital Database for Screening Mammography (DDSM) [18]. The DDSM database was created by Massachusetts General Hospital in the University

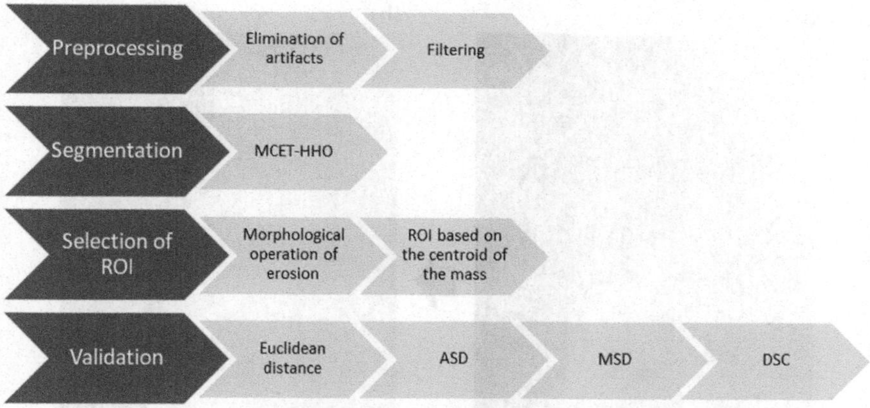

Fig. 2 Flowchart of the methodology followed

of South Florida and Sandia National laboratories and contains 2620 cases of digital mammograms compressed with lossless JPEG encoding which is used in the work [1, 8, 44].

Each case contains four images, two projections (craniocaudal (CC) and mediolateral oblique (MLO)) per side (left and right), also it contains extra information about the case (date of study, patient age, breast density, type of pathology, number of anomalies, etc.) and information about the image (file name, image type, scan date, scanner type, pixels per line, bits per pixel, lesion location, resolution, etc.). All information contained in the DDSM was provided by expert doctors.

In this work, are used 4 cases of patients who presented mammograms with cancerous tumors on one side of their breasts (left or right) with their two projections (CC and MLO).

4.2 *Preprocessing*

Preprocessing is an essential and useful step, which is applied before any image segmentation technique to achieve reliable and adequate precision, improving the contrast of the image and highlighting the edges of it [39]. The proposed preprocessing method is composed of two stages, elimination of artifacts and filters.

4.2.1 Elimination of Artifacts

It is necessary to isolate the breast for the other stages of the methodology. The existence of undesirable artifacts or structures in mammography (noise, edges, marks) is not relevant and affects the correct functioning of the proposed methodology.

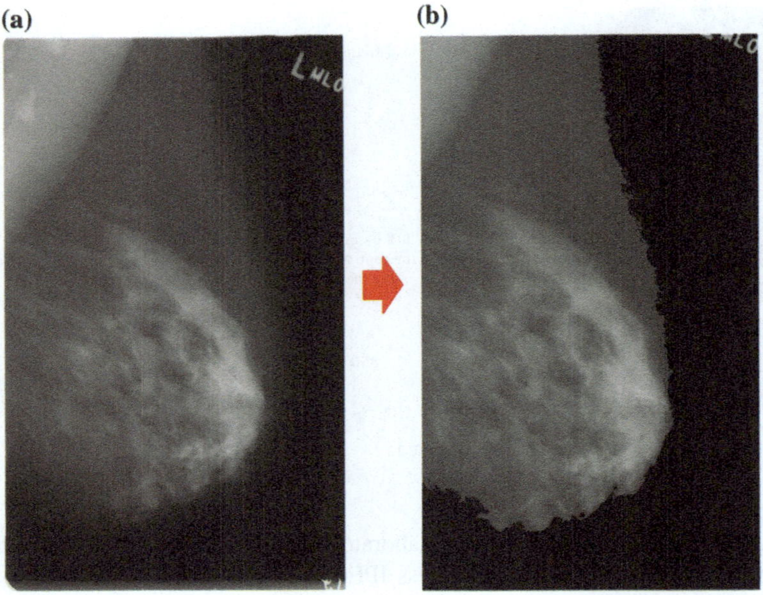

Fig. 3 Background removal of the image, **a** original image, **b** image after background removal

First, the size of the mammograms is resized because the images of the DDSM database have an average size of 4000 × 6000 pixels. In this work, the images are reduced by 70% of the original size to decrease the computation time. In the literature there are some works where they adopted this and showed that resizing does not cause negative results in the processing of mammograms [14, 33, 40].

Next, 30 pixels away from the side, top and bottom edges are removed. After that, a background removal is done, as shown in Fig. 3. In this step, a global mammography threshold is calculated and each pixel value of the image is compared using a 3 × 3 kernel with the global threshold. If three of the neighbors are less or equal to this threshold, the pixel is replaced by the color black (value of pixel 0).

Then, a vertical line adjustment strategy is applied to remove the external artifacts from mammograms, such as labels, markers, scratches and tapes with adhesive, as shown in Fig. 4. This method is implemented using a vertical line in the center of the image, separating the mammography on two sides (a and b), which moves a column to the right if the image is from the left breast and to the left if it is from the right breast, until the deviation standard of the whole column is less or equal to 0.04. When the regions a and b of the image are defined, the pixels of the region where no useful mammography information is found are converted to black (pixel value 0).

Finally, only the breast and part of the pectoral muscle were maintained, the pixels of these tissues have very similar high intensities. Due to this, it is necessary to eliminate the information of the pixels of the part of the muscle, and for this, the methodology of this chapter proposes a horizontal line adjustment strategy, as shown in Fig. 5. This method is implemented using a horizontal line in the top of the image,

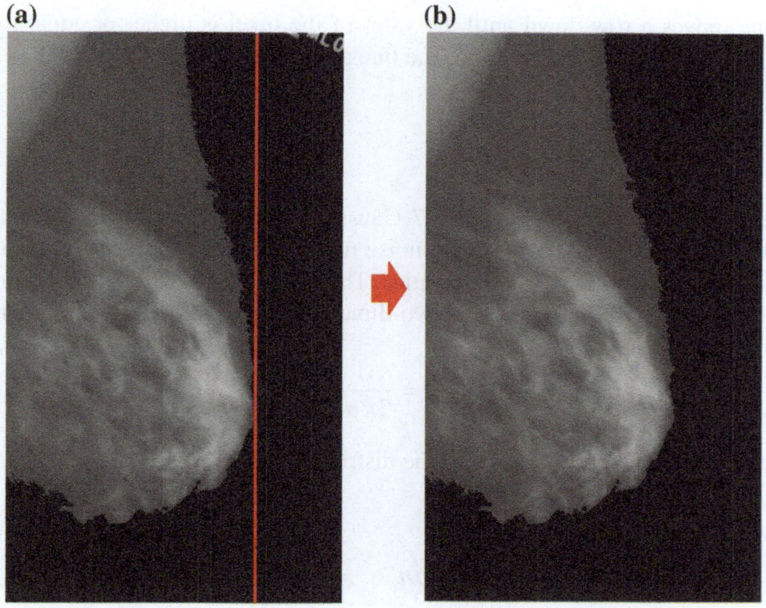

Fig. 4 Vertical line adjustment of the image, **a** vertical line located to the right on a left breast, **b** image after vertical adjustment

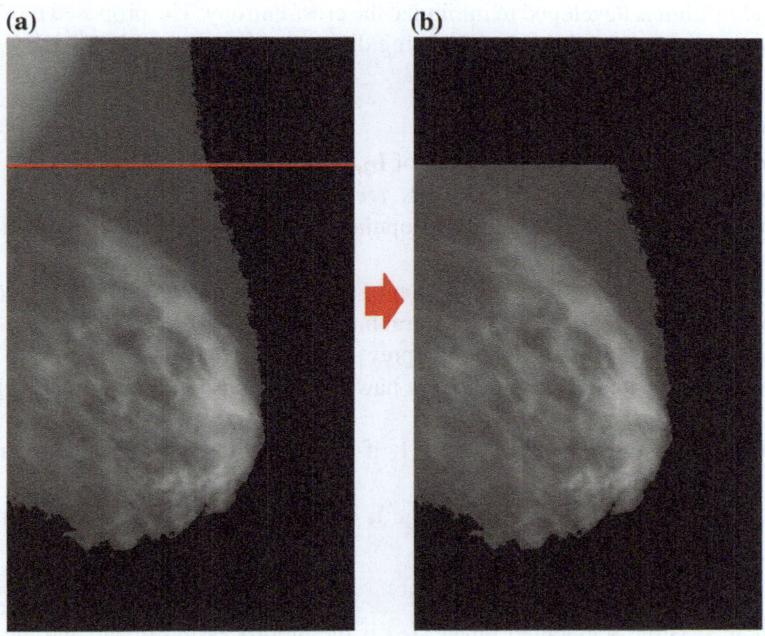

Fig. 5 Horizontal line adjustment of the image, **a** horizontal line located on the top of the image, **b** image after horizontal adjustment

this line moves a row down until the valor of the pixel is higher or equal to 100. Then, the pixels of the region above the line are replaced by the value of 0.

4.2.2 Filtering

In the proposed methodology, a 7×7 Gaussian filter is applied, which is a non-uniform low pass filter to eliminate noise and soften the images to improve the contrast of tissues with different densities. The Gaussian filter for a pixel (i, j) that is used in the proposed system uses a two-dimensional Gaussian distribution function is shown in Eq. 20.

$$\mathbf{G}(i, j) = \frac{1}{2\pi\sigma^2} e^{\frac{-i^2+j^2}{2\sigma^2}} \tag{20}$$

where σ is the standard deviation of the distribution.

4.3 MCET-HHO Segmentation

The main objective of the image segmentation is to identify specific ROIs that contain significant information in the breast, such as malignant masses. For this approach, the HHO algorithm is developed to minimize the cross entropy. The proposed method is simple and easy to implement, presenting different steps of the MCET-HHO algorithm below:

Step 1: Read the image $\mathbf{I_{Gr}}$.
Step 2: Calculate the histogram h_{Gr} of $\mathbf{I_{Gr}}$.
Step 3: Initialize the HHO parameters: $iter_{\max}$, N.
Step 4: Initialize the location of a population of Harris hawks $\mathbf{X}$ of N random particles with nt dimensions.
Step 5: Evaluate the objective function (f_{cross}) with Eq. 5 for each element of $\mathbf{X}$.
Step 6: Set X_{rabbit} as the location of rabbit (best location).
Step 7: Calculate E (the energy of a prey) with Eq. 9 for each hawk $\mathbf{X}_i$.
Step 8: Update the location of Harris hawks $\mathbf{X}$ depending of the energy value of his prey.
Step 9: The t index is increased in 1, if the stop criteria ($t \geq Iter_{\max}$) are not satisfied jump to step 5.
Step 10: Generate the segmented image $\mathbf{I_s}$ with the thresholds values contained in X_{rabbit}.

Where $\mathbf{I_s}$ is the segmented image, $\mathbf{I_{Gr}}$ is the preprocessing image and the stopping criterion for the MCET-HHO algorithm is the maximum number of iterations ($iter_{\max}$) and is set to 250 and the population size (N) of 30.

After obtaining the optimal values found by the MCET-HHO algorithm it is possible to obtain a series of levels that allow to thresholdize the different levels of gray that the image presents, in order to facilitate the identification of regions of interest that could go unnoticed in another way. Therefore, the breast is segmented using the highest value obtained by these optimal values (**th**) using the bilevel threshold segmentation to obtain a mass in the foreground (pixel values of 1) and the background (pixel values of 0).

4.4 Selection of ROI

Then, erosion is applied, which is a morphological operation that eliminates the pixels that do not belong to the object. This means that if some of the pixels next to the study pixel do not belong to the object, then the study pixel does not belong to the object either. For this work, a disc with a radius of 3 pixels was used as structuring object. Finally, there are the coordinates of the center of the ROI, which are the centroid of the mass obtained by eroding the pixels of the bilevel image. A square of 300×300 pixels is obtained as an automatic ROI.

4.5 Validation

A set of metrics was applied in order to compare the accuracy of the ROI obtained automatically against the ROI presented by the DDSM database. The first assessed metric is the euclidean distance (d). The euclidean distance between the coordinates of the center of the ROI of the database (A) and the center of the automatic ROI (B) is the length of the line that connects them, and is defined by:

$$d = \sqrt{(x_B - x_A) + (y_B - y_A)} \tag{21}$$

Another of the metrics used is the average symmetric distance (ASD), defined as the average euclidean distance between two surfaces, where 0 indicates a perfect match, by:

$$\mathrm{ASD}(\mathbf{A}, \mathbf{B}) = \frac{1}{|T_A + T_B|} \left(\sum_{\mathbf{P_A} \in T_A} d(\mathbf{P_A}, T_B) + \sum_{\mathbf{P_B} \in T_B} d(\mathbf{P_B}, T_A) \right) \tag{22}$$

where $d(P_A, T_B)$ and $d(P_B, T_A)$ indicate the shortest euclidean distance between each point of the perimeter of the ROI database (**A**) to the perimeter of the ROI obtained by the proposed methodology (**B**), and vice versa. T_A represents the number of coordinates of the perimeter of the ROI **A** and T_B the number of points of the ROI **B** [3, 7].

Maximum symmetric distance (MSD) is the metric used to indicate the maximum distance (or Hausdorff distance) between two contours **A** (ROI of the database) and **B** (automatic ROI) [3, 7], calculated as:

$$\mathrm{MSD}(\mathbf{A}, \mathbf{B}) = \max\left(\max(d(P_A, T_B)), \max(d(P_B, T_A))\right) \tag{23}$$

The last metric is the dice similarity coefficient (DSC), it is the index most widely used to measure the coincidence between 2 masses, calculated as:

$$\mathrm{DSC}(\mathbf{A}, \mathbf{B}) = \frac{2(\mathbf{A} \cap \mathbf{B})}{|T_A + T_B|} \tag{24}$$

where the DSC varies from 0 to 1, and 1 indicates a perfect overlap.

5 Results and Discussion

This section provides the experimental results of the methodology presented in this work based on different metrics that allow evaluate the efficiency and precision of the proposed algorithm.

In Figs. 6, 7, 8 and 9 are shown the results of the methodology applied to both views, CC and MLO, of the mammographies used for this work. In (a), the original images with their respective ROIs are observed. Then, in (b) the images were submitted to the preprocessing step, where it is possible to note that the background is uniformly black and the artifacts were removed. The next step, (c), present the images segmented, being possible to identify the multilevels of gray that are obtained from the selected thresholds. The selection of the ROIs is shown in (d), where the blue square is the representation of the ROI calculated and the red square, as well as the region bordered with a red dotted line, are the original ROIs. It is important to mention that one of the limitations of the calculated ROIs is that they are generated from a fixed size, while the original ROIs are adaptive to the different shapes and sizes of the tumors.

Figure 6 shows the image of a right breast with presence of a malignant tumor in the upper quadrants. In section (d), the comparison of the ROIs, the calculated and the original, presents similarity in the identification of the tumor, since the blue square accurately surrounds the region where the tumor is most evident in both views, based on the change that occurs in the levels of gray.

In Fig. 7 is shown the image of a left breast with presence of a malignant tumor being part of the upper and lower quadrants. The overlap that the ROIs present allows to observe the significant performance of the calculated ROI; however, the precision shown by the blue square in comparison with the red square can be affected because it does not have a dynamic behavior when calculated.

Then, in Fig. 8 is presented the image of a right breast with presence of a malignant tumor in the lower quadrants. After applying the different processing techniques, the

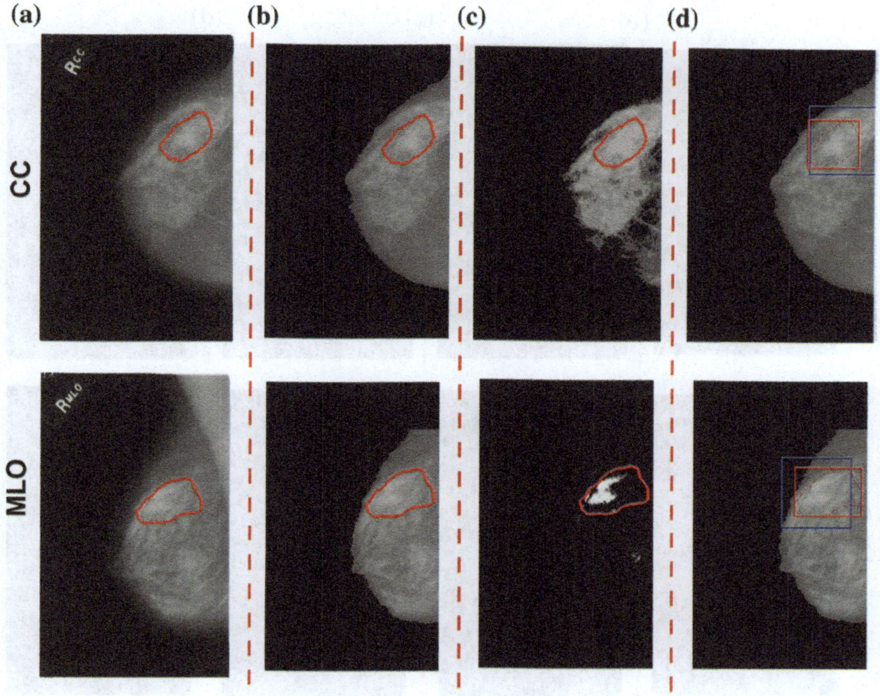

Fig. 6 CC and MLO views of mammography 1, **a** original images with the ROIs containing a malignant tumor, **b** preprocessed images, **c** segmented images, **d** contrast between the selected ROI (blue) and the original ROI (red)

ROI calculated presents a significant overlap with the original ROI, differing only in the shape, since the original ROI, being dynamic, acquires a rectangle shape, while the calculated ROI remains as a square. However, it should be noted that the blue square respects the size of the tumor in the image based on the change shown by shades of gray, without leaving the margin observed in the red rectangle.

Figure 9 shows the image of a left breast with presence of a malignant tumor in the upper quadrants. The calculated and the original ROIs overlap almost completely, both presenting a very similar square shape. The blue square presents a highly significant behavior, identifying with great precision the region where the tumor is located and also, matching with the margins of the red square.

Table 1 shows the average values of d, ASD, MSD and DSC obtained for each image in both views, CC and MLO, as a result of 35 iterations, as well as its standard deviation (SD). These values were calculated based on the comparison between the original ROI and the ROI obtained through the algorithm. In addition, the average time it takes the algorithm to calculate the ROI of the image is also included.

As described before, the d value represents the Euclidean distance between the coordinates of the center of the original ROI and the center of the calculated ROI, where is possible to observe that the most of the images obtained similar average

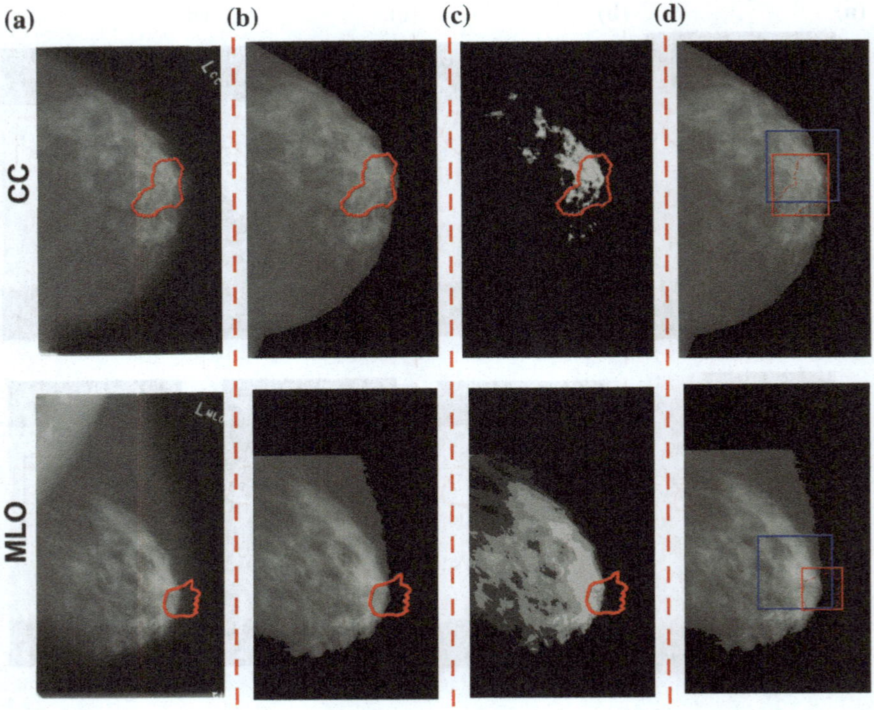

Fig. 7 CC and MLO views of mammography 2, **a** original images with the ROIs containing a malignant tumor, **b** preprocessed images, **c** segmented images, **d** contrast between the selected ROI (blue) and the original ROI (red)

and SD values between each other, which means that the centers of both regions of interest are close, so the identification of the location of the ROI through the algorithm obtains a result similar to the original. In addition, the SDs have low values, allowing to know that the behavior of the algorithms has robustness when identifying the center of the ROIs. In the image 2 are present the highest values of d, especially in the MLO view. According to the Figure, this may be due to the fact that tissue that has similar gray levels is found around the tumor, making it difficult for the algorithm to locate the center of the region that contains the tumor. In the MLO view this becomes more evident because the ROI is decreased considerably due to its visibility from this perspective.

On the other hand, in the image 4 it is observed a significantly higher value in the CC view than in the MLO view. It is worth highlighting the results of this image, which presents the particularity that on its contour, in the upper quadrant, a small white mass can be observed only in the CC view. Due to the high levels of gray that the pixels present within the mass, it is caused that in some iterations the algorithm presents confusion and locates the section where the mass is located as the ROI, while in other iterations the metaheuristic behavior of the algorithm allows correctly locate the ROI, thus significantly affecting the values of the metrics on this image.

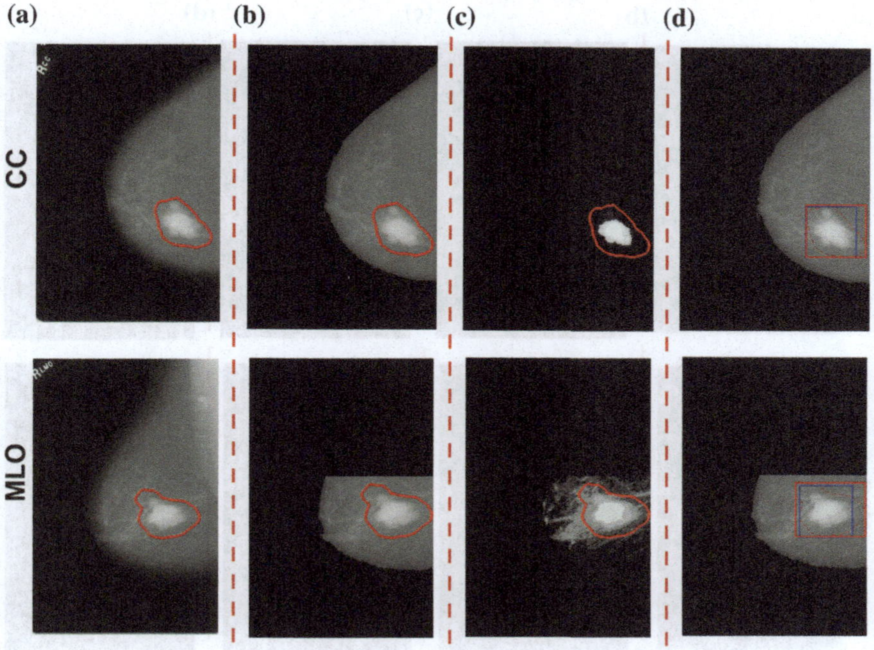

Fig. 8 CC and MLO views of mammography 3, **a** original images with the ROIs containing a malignant tumor, **b** preprocessed images, **c** segmented images, **d** contrast between the selected ROI (blue) and the original ROI (red)

Therefore, this view of the image presents this high d value because in each iteration the center of the ROI can be presented in any of the two locations.

The ASD values obtained allow to know that for every image, the distance between the surface of the original ROI and the surface of the ROI calculated is not significant, since 0 indicates a perfect match. It should be noted that for the approach of this work, a ROI of static size is used (based on the image presented by the tumor with the greatest surface area), unlike the ROI with dynamic behavior of the database, which can cause the difference in surfaces to be greater if the ROI should cover a small tumor.

However, most of the images present lower values in the CC view than in the MLO, except for the image 4, where the MLO view present a significant improvement. This can happen due to the mixing of gray levels that occurs when observing the breast from another position. When acquiring an image with MLO view, the splicing of the tissues can make its interpretation complex, even for specialists, since the tumor can take gray values similar to those of the tissue, as shown in Fig. 7c, where only two gray levels are presented for the CC view, while four are shown in the MLO view. Because of this, it becomes confusing for the algorithm to know which is the surface that contains the tumor shape, being possible to include healthy tissue inside it. This at the same time causes the calculated ROI to cover an area larger than what it should really identify, increasing the distance between it and the original ROI.

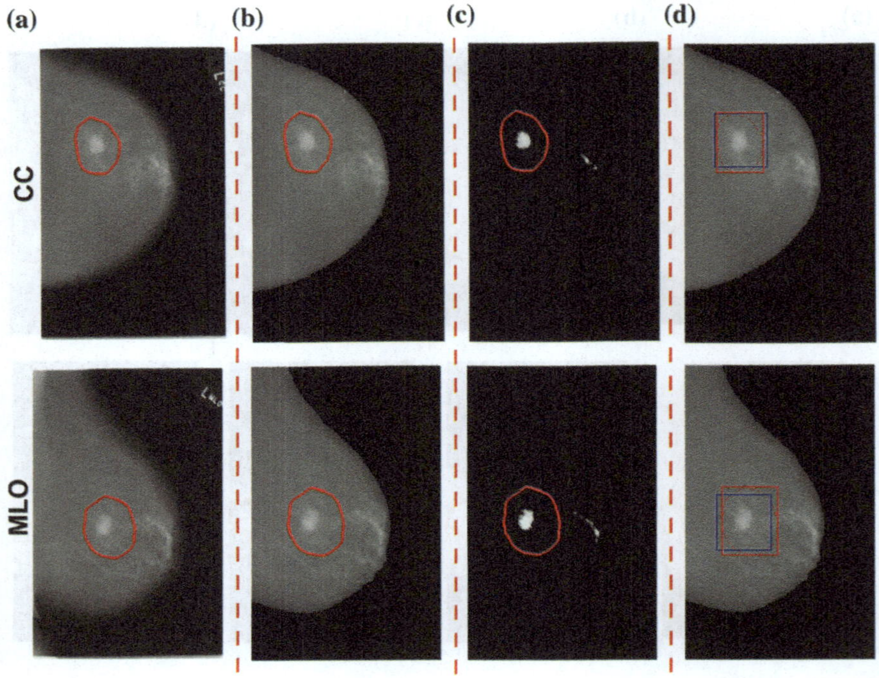

Fig. 9 CC and MLO views of mammography 4, **a** original images with the ROIs containing a malignant tumor, **b** preprocessed images, **c** segmented images, **d** contrast between the selected ROI (blue) and the original ROI (red)

Table 1 Average values of the metrics used for the validation

	View	*d* (mm)	ASD (mm)	MSD (mm)	DSC	Time (s)
1	CC	1.828 (3.612)	2.002 (1.224)	4.438 (3.966)	0.707 (0.008)	53.831 (9.660)
	MLO	2.490 (3.223)	2.284 (1.184)	3.807 (0.758)	0.672 (0.008)	73.871 (3.287)
2	CC	4.012 (3.956)	2.606 (1.180)	5.529 (3.277)	0.629 (0.008)	69.949 (2.660)
	MLO	6.219 (2.365)	4.231 (1.475)	10.925 (2.999)	0.361 (0.011)	34.428 (0.632)
3	CC	1.431 (1.989)	0.792 (0.446)	2.812 (1.696)	0.900 (0.002)	138.123 (3.366)
	MLO	1.298 (1.027)	1.369 (0.444)	3.703 (1.168)	0.834 (0.003)	144.780 (3.882)
4	CC	6.034 (159.898)	3.609 (83.496)	6.712 (153.737)	0.620 (0.405)	92.998 (23.583)
	MLO	1.903 (1.626)	1.694 (0.517)	2.984 (3.041)	0.790 (0.003)	128.395 (3.729)

In the image 4, the results of the CC view are again affected, where an important change is presented in the ASD in comparison with the value obtained in its MLO view and also, a significantly high SD.

The next metric taken into account is the MSD, which allows to indicate the maximum distance between two contours. As in the case of the ASD, all the images obtained low values, indicating a short distance between the contours of both ROIs. The highest MSD values are presented by image 2, which is evident if seen in Fig. 7, where the contour of the calculated ROI has a significant distance from the contour of the original ROI due to its static size, since the size of the tumor is significantly smaller than the fixed ROI size. The obtained SDs present small values in their majority, except again for the case of the image 4, showing the little variation that there is in each iteration for the calculation of the ROI and verifying the robustness of the behavior of the algorithm.

The values of DSC, which allow to measure the overlapping in a range from 0 to 1, show significant values higher than 0.6 in most of the images, by exception of the image 2, which, as mentioned above, is affected by the limitation related to the dynamism of the calculated ROI, presenting a DSC of 0.361.

The highest DSC values are obtained by image 3 for both views, getting an almost complete overlap, as can be observed in Fig. 8, where ROIs overlap in most of the contour. An important point that allows the adequate behavior of the algorithm on this image is that the region where the tumor is presents gray values that are clearly distinguished from the rest of the breast, facilitating the calculation of the ROI and allowing its high concordance with the original ROI.

Finally, the time it takes the algorithm to develop the calculation of the ROI is presented to allow comparison in the variations it may have depending on the information contained in the image. As it is possible to observe, the time required to calculate the ROI presents considerable changes between each image, however, despite the fact that the images present complex information, due to the heterogeneity that can occur in the levels of gray when superimposing the tissue of the breast and the tumor, the average processing time does not exceed 145 s. So it is shown that the computational cost that the algorithm takes to carry out its procedure, based on the processing time, varies depending on the content of the image without affecting the results linearly, since the times obtained can not be relate directly to the values of the calculated metrics.

6 Conclusions

According to the results obtained, it is shown that the automatic ROI has a significantly high overlap with the ROI of the database, the euclidean distance between the coordinates of the center of both ROIs is small, in addition to the average symmetric distance and the maximum symmetrical distance presents significantly small values. Therefore, a tool capable of identifying possible ROIs where an abnormal or malignant mass is found in the images is obtained in this work, with the aim of the early detection of breast cancer.

Although this chapter does not intend to provide a diagnostic method for breast cancer by itself, the automatic detection of ROI using the MCET-HHO method contributes to the improvement of digital mammography to facilitate the work of health professionals in the diagnosis and breast cancer monitoring as a CAD tool.

7 Future Work

For the future development of the submitted proposal, the combination of the contribution if this work with a superior system, such as artificial neural networks (ANN), is proposed, in order to support the cancer diagnosis procedure. To achieve this, a large set of digital mammograms will be included for the development of an ANN architecture, obtained from the DDSM database, providing an automatic tool to make a supportive diagnosis and thus improve the early diagnosis of breast cancer that occurs due to a bad interpretation of the images.

References

1. R. Agarwal, O. Diaz, X. Lladó, M.H. Yap, R. Martí, Automatic mass detection in mammograms using deep convolutional neural networks. J. Med. Imaging **6**(3), 031,409 (2019)
2. M.A. Al-antari, M.A. Al-masni, M.T. Choi, S.M. Han, T.S. Kim, A fully integrated computer-aided diagnosis system for digital x-ray mammograms via deep learning detection, segmentation, and classification. Int. J. Med. Inform. **117**, 44–54 (2018)
3. E.R. Arce-Santana, A.R. Mejia-Rodriguez, E. Martinez-Peña, A. Alba, M. Mendez, E. Scalco, A. Mastropietro, G. Rizzo, A new probabilistic active contour region-based method for multi-class medical image segmentation. Med. Biol. Eng. Comput. **57**(3), 565–576 (2019)
4. I. Bankman, *Handbook of Medical Image Processing and Analysis* (Elsevier, 2008)
5. M. Bari, A. Ahmed, S. Naveed et al., Lungs cancer detection using digital image processing techniques: a review. Mehran Univ. Res. J. Eng. Technol. **38**(2), 351–360 (2019)
6. R. Blanks, R. Given-Wilson, R. Alison, J. Jenkins, M. Wallis, An analysis of 11.3 million screening tests examining the association between needle biopsy rates and cancer detection rates in the English nhs breast cancer screening programme. Clin. Radiol. (2019)
7. S. Broggi, E. Scalco, M.L. Belli, G. Logghe, D. Verellen, S. Moriconi, A. Chiara, A. Palmisano, R. Mellone, C. Fiorino et al., A comparative evaluation of 3 different free-form deformable image registration and contour propagation methods for head and neck MRI: the case of parotid changes during radiotherapy. Technol. Cancer Res. Treat. **16**(3), 373–381 (2017)
8. K.H. Cha, N. Petrick, A. Pezeshk, C.G. Graff, D. Sharma, A. Badal, A. Badano, B. Sahiner, Reducing overfitting of a deep learning breast mass detection algorithm in mammography using synthetic images, in *Medical Imaging 2019: Computer-Aided Diagnosis*, vol. 10950. (International Society for Optics and Photonics, 2019), p. 1095004
9. E. Cuevas, M. Cienfuegos, D. ZaldíVar, M. Pérez-Cisneros, A swarm optimization algorithm inspired in the behavior of the social-spider. Exp. Syst. Appl. **40**(16), 6374–6384 (2013)
10. E. Cuevas, A. González, D. Zaldívar, M. Pérez-Cisneros, An optimisation algorithm based on the behaviour of locust swarms. Int. J. Bio-Inspired Comput. **7**(6), 402–407 (2015)
11. E. Cuevas, V. Osuna, D. Oliva et al., *Evolutionary Computation Techniques: A Comparative Perspective*, vol. 686 (Springer, 2017)
12. M.P. De Albuquerque, I.A. Esquef, A.G. Mello, Image thresholding using Tsallis entropy. Pattern Recogn. Lett. **25**(9), 1059–1065 (2004)
13. M.A. Díaz-Cortés, N. Ortega-Sánchez, S. Hinojosa, D. Oliva, E. Cuevas, R. Rojas, A. Demin, A multi-level thresholding method for breast thermograms analysis using dragonfly algorithm. Infrared Phys. Technol. **93**, 346–361 (2018)
14. J.O.B. Diniz, P.H.B. Diniz, T.L.A. Valente, A.C. Silva, A.C. de Paiva, M. Gattass, Detection of mass regions in mammograms by bilateral analysis adapted to breast density using similarity indexes and convolutional neural networks. Comput. Methods Program. Biomed. **156**, 191–207 (2018)

15. M. Dorigo, G. Di Caro, Ant colony optimization: a new meta-heuristic, in *Proceedings of the 1999 Congress on Evolutionary Computation-CEC99 (Cat. No. 99TH8406)*, vol. 2 (IEEE, 1999), pp. 1470–1477
16. M.A. El Aziz, A.A. Ewees, A.E. Hassanien, Whale optimization algorithm and moth-flame optimization for multilevel thresholding image segmentation. Exp. Syst. Appl. **83**, 242–256 (2017)
17. L. He, S. Huang, Modified firefly algorithm based multilevel thresholding for color image segmentation. Neurocomputing **240**, 152–174 (2017)
18. M. Heath, K. Bowyer, D. Kopans, R. Moore, W.P. Kegelmeyer, The digital database for screening mammography, in *Proceedings of the 5th International Workshop on Digital Mammography* (Medical Physics Publishing, 2000), pp. 212–218
19. A.A. Heidari, S. Mirjalili, H. Faris, I. Aljarah, M. Mafarja, H. Chen, Harris Hawks optimization: algorithm and applications. Future Gener. Comput. Syst. **97**, 849–872 (2019)
20. E.L. Henriksen, J.F. Carlsen, I.M. Vejborg, M.B. Nielsen, C.A. Lauridsen, The efficacy of using computer-aided detection (CAD) for detection of breast cancer in mammography screening: a systematic review. Acta Radiol. **60**(1), 13–18 (2019)
21. S. Hinojosa, D. Oliva, E. Cuevas, M. Pérez-Cisneros, G. Pájares, Real-time video thresholding using evolutionary techniques and cross entropy, in *2018 IEEE Conference on Evolving and Adaptive Intelligent Systems (EAIS)* (IEEE, 2018), pp. 1–8
22. M. Hmida, K. Hamrouni, B. Solaiman, S. Boussetta, Mammographic mass segmentation using fuzzy contours. Comput. Methods Program. Biomed. **164**, 131–142 (2018)
23. J.N. Kapur, P.K. Sahoo, A.K. Wong, A new method for gray-level picture thresholding using the entropy of the histogram. Comput. Vision Graph. Image Process. **29**(3), 273–285 (1985)
24. J. Kennedy, Particle swarm optimization, in *Encyclopedia of Machine Learning* (2010), pp. 760–766
25. V. Kovalevsky, Image segmentation and connected components, in *Modern Algorithms for Image Processing* (Springer, 2019), pp. 167–186
26. C.H. Li, C. Lee, Minimum cross entropy thresholding. Pattern Recogn. **26**(4), 617–625 (1993)
27. J. Lian, Z. Yang, W. Sun, Y. Guo, L. Zheng, J. Li, B. Shi, Y. Ma, An image segmentation method of a modified SPCNN based on human visual system in medical images. Neurocomputing **333**, 292–306 (2019)
28. C. Liu, W. Liu, W. Xing, A weighted edge-based level set method based on multi-local statistical information for nocoisy image segmentation. J. Visual Commun. Image Represent. **59**, 89–107 (2019)
29. A.G. Mathioudakis, M. Salakari, L. Pylkkanen, Z. Saz-Parkinson, A. Bramesfeld, S. Deandrea, D. Lerda, L. Neamtiu, H. Pardo-Hernandez, I. Solà et al., Systematic review on women's values and preferences concerning breast cancer screening and diagnostic services. Psycho-Oncology (2019)
30. H. Min, S.S. Chandra, S. Crozier, A.P. Bradley, Multi-scale sifting for mammographic mass detection and segmentation. Biomed. Phys. Eng. Exp. (2019)
31. B. Mughal, N. Muhammad, M. Sharif, Adaptive hysteresis thresholding segmentation technique for localizing the breast masses in the curve stitching domain. Int. J. Med. Inform. **126**, 26–34 (2019)
32. T. Nayak, N. Bhat, V. Bhat, S. Shetty, M. Javed, P. Nagabhushan, Automatic segmentation and breast density estimation for cancer detection using an efficient watershed algorithm, in *Data Analytics and Learning* (Springer, 2019), pp. 347–358
33. O.P.S. Neto, O. Carvalho, W. Sampaio, A. Corrêa, A. Paiva, Automatic segmentation of masses in digital mammograms using particle swarm optimization and graph clustering, in *2015 International Conference on Systems, Signals and Image Processing (IWSSIP)* (IEEE, 2015), pp. 109–112
34. D. Oliva, M.A. Elaziz, S. Hinojosa, Image segmentation using Kapur's entropy and a hybrid optimization algorithm, in *Metaheuristic Algorithms for Image Segmentation: Theory and Applications* (Springer, 2019), pp. 85–99

35. D. Oliva, M.A. Elaziz, S. Hinojosa, Multilevel thresholding for image segmentation based on metaheuristic algorithms, in *Metaheuristic Algorithms for Image Segmentation: Theory and Applications* (Springer, 2019), pp. 59–69
36. D. Oliva, S. Hinojosa, V. Osuna-Enciso, E. Cuevas, M. Pérez-Cisneros, G. Sanchez-Ante, Image segmentation by minimum cross entropy using evolutionary methods. Soft Comput. 1–20 (2017)
37. G. Pei, Y. Zhang, *Digital Orthopedics* (Springer, 2019)
38. M. Posso, J. Louro, M. Sánchez, M. Román, C. Vidal, M. Sala, M. Baré, X. Castells, Study group B, et al., Mammographic breast density: how it affects performance indicators in screening programmes? Eur. J. Radiol. **110**, 81–87 (2019)
39. T. Sadad, A. Munir, T. Saba, A. Hussain, Fuzzy c-means and region growing based classification of tumor from mammograms using hybrid texture feature. J. Comput. Sci. **29**, 34–45 (2018)
40. W.B. de Sampaio, A.C. Silva, A.C. de Paiva, M. Gattass, Detection of masses in mammograms with adaption to breast density using genetic algorithm, phylogenetic trees, LBP and SVM. Exp. Syst. Appl. **42**(22), 8911–8928 (2015)
41. E. Seeram, Digital image processing concepts, in *Digital Radiography* (Springer, 2019), pp. 21–39
42. P. Shi, J. Zhong, A. Rampun, H. Wang, A hierarchical pipeline for breast boundary segmentation and calcification detection in mammograms. Comput. Biol. Med. **96**, 178–188 (2018)
43. R. Storn, K. Price, Differential evolution-a simple and efficient heuristic for global optimization over continuous spaces. J. Glob. Opt. **11**(4), 341–359 (1997)
44. N. Thome, S. Bernard, V. Bismuth, F. Patoureaux, et al., Multitask classification and segmentation for cancer diagnosis in mammography (2019)
45. G. Valvano, G. Santini, N. Martini, A. Ripoli, C. Iacconi, D. Chiappino, D. Della Latta, Convolutional neural networks for the segmentation of microcalcification in mammography imaging. J. Healthc. Eng. (2019)
46. X.S. Yang, Firefly algorithms for multimodal optimization, in *International Symposium on Stochastic Algorithms* (Springer, 2009), pp. 169–178
47. P.Y. Yin, Multilevel minimum cross entropy threshold selection based on particle swarm optimization. Appl. Math. Comput. **184**(2), 503–513 (2007)

Cancer Cell Prediction Using Machine Learning and Evolutionary Algorithms

Karla Avila-Cardenas and Marco Pérez-Cisneros

Abstract Cancer is a disease that affects the global population indistinctly. It is considered the second cause of death in the world. Early detection can reduce cancer mortality. However, instruments and equipment for diagnostic are often expensive and insufficient. This makes doctor's work becomes complex and often, cancer patients do not receive a diagnosis until the disease is advanced. Machine Learning (ML) has shown to be useful for classification and prediction problems but it faces some limitations. Mainly, because it depends on the quality of the information. Over the years, different mechanisms have been developed to solve them, but there is not any mechanism that eliminates all ML difficulties. Because of that, this area remains open to new promising discoveries and ideas. Evolutionary Algorithms (EAs) have demonstrated to be useful for solving optimization problems in a heuristic way. This chapter presents a comparative study related to the prediction of cancer cells based on Machine Learning and Evolutionary Algorithms. As well as, a brief introduction of machine learning and evolutionary technics is presented. Also, the procedures' implementation and performance are described. The results obtained show that the AEs can support a ML method, guiding the learning process.

Keywords Cancer prediction · Artificial neuronal net · Multiple linear regression · Evolutionary technics · Predictive analytics · Optimization

1 Introduction

Cancer is a disease that affects the global population indistinctly. It is considered the second cause of death in the world. In 2018, they were estimated 609,640 deaths from this disease and 1,735,350 new cases. Early detection can reduce cancer mortality because it increases the likelihood of treatments effectiveness [1, 2]. Unfortunately,

K. Avila-Cardenas (✉) · M. Pérez-Cisneros
Departamento de Electrónica, Universidad de Guadalajara, CUCEI, Guadalajara, Jalisco, Mexico
e-mail: karla.acardenas@alumnos.udg.mx

M. Pérez-Cisneros
e-mail: marco.perez@cucei.udg.mx

D. Oliva and S. Hinojosa (eds.), *Applications of Hybrid Metaheuristic Algorithms for Image Processing*, Studies in Computational Intelligence 890,
https://doi.org/10.1007/978-3-030-40977-7_16

instruments and equipment for diagnostic are often expensive and, therefore, sometimes they are not available. This makes doctor's work become complex and often, cancer patients do not receive a diagnosis until the disease is advanced.

However, with the advances in network infrastructure, data storage, and processing capacity, different kind of devices have emerged and with them, new areas of knowledge have surged. One of these areas is The Artificial Intelligent (AI) [3]. There are sundry AI technics, but machine learning has become one of the most accepted areas because it is used in popular software for marketing, entertainment, and engineering [4].

In fact, some scientific disciplines, such as bioinformatics [5, 6], medical research [7, 8] and image processing [9, 10] have used ML method for cancer classification and the results obtained from this kind of research works have shown that machine learning can help in the medical diagnosis. Of course, the information obtained is not the final diagnosis. It is very important to clarify that, because even though usually the results support the expert opinion could be possible that this affirmation was not true due ML has some limitations, like: overfitting, the available computational resources and the quality of the data used through the training process.

The Artificial Neuronal Networks (ANN) are one of the most common ML technique using nowadays. Furthermore, ANN have shown to be useful for classification and prediction problems but, as can be deduced, the ANN faces the same limitations as machine learning.

Over the years, different mechanisms have been developed to solve the previously mentioned problems but usually, when a difficulty is eliminated, a new one arrives. A mechanism that eliminates all these difficulties has not been found yet. Therefore, this area remains open to new promising discoveries and ideas.

It should be mentioned that one of the indispensable stages of neuronal networks is to assign the appropriate values to the weight matrix that will be modified throughout the training process to generate a learning model. Unfortunately, this step of the process has been diminished because in some cases it is difficult to define manually a "good" set of weight values.

On the other hand, Evolutionary Algorithms (EA) are a heuristic search processes use for optimization (minimization or maximization of an objective function). EAs have demonstrated to be useful for solving engineering problems [11, 12], image processing [13, 14], control problems [15, 16], and others that depend on the identification of "good" values, known as the optimum value. Because of that, this work is presented a comparative study related to the prediction of cancer cells based on machine learning and evolutionary algorithms.

Also, a brief introduction of machine learning and evolutionary technics is mentioned in Sect. 2. The procedures' implementation and performance are presented in part 3. The results obtained are deeply discussed in Sect. 4. And in the final section of this character, some important conclusions are presented.

2 Preliminary Concepts

In this section, some concepts used throughout this work are described. Such as machine learning, types of learning, artificial neural nets, evolutionary techniques, and so others.

2.1 *Machine Learning*

As mentioned earlier, machine learning is a kind of AI [17]. It gets a model based on some input data. As shown in (Fig. 1), ML can be treated as a black box that returns a model created by the generalization of the input data. This generalization is made base on the type of learning that it is used. It is important to mention that there are different forms in which ML can be classified. For example, Nilsson [18] catalog the ML techniques depending on the hypothesis in which they are based: Statistics and Brain Models. However, the common classification for ML methods is based on three types of learning: Supervised, Unsupervised and Reinforcement [19].

- Supervised learning (SL)
  In the supervised learning, the new knowledge is obtained by a similar process as human learn things: solving exercise problems. Ones the exercise is solved, the result is compared with the correct answer. If the result is wrong, the knowledge generated is modified. It should be said that this is an iterative process which ends by a stop standard, usually, when the result has a minimal difference with the correct answer. This kind of learning can be approached to non-parametric methods, discriminant functions, and model-based approaches. Normally, this kind of algorithms used input data format as follow: $[Input, CorrectOuput]$.
- Unsupervised learning (UL)
  In contrast, unsupervised learning treats new learning analytically, i.e. it is used to figure out the data characteristics. Normally, this kind of algorithms used input data format as $[Input]$.

Fig. 1 Machine learning training process

- Reinforcement learning (RL)
  This kind of learning is based on. It uses a set of inputs, some outputs and a grade for that outputs as training data. Commonly, it is employed when a live interaction is required, such as control or game plays [17]. The RL input format: $[Input, SomeOuput, Grade]$.

Each of these approaches has a set of characteristics that make them suited to different types of applications. For example, unsupervised learning can be used in clustering or reduction of dimensionality. On the other hand, supervised learning can be distinguished, mainly, two types of techniques: classification and regression.

2.1.1 Classification

Classification is a way for discriminating between a finite number of classes. Usually, most of the ML methods in supervised learning focus on classification. One of the principal methods for this is the Artificial Neural Networks (ANNs).

Artificial Neural Networks (ANNs)

The ANNs are based on how the brain works. Essentially, the ANN develops associations between a set of nodes (neurons) to generate internal learning rules that guide a learning process [20]. The main structure in this architecture is de perceptron, the simplest model for a neuron that can be observed in Fig. 2. This ML technique was introduced in the middle 40s and it has had different periods of reborn and death, but actually is one of the principal ML methods because of its easy learning (by heuristic learning), easy data and knowledge representation, parallelism, error tolerance, and nodes connection.

Where x_i is the input data, w_i are the weights and b is the bias. Also, it is important to mention that the ANN process is divided into two phases: training and testing. The training part is the most important step in the ANN. It consists of a set x_i where its format is equal to the mentioned in supervised learning section. The training data is introduced to obtain the output of the net. Then the error is calculated to adjust a set of weights w_n. Usually, this procedure is repeated until a number of iterations, called epoch, is reached. The second part consists on a set with the same format mentioned

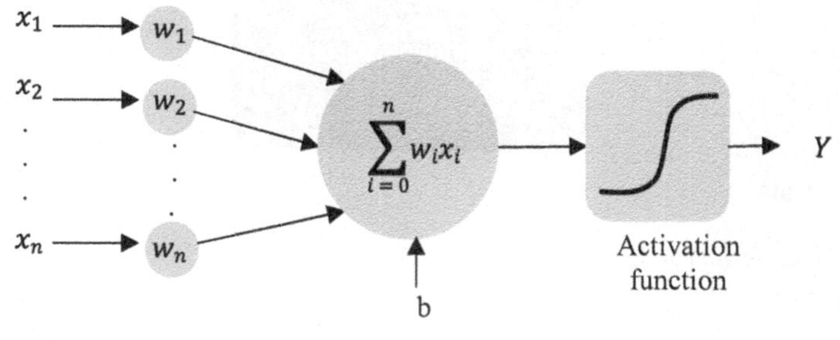

Fig. 2 Simple perceptron

previously, but with different data than the previously used in the training part to evaluate the network performance and if it is necessary re-train the net.

The procedure for modifying the weights based on the given data is named: learning rule. The Delta Rule is the representative learning rule for a single-layer neuronal net and can be described by (1).

$$w_{ij} = w_{ij} + \alpha e_i x_j \tag{1}$$

where x_j is the output from the node j, the error of the output node i is defined as e_i, w_{ij} is the weight between the output and input nodes, and α is the learning rate between $0 \leq \alpha \leq 1$.

Based on this information, it is possible to define the following steps for the training process using the delta rule.

1. Initialize the weight with adequate values. Commonly, a random initialization is used.
2. Enter the training data to the neural network and obtain the output.

$$y_i = \varphi(wx + b) \tag{2}$$

   where $\varphi(v)$ is the activation function.
3. Calculate the error of the output y_i using the correct output d_i

$$e_i = d_i - y_i \tag{3}$$

4. Calculate the weight updates, using the delta rule.

$$\Delta w_{ij} = \alpha e_i x_j \tag{4}$$

   where α is the learning rate, x_j is the output from the node j, e_i is the error of the output node i.
5. Adjust the weights.

$$w_{ij} = w_{ij} + \Delta w_{ij} \tag{5}$$

6. Perform steps 2–5 for all training data.
7. Repeat steps 2–6 until the stop criteria is reached.

2.1.2 Regression

Regression is a way for fitting data to a model through adjusting a curve to the data and it is, commonly, used to predict the behavior of systems or analyze experimental information. There are two main approaches for the regression: Simple Linear Regression (SLR) and Multiple Linear Regression (MLR).

The first one focuses on the relation between a dependent variable y (also named as a response) and an independent variable x (called the predictor). It is based on Eq. (6) but, it is needed to say that there are different ways to make linear regression [21]. Also, it should be mentioned that α is the ordinate at the origin, namely, the y value obtained when $x = 0$; β is the slope of the straight, and ε is the error [22].

$$y = \alpha + \beta x + \varepsilon \tag{6}$$

On the other hand, the multiple linear regression is the process done for multiple dimensions and produces a model that can be used for prediction [4]. In the next section, this topic will be explained in detail.

Multiple Linear Regression (MLR)

Multiple linear regression is a regression model based on more than one regression variable. This model can be represented by (7).

$$y = \beta_0 + \beta_1 x_1 + \beta_2 x_2 + \cdots + \beta_k x_k + \varepsilon \tag{7}$$

where $x_i | i = 1, 2, \ldots, k$ are the regression variables, $\beta_i | i = 1, 2, \ldots, k$ are the regression coefficients which represent the expected change in the response y when the other regression variables stay constant and ε is the expected error [23].

Usually, for most of the real problems, the regression coefficients are unknown so, they should be estimated with sample data. To do this, it is possible to use the Least Squares Method (LSM), which can be defined by (8) and (9). Where y_i are the answers, β_j are the regression coefficients and x_{ij} are the regression variables.

$$S(\beta_0, \beta_1, \ldots, \beta_k) = \sum_{i=1}^{n} \varepsilon_i^2 \tag{8}$$

$$\sum_{i=1}^{n} \varepsilon_i^2 = \sum_{i=1}^{n} \left(y_i - \beta_0 - \sum_{j=1}^{k} \beta_j x_{ij} \right)^2 \tag{9}$$

2.2 Optimization and Evolutionary Techniques

Optimization is the methodology used to find the best among available alternatives (feasible solutions) that yields the best quality (minimum or maximum) measured by an objective function. Generally, an optimization problem can be defined as (10).

$$\max_{x \in X} f(x) \quad \text{or} \quad \min_{x \in X} f(x) \tag{10}$$

where x is the decision variable, X is the feasible region and f is the objective function.

Optimization has become popular as it allows solving a wide variety of problems in different fields of science. The classic optimization methods do not provide a satisfactory result for problems which are not differentiable or unimodal. For this type of to, Evolutionary Algorithms (EAs) are the answer. Because they perform a search process in a heuristic way. Often, evolutionary methods are inspired by phenomena and processes present in nature. It should be mentioned that due to the above, a large number of EAs have been developed since the begging of this area. Through this section, some EAs are discussed to introduce as previous concepts for this work and to help the reader become familiar with this area.

2.2.1 Differential Evolution (DE)

This algorithm was proposed by Storn and Price [24]. It was introduced as a novel direct search method that covered the main limitations of the AEs in those moments. Such as the ability to handle non-differentiable, nonlinear and multimodal cost functions, parallelizability, ease of use; i.e. few control variables to steer the minimization, and good convergence properties.

Basically, the DE has three principal processes: Mutation, Crossover, and Selection. These processes' names come from the Genetic Algorithm (GA) concepts because GA is considered as father of the EAs. Next, the DE steps are described in detail.

- Initialization.
  The population vector is defined as:

$$x_{i,G}\big|i = 1, 2, \ldots, NP \tag{11}$$

  where G is the index for each generation and NP is the population size. A uniform probability distribution is used to set initial values $x_{i,G}$.
- Mutation.
  On this step, a new vector $v_{i,G}$ is generated adding the weighted difference between two population vectors to a third vector. This is represented by the next Eq. (12).

$$v_{i,G} = r_0 + F \cdot (r_1 - r_2) \tag{12}$$

  where r_0, r_1 and r_2 are randomly selected vectors from the population and $F \in (0, 1)$ is the scaling factor [25]. Figure 3 allows to understand this idea more clearly.
- Crossover.
  The vector generated in the previous step is mixed with the original population vector to produce a trial vector (Fig. 4). *CR* is the crossover probability.

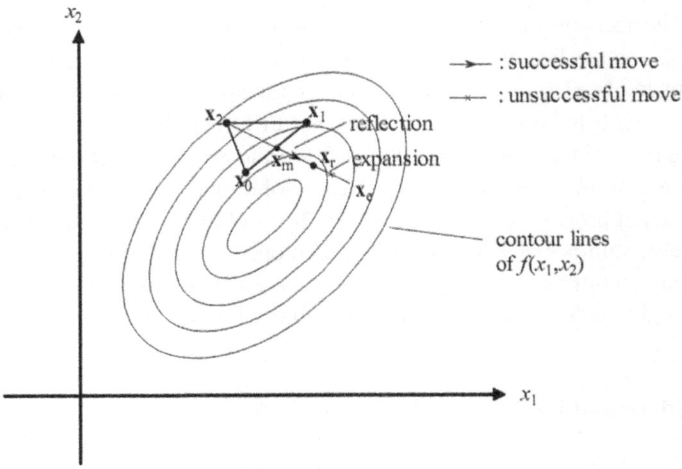

Fig. 3 Geometric mutation representation

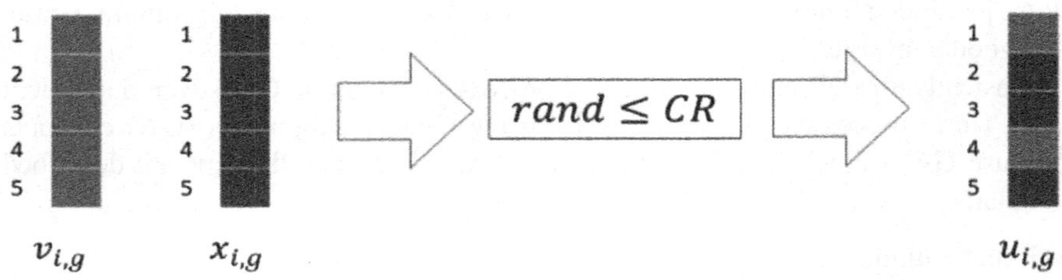

Fig. 4 Crossover process

- Selection.
  If the trial vector yields a lower cost function value than the target vector, the trial vector replaces the target vector in the following generation.
  It should be mention, that over the years, this algorithm has experienced several promising modifications [26–28], but it has maintained as the main steps the concepts mentioned above.

2.2.2 Particle Swarm Optimization (PSO)

PSO is one of the most popular EAs since it was proposed by Kennedy and Eberhart [29]. The impact produced by PSO in the scientific research was cased because PSO is a population-based optimization technique inspired by social behavior of bird flocking/roosting or fish schooling. It is considered the father of a new branch in the EAs, called the swarm intelligence. This algorithm introduced the notion of a social network in which each member of the swarm has an active communication with the other members of the group, they share discoveries and experiences [30].

The process of this algorithm is shown in Fig. 5. The steps 1–4 are made for all particles in all dimensions. As it is visible, the initialization step is necessary for all EAs and usually, a uniform distribution is used.

Firstly, a new speed is generated using (13). This speed determines how the particles are displaced. It should be said that v_i is the previous speed. rand1 and rand2 are random numbers with a uniform distribution. P of the best local positions for each particle. G is the best global particle and X_i are the particles' position.

Fig. 5 PSO algorithm

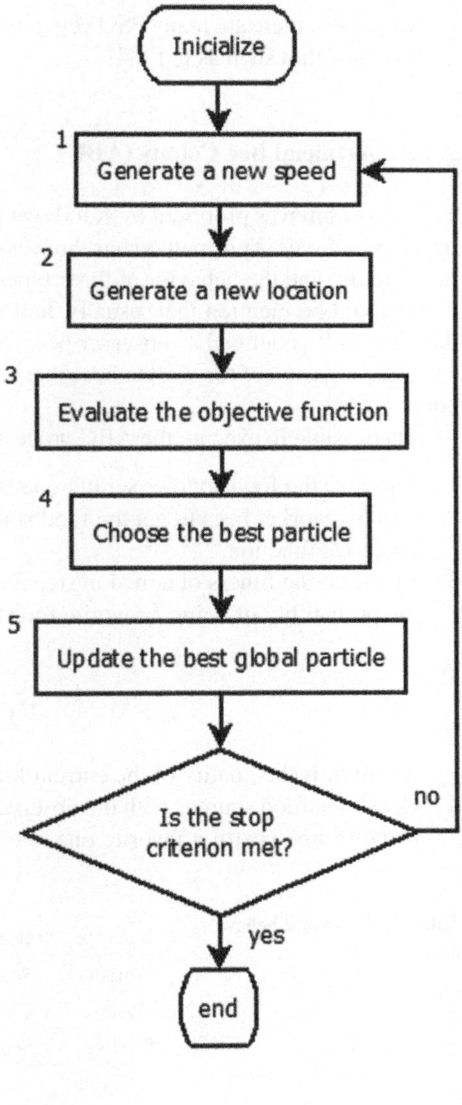

$$v_{i+1} = v_i + rand1 \cdot (P - X_i) + rand2(G - X_i) \tag{13}$$

Then, a new position is created based on the actual particles' position and the speed calculated previously. As shown in (14).

$$x_{i+1} = x_i + v_{i+1} \tag{14}$$

At once, the new position was created, it must be evaluated by the objective function. Next step is to update the best local positions for each particle P and the best global particle G. The PSO procedure is iterative and ends when a stop criterion is reached.

Nowadays, there are many PSO versions in the literature that had shown the power of this algorithm such as [31–34].

2.2.3 Artificial Bee Colony (ABC)

This algorithm was proposed by Karaboga [35]. It is based on how a bee colony can find food sources. This method has three essential components: the food sources, the food quality and the behavior of three types of bee, shown in Table 1.

It should be mention that, usually, half of the hive is defined as worker bee and the other half is defined as observer bee. On the other hand, the explorer bees are a selected subgroup of the worker bees that have left their food sources to explore new zones [25].

It is possible to execute the ABC process by applying the following instructions:

1. Initialize the food sources setting one explorer bee in each of them.
2. Use the worker bees to get the food sources quality by evaluating them into the objective function.
3. Based on the fitness obtained in step 2, get the favorite food sources calculating the probability of being a favorite food source for each one.

$$p_i = \frac{fit}{\sum_{i=1}^{N} fit_i} \tag{15}$$

   where fit is the quality of the current food source.
4. Check the food sources with the observer bees. Based on a finite number of tries; if the source is still a favorite one, the worker bees will continue exploiting it.

Table 1 Types and behavior of bee in ABC

Bee type	Behavior
Worker	Searches food surrounding a known food source
Observer	Choices the best food sources
Explorer	Searches new food sources

Otherwise, the explorer bees will find new food sources using the (16).

$$v_{i,j} = x_{i,j} + \phi_{i,j}\left(x_{i,j} - x_{k,j}\right) \tag{16}$$

where $i, k \in \{1, 2, \ldots, N\}|i \neq k$, $j \in \{1, 2, \ldots, d\}$ and $\phi_{i,j} \in [-1, 1]$.

2.2.4 Flower Pollination Algorithm (FPA)

The Flower Pollination Algorithm is inspired by the flowers' proliferation role in plants. It was proposed by Yang in 2012 [36]. This metaheuristic method has shown to be useful for different kind of problems, such as [37–40].

Pollination is the process in which the pollen is transfer to the female gamete of the plant. There are two principal classifications for it: the first one is based on pollination types and the second one is based on the pollinators. On Table 2 the main characteristics of each one are described.

The FPA takes the following principles for ensuring the quality of the search creating an adequate mix between exploitation and exploration [41].

- Biotic, cross-pollination acting as global search via Lévy flight.
- The local search is performed by abiotic and self-pollination process.
- Flower constancy can be involved due to the similarity of two flowers.
- The switching between local and global search is made by a random probability.

The FPA process can be defined by the following steps [42]:

i. Initialize parameter with switching probability
ii. Generate an initial population of flowers randomly
iii. Evaluate the population and find the best solution
iv. For each flower, based on a probabilistic acceptation, do global (17) or local (18) pollination.

$$x_i^{t+1} = x_i^t + \gamma L(x_i^t - gbest) \tag{17}$$

Table 2 Types of pollination

Pollination types	
Cross-pollination	The pollen is transfer from one flower to another in a different plant
Self-pollination	The pollen is transfer from one flower to another in the same plant
Pollinators	
Biotic-pollination	The pollen is transfer by animal or insect that visit the flower for sipping nectar
Abiotic-pollination	The pollen is transfer by the wind, diffusion in water or gravity

where x_i^t is the solution i at time t, the current best solution is represented by $gbest$, γ is a scaling factor, and L is the Lévy flight step.

$$x_i^{t+1} = x_i^t + \varepsilon\left(x_j^t - x_k^t\right) \tag{18}$$

where x_i^t is the solution i at time t, x_j^t and x_k^t are randomly selected solutions, and ε is a random number between [0, 1].

v. Evaluate new solutions.
vi. Update the solutions and the current best.

2.2.5 Gravitational Search Algorithm (GSA)

It was proposed by Rashedi et al. [43]. Unlike the previous algorithms that are bio-inspired, the GSA is inspired in a physical phenomenon: the Newtonian gravity and the laws of motion.

Newton's gravitational force focus on how gravity acts between separated particles without any intermediary and without any delay so that, each particle attracts every other particle.

In physics, there are three types of masses, such as active gravitational mass (M_a), passive gravitational mass (M_p) and inertial mass (M_i). The first one is a measure of the strength of the gravitational field due to a particular object. The Next one, is related to the strength of an object's interaction with the gravitational field. And the last one is a measure of an object resistance to changing its state of motion when a force is applied.

The GSA is based on the previous data such that it can be defined as an isolated system of masses and obey the gravity and motion laws. The algorithm can be described in the next steps.

1. Define the initial $G(t)$ value and position of the masses, randomly.
2. Calculate $best(t)$, $worst(t)$ and $M_i(t)$ using the following equations.

$$best(t) = \min_{j \in \{1,\ldots,N\}} fit_j(t) \tag{19}$$

$$worst(t) = \max_{j \in \{1,\ldots,N\}} fit_j(t) \tag{20}$$

$$m_i(t) = \frac{fit_i(t) - worst(t)}{best(t) - worst(t)}$$
$$M_i(t) = \frac{m_i(t)}{\sum_{j=1}^{N} m_j(t)} \tag{21}$$

It should be mentioned that $\mathrm{fit}_j(t)$ represents the fitness value and initially, the gravitational masses are equal to inertia mass.

3. Calculate the force acting between masses by (22).

$$F_{ij}^{d}(t) = G(t)\frac{M_{pi}(t) \times M_{aj}(t)}{R_{ij}(t) + \varepsilon}\left(x_{j}^{d}(t) - x_{i}^{d}(t)\right) \tag{22}$$

where M_{aj} is the active gravitational mass related to the agent j, M_{pi} is the passive gravitational mass related to the agent i, $G(t)$ is gravitational constant at time t that will be reduced with time, ε is a small constant, and $R_{ij}(t)$ is the Euclidian distance between the agents i and j.

4. Calculate the velocity (23) and acceleration (24).

$$v_{i}^{d}(t+1) = rand_{i} \times \left(v_{i}^{d}(t) + a_{i}^{d}(t)\right) \tag{23}$$

$$a_{i}^{d}(t) = \frac{F_{i}^{d}(t)}{M_{ii}(t)} \tag{24}$$

where $v_{i}^{d}(t)$ is the velocity in the previous time, M_{ii} is the inertia mass of the particle i, and $F_{i}^{d}(t)$ gives a stochastic characteristic to the GSA and can be defined by the following equation.

$$F_{i}^{d}(t) = \sum_{j=1, j\neq i}^{N} rand_{j} F_{ij}^{d}(t) \tag{25}$$

5. Update agents' positions.

$$x_{i}^{d}(t+1) = x_{i}^{d}(t) + v_{i}^{d}(t+1) \tag{26}$$

2.2.6 Cuckoo Search Algorithm (CS)

This metaheuristic algorithm is based on the cuckoo reproduction strategy and the Lévy flights. Firstly, some species leave their eggs in common nests. If a host bird discovers the eggs are not their own, they will throw the eggs away or abandon the nest. On the other hand, Lévy flight provide a random walk while the random step length is drawn from a Lévy distribution. This permits an efficient exploration of the search space.

The CS can be described by Fig. 6. Where the initial population of n the host nest is made randomly. A cuckoo is selected and evaluated then it is compared with the chosen nest, using their fitness and if it is better, it will replace the chosen nest.

Is should be mentioned that based on a probability p_a, a fraction of worse nests is replaced with new solutions that can be defined using (27).

$$x_{i}^{t+1} = x_{i}^{t} + \alpha \oplus L\acute{e}vy(\lambda) \tag{27}$$

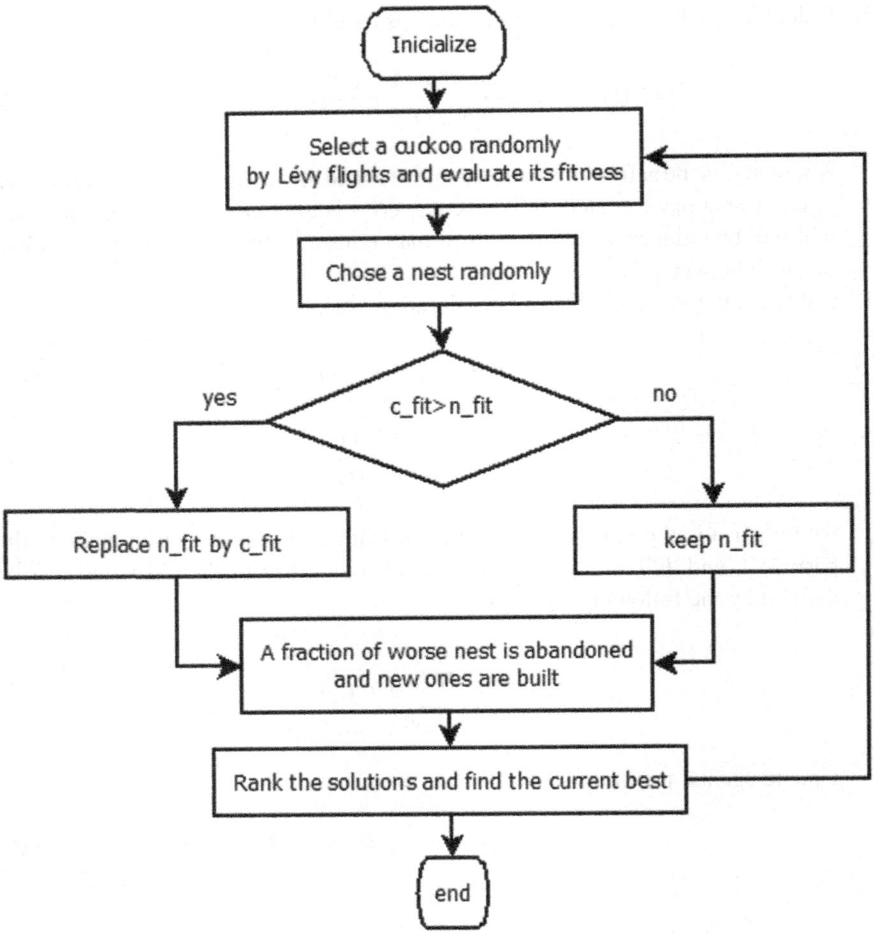

Fig. 6 Cuckoo search algorithm

3 Problem Formulation

Studying Cancer has been an important issue for humanity for a long time. Through the years, new information about cancer has appeared. Especially now, that it is possible to store a huge quantity of information thanks to technological advances. One of the known sets of cancer data is the Breast Cancer Wisconsin Data Set [44]. It was made from 699 clinical reports of tumor cases, in which 458 records are benign tumors and 241 malignant. In Table 3 Cancer Data Set content is described in detail.

Some researchers have used this data set for identifying the benign and malign tumors using machine learning [45]. As mentioned previously, machine learning has become a promising area as it has proven to be useful for classification and prediction problems that have been applied in various areas such as marketing, finance,

Table 3 Cancer data set content

Attribute information	Range
Clump thickness	1–10
Uniformity of cell size	1–10
Uniformity of cell shape	1–10
Marginal adhesion	1–10
Single epithelial cell size	1–10
Bare nuclei	1–10
Bland chromatin	1–10
Normal nucleoli	1–10
Mitoses	1–10
Class	(2 for benign, 4 for malignant)

engineering, and so others. However, ML faces some limitations. Mainly, because it depends on the quality of the information, the epochs used in the training process, the weights adequate initialization and so others [17].

On the other hand, evolutionary algorithms have shown their effectiveness in several engineering problems [46–48] because they find an adequate solution though a search strategy that combines stochastic and deterministic procedures.

Based on the qualities that characterize each of these two areas, recently, researchers have focused their efforts on studying and making new proposals that combine the advantages of different algorithms in both areas, generating several approaches. One of these approaches uses evolutionary techniques for supporting the learning process in machine learning. In [49] various algorithms that use this approach are presented. One of them shows a neuroevolutionary approach, which can adaptively learn a network structure and size appropriate to the task. Other, performed a search of a deep neural network (DNN) using Genetic Algorithms (GAs). In [50] the ABC algorithm is used for training a Feed-Forward Neuronal Network. These items have in common the objective to strengthen the learning process with an evolutionary method. As it can be seeing, this area remains open to new promising discoveries and ideas.

The present work shows a comparative study related to the prediction of cancer cells based on machine learning helped by evolutionary algorithms through the learning process. The regression coefficients obtained from an evolutionary technique to fit the initial ANN weights to the training data to support the learning process. On this section, all the procedures performed are described in detail.

Generally, the strategy used for this work can be seen in Fig. 7 and consists of three main stages: pre-processing of data, multiple linear regression using evolutionary algorithms and the artificial Neural Network process.

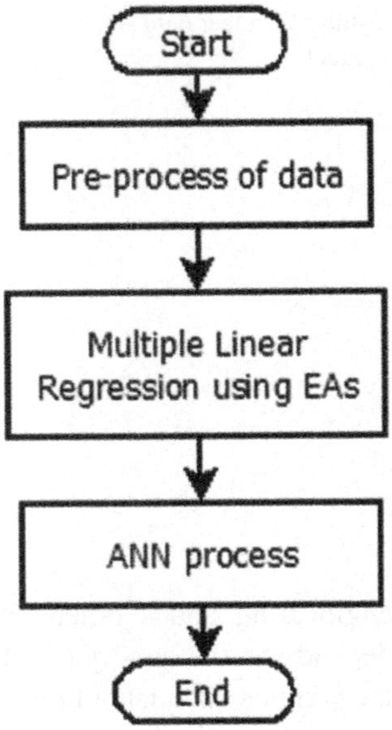

Fig. 7 Strategy stages

3.1 *Pre-process of Data*

One of the most important parts in the data analysis is the pre-process of data because sometimes the data of a set have errors, gaps, etc. The following issues were focused on this step.

- Checking that all records are complete.
  As mentioned in Table 3, each pattern has nine attributes in a range between 1 and 10.
- Assigning new ranges for classes.
  The original range for the classes in the data set is two for benign, four for malignant. To facilitate the handling of data throughout the complete process, the classes range was changed to zero (0) for benign and one (1) for malignant.
- Dividing data for training and testing.
  Usually, when data has a high dimension, the resulting model cannot be seen intuitively. So, a different method, called validation, should be used to determinate whether the model is overfitting or not. To prepare the data for the next steps, the Cancer data set is divided randomly into two parts: training and testing data. In the next table (Table 4), these types of data are described in detail.

Table 4 Data types

Data	Percentage (%)	Number of records
Training data	60	419
Testing data	40	280

3.2 Multiple Linear Regression Using EAs

As mentioned in Sect. 2.1.2, a regression is a way for fitting data to a model by adjusting a curve to the data. For models based on more than one regression variable, the multiple linear regression is used. Based on (7) it is possible to define the following Eq. (28) for this work.

$$y = \beta_0 + \beta_1 x_1 + \beta_2 x_2 + \cdots + \beta_8 x_8 + \beta_9 x_9 + \varepsilon \tag{28}$$

where $\beta_j | j = 0, 1, \ldots, 9$ are the regression coefficients, $x_i | i = 1, 2, \ldots, 9$ are the cancer data set attributes, excepting the class field, and ε is assumed as zero. Then it is feasible to think about using the least-squares method (check Sect. 2.1.2). That, for this work, is defined in (28).

$$S(\beta_0, \beta_1, \ldots, \beta_k) = \sum_{i=1}^{419} \left(y_i - \beta_0 - \sum_{j=1}^{9} \beta_j x_{ij} \right)^2 \tag{29}$$

where, as mentioned previously, β_j are the regression coefficients, x_i are the cancer data set attributes and y_i is the correct answer. It should be mentioned that the first sum is made for the 419 records contained in the training data set. On the other hand, the second sum takes the 9 attributes contained in each record.

Remembering Sect. 2.2, evolutionary algorithms have been used for solving optimization problems because they can find an adequate solution through a heuristic search strategy that combines stochastic and deterministic procedures. An optimization problem can be defined as follows:

$$\min_{-1000 \leq \beta_0, \beta_1, \ldots, \beta_k \leq 1000} S(\beta_0, \beta_1, \ldots, \beta_k) \tag{30}$$

where $S(\beta_0, \beta_1, \ldots, \beta_k)$ was defined in (7) and β_j is subject to the next condition $-1000 \leq \beta_j \leq 1000$. Now, it is viable to transfer this optimization problem through an evolutionary technique to estimate the regression coefficients.

3.3 Artificial Neural Network Process

On the other hand, ANNs are a way for discriminating between a finite number of classes. In the problem described in this section, the main objective is to classify between two classes: benign and malignant. Because of that, this ML technique is used. The structure of the ANN used in this work is shown in Fig. 8.

Where x_i given that $1 \leq i \leq 9$ are the record attributes, w_i are the weights. Returning to the steps for the training process using the delta rule in Sect. 2.1.1 and adapting it to this net, the following algorithm is used.

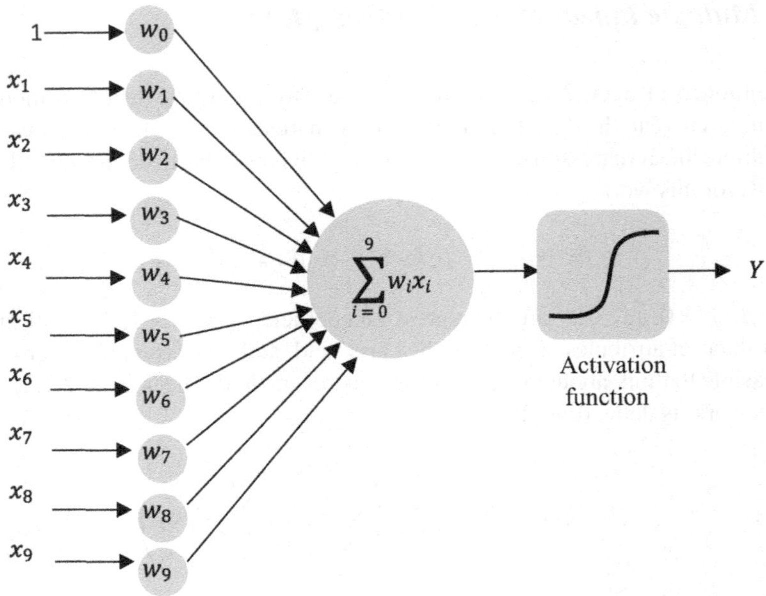

Fig. 8 Structure ANN used

1. Initialize the weight with adequate values obtained from the MLR made by evolutionary algorithms.
2. Enter the training data to the neural network and obtain the output using the Sigmoid function.

$$\varphi(v) = \frac{1}{1 + e^{-v}} \tag{31}$$

3. Calculate the error of the output.
4. Calculate the weight updates, using the delta rule.
5. Adjust the weights.
6. Perform steps 2–5 for the 419 training records.
7. Repeat steps 2–6 until the stop criteria is reached.

4 Results

In this section, the details related to the testes made are described. Also, the results obtained in each case are discussed. It should be mentioned that all the statistics were obtained through 20 repetitions of each test. To evaluate the performance of the previous hypothesis, it was made an experiment based on the comparison of two tests:

- Test 1: the classification made by the original ANN (without evolutive help) as described in Sect. 2.1.1.
- Test 2: the classification made by the ANN supporting with evolutionary algorithms mentioned in Sect. 2.2 (ABC, CS, DE, FPA, GSA, PSO). The process for this test is the same as described in Fig. 6 but it should be mentioned that the process is made for each evolutionary algorithm.

It should be mentioned that, in both testes, ones the training process has finished the validation process begins executing the following steps.

1. Enter the testing data to the neural network.
2. Calculate the output using Eq. (2).
3. Round the value obtained in the previous step to get a discrete value.
4. Compare the generated output with the correct answer and make the corresponding statistics.
5. Repeat steps 1–4 for the 280 testing records.

The general parameters used for all experiments are shown in Tables 5 and 6.

Once the tests were implemented, come up the dude about how can the performance be evaluated? One of the accepted evaluations of ML methods is the confusion matrix in which the following statistics can be analyzed.

- The accuracy: the relationship between correct classifications and all classifications.
- Specificity: measures the proportion of actual negatives that are correctly identified.
- Precision: measures the correct classification.
- Sensibility: measures the proportion of actual positives that are correctly identified.

Table 5 Test 1 parameters

Parameter	Value
Training epochs	100,000
Learning rate	0.1
Bias	1

Table 6 Test 2 parameters

Parameter	Value
No. of particles	50
No. iterations	3000
Training epochs	100,000
Learning rate	0.1
Bias	1

In Table 7 these measurements are presented. It is important to mention that these data are the mean obtained from the 20 repetitions in each case for each test. Based on them, it can be said that for all EAs in Test 2 the obtained results give an improvement to the traditional learning process in the ANN. Also, can be noticed that the ABC and CS algorithms have a better performance than the other EAs.

Based on the confusion Matrix data, essentially, with the sensibility and specificity, a Receiver Operating Characteristic (ROC Curve) can be obtained. It is presented in Fig. 9 that illustrates, graphically, the performance for each experiment.

To present a deeper perspective to this analysis, in Table 8 the accuracy results are shown in detail. With this information, it can be seen the support given by the EAs to the learning process. Based on all the information presented, a deeper study of each EA can be done. For example, the results generated by PSO and FPA are similar to

Table 7 Confusion matrix data

	Accuracy	Specificity	Precision	Sensibility
Without evolutive	0.9503	0.9592	0.9179	0.9382
ABC	**0.9660**	**0.9795**	**0.9606**	**0.9437**
CS	**0.9643**	**0.9756**	**0.9519**	**0.9438**
DE	**0.9660**	**0.9753**	**0.9524**	**0.9496**
FPA	**0.9602**	**0.9711**	**0.9462**	**0.9409**
GSA	**0.9650**	**0.9798**	**0.9608**	**0.9395**
PSO	**0.9588**	**0.9657**	**0.9327**	**0.9481**

Bold numbers are used to emphasize the evolutionary algorithms results

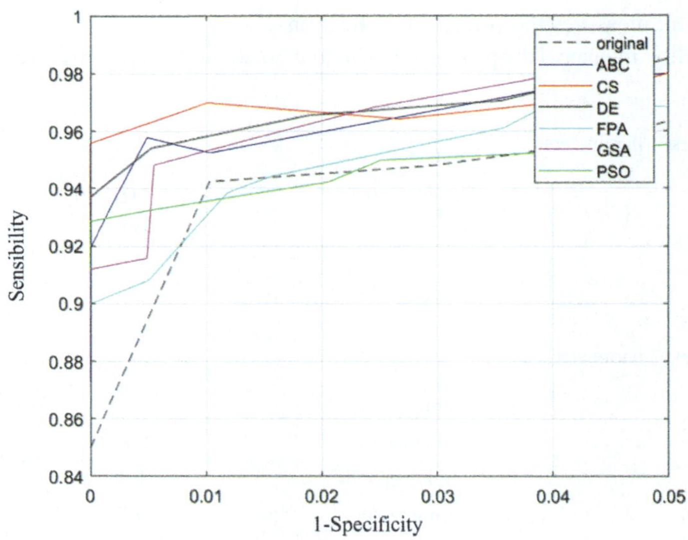

Fig. 9 Roc curve

Table 8 Accuracy results

	Mean	Maximum	Minimum	Mode
Without evolutive	0.9503	0.9632	0.9264	0.9554
ABC	**0.9660**	**0.9766**	**0.9599**	**0.9656**
CS	**0.9643**	**0.9833**	**0.9532**	**0.9571**
DE	**0.9660**	**0.9799**	**0.9565**	**0.9666**
FPA	**0.9602**	**0.9766**	**0.9431**	**0.9662**
GSA	**0.9650**	**0.9833**	**0.9532**	**0.9659**
PSO	**0.9588**	**0.9833**	0.9264	**0.9666**

Bold numbers are used to emphasize the evolutionary algorithms results

the original test. Maybe, it can be said that they tie with the traditional ANN process. On the other hand, the results generated by ABC, DE, and GSA are higher than the results obtained by the traditional ANN process.

Another accepted evaluation for ML methods is the analysis of the Mean Squared Error (MSE) that is calculated by (32).

$$MSE = \frac{1}{n}\sum_{i=1}^{n}(\hat{y}_i - y_i)^2 \quad (32)$$

In Table 9 the statistics related to the MSE are shown. It can be noticed that all AEs test, minimized the MSE resultant. Considering all the MSE data it can be said that the GSA and CS are the algorithms that give a promising general result. But, it should b mentioned that, in general, all the EAs results have been minimized compared with the original procedure.

Table 9 MSE results

	Mean	Maximum	Minimum	Mode
Without evolutive	0.0494	0.0736	0.0301	0.0468
ABC	0.0651	0.4010	0.0234	0.0334
CS	**0.0356**	**0.0468**	**0.0167**	**0.0435**
DE	0.0489	0.3340	0.0201	0.0401
FPA	0.0398	0.0569	0.0234	0.0334
GSA	**0.0365**	**0.0680**	**0.0167**	**0.0368**
PSO	0.0411	0.0703	0.0334	0.0334

Bold numbers are used to emphasize the minimum MSE obtained results

5 Conclusions

Cancer is considered the second cause of death in the world. Early detection can reduce cancer mortality, but, frequently, cancer patients do not receive a diagnosis until the disease is advanced. Thanks to technological advances, it is possible to analyze cancer data using machine learning. However, it faces some limitations. Mainly, because it depends on the quality of the information, the epochs used in the training process, the weights adequate initialization and so others. On the other hand, evolutionary algorithms have demonstrated to be useful for solving optimization problems in a heuristic way. This character had shown, based on the results presented previously, that using the regression coefficients obtained from an evolutionary technique to fit the initial ANN weights to the training data can support the learning process, improve the classification performance and minimize the mean square error.

Also, this work focused on studying and making new proposals that combine the advantages of different algorithms taking the on the qualities that characterize the machine learning methods and the evolutionary techniques, on the approach that uses the evolutionary techniques for supporting the learning process in machine learning.

In such a way, as future work, it is expected that this study will serve as an inspiration for other researchers and students to focus more on this type of algorithms that allow improving current techniques by means of combining them with other techniques in order to take the main qualities of each one to potentially a heuristic or a learning process. As well as, to improve the possible results in various applications that provide a solution for the problems of today's society.

References

1. M. Plummer, C. de Martel, J. Vignat, J. Ferlay, F. Bray, S. Franceschi, Global burden of cancers attributable to infections in 2012: a synthetic analysis. Lancet Glob. Heal. **4**(9), e609–e616 (2016)
2. Estadísticas del cáncer—National Cancer Institute (2018). https://www.cancer.gov/espanol/cancer/naturaleza/estadisticas. Accessed 09 May 2019
3. P. Ponce Cruz, *Inteligencia Artificial con Aplicaciones a la Ingeniería* (Mexico, DF, Alfaomega, 2010)
4. M. Paluszek, S. Thomas, *MATLAB Machine Learning* (2016)
5. D. Kaladhar, B. Chandana, P. Kumar, Predicting cancer survivability using classification algorithms. Int. J. Res. Rev. Comput. Sci. **2**(2), 340–343 (2011)
6. J.A. Cruz, D.S. Wishart, Applications of machine learning in cancer prediction and prognosis. Cancer Inform. **2**, 59–77 (2006)
7. A. Raad, A. Kalakech, M. Ayache, in *The 13th International Arab Conference on Information Technology ACIT Breast Cancer Classification Using Neural Network Approach: MLP and RBF*, 10–13 Dec 2012, p. 15–19
8. D. Fehr et al., Automatic classification of prostate cancer Gleason scores from multiparametric magnetic resonance images. Proc. Natl. Acad. Sci. **112**(46), E6265–E6273 (2015)
9. B.M. Wise. J.M. Shaver, in *Detection of Cervical Cancer from Evoked Tissue Fluorescence Images Using 2- and 3-Way Methods,* vol. 1087210 (2019), p. 35

10. M.Z. Alom, C. Yakopcic, M.S. Nasrin, T.M. Taha, V.K. Asari, Breast cancer classification from histopathological images with inception recurrent residual convolutional neural network. J. Digit. Imaging (2019)
11. A.W. Mohamed, A novel differential evolution algorithm for solving constrained engineering optimization problems. J. Intell. Manuf. **29**(3), 659–692 (2018)
12. P. Díaz, M. Pérez-Cisneros, E. Cuevas, O. Camarena, F.A.F. Martinez, A. González, A swarm approach for improving voltage profiles and reduce power loss on electrical distribution networks. IEEE Access **6**, 49498–49512 (2018)
13. K.G. Dhal, S. Das, A dynamically adapted and weighted bat algorithm in image enhancement domain. Evol. Syst. **10**(2), 129–147 (2018)
14. K.G. Dhal, A. Das, S. Ray, J. Gálvez, S. Das, *Nature-Inspired Optimization Algorithms and Their Application in Multi-thresholding Image Segmentation*. No. 0123456789 (Springer, Netherlands, 2019)
15. T. Bui, S.D. Stoller, J. Li, Greedy and evolutionary algorithms for mining relationship-based access control policies. Comput. Secur. **80**, 317–333 (2019)
16. H. Yoshida, D. Azuma, Y. Fukuyama, Dependable parallel canonical differential evolutionary particle swarm optimization for voltage and reactive power control. IFAC-PapersOnLine **51**(28), 167–172 (2018)
17. P. Kim, *MATLAB Deep Learning. With Machine Learning, Neuronal Networks And Artificial Intelligence* (2017)
18. N.J. Nilsson, *Introduction to Machine Learning an Early Draft of Proposed Textbook* (2005)
19. G. Englebienne, *Machine Learning Pattern Recognition Lecture Notes* (2013)
20. R.E. Bello Peréz, Z.Z. García Valdivia, M.M. García Lorenzo, A. Reynoso Lobato, *Aplicaciones de la inteligencia artificial*. México (2002)
21. S. Chatterjee, A.S. Hadi, *Regression Analysis by Example,* 5th edn. (2012)
22. M.C. Carollo, *Regresión lineal simple* (2011), p. 1–31
23. D.C. Mongomery, E.A. Peak, G.G. Vining, *Introducción al analisis de regresión lineal* (Grupo Editorial Patria, México, 2011)
24. R. Storn, K. Price, Differential evolution—a simple and efficient heuristic for global optimization over continuous spaces. Australas. Plant Pathol. **38**(3), 284–287 (1995)
25. E.V. Cuevas, J.V. Osuna, D.A. Oliva, M.A. Diaz, *OPTIMIZACIÓN. Algoritmos programados con MATLAB* (Alfaomega, Ciudad de México, 2016)
26. A.K. Qin, P.N. Suganthan, in *Self-adaptive Differential Evolution Algorithm for Numerical Optimization* (2005), p. 1785–1791
27. S. Das, A. Abraham, A. Konar, Automatic clustering using an improved differential evolution algorithm. IEEE Trans. Syst. Man, Cybern.—Part A Syst. Humans **38**(1), 218–237 (2007)
28. J. Liu, J. Lampinen, A fuzzy adaptive differential evolution algorithm. Soft. Comput. **9**(6), 448–462 (2005)
29. J. Kennedy, R. Eberhart, B. Gov, Particle swarm optimization. Encycl. Mach. Learn. 760–766 (1995)
30. X.-S. Yang, *Engineering Optimization. An Introduction with Metaheuristic Applications* (Wiley, United States of America, 2010)
31. M. Pluhacek, A. Viktorin, R. Senkerik, T. Kadavy, I. Zelinka, Extended experimental study on PSO with partial population restart based on complex network analysis. Log. J. IGPL (2018)
32. N.P. Holden, A.A. Freitas, *A Hybrid PSO/ACO Algorithm for Classification* (2007), p. 2745
33. A. Modiri, K. Kiasaleh, Modification of real-number and binary PSO algorithms for accelerated convergence. IEEE Trans. Antennas Propag. **59**(1), 214–224 (2011)
34. H. Fan, A modification to particle swarm optimization algorithm. Eng. Comput. **19**(7–8), 970–989
35. D. Karaboga, *An Idea Based on Honey Bee Swarm for Numerical Optimization* (Kayseri Turkey, 2005)
36. X.-S. Yang, Flower pollination algorithm for global optimization, in *Unconventional Computation and Natural Computation* (2012), p. 240–249

37. J. Gálvez, E. Cuevas, O. Avalos, Flower pollination algorithm for multimodal optimization. Int. J. Comput. Intell. Syst. **10**(1), 627 (2017)
38. R. Salgotra, U. Singh, A novel bat flower pollination algorithm for synthesis of linear antenna arrays. Neural Comput. Appl. **30**(7), 2269–2282 (2018)
39. J. Senthilnath, S. Kulkarni, S. Suresh, X.S. Yang, J.A. Benediktsson, FPA clust: evaluation of the flower pollination algorithm for data clustering. Evol. Intell. (0123456789) (2019)
40. X. Yang, *Nature-Inspired Algorithms and Applied Optimization*, vol. 744 (2018)
41. X.S. Yang, M. Karamanoglu, X. He, Flower pollination algorithm: a novel approach for multiobjective optimization. Eng. Optim. **46**(9), 1222–1237 (2014)
42. M. Abdel-Basset, L.A. Shawky, Flower pollination algorithm: a comprehensive review. Artif. Intell. Rev. 1–25 (2018)
43. E. Rashedi, H. Nezamabadi-pour, S. Saryazdi, GSA: a gravitational search algorithm. Inf. Sci. (Ny) **179**(13), 2232–2248 (2009)
44. W.H. Wolberg, *UCI Machine Learning Repository: Breast Cancer Wisconsin (Original) Data Set* (1992)
45. E. Yuan, *Logistic Regression* (2014). http://eric-yuan.me/logistic-regression/. Accessed 28 May 2019
46. D. Oliva, *Studies in Computational Intelligence 825 Metaheuristic Algorithms for Image Segmentation : Theory and Applications*
47. Y. Zheng, J. Qu, Y. Zhou, An improved PSO clustering algorithm based on affinity propagation 2 an overview of particle swarm optimization. **12**(9), 447–456 (2013)
48. M.R. Alrashidi, S. Member, A survey of particle swarm optimization applications in electric power systems. **13**(4), 913–918 (2009)
49. H. Iba, *Evolutionary Approach to Machine Learning and Deep Neural Networks* (2018)
50. V. Torra, Y. Narukawa, Y. Yoshida, *Modeling Decisions for Artificial Intelligence* (Berlin, 2007)

Metaheuristic Approach of RMDL Classification of Parkinson's Disease

V. Kakulapati and D. Teja Santhosh

Abstract Today, many people suffering from Parkinson's disease (PD), chronic and progressive illness throughout the world. This disease is easy to identify, patients coming with tremors, slowness of movement and freezing-of-gait. In this work we study computerized analysis of PD and illustrate a method to classify the diverse features of deep brain surgical images using the machine learning methods. The feature of deep brain can be analyzed, and their features can recognize with the help of image processing methods. In this work we implemented Random Multimodal Deep Learning (RMDL): a new ensemble, deep learning approach for classification. By utilizing RMDL method we achieve to getting better robustness and precision through ensembles of deep learning methods. Which can allow a variety data as input such that text, video, images, and symbolic. In this work, RMDL shows test results for images of PD brain images. The obtained results generate consistently enhanced performance than model methods over a broad range of data types and classification problems.

Keywords Deep learning · Classification · RMDL · Brain · Performance · Text · Metaheuristic · Swarm optimization

1 Introduction

Human being nervous system is are two major divisions-one is the central nervous system (CNS) and second is the peripheral nervous system (PNS). In CNS, the brain and spinal cord are present. In PNS nerves, and ganglia is present. About 100 billion neurons are present in an adult brain, which connected to CNS of the adult body to control the muscles and total organs. There is a malfunction representing

V. Kakulapati (✉) · D. Teja Santhosh
Sreenidhi Institute of Science and Technology, Yamnampet, Ghatkesar, Hyderabad, Telangana 501301, India
e-mail: vldms@yahoo.com

D. Oliva and S. Hinojosa (eds.), *Applications of Hybrid Metaheuristic Algorithms for Image Processing*, Studies in Computational Intelligence 890,
https://doi.org/10.1007/978-3-030-40977-7_17

as neurodegenerative disease which represents the disorders of nervous system by affecting the nervous system, and they make the nervous system to dysfunction by progressively affecting it. Due to the lack of regularity, the main cause of these dysfunctions cannot identify at the earliest stage of life, and the disease remains unknown. The main cause of this unknown identification, due to genetic nature of adult nervous system, environmental factors of the disease and the aging of body may be as a natural process.

Existing medical treatments for the neurodegenerative disease are very rare, and they are aimed to reduce the symptoms to improve the quality of life and extend the normal adult life expectancy. As age is one of the factors of this disease attack, the expectation has reached, as by 2030 these types of neurodegenerative disease may further increase more. The two common types of neurodegenerative brain diseases are Parkinson's disease (PD) and Alzheimer's disease (AD). The PD clinically makes the adult body bradykinesia: rest tremor, muscular rigidity, and postural instability. PD is effected over 5 million people, among 92% of patients with this disease develops dementia. The AD, on the other hand, within 1 and 8 years, is also a type of dementia which affects 8 million people.

Knowing the difference between PD and AD, because of brain disorders and diagnose them at the earliest, is a challenging task in the clinical grounds, as both PD and AD has to have overlapping pathological and clinical similar features. Like these, around 30% of the patient diagnosed through early detection of PD, at the time of treatment, as the later treatment begins, the patients found with other similar parkinsonian diseases.

To overcome these limitations, the advanced technology made used [1, 2], through the developed computer-aided algorithms, which stores a large data of patient information to process and diagnose. These algorithms help the physician by working among them and reduce the risk to be caused to the patient in future. And the large data storage of patient information makes the algorithms to have large cases of PD information to diagnose at the early stage with the patient's history at the doctor. The algorithms with combined clinical imaging methods [3, 4] are able to classify [5] the patterns of PD disease symptoms where the neuronal dysfunction is present and provides the information about the PD treatment [6–15] based on the choice and appropriate availability of treatment at the clinic which is followed by the patient needs.

The objectives of this research work are:

1. Study the PD and provide the hybrid metaheuristic approach for Random Multimodal Deep Learning (RMDL) method in the analysis of DICOM image data.
2. Provide the information about the sample of disease collected from the patient, through a classifier, to sample either PD, by training with the large database of patient's information.
3. Features are classified based on the subject of PD disease.

This chapter organized as follows: In Sect. 2, the base of this research work presented. In Sect. 3, the approach used in the DICOM image feature classification shown. The next Sect. 4, presents the experiments done and the results obtained. Finally in the last Sect. 5, the findings were given and presented the conclusions of this work.

2 State of the Art

In this section, the aspects of the areas, in which research chapter outlines presented. They are neurodegenerative brain disease PD and clinical image processing.

2.1 *Neurodegenerative Brain Diseases PD*

It is an incurable disease which links with age, through a progressive loss of function of nervous system in the adult brain. Due to this PD, the sending and receiving impulses through nerves do not reproduce among them, because of this, the damaged nerves cannot replaced. These changes in brain synaptic activity are accomplished by change in glucose levels for consumption, in the distant brain regions through differentiation, also vice versa may happen in glucose levels. And also lead to development of irregular activity patterns in the metabolic activity. By these, with the movement or mental functioning, are the causes for parkinsonian syndromes and dementia can arise respectively. As this disease can be identified at the early stage, but could not identify the features, when it is overlapped with the other features of disease like AD, because PD disease features changes due to the course of time in adult body. The main characteristics of PD are, this disease caused due to the presence of loss of dopamine producing nerve cells that emanate from the substantia nigra pars compacta in the midbrain, and project to the striatum, that regulates the movement. Based on the patient medical history and through neurological tests, from the medical database of person to person, and the timeline of the disease occurrences, through progressive growth, it is effected through the presence of multisystem disorders with motor and non-motor symptoms, because this PD will begin gradually and may become worse over the timeline of clinical observation.

2.2 *Clinical Image Processing*

Through the clinical image processing the distinguishing of PD neurodegenerative disorders is possible, through the automated image processing method, the medical image dataset can lead to analyse the patient clinical observations possible. By the approaches of image processing, the subjective evaluation through analysis and objective evaluation through results, having a large patient's history data storage can make a possible patient observation in the clinical environment gives a diverse PD change in time characteristics.

Use of learning computerized analysis of PD, by integrating the image data processing, movement to computerized methods to statistical analysis by maximizing the patient's information through the data extraction is made possible. The results of the computerized analysis help the clinic observers to differentiate between the several PD conditions through the compared diagnose reports and through the overlapped PD features. The process of computerized analysis through image processing of PD disease diagnose is mainly based on: by studying of features of the disease, by extracting the patterns of disease, classify the disease features and patterns, through acquiring the PD disease images and its features, comparing and extracting the PD disease, selecting the features with which matches to PD and classify the matches features to the disease with clinical diagnose data. Using DICOM image features of brain imaging is performed, by statistical data mapping at the initial stage. Through a component analysis the PD characteristics were patterned from the disease data to disease features. These components occupied for marking to observe the location of PD disease percentage among the dataset selected. The percentage score through subjective and objective compared, and the result of the above said analysis helps in PD reorganization.

In this chapter, a Random Multimodal Deep Learning (RMDL) approach is performed to classify the PD disease through image classification. Through RMDL, the DICOM image of PD disease is separated in to number of layers and nodes, which contains multi Deep Neural Networks (DNN), Deep Recurrent Neural Networks (RNN), and Deep Convolutional Neural Networks (CNN), These all models are trained with clinical dataset to predict and calculate for majority votes of classified features. The flowchart for the interaction of the three NN shown in Fig. 1.

DNN generates the PD features through connect-and-hide, the processing patient image information learned by dataset and these learned image set classified into layers through multi-connection and these layers receives connection from the previous and present layer, by providing connections to the next layer of connection in hidden layer part. These connections have input and output feature space, where input feature space holds the hidden layers and first layers connected and output feature space holds the number of feature classes for multi-class classification and binary

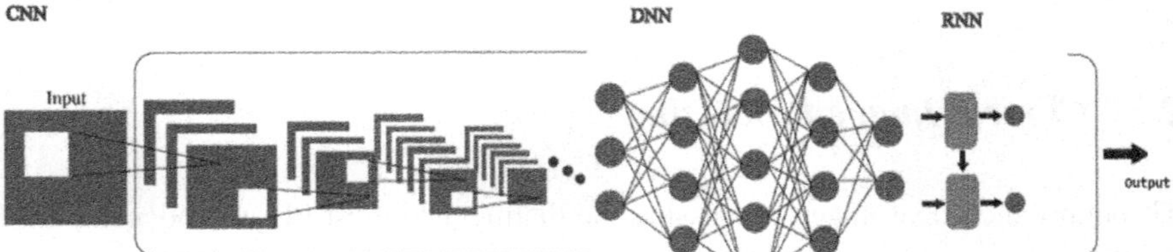

Fig. 1 The interaction of the three NN

classification only one output observed. The feature weights of the DNN are considered in a sequence and classifies them for semantic analysis through connection structures of the clinical dataset provided through RNN. The structures resulted are processed for image tensor, through convolution process the set of kernels deployed in the image tensor. These convolution layers are the feature maps of PD disease by the clinical dataset, results in a group of mapped features through pooling process, by selecting maximum elements of features from the pooling window. The fully connected structure results through the back-propagation process by varied PD feature detection accuracy.

3 Method

This section focus on the steps followed for the implementation of proposed work that has applied to perform feature extraction, classification of DICOM images. In Fig. 2, the flow diagram of the process that followed in the proposed work given.

3.1 DICOM Data

Subject PD images collected from clinical hospitals. In this research work, 50 patients samples and are diagnosed through clinical data. In the dataset, there are variations in gender, duration of the disease and age of PD patients. In Fig. 3, few samples of PD dataset images shown.

3.2 Feature Extraction

Extract the features of PD images, this chapter proposes a PD feature extraction method and is based on the % if PD intensity. The proposed approach is summarized as follows in Algorithm 1:

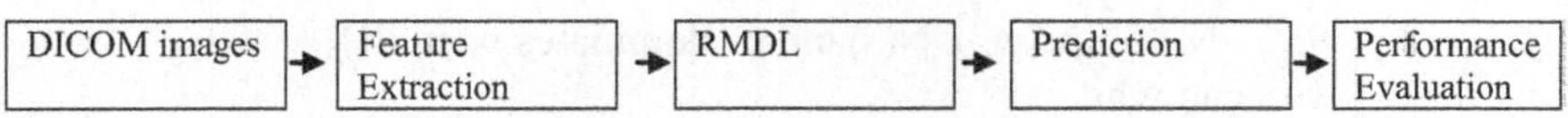

Fig. 2 Proposed workflow

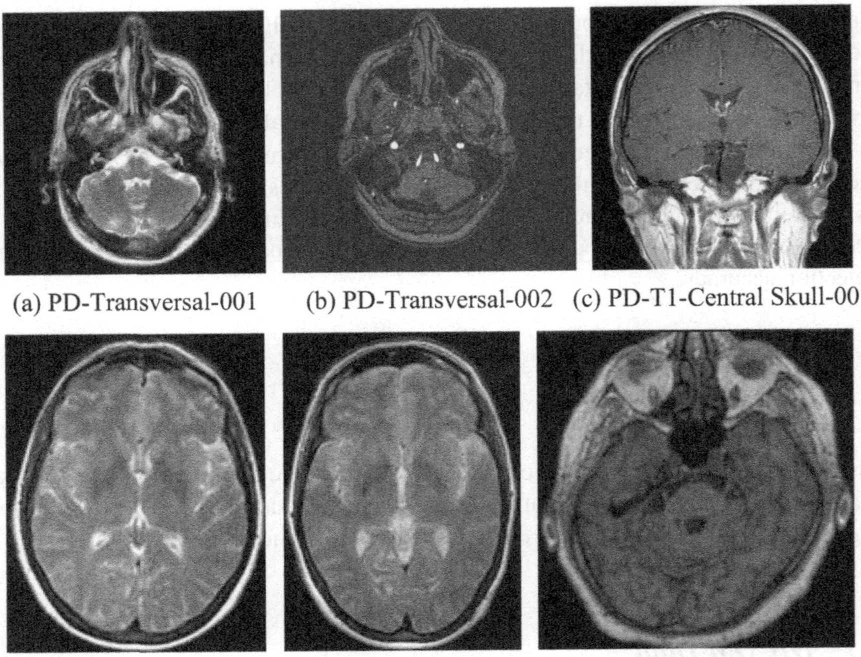

(a) PD-Transversal-001 (b) PD-Transversal-002 (c) PD-T1-Central Skull-001

(d) PD-Axial-001 (e) PD-Axial-003 (f) PD-T1-Central Skull-003

Fig. 3 DICOM images of different PD datasets with which the proposed work worked

```
Algorithm 1: PD CNN Feature Extraction Method
Input: PDi {set of PD images},
N(.){set of native structures}, pi(positions)
PD<--{Npi}(set of non-native positions)
Ns<-{pi'| pi' ϵ N(s)and pi' dominates pi' in PD}
while Ns≠null do
    for all pi' ϵ Ns do
        PD<---update (pi',PD)
    end for
   pi'<--choice PD(x) where xϵ Ns
   pi <-- pi'
   Ns<-- {pi'| pi' ϵ N(s)and pi' dominates pi' in PD}
end while
return PD
```

The PD Feature Extraction Method describes steps for PD image feature extraction the multi-native structures of PD are differentiated by positions of the PD compared dataset images. In this multi-native PD content, the results of non-native positions are evaluated by dominance PD relationship, by comparing the region of PD in each position of dataset taken. The set of non-native structures of ND starts with the initial

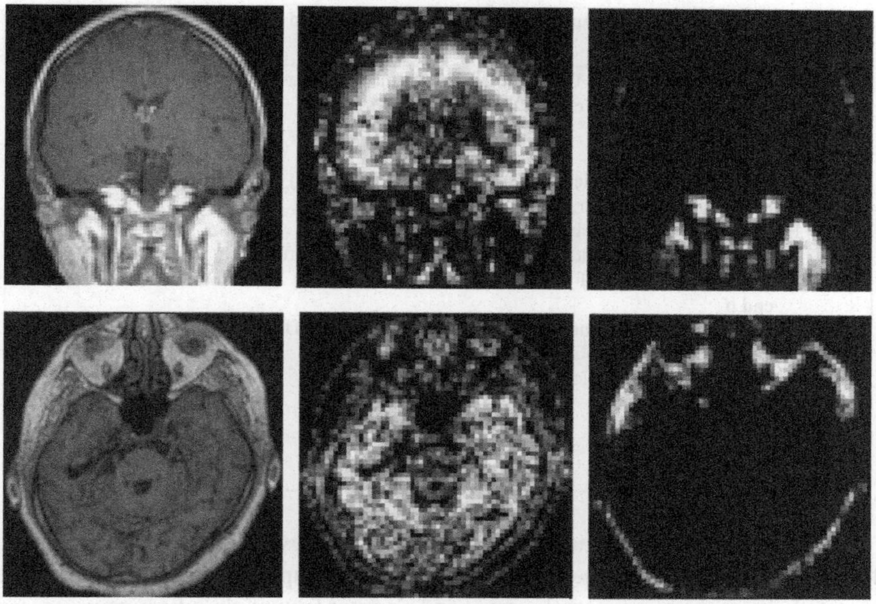

(a) Trained Image (b) Native Positions (c) Non-native Positions

Fig. 4 Some training images, with the features

null value, and at each iteration a $p_i^{'}$ solution of PD results uniformly. The natives of $p_i^{'}$ are generated regularly and the values of $p_i^{'}$ incremented results in non-dominance solutions. The initial p_i solution value becomes $p_i^{'}$ as the PD features extracted and this process repeats until a $p_i^{'}$ positions of PD are selected, even though the dominance of $p_i^{'}$ is considered. The trained image with its proposed feature extracted images shown in Fig. 4.

3.3 RMDL

In this, the investigation of PD effects performed through the hybrid metaheuristics in the implementation of classifier ensembles, which performed on PD DICOM datasets. In this classification, the performance of multi-features and multi-attributes in the selection of PD disease detection of classifier ensembles assessed. Along with this, the separate and similar PD feature objectives were assessed, to get all the possibilities of used classifier ensemble considered. The main steps of the proposed classifier presented in Algorithm 2. During the observations of Algorithm 2, this algorithm applicable to both the separate and similar PD features.

Algorithm 2: PD RMDL metaheuristic classifier ensemble method

Input: Dataset (Dset), Features Objective (FOb), Number of Classifiers (NoC), Attributes(Attr), Classifiers (Clasf), Number of Features (NoF)

```
NoF<--O and NoC<--O
  while StartCondition do
    Attr, Clasf, FOb=Classifier(Dset,Attr,NoF,Clasf,FOb,NoC)
    Ensb<--PDEnsemble(Attr, Clasf, FOb)
    Evalt<--PDEvaluate (Ensb,FOb)
      if (Ensb not ∈ NoF) and (Condition(Ensb,NoF,Evalt)=True) then
        Includes (Ensb, NoF)
      end if
      if (Ensb not ∈ NoC) and (Condition(Ensb,NoC,Evalt)=True) then
        Includes (Ensb, NoC)
        NoC<--PDEnsemble(FOb,NoF,Clasf)
      end if
  end while
return NoF
return NoC
```

In the RMDL, the RNN and DNN performed. As these are robust and have made this work result-oriented, these incorporated in this chapter. Once the features were selected, the ensemble system designed and a proper feature classification processed through the stacking of features in classifier. The stacking method trains the classifier with the datasets of PD disease and results the classification of PD disease accurately.

3.4 Prediction

For the proposed work analysis, 50 PD DICOM datasets used from clinical data. These data describe PD in DICOM format and the calculations performed in MatLab environment using the machine learning packages. Firstly, the descriptive analysis conducted, where the study of PD data visualization with different samples performed, results in finding the similar features of PD from the stored dataset. Validation test is performed using linear regression on the trained and feature datasets. The accuracy of the validation set gives an important indicator for the proposed work. The main steps of the proposed predictor presented in Algorithm 3. During the observations of Algorithm 3, this algorithm applies to decision strategy of PD features.

Algorithm 3: PD Probabilistic prediction method

```
Input: NoC, set of features S, validation V, Prediction P_PD, i is the number of input PD features,
for each feature FOb defined by the rows of NoF do
  Let Attr=x, NoF=y, S_i={(x,NoC(i,y)):(x,y)ϵS}
  Let d_i=Dset(S_i)
   while StartCondition do
     for S≤1 do
       if NoC<S
         V<--O
       end if
     end for
     for S≥1 do
       if NoC>S
         outcomeprob_d(x,y)=2 P_PD i[NoC(i,y)bi(x)+(1-NoC(i,y))(1-d_i(x))]-1
         setofpred_b(x,y)= P_PD (x,y)~Dset S(V(y),x)
         V_d(f,D)= P_PD (x,y)~Dset (V(y)-x)^2
       end if
     end for
return {di}
```

3.5 *Performance Evaluation*

The classifier performance made by 10-fold validation of 5 repetitions, at each round the dataset of stored samples randomly partitioned into five equal-sized subsets of PD. Among these 5, one set used for training the PD data and the remaining four are used for validation of PD data to test the model of the classifier proposed. This processed is repeated for 10 times, for each running fold 10 subsets are validated, by performing multiple subsets validation, the 10-fold validation achieves, which is more accurate estimation of the proposed model quality, through the reduced risk of time consumption, which results in average validation runs over the folds.

4 Experiments and Results

In this section, the experimental framework of the proposed hybrid metaheuristics approaches work and the simulated results established. All data were collected through clinical diagnose and web results with Unified Parkinson's Disease Rating Scale (UPDRS) and Hoehn and Yahr scale (H&Y), the summary of data set for the study of this work tabulated in Table 1.

During the experimental framework, the different types of PD disease subjects taken into account, by taking the 3 phases in feature extraction process. First, using the feature subjects through PD CNN method, then during second, RMDL method

Table 1 Collected PD data summary

Subject	H&Y	Count	Gender (M/F)	Age	Duration of illness (years)	UPDRS
PD	1	19	10/9	60.5	4	11.7
PD	2	28	16/12	61.8	3	22.8
PD	3	3	2/1	54.2	1	29.5

Table 2 Classification of PD features by center

Source	Subjects	Experiment	Class 1	Class 2	Classification
DICOM	Transversal	E1	√	–	Linear
DICOM	T1-Central Skull	E2	√	√	Vector
DICOM	Axial	E3	√	√	Vector

is used in the classified PD data. Finally, both the methods with prediction is applied to combine the classified PD features. The experiments and classifications tabulated in Table 2.

The performance of the proposed work is evaluated using an objective measure based. The result images shown in Fig. 5.

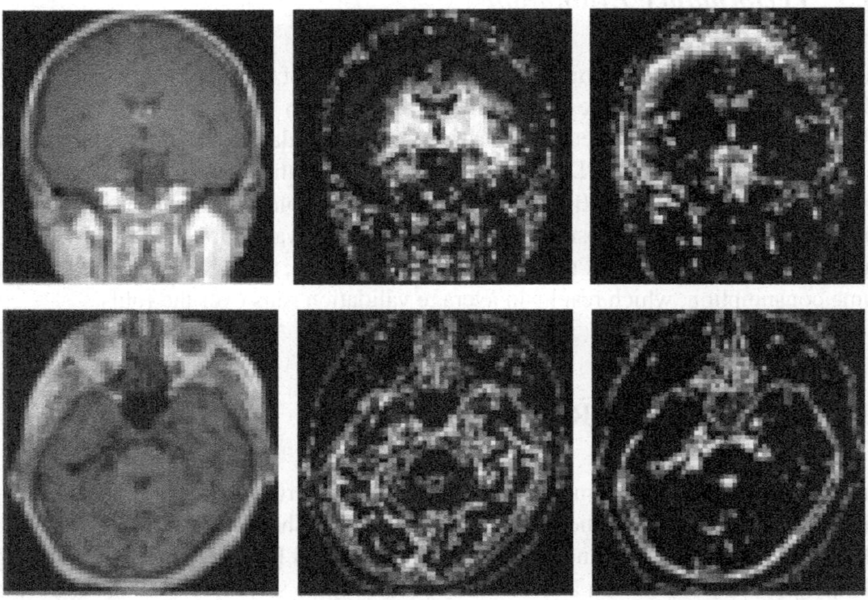

(a) Trained Image (b) Class 1 resulting image (c) Class 2 resulting image

Fig. 5 Resulting images of the proposed work

Table 3 Definitions of four labels of classifier

Label	Definition
TP	Feature extracted as class 1 and have classified as class 1
FN	Feature extracted as class 2 and have classified as class 1
FP	Feature extracted as class 1 and have classified as class 2
TN	Feature extracted as class 2 and have classified as class 2

The evaluation measure indicators performed in this proposed work described below.

1. $Sensitivity = TP/TP + FN$
2. $Specificity = TN/FP + TN$

where TP is True Positive, FN is False Negative, FP is False Positive, and TN is True Negative. The description of these four labels given in Table 3.

The objective measure performance of the proposed with the existing works presented in Table 4 with the experiments performed.

From the experiments, E1, E2 and E3 in Table 2, the performance of proposed work which given in Table 4, cam conclude that the PD DICOM image datasets used in this chapter are analyzed and distinguished between the PD across the center. Also by the results the performance measure of proposed classifier is better than the compared.

Table 4 Comparative results of the proposed RMDL method

Subject	Experiment	Classifier	Sensitivity (%)	Specificity (%)
PD	E1	SVM [16]	68.89	78.11
		Proposed	68.90	79.22
		GMLVQ [17]	68.59	81.07
PD	E2	SVM [16]	94.84	95.14
		Proposed	92.52	95.22
		GMLVQ [17]	88.28	95.58
PD	E3	SVM [16]	84.35	84.69
		Proposed	85.66	84.22
		GMLVQ [17]	82.86	83.05

5 Conclusions

The set of non-native positions and native positions have distinguished from the trained and learned dataset. These positions features are extracted to apply for RMDL ensemble. Used 5 fold classifier, the 100 K features were extracted and predicted the required features through decision strategy. From the results observed in Tables 2 and 4, the features and classifier level distinguishing achieved. And the features are extracted based on the samples of DICOM dataset. The classified features are varied from center with different discriminative features, to build the classifier ensemble. Compared with SVM and GMLVQ, the proposed work shown better results. The proposed work can used in the future works of vision feature extraction process during the multi-cross validation analysis.

References

1. A. Mozaffari, A. Fathi A, A natural-inspired optimization machine based on the annual migration of salmons in nature (2013). Available from: https://arxiv.org/ftp/arxiv/chapters/1312/1312.4078.pdf. Accessed 2 Apr 2015
2. M. Singh, A. Bhasin, Surender, Using meta-heuristic approaches in web document clustering in web search. Int. J. Pure Appl. Math. **119**(15), 2853–2861 (2018)
3. K. Morgen, G. Sammer, L. Weber, B. Aslan, C. Müller, G.F. Bachmann, D. Sandmann, M. Oechsner, D. Vaitl, M. Kaps, I. Reuter, Structural brain abnormalities in patients with Parkinson disease: a comparative voxel-based analysis using T1-weighted MR imaging and magnetization transfer imaging. AJNR Am. J. Neuroradiol. **32**(11), 2080–2086 (2011). https://doi.org/10.3174/ajnr.A2837. Epub 11 Nov 2011
4. R. Helmich, D. Vaillancourt, D. Brooks, The future of brain imaging in Parkinson's disease. J. Parkinson's Dis. **8**, S47–S51 (2018). https://doi.org/10.3233/JPD-181482
5. K. Kowsari, M. Heidarysafa, D.E. Brown, K.J. Meimandi, L.E. Barnes (2018) RMDL: Random Multimodel Deep Learning for classification, in ICISDM '18, 9–11 April 2018, Lakeland, FL, USA. arXiv:1805.01890v2 [cs.LG] 31 May 2018
6. H. Braak, U. Rub, E.N. Jansen Steur et al., Cognitive status correlates with neuropathologic stage in Parkinson disease. Neurology **64**, 1404–1410 (2005)
7. P.M. Mattila, M. Roytta, H. Torikka et al., Cortical lewy bodies and alzheimer-type changes in patients with Parkinson's disease. Acta Neuropathol **95**, 576–582 (1998)
8. S.N. Gomperts, D.M. Rentz, E. Moran et al., Imaging amyloid deposition in lewy body diseases. Neurology **71**, 903–910 (2008)
9. E.J. Burton, I.G. McKeith, D.J. Burn et al., Cerebral atrophy in Parkinson's disease with and without dementia: a comparison with Alzheimer's disease, dementia with Lewy bodies and controls. Brain **127**(pt 4), 791–800 (2004)
10. A. Nagano-Saito, Y. Washimi, Y. Arahata et al., Cerebral atrophy and its relation to cognitive impairment in Parkinson disease. Neurology **64**, 224–229 (2005)
11. C. Summerfield, C. Junque, E. Tolosa et al., Structural brain changes in Parkinson disease with dementia: a voxel-based morphometry study. Arch. Neurol. **62**, 281–285 (2005)
12. N. Pyatigorskaya, B. Magnin, M. Mongin, L. Yahia-Cherif, R. Valabregue, D. Arnaldi, C. Ewenczyk, C. Poupon, M. Vidailhet, S. Lehericy, Comparative study of MRI biomarkers in the substantia nigra to discriminate idiopathic Parkinson disease. AJNR Am. J. Neuroradiol. **39**, 1460–1467 (2018)

13. P.J. Planetta, E. Ofori, O. Pasternak, R.G. Burciu, P. Shukla, J.C. DeSimone, M.S. Okun, N.R. McFarland, D.E. Vaillancourt, Free-water imaging in Parkinson's disease and atypical parkinsonism. Brain **139**, 495–508 (2016)
14. J. Schulz, G. Pagano, J.A. Fernandez Bonfante, H. Wilson, M. Politis, Nucleus basalis of Meynert degeneration precedes and predicts cognitive impairment in Parkinson's disease. Brain **141**, 1501–1516 (2018)
15. R.G. Burciu, E. Ofori, D.B. Archer, S.S. Wu, O. Pasternak, N.R. McFarland, M.S. Okun, D.E. Vaillancourt, Progression marker of Parkinson's disease: a 4-year multi-site imaging study. Brain **140**, 2183–2192 (2017)
16. L. Anderlucci, L. Guastadisegni, C. Viroli, Classifying textual data: shallow, deep and ensemble methods, pp. 1–16. arXiv:1902.07068v1 [cs.CL] 18 Feb 2019
17. S. Saralajew, L. Holdijk, M. Rees, T. Villmann, Prototype-based neural network layers: incorporating vector quantization, pp. 1–19. arXiv:1812.01214v2 [cs.LG] 18 Jan 2019

Fuzzy-Crow Search Optimization for Medical Image Segmentation

A. Lenin Fred, S. N. Kumar, Parasuraman Padmanaban, Balazs Gulyas and H. Ajay Kumar

Abstract Fuzzy C Means (FCM) clustering technique is widely used in medical image segmentation. The classical FCM is sensitive to noise and the objective function often gets trapped in local optima. In this work, we have combined FCM with various optimization techniques to find the optimum cluster centres. The Artificial Bee Colony (ABC), Firefly, Cuckoo and Simulated Annealing (SA) are the widely used optimization techniques in segmentation as well as for classification problems. Here we are proposing the Crow Search (CS) Optimization for the image segmentation using FCM. The CS optimization was found to produce satisfactory and promising results when compared with ABC, Firefly, Cuckoo and SA algorithm in the segmentation of abdomen CT images using FCM. In addition, a thorough investigation has been made in choosing the appropriate validation functions for the clustering techniques. The algorithms have been developed in Matlab 2015a and tested on CT abdomen datasets. Finally, the hardware implementation of FCM-CS algorithm was done in Raspberry Pi B+ model embedded board.

A. Lenin Fred · H. Ajay Kumar
Mar Ephraem College of Engineering and Technology, Elavuvilai 629171, Tamil Nadu, India
e-mail: leninfredip@gmail.com

H. Ajay Kumar
e-mail: ajayhakkumar@gmail.com

S. N. Kumar (✉)
Amal Jyothi College of Engineering, Kanjirappally 629001, Kerala, India
e-mail: appu123kumar@gmail.com

P. Padmanaban · B. Gulyas
Nanyang Technological University, Nanyang Drive, Singapore 636921, Singapore
e-mail: ppadmanabhan@ntu.edu.sg

B. Gulyas
e-mail: balazs.gulyas@ntu.edu.sg

D. Oliva and S. Hinojosa (eds.), *Applications of Hybrid Metaheuristic Algorithms for Image Processing*, Studies in Computational Intelligence 890,
https://doi.org/10.1007/978-3-030-40977-7_18

1 Introduction

Image segmentation is the method of taking out of the desired region of interest. The thresholding, region growing, and edge detection is classical segmentation algorithms; however, the classical techniques fail in many cases for images with complex ROI [1]. Fuzzy clustering is a soft segmentation model that is widely used in medical and remote sensing applications. The classical FCM initializes the cluster centroids randomly and often stuck in local optima. Some of the related works of FCM in combination with the optimization algorithms are as follows; the issues in conventional FCM are cluster centroid initialization and the objective function trapping in the local minima. A noise probability term (Pn) is [2] introduced in the FCM algorithm for the segmentation of brain tissue in MR images, ABC classifies the pixels into noisy and normal based on the value of Pn and produces good segmentation accuracy, when compared with the classical FCM algorithm. Compared with other metaheuristic algorithms like Particle Swarm Optimization (PSO), Genetic Algorithm (GA), Differential Evolution (DE); Artificial Bee Colony (ABC) algorithm produces good results [2]. Ayat Alrusan [3] apply ABC algorithm in the FCM segmentation technique for the MR images, the performance of ABC-FCM was superior when compared with the conventional FCM approach. The ABC-FCM algorithm [4] was applied for the segmentation of tumor in MR brain images, the fitness function of ABC algorithm determines the threshold value for tumor delineation and the result was refined by FCM. Dervis Karaboga et al. [5] performed a comprehensive survey on ABC algorithm variants and its applications in various domains. The global-local search capacity of the ABC algorithm was improved by a hybrid technique comprising of ABC and GA algorithms [6].

An improved discrete Binary ABC algorithm was developed that produces good results than FCM, PSO-FCM, GA-FCM, K-means and discrete ABC techniques [7]. Gbest-guided Artificial Bee Colony (GABC) is an improvement of ABC algorithm and was utilized in FCM, better convergence rate with the global optimum solution than the ABC algorithm [8]. The fuzzy membership function uses ABC to search for optimum cluster centre and generates efficient segmentation result for grayscale synthetic, medical and texture images [9]. Similarly, in [10], the authors reported that ABC-FCM was efficient for segmentation of MR images with noise when compared with FCM and Possibilistic C-Means Algorithm. The firefly algorithm was used for the initialization of cluster centre in K-means segmentation and produces better results than traditional FCM, FCM-PSO algorithm [11]. In [12], the firefly algorithm was utilized in FCM for the segmentation of lesions in the multiple sclerosis MR images. Similarly, in [13], the FCM algorithm performance was improved by the incorporation of the firefly algorithm for the segmentation of simulated and real-time MR brain images.

The variants of firefly algorithm (FA) was applied for multi thresholding segmentation, Brownian distribution based FA performance was better than classical FA and levy flight FA [14]. In [15], the FCM based on FA was found to be robust in the segmentation of lung nodules in CT images. The FCM segmentation result was improved by a hybrid optimization technique comprising of FA and GA algorithm with less computation time [16]. The chaotic simulated annealing based FCM produces global optimum value for the objective function of FCM with less computation time and good segmentation accuracy on grayscale benchmark images [17].

The Cuckoo optimization is used in the feature extraction in mammogram images by coupling with Ant Colony Optimization (ACO), Cuckoo search is employed in the local search of ACO to speed up the process [18]. The Cuckoo optimization is employed in FCM for the segmentation of blood vessels in CT images of the liver [19]. A hybrid segmentation algorithm comprising of FCM, Random walker and Cuckoo optimization was employed for the automatic liver tumor segmentation [20]. Saptarshi Sengupta et al. proposed Quantum behaved Particle Swarm optimization-based FCM for the data clustering of UCI Machine Learning Repository. The outcomes were superior when compared with the PSO, K- means and QPSO K-means algorithms [21]. The crow search optimization was found to be effective in the optimal PV and capacitor placement for loss drop in the electrical distribution systems [22]. The chicken swarm optimization algorithm when coupled with the FCM was found to produce efficient results when compared with the classical FCM, FCM-ABC and FCM-PSO algorithms [23]. The global search capacity of the PSO algorithm was improved by coupling with the simulated annealing and the hybrid optimization was incorporated in the FCM algorithm [24].

The Grey Level Co-occurrence Matrix (GLCM) feature extraction was utilized in the cluster number selection and Cluster Number Adaptive Fuzzy c-Means (CNAFCM) image segmentation algorithm yield efficient results for synthetic and natural images from Berkeley database [25]. The K-prototype clustering algorithm is widely used in data mining and the CS algorithm was incorporated, that produces global optimum solution [26]. The CS algorithm was found to be robust in the power system applications like load flow analysis [27]. The fuzzy PSO generates efficient segmentation results for MR brain images even in the presence of noise and intensity inhomogeneity [28]. The classical FCM often fails in the optimal selection of cluster centre and in this proposed research work, various optimization algorithms are coupled with FCM for the segmentation of abdomen CT images. The hybrid FCM and Crow optimization algorithm generates efficient results and analysis has been made by validation metrics for the selection of cluster number. The materials and methods highlight the data acquisition; classical FCM, Crow optimization algorithm and finally results and discussions are drawn.

2 Materials and Methods

2.1 Data Acquisition

The ten DICOM CT data sets of the abdomen are used for the algorithm analysis. The images have been attained from Optima CT machine. The Metro Scans and

Research Laboratory supports the research work by providing medical data sets and the expert radiologist helps in the identification of anomalies.

2.2 Overview

Fuzzy C Means (FCM) clustering technique gains much importance in medical image segmentation [28]. The classical FCM is sensitive to noise and the objective function often gets trapped in local optimal value. In this research work, FCM was coupled with various optimization techniques to find the optimum cluster centres [29]. The Artificial Bee Colony (ABC), Firefly, Cuckoo and Simulated Annealing (SA) are the widely used optimization techniques in segmentation as well as for classification problems. Here, the Crow Search (CS) Optimization [30] is proposed for the image segmentation using FCM. The CS optimization was found to produce satisfactory and promising results when compared with ABC, Firefly, Cuckoo and SA algorithm in the segmentation of abdomen CT images using FCM. In addition, a detailed investigation has been made in choosing the appropriate validation functions for the clustering techniques.

The classical FCM often fails in the optimal selection of the cluster centre and in this proposed research work, various optimization algorithms are coupled with FCM for the initialization of cluster centroids [31]. The Crow search optimization algorithm, when coupled with FCM generates efficient result and analysis has been made by validation metrics for the selection of cluster number.

2.3 Fuzzy C Means Clustering

The FCM relies on fuzzy sets to generate a membership matrix while minimizing objective function [22]. The proposed algorithm in this work comprises of two stages. The first stage involves the cluster centre estimation by optimization algorithm and the second stage is the segmentation by clustering using the cluster centres determined in the previous stage. The optimization algorithm replaces random initialization of cluster centres with optimized values.

The objective function of the FCM algorithm is defined as the sum of distances between the pattern and cluster centres and is represented in Eq. 1.

$$J = \sum_{i=1}^{P} \sum_{j=1}^{C} U_{i,j}^{f} d\left(y_i, v_j\right) = \sum_{i=1}^{P} \sum_{j=1}^{C} U_{i,j}^{f} \left\| y_i, v_j \right\|^2 \quad (1)$$

where C is the number of clusters and in the context of image processing, it represents the number of segmented regions; P is the number pixels; f is the fuzzifier intensity greater than 1; U_{ij} is the membership function of the attribute y_i to cluster j.

The U_{ij} satisfy the following conditions represented in Eq. 2.

$$U_{ij} \in [0, 1]; \sum_{j=1}^{C} U_{ij} = 1 \tag{2}$$

U_{ij} is defined as the fuzzy membership of the pixel y_i in the jth cluster; v_j is the jth cluster centre.

The membership function and cluster centre are updated as follows represented in Eqs. 3 and 4.

$$U_{ij} = \frac{1}{\sum_{c=1}^{C}\left(\frac{D_{ij}}{D_{ic}}\right)^{\frac{2}{f-1}}} = \frac{1}{\sum_{c=1}^{C}\left(\frac{\|y_i - v_j\|^2}{\|y_j - v_c\|^2}\right)^{\frac{1}{f-1}}} \tag{3}$$

and

$$V_i = \frac{\sum_{i=1}^{N} U_{ij}^{f} y_i}{\sum_{i=1}^{N} U_{ij}^{f}} \tag{4}$$

The FCM algorithm was coupled with an optimization algorithm to find the near-optimal cluster centre. Though the FCM algorithm has been combined with many optimization techniques, the proposed research work employs Crow search optimization algorithm. After the generation of optimal cluster centres, pixels are assigned to the nearest cluster. Hence, the segmented image is obtained.

2.4 Crow Search Optimization

The Crow search algorithm is a bio-inspired optimization technique that relies on the clever traits of crows The CSO was validated on benchmark functions and efficient results were produced, when compared with the PSO and GA algorithm [32]. The crow search optimization was found to be efficient in breast cancer detection when coupled with the Intuitionistic Fuzzy Clustering algorithm. The CSO was found to be efficient in segmentation when compared with the classical optimization algorithms like PSO [33]. The crow search optimization was coupled with the cross-entropy thresholding technique for the segmentation of MR brain images [34]. The CSO was also found to be efficient in the early diagnosis of Parkinson's disease based on handwriting exams [35]. The CS is a population-based optimization technique that can be applied to engineering problems. The principle of CS is based on the biological trait of the crow; the excess food is stored in hidden places and will be retrieved when the food is required. The objective of the optimization algorithm is to determine the optimum solution for engineering design problems that comprises a large number of decision variables. In real-world problems, the designer will be interested in the global optimum value of the objective function rather than the local optimum value determined by the conventional optimization techniques. The underlying principle of Crow Search (CS) optimization is based on its flock behaviour. Crows have the capacity to store the information in their memory regarding where they hide their food and they follow each other to do thievery for food.

The steps of CS optimization are summarized as follows

Step 1: The optimization problem is defined and set the parameters of CS optimization algorithm

The parameters that can be adjusted to yield a refined result are flock size (N), the maximum number of iterations ($iter_{max}$), flight length (fl) and awareness probability (AP).

Step 2: Initialization of crow's position and memory.

The position of crow's *i* in the search space is defined by a vector represented in Eq. 5.

$$Y^{i,iter} = \left[Y_1^{i,iter}, Y_2^{i,iter}, \ldots Y_d^{i,iter}\right] \tag{5}$$

where *iter* represents the number of iterations.

The solution space is of 'd' dimension with N crows randomly positioned.

In the matrix form, the crow's position is represented as follows in Eq. 6.

$$Crows = \begin{bmatrix} Y_1^1 & Y_2^1 & Y_1^1 \\ Y_1^2 & Y_2^2 & Y_1^2 \\ Y_1^N & Y_2^N & Y_d^N \end{bmatrix} \tag{6}$$

In the initial iteration stage, the crow has no experience, the assumption is that crow's hide the food in initial positions. The matrix for memory is written as follows represented in Eq. 7.

$$Memory = \begin{bmatrix} m_1^1 & \cdots & m_d^1 \\ \vdots & \ddots & \vdots \\ m_1^N & \ldots & m_d^N \end{bmatrix} \tag{7}$$

Step 3: Evaluate the fitness function

The quality of all crow's location is determined by incorporating the decision variables values into the objective function.

Step 4: Generation of the newly updated position

Suppose, when a crow i wants to generate a new position, it randomly selects any one of the flock member (for example j) and follows it, where the food is being hidden.

There are two possible cases

Case i: The crow *i* will reach the location of food storage by crow *j*, when it does not realize that crow *i* is subsequently as

The new location of crow *i* is determined as follows represented in Eq. 8.

$$Y^{i,iter+1} = Y^{i,iter} + P_i \times fl^{iter} \times \left(m^{j,iter} - Y^{i,iter}\right) \tag{8}$$

where P_i represents a random number with uniform distribution and *fl* is the flight length of the crow *i* at iteration iter.

Case ii: The crow *i* will be fooled by crow *j*, when it realizes that crow *i* is following it and it will fly into a random location in solution space.

The mathematical representations of the two cases are represented as follows in Eq. 9.

$$Y^{i,iter+1} = \begin{cases} Y^i + P_i \times fl^{iter} \times \left(m^{j,iter} - Y^{i,iter}\right) P_j \geq AP \\ a\ random\ position\ otherwise \end{cases} \tag{9}$$

Step 5: Estimate the fitness value of the updated new position

The estimation of the fitness function is done for each new position of the crow.

Step 6: Evaluate the feasibility of updated new positions

The new position of all crow is evaluated; updation of position is done, if the current position is feasible, otherwise, it will stay in the old position.

Step 7: Updation of crow memory

The crow updates its memory in a new position when the fitness function value of the updated new position is better than the memorized position. The expression for crow memory updation is represented in Eq. 10.

$$m^{i,iter+1} = \begin{cases} Y^{i,iter+1} \times f^{i,iter+1}\ is\ better\ than\ f\left(m^{i,iter}\right) \\ m^{i,iter}\ otherwise \end{cases} \tag{10}$$

Step 8: Evaluation of the termination criterion

The steps 4–7 are repeated for a maximum number of iterations and the best position of the memory in terms of the objective function is considered as the solution of the optimization problem.

The CS optimization was used for the determination of cluster centres for the FCM algorithm. The CS algorithm-generated cluster centres are used for the clustering process instead of the randomly initialized cluster centres by the classical FCM. Once the initialization of the cluster centre is provided, for every iteration based on the fuzzy membership function, cluster centres are recalculated. The goal is to minimize the objective function by choosing an ideal combination of cluster centres. The FCM algorithm terminates when there is no more change in the cluster centres.

2.5 FCM-Crow Optimization Segmentation Algorithm

In this research work, the crow search optimization was employed in the FCM algorithm for optimum cluster centres value selection. The crows' searches in the environment and search space of optimization are the environment and are depicted in Fig. 1.

The biological characteristics of crow based on their habit of hiding surplus food in certain places and recovering the food upon need were used for the formulation of the optimization algorithm. The crows will try to find the place, where the food was stored by a crow. The crow is an intelligent bird and the brain size is relatively large when compared with the body size. The crow will try to fool another crow following it to discover the hidden food.

The Fig. 1a shows the case of flight length greater than one and Fig. 1b shows the case of flight length less than one. In the environment, each position of the crow represents the feasible solution. The objective function (fitness value) represents the quality of food source and the global solution represents the best food source of the environment. The two cases of the crow search optimization, $fl < 1$ and $fl > 1$ are depicted above in Fig. 1. The crow i can go to any position in the dashed line.

The four modules of FCM—CS algorithm are as follows; (i) Initialization stage, (ii) Fitness function evaluation, (iii) Memory updation stage, (iv) Segmentation stage. In the first stage, the parameters of the crow search optimization and FCM algorithm are initialized. The parameters of the crow optimization are flight length (fl) and Awareness Probability (AP).

In the second stage, the fitness value and the new location of the crows are determined. The crow memory updation takes place in the third stage and in the fourth stage, cluster centroids are initialized by crow search optimization for FCM segmentation.

The cluster centre is expressed as follows in Eq. 11.

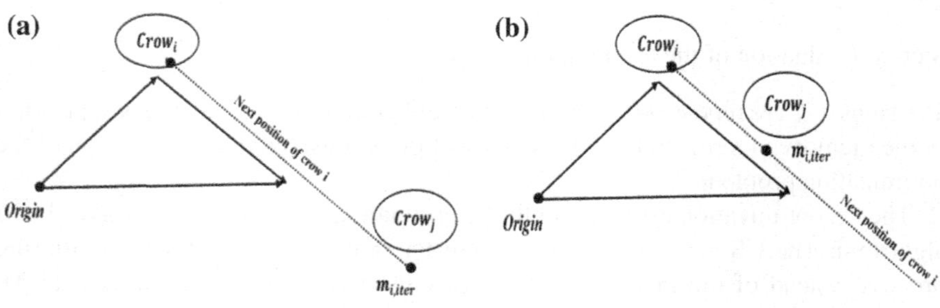

Fig. 1 Crow's position corresponding to two cases of flight length

$$V_i = \frac{\sum_{i=1}^{N} U_{ij}^{f} y_i}{\sum_{i=1}^{N} U_{ij}^{f}} \tag{11}$$

where y_i represents the position of the crow.

The membership function is expressed as follows in Eq. 12.

$$U_{ij} = \frac{1}{\sum_{c=1}^{C}\left(\frac{\|y_i - v_j\|^2}{\|y_j - v_c\|^2}\right)^{\frac{1}{f-1}}} \tag{12}$$

The global solution was obtained, when the predefined iterations are over. The new position of the crow is determined and stages ii, iii and iv are repeated until the objective function is minimized with global minima.

2.6 Tuning of Parameters for Optimization Algorithms

In this work, the FCM was coupled with optimization algorithms like Cuckoo, Simulated annealing, Firefly, Artificial Bee Colony and Crow for the segmentation of grayscale CT/MR images. The population size and the maximum number of iterations are common parameters to be specified for all optimization algorithms. Apart from the above-said parameters, based on the unique characteristics of each algorithm, tuning of other parameters is needed and sometimes it is a time-consuming process.

In crow search algorithm, only two parameters have to be tuned; flight length and awareness probability. The CS optimization has memory in which the good solutions are memorized. During each iteration, each crow randomly selects one of the flock and moves towards its food hiding place. The tuning of parameters plays a prominent role in the optimization algorithms and less number of parameters will be ideal for the user. Less number of control parameters makes a metaheuristic algorithm an efficient one for the arrival of global minima. Few parameters make an algorithm user-friendly with less complexity and improve the generosity. The SA algorithm requires the proper setting of temperature since it influences the convergence rate. The low-temperature setting will cause the algorithm to stuck in local optima and high-temperature setting will pass the optima several times, in the proposed FCM-SA segmentation initial temperature value of 500 with 250 iterations were used. For many optimization problems, it has been proved that Particle Swarm Optimization (PSO) can outperform the Genetic Algorithm (GA).

The cuckoo optimization algorithm was found to be efficient than PSO and GA in the case of testing on benchmark test function. In the cuckoo optimization algorithm, random step size is derived from the levy distribution and it can be changed based on the application. The step size (α) in cuckoo algorithm plays a vital role since a low value will limit the search process and high value will cause the new cuckoo egg position outside the boundary, thereby convergence rate is poor. The probability of an abandoned nest (Pa) is set to 0.25 in many applications and in the proposed FCM-Cuckoo segmentation; Pa = 0.25, $\lambda = 1.5$ (levy distribution coefficient) and $\alpha = 0.25$ are taken.

The convergence rate of the ABC algorithm is poor and often in many applications, it gets stuck in local minima. The pheromone evaporation parameter (ρ) in local and global update ranges from (0, 1) and the stagnation limit is a positive integer (L). The number of employed bees (Ne) is 10–30% of the total population. In the proposed FCM-ACO segmentation; $\rho = 1$, L = 10 and Ne = 30% of population are used. The parameters of the Firefly algorithm are attractive coefficient (β_0), the light absorption coefficient (γ) and the step parameter (α). The firefly algorithm also has the issue of getting into local minima and in the proposed FCM-Firefly segmentation; $\beta_0 = 1$, $\gamma = 1$ and $\alpha = 0.25$ are used.

The CS optimization was found to be effective for a wide number of engineering problems like three bar truss, pressure vessel, tension/compression string, welded beam, gear train and Belleville spring [33]. In CS optimization, flight length and awareness probability are the two parameters to be tuned and for all the medical datasets, fl = 2 and AP = 0.1 give satisfactory results. The population (N = 10) and the number of iteration (iter = 50) are set to a common value for Cuckoo, Firefly, ABC, SA, and CS-based FCM segmentation algorithms.

In this research work, the random initialization of cluster centroids was replaced by optimization algorithms. The choice of cluster number was made by the cluster validation metrics.

3 Results and Discussion

The FCM-CS optimization approach was proposed in this work for the segmentation of anomalies on abdomen CT images. Some of the widely used optimization algorithms in the FCM clustering are Artificial Bee colony, Simulated annealing, Firefly and Cuckoo and hence the proposed FCM-CS optimization work was validated by comparing with the results of above said optimization techniques. The Cluster Validity Indexes (CVIs) plays a vital role in the evaluation of centroid-based clustering segmentation algorithms.

The terminologies used for describing CVIs are expressed below.

N: Number of data objects for clustering
m: fuzziness measure
y_i: the ith data item (, $1 \leq i \leq N$)
K: the count of clusters
C_k: the Kth $1 \leq k \leq K$ cluster
$|C_k|$: the number of data objects in the Kth cluster
v_k: the centroid of the kth cluster.
$\hat{v}$: the centroid of all data objects

$\|x - y\|$: Euclidean distance measure between data objects
F_{ik}: fuzzy membership degree of y_i belonging to C_k

The expression for centroid is expressed in Eq. 13 as follows

$$\hat{v} = \frac{1}{N}\sum_{i=1}^{N} \chi_i \tag{13}$$

The partition coefficient (PC) and partition entropy (PE) are widely used indexes for the evaluation of clustering validity [23, 24]. The PC determines the mean strength of belongingness of data and PE is the entropy measurement determined by taking the logarithm of the PC. The PC and PE are based on the membership values F_{ik} of the fuzzy partition of dataset [23]. The best and optimal partition is achieved by maximizing PC (or minimizing PE). The PC and PE focus only on the compactness and variations with cluster and they are expressed as follows in Eqs. 14 and 15.

$$PC(K) = \frac{1}{N}\sum_{k=1}^{K}\sum_{i=1}^{N} F_{ik}^2 \tag{14}$$

$$PE(K) = \frac{1}{N}\sum_{k=1}^{K}\sum_{i=1}^{N} F_{ik}\log_2(F_{ik}) \tag{15}$$

The Calinski-Harabasz Index (CHI) [34] represents the spatial relationship on pixels through agglomerative and the divisive method; the maximum value indicates the best clustering result and is expressed in Eq. 16.

$$CHI(K) = \frac{W_{1K}}{K-1}\bigg/\frac{W_{2K}}{N-K} \tag{16}$$

where

$$W_{1K} = \sum_{k=1}^{K} |C_k| \, \|v_k - \hat{v}\|$$

$$W_{2K} = \sum_{k=1}^{K}\sum_{\chi_i \epsilon C_k} \|y_i - v_k\|^2$$

The Davies Bouldin Index (DBI) [25] is a function of the ratio of sum within-cluster scatter to between cluster separations; the minimum value indicates the best clustering result and is expressed in Eq. 17.

$$DBI(K) = \frac{1}{K}\sum_{k=1 \; j\neq k}^{k\max} \frac{S_j + S_k}{\|v_j - v_k\|^2} \tag{17}$$

where Sj is the average Euclidean distance of the vectors in class j to the centroid of class k.

The Xie and Beni index (XBI) is defined as the ratio of the fuzzy within-cluster sum of squared distances to the product of the number of elements and the minimum distance between the cluster separations. The minimum value of XBI indicates the best clustering result [26] and is expressed in Eq. 18.

$$XBI(k) = \frac{\sum_{k=1}^{K} \sum_{i=1}^{N} F_{ik}^{K} \| y_i - v_k \|^2}{N.\min_{i \neq j} \| y_i - v_j \|^2} \tag{18}$$

The Fukuyama and Sugeno Index (FSI) is based on the compactness and separation. The first term in the equation represents the compactness measure and the second term represents the degree of separation. The minimum value of FSI indicates the best clustering result [26] and is expressed in Eq. 19.

$$FSI(k) = \sum_{k=1}^{K} \sum_{i=1}^{N} F_{ik}^{m} \| y_i - v_k \|^2 - \sum_{k=1}^{K} \sum_{i=1}^{N} F_{ik}^{m} \| v_k - \hat{v} \|^2 \tag{19}$$

The SC Index (SCI) determines the compactness ratio by combining the two functions, $SF_1(K)$ and $SF_2(K)$; the maximum value indicates the best clustering result [27] and is expressed in Eq. 20.

$$SCI(K) = SF_1(k) - SF_2(k) \tag{20}$$

$$SF_1(k) = \frac{\frac{1}{k} \sum_{k=1}^{K} \| v_k - \hat{v} \|}{\sum_{k=1}^{K} \left(\sum_{i=1}^{N} F_{ik}^{m} \| y_i - v_k \|^2 \Big/ \sum_{i=1}^{N} F_{ik} \right)} \tag{21}$$

$$SF_2(k) = \frac{\sum_{k=1}^{k-1} \sum_{j=k+1}^{K} \left(\sum_{i=1}^{N} \left(\min(F_{ik}, F_{ij})^2 \right) \Big/ n_{ij} \right)}{\left(\sum_{i=1}^{N} \max_{1 \le k \le K} F_{ik}^2 \right) \Big/ \left(\sum_{i=1}^{N} \max_{1 \le k \le K} F_{ik} \right)} \tag{22}$$

where

$$n_{ij} = \sum_{i=1}^{N} \min(F_{ik}, F_{ij})$$

The CS Index (CSI) estimates the ratio of compactness separation of the data objects and the centroids; the minimum value indicates the best clustering result [27] and is expressed in Eq. 23.

$$CSI(K) = \frac{\sum_{k=1}^{K}\left(\frac{1}{|C_k|}\left(\sum_{\chi_j \in C_k} \max_{\chi_j \in C_k} \left\| y_j - y_i \right\|\right)\right)}{\sum_{j=1}^{K} \min_{i \neq j} \left\| V_i - V_j \right\|} \quad (23)$$

The Partition Coefficient and Exponential Separation (PCAES) Index [27] explores the normalized partition coefficient and exponential separation for each cluster and the maximum value indicates the best clustering result and is expressed in Eq. 24.

$$PCAES(K) = \sum_{k=1}^{K}\sum_{i=1}^{N} \frac{F_{ik}^2}{F_M} - \exp\left(\frac{-\min_{i \neq j} \|v_k - v_h\|^2}{\beta_T}\right) \quad (24)$$

where

$$F_M = \min_{1 \leq k \leq K} \left\{ \sum_{i=1}^{N} F_{ik}^2 \right\}$$

$$\beta_T = \frac{1}{K} \sum_{k=1}^{K} \left\| v_k - \hat{v} \right\|^2$$

The Pakhira-Bandyopadhyay-Maulik (PBMF) Index is derived from the fuzzy variant of PBM Index. It is based on the compactness of clusters and maximum separation between clusters. On varying the number of clusters, the maximum value of PBMF indicates the best cluster value [35] and is expressed in Eq. 25.

$$PBMF(K) = \frac{\max_{j \neq k} \left\| v_j - v_k \right\| \times \sum_{i=1}^{N} F_{i1} \|y_i - v_1\|}{\sum_{k=1}^{K} \sum_{i=1}^{N} F_{ik}^m \|y_i - v_k\|} \quad (25)$$

The WL Index (WLI) is a novel metric for the validation of clusters and it is based on the fuzzy weighted distance and fuzzy cardinality of the clusters. The minimum value of WLI indicates the best clustering result [27] and is expressed in Eq. 26.

$$WLI = \sum_{k=1}^{K} \left[\frac{\sum_{i=1}^{N} F_{ik}^2 \|y_i - v_k\|^2}{\sum_{i=1}^{N} F_{ik}} \right] \quad (26)$$

The FCM with various optimization algorithms is executed for 12 runs per selected slice in each dataset with the cluster number varied from 3 to 6. The average of objective function value for the optimization algorithms coupled with the FCM for the selective slice from each dataset is tabulated in Table 1.

Table 1 Objective function value of optimization algorithms

Dataset ID	Cluster value	FCM-ABC	FCM-Cuckoo	FCM-Firefly	FCM-Crow	FCM-SA
1	3	3.6309	3.5383	4.0457	3.5340	3.5346
	4	2.0739	2.0699	2.4934	2.0059	2.0245
	5	1.5778	1.5254	2.2278	1.5171	1.5259
	6	1.1276	0.9927	1.3895	1.0173	1.0355
2	3	3.4905	3.4724	4.7723	3.4748	3.4693
	4	2.1827	2.1534	2.3199	2.1205	2.1404
	5	1.2865	1.4160	1.4195	1.4467	1.4612
	6	1.0383	0.9394	1.1479	0.9931	1.0467
3	3	3.3065	3.2868	3.9216	3.2849	3.2851
	4	2.0327	1.9261	1.9346	1.9281	2.1421
	5	1.1589	1.1707	1.4036	1.0985	1.4292
	6	1.0558	0.8586	1.0664	0.8422	0.9732
4	3	3.3352	3.2790	3.4166	3.2757	3.2847
	4	1.6983	1.6314	2.1998	1.6117	1.6668
	5	1.0495	0.9754	1.1336	1.0210	1.0990
	6	0.7034	0.6628	0.8419	0.6453	0.6427
5	3	6.9542	6.8998	7.5332	6.9000	6.8981
	4	4.7670	4.6337	4.6325	4.6766	4.6798
	5	3.2224	2.8412	3.2972	2.8135	3.3307
	6	2.0860	1.7209	2.9164	1.7109	1.7250
6	3	3.4837	3.4411	4.1736	3.4642	3.4410
	4	2.3319	2.2337	2.8066	2.3666	2.4336
	5	1.5357	1.4278	2.1472	1.5468	1.5820
	6	1.2192	1.0251	1.4459	1.0391	1.0556
7	3	4.0891	4.0360	4.6214	4.0312	4.0310
	4	2.2407	2.1730	2.6480	2.3238	2.3051
	5	1.4776	1.4046	1.9467	1.4730	1.9665
	6	1.0398	1.0665	1.1384	1.0139	1.2278
8	3	6.7511	6.7300	6.9227	6.7207	6.7179
	4	4.2555	4.0053	4.9812	3.9644	3.9674
	5	2.8004	2.5351	2.7705	2.6393	2.8946
	6	1.7121	1.6909	1.8181	1.6733	1.8851
9	3	2.8345	2.8137	2.9093	2.8142	2.8145
	4	1.7840	1.7000	2.2967	1.7102	1.7093
	5	1.2982	1.1714	1.3539	1.1966	1.2694

(continued)

Table 1 (continued)

Dataset ID	Cluster value	FCM-ABC	FCM-Cuckoo	FCM-Firefly	FCM-Crow	FCM-SA
	6	0.7744	0.8022	0.8945	0.7466	0.8940
10	3	3.2133	3.1777	4.3922	3.1778	3.1757
	4	2.0759	1.9054	2.3913	1.8881	2.1681
	5	1.0712	1.0738	1.7428	1.0277	1.1024
	6	0.9547	0.8733	1.0798	0.8627	0.9806

Table 1 shows that the Crow optimization algorithm produces good results and in many cases, the objective function was minimized when compared with the other optimization algorithms.

The FCM-CS optimization results are taken further for analysis and for a typical slice in dataset ID1, the clustering result performance validation are depicted in Figs. 2, 3, 4, and 5. The tabulation of cluster validity functions is represented as CVI +/−. The metrics with "+" sign value should be high and "−" value should be low for efficient clustering result. Figures 6 and 7 depict the FCM-CS optimization results for the datasets (ID1–ID10). The choice of the cluster number is a vital one and for deciding that, Cluster Validity metrics are used.

The cluster validity functions are determined for FCM with various optimization algorithms and cluster number is decided from their respective values. The CVI's for various cluster values in the first run corresponding to selective slice in each dataset are represented in Tables 2 and 3. For the dataset ID1, the CVI's PC^+, PE^-, CHI^+, DBI^-, FSI^-, CSI^-, $PBMF^+$, XBI^- favours a cluster number of 3, CVI's SCI^+, WLI^-, $PCAES^+$ index favours a cluster number of 6. For each input image with various cluster number ranging from 3, 4, 5, 6; 12 runs of execution are made for FCM-CS optimization algorithm and CVI's are determined.

Tables 2 and 3 below depicts the counts of cluster number given by a metric during the 12 runs of a selective slice from each dataset.

Tables 4 and 5 were deduced from Tables 2 and 3 by choosing the maximum cluster number for each CVI result corresponding to 12 runs.

Table 6 below depicts the actual cluster number specified by a metric based on the maximum count during execution. The second row of Table 6 represents the possible cluster values decided by the CVI and third row represents the actual cluster value chosen.

The CVI's value helps to choose the appropriate cluster number for an application. Moreover, since the optimization algorithm is used; a definite number of program executions have to be made for validation. The judgment cannot be done with a few CVI's; hence in this research work, nearly 11 CVI's are used.

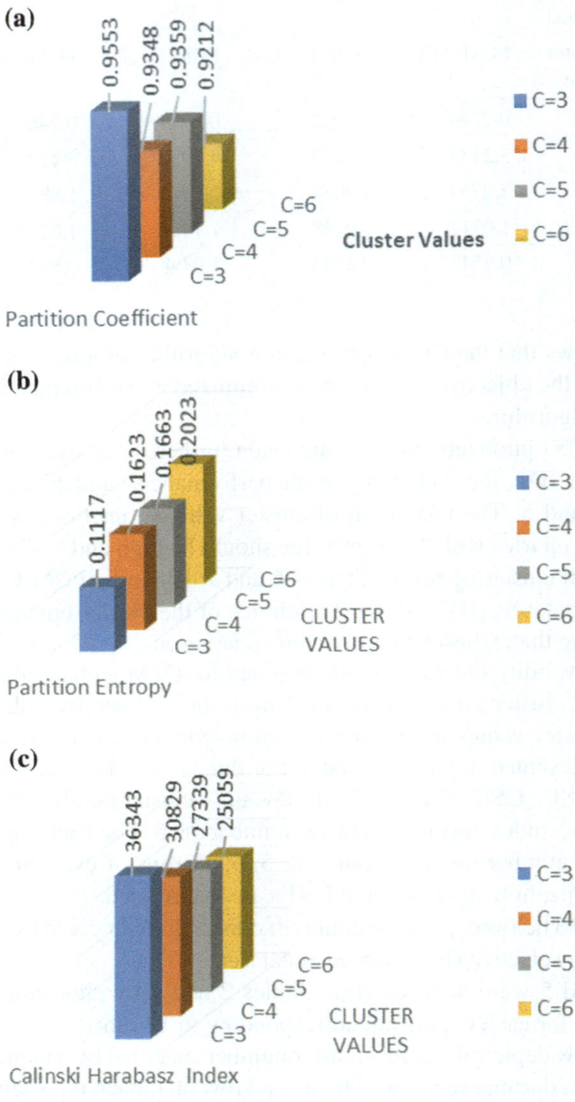

Fig. 2 Cluster validity index plot for the FCM-CSO in the first run for the image from data set ID1; **a** partition coefficient, **b** partition entropy, **c** Calinski Harabasz index

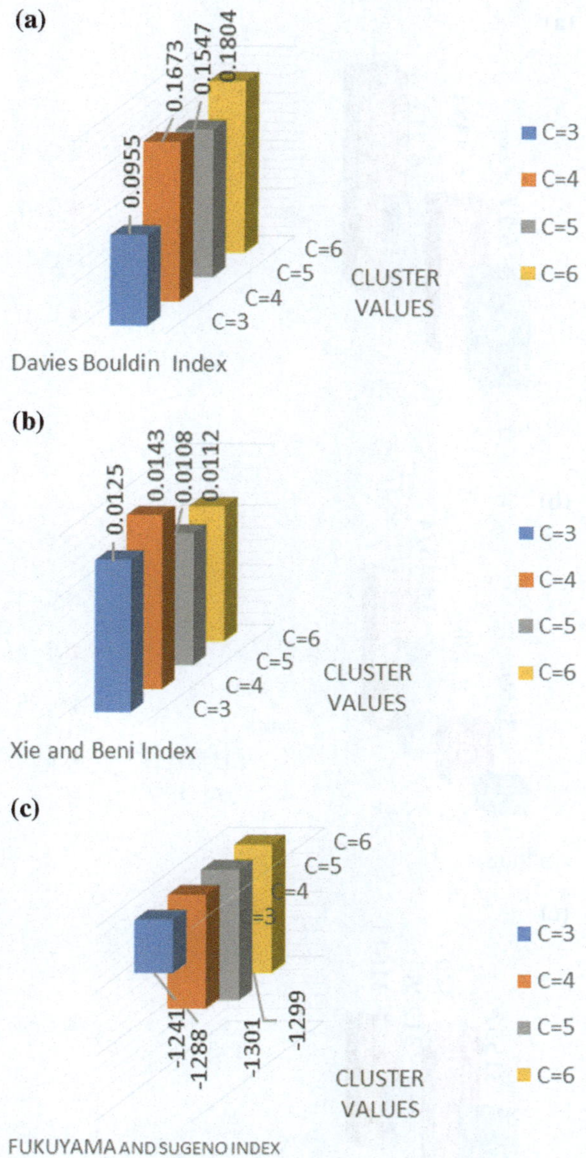

Fig. 3 Cluster validity index plot for the FCM-CSO in the first run for the image from data set ID1; **a** David Bouldin index, **b** Xie and Beni index, **c** Fukuyama and Sugeno index

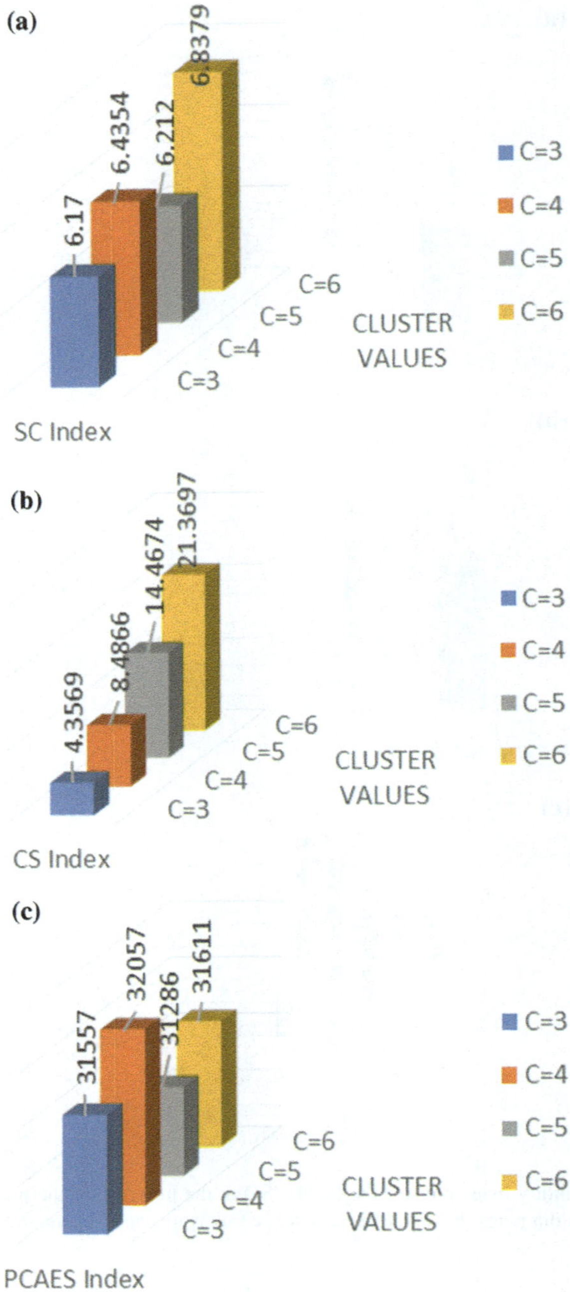

Fig. 4 Cluster validity index plot for the FCM-CSO in the first run for the image from dataset ID1; **a** SC index, **b** CS index, **c** PCAES Index

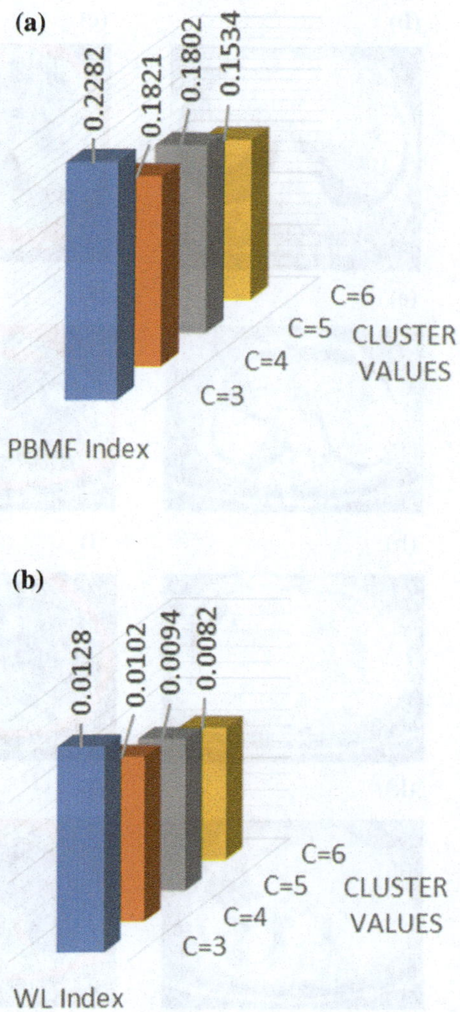

Fig. 5 Cluster validity index plot for the FCM-CSO in the first run for the image from data set ID1; **a** PBMF index, **b** WL index

4 Hardware Implementation

The hardware implementation of FCM-CS segmentation algorithm was done in Raspberry Pi B+ model embedded board. Initially, for the comparative analysis, the algorithms are developed in Matlab 2015a. The specifications of the Raspberry Pi B+ model embedded board are as follows; ARM11 processor, 512 GB RAMS, 700 MHz, 2 USB and an Ethernet port.

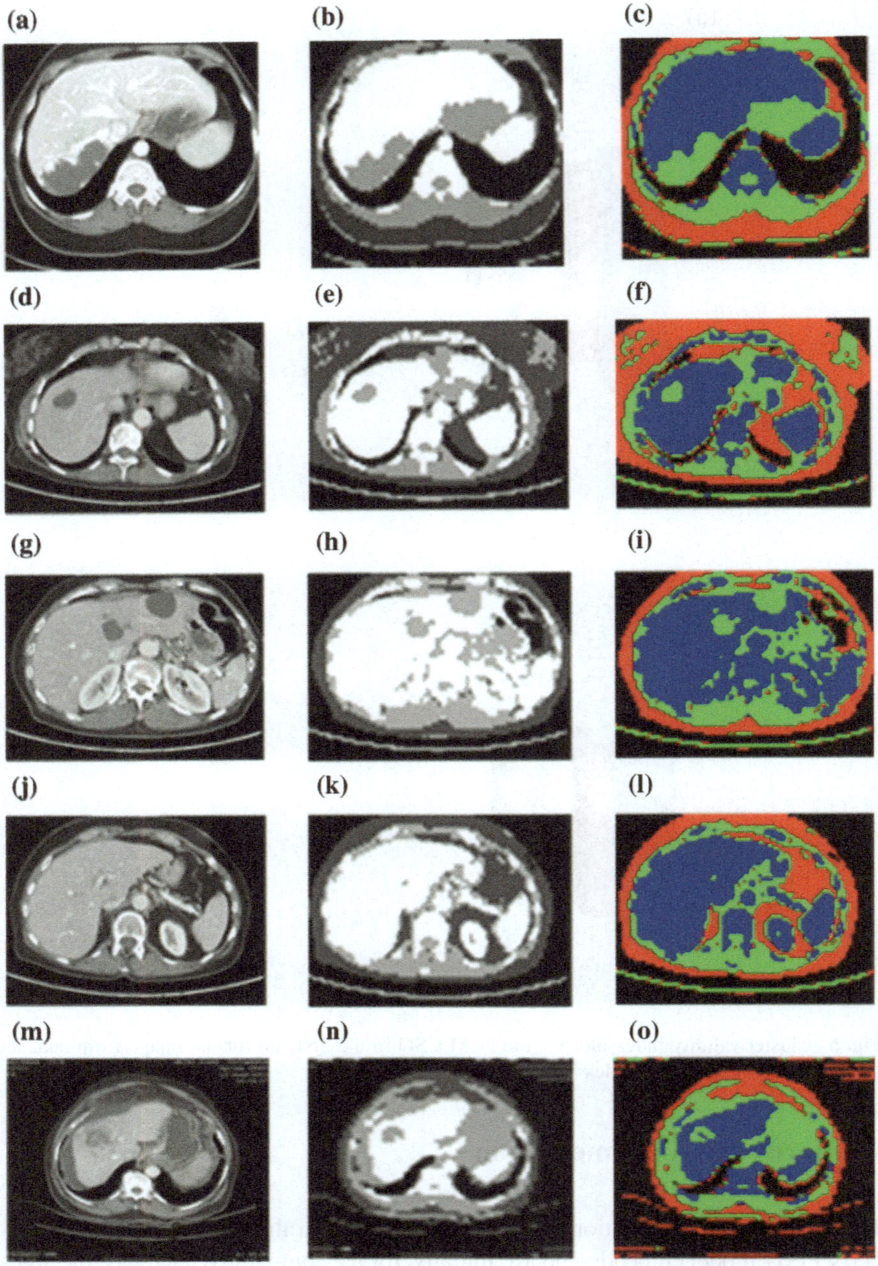

Fig. 6 The first column depicts input CT images corresponding to data set ID 1, 2, 3, 4, 5; second and third column depicts the FCM-CS optimization algorithm result

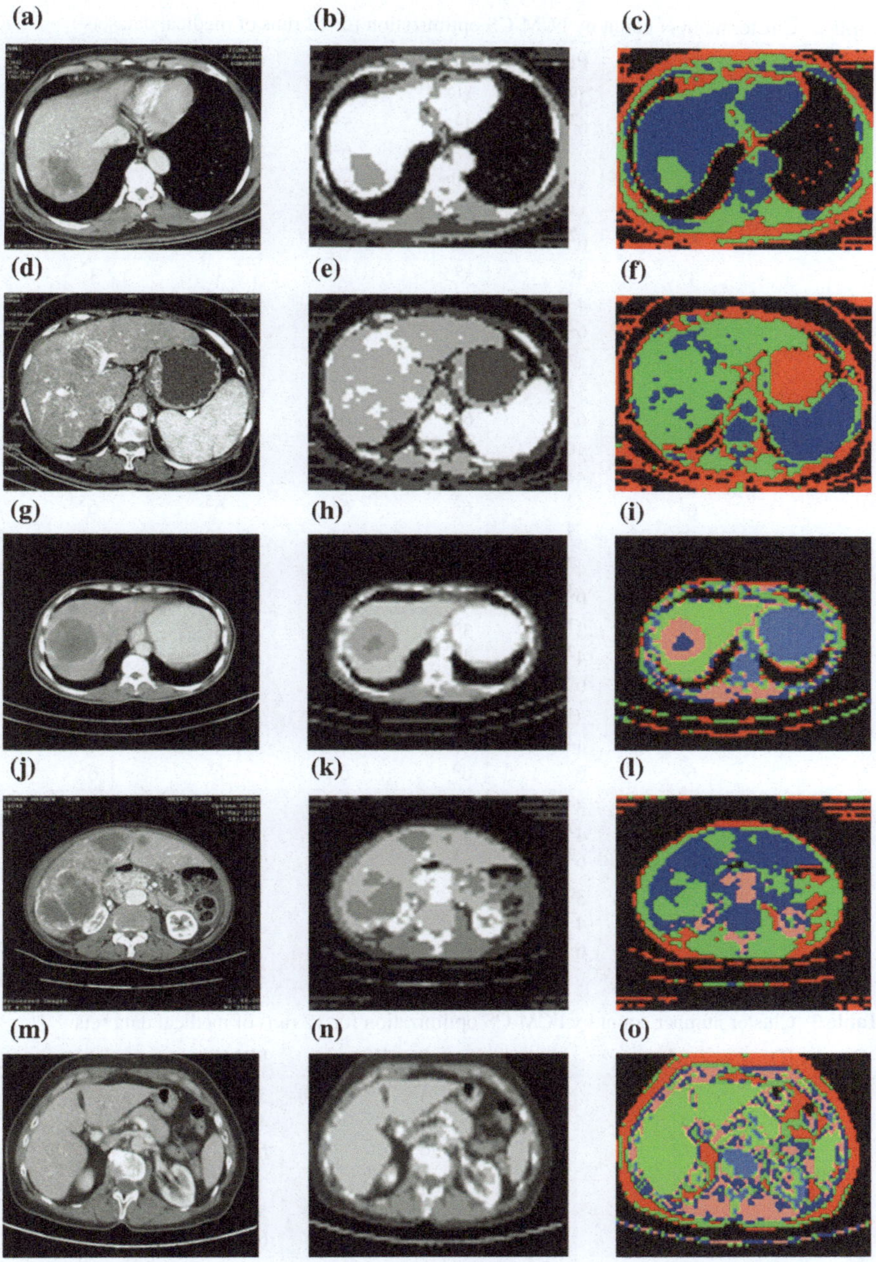

Fig. 7 The first column depicts input CT images corresponding to data set ID 6, 7, 8, 9, 10; second and third column depicts the FCM-CS optimization algorithm result

Table 2 Cluster number count by FCM-CS optimization for 12 runs of medical datasets

Dataset ID	PC⁺	PE⁻	CHI⁺	DBI⁻	XBI⁻	FSI⁻
1	3^{9} 6^{3}	3^{9} 6^{3}	3^{10} 4^{2}	3^{9} 5^{1} 6^{2}	3^{5} 5^{3} 6^{4}	3^{9} 6^{3}
2	3^{9} 5^{1} 6^{2}	3^{9} 5^{1} 6^{2}	3^{10} 6^{2}	3^{9} 5^{1} 6^{2}	3^{3} 5^{2} 6^{7}	3^{2} 5^{3} 6^{7}
3	3^{8} 4^{2} 6^{2}	3^{8} 4^{2} 6^{2}	3^{9} 4^{2} 6^{1}	3^{5} 4^{5} 6^{2}	3^{3} 4^{3} 6^{6}	3^{2} 4^{3} 6^{7}
4	3^{2} 5^{2} 6^{8}	3^{2} 5^{3} 6^{7}	3^{6} 5^{3} 6^{3}	3^{2} 5^{3} 6^{7}	3^{4} 5^{2} 6^{6}	3^{4} 5^{2} 6^{6}
5	3^{8} 5^{2} 6^{2}	3^{8} 5^{2} 6^{2}	3^{7} 5^{3} 6^{2}	3^{4} 5^{6} 6^{2}	3^{4} 5^{3} 6^{5}	3^{3} 5^{4} 6^{5}
6	3^{7} 4^{2} 6^{3}	3^{7} 4^{2} 6^{3}	3^{8} 4^{2} 6^{2}	3^{3} 4^{2} 6^{7}	3^{3} 4^{3} 6^{6}	3^{3} 4^{2} 6^{7}
7	3^{8} 4^{2} 6^{2}	3^{7} 4^{2} 6^{3}	3^{8} 4^{2} 6^{2}	3^{8} 4^{2} 6^{2}	3^{2} 4^{2} 6^{8}	3^{2} 4^{2} 6^{8}
8	3^{3} 4^{2} 6^{7}	3^{4} 4^{2} 6^{6}	3^{4} 4^{2} 6^{6}	3^{2} 4^{3} 6^{7}	3^{3} 4^{3} 6^{6}	3^{2} 4^{3} 6^{7}
9	3^{3} 4^{6} 6^{3}	3^{3} 4^{7} 6^{2}	3^{6} 4^{3} 6^{3}	3^{3} 4^{6} 6^{3}	3^{3} 4^{3} 6^{6}	3^{2} 4^{3} 6^{7}
10	3^{6} 4^{2} 6^{4}	3^{6} 4^{2} 6^{7}	3^{5} 4^{3} 6^{4}	3^{2} 4^{8} 6^{2}	3^{3} 4^{3} 6^{6}	3^{3} 4^{4} 6^{5}

Table 3 Cluster number count by FCM-CS optimization for 12 runs of medical data sets

Dataset ID	SCI⁺	CSI⁻	PCAES⁺	PBMF⁺	WLI⁻
1	3^{4} 4^{1} 6^{7}	3^{9} 4^{1} 6^{2}	3^{2} 4^{3} 6^{7}	3^{9} 4^{1} 6^{2}	3^{3} 5^{4} 6^{5}
2	3^{1} 5^{3} 6^{8}	3^{8} 5^{1} 6^{3}	3^{6} 5^{3} 6^{3}	3^{6} 5^{5} 6^{1}	3^{4} 5^{3} 6^{5}
3	3^{3} 4^{3} 6^{6}	3^{7} 4^{2} 6^{3}	3^{7} 4^{2} 6^{3}	3^{3} 4^{3} 6^{6}	3^{6} 4^{3} 6^{6}

(continued)

Table 3 (continued)

Dataset ID	SCI$^+$	CSI$^-$	PCAES$^+$	PBMF$^+$	WLI$^-$
4	3^3 5^4 6^5	3^7 5^2 6^3	3^3 5^2 6^7	3^3 5^3 6^6	3^3 5^3 6^6
5	3^4 5^3 6^5	3^6 5^3 6^3	3^5 5^4 6^3	3^2 5^7 6^3	3^3 5^4 6^5
6	3^3 4^3 6^6	3^7 4^2 6^3	3^8 4^2 6^2	3^7 4^3 6^2	3^3 4^3 6^6
7	3^3 4^2 6^7	3^8 4^2 6^2	3^3 4^3 6^6	3^8 4^2 6^2	3^3 4^2 6^7
8	3^2 4^3 6^7	3^6 4^3 6^3	3^3 4^6 6^3	3^6 4^3 6^3	3^3 4^3 6^6
9	3^4 4^2 6^6	3^7 4^3 6^2	3^3 4^7 6^2	3^3 4^6 6^3	3^2 4^3 6^7
10	3^3 4^3 5^6	3^7 4^3 6^2	3^2 4^8 6^2	3^2 4^8 6^2	3^3 4^3 6^6

Table 4 Cluster number decision by cluster validity metrics

Dataset ID	PC$^+$	PE$^-$	CHI$^+$	DBI$^-$	XBI$^-$	FSI$^-$
1	3	3	3	3	3	3
2	3	3	3	3	6	6
3	3	3	3	3	6	6
4	6	6	3	6	6	6
5	3	3	3	5	6	6
6	3	3	3	6	6	6
7	3	3	3	3	6	6
8	6	6	6	6	6	6
9	4	4	3	4	6	6
10	3	6	3	4	6	6

For hardware implementation, the FCM-CS segmentation algorithm in Matlab code was converted into java and incorporated in the hardware. The prediction-based compression algorithm was also incorporated in the hardware for data transfer through cloud network. Prior to segmentation and compression, the preprocessing was performed. The Hardware implementation of the FCM-CS segmentation algorithm is depicted in Fig. 8.

Table 5 Cluster number decision by cluster validity metrics

Dataset ID	SCI$^+$	CSI$^-$	PCAES$^+$	PBMF$^+$	WLI$^-$
1	3	3	6	3	6
2	6	3	3	3	6
3	6	3	3	6	6
4	6	3	6	6	6
5	6	3	3	5	6
6	6	3	3	3	6
7	6	3	6	3	6
8	6	3	4	3	6
9	6	3	4	4	6
10	5	3	4	4	6

Table 6 Final cluster number decision by cluster validity metrics

Dataset ID	1	2	3	4	5	6	7	8	9	10
Possible cluster values	3^9 6^2	3^7 6^4	3^6 6^5	3^9 6^2	3^5 6^4 5^2	3^6 6^5	3^6 6^5	3^2 6^8	3^2 4^5 6^4	3^3 4^3 6^4 5^1
Chosen cluster value	3	3	3	3	3	3	3	6	4	6

5 Conclusion

This paper proposes an FCM-CS optimization algorithm for the segmentation of abdomen CT images. The cluster validity functions are used for the optimal selection of cluster number. The FCM-CS optimization algorithm generates robust results. The proposed CS optimization results were compared with the widely used optimization algorithms in clustering like ABC, SA, Firefly, and Cuckoo techniques. The CS is a population-based algorithm with two tunable parameters (flight length and awareness probability). The awareness probability is to control the diversity of algorithm; few parameters have to be tuned when compared with the other widely used optimization algorithms. The hardware implementation of FCM-CS algorithm was implemented by Raspberry Pi B+ model embedded board. Our future work will be the improvement in performance of CS optimization by replacing the random variable with chaotic functions. The aim of this research work is not to state that FCM-CS optimization could beat conventionally using optimization techniques in clustering, but to depict that the CS optimization can be considered as an efficient alternative for other classical optimization techniques in the FCM segmentation process.

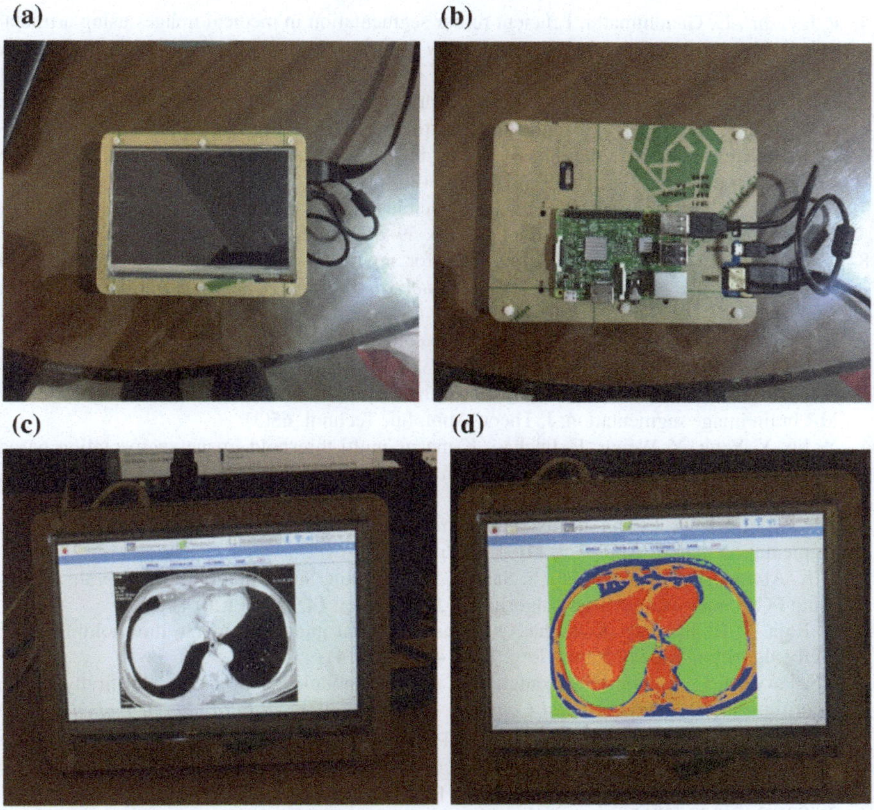

Fig. 8 Hardware implementation of the FCM-CS segmentation algorithm

Acknowledgements The authors would like to acknowledge the support provided by Nanyang Technological University under NTU Ref: RCA-17/334 for providing the medical images and supporting us in the preparation of the manuscript.

References

1. S.N. Kumar, A. Lenin Fred, S. Muthukumar, H. Ajay Kumar, P. Sebastian Varghese, A voyage on medical image segmentation algorithms. Biomed. Res. 1–13 (2017)
2. M. Shokouhifar, S. Abkenar, An artificial bee colony optimization for MRI fuzzy segmentation of brain tissue, in *2011 International Conference on Management and Artificial Intelligence, IPEDR*, vol. 6. (IACSIT Press, Bali, Indonesia, 2011)
3. A. Alrosan, N. Norwawi, W. Ismail, W. Alomoush, Artificial bee colony based fuzzy clustering algorithms for MRI image segmentation, in *Proceedings of the International Conference on Advances in Computer Science and Electronics Engineering—CSEE*, pp. 225–228

4. P. Jayashri, D. Gandhimathi, Efficient tumor segmentation in medical images using artificial bee colony optimization algorithm and fuzzy c-means clustering. Int. J. Adv. Res. Comput. Sci. Manage. Stud. **2**(8), 98–101 (2014)
5. D. Karaboga, B. Gorkemli, C. Ozturk, N. Karaboga, A comprehensive survey: artificial bee colony (ABC) algorithm and applications. Artif. Intell. Rev. **1**, 1–37 (2014)
6. C. Ozturk, E. Hancer, D. Karaboga, A novel binary artificial bee colony algorithm based on genetic operators. Inf. Sci. **10**(297), 154–170 (2015)
7. C. Ozturk, E. Hancer, D. Karaboga, Dynamic clustering with improved binary artificial bee colony algorithm. Appl. Soft Comput. **31**(28), 69–80 (2015)
8. X. Song, G. Li, L. Luo, Optimizing FCM for segmentation of image using Gbest-guided artificial bee colony algorithm, in *2015 11th International Conference on Natural Computation (ICNC)*, IEEE, pp. 764–768, 15 Aug 2015
9. A. Bose, K. Mali, Fuzzy-based artificial bee colony optimization for gray image segmentation. SIViP **10**(6), 1089–1096 (2016)
10. P. Krishnan, P. Ramamoorthy, Fuzzy clustering based ant colony optimization algorithm for MR brain image segmentation. J. Theor. Appl. Inf. Technol. **65**(3)
11. Y. Jie, Y. Yang, Y. Weiyu, F. Jiuchao, K-means multi-threshold image segmentation based on firefly algorithm, in *3rd International Conference on Multimedia Technology (ICMT-13)*, (Atlantis Press, 2013), pp. 134–142
12. M.K. Alsmadi, A hybrid firefly algorithm with fuzzy-c mean algorithm for MRI brain segmentation. Am. J. Appl. Sci. **11**(9), 1676–1691 (2014)
13. W.K. Alomoush, S.N. Abdullah, S. Sahran, R.I. Hussain, Segmentation of MRI brain images using FCM improved by firefly algorithms. J. Appl. Sci. **14**, 66–71 (2014)
14. N. Raja, V. Rajinikanth, K. Latha, Otsu based optimal multilevel image thresholding using firefly algorithm. Model. Simul. Eng. **1**(2014), 37 (2014)
15. S.S. Parveen, C. Kavitha, Segmentation of CT lung nodules using FCM with firefly search algorithm, in *2015 International Conference on Innovations in Information, Embedded and Communication Systems (ICIIECS), IEEE*, pp. 1–6, 19 Mar 2015
16. M. Haghdoost, A.H. Abadi, S.J. Abedini, A new method of fuzzy clustering by using the combination of the firefly algorithm and the particle swarm optimization algorithm. WALIA J. **31**(S3), 245–252 (2015)
17. Q. Yang, Z.Q. Wang, Y. Xu, Fuzzy c-means image segmentation algorithm based on chaotic simulated annealing. Appl. Mech. Mater. **624**, 536–539 (2014)
18. J.B. Jona, N. Nagaveni, Ant-cuckoo colony optimization for feature selection in digital mammogram. Pak. J. Biol. Sci. PJBS **17**(2), 266–271 (2014)
19. B.F. Nia, A. Branch, A hybrid automatic segmentation method of blood vessels in CT scan images of liver. J. Med. Imaging Health Inform. **7**(4), 799–804 (2017)
20. M. Moghbel, S. Mashohor, R. Mahmud, M.I.B. Saripan, Automatic liver tumor segmentation on computed tomography for patient treatment planning and monitoring. EXCLI J. **15**, 406 (2016)
21. J.C. Bezdek, R. Ehrlich, W. Full, FCM: the fuzzy c-means clustering algorithm. Comput. Geosci. **10**(2–3), 191–203 (1984)
22. A. Askarzadeh, A novel metaheuristic method for solving constrained engineering optimization problems: crow search algorithm. Comput. Struct. **169**, 1–12 (2016)
23. K.L. Wu, M.S. Yang, A cluster validity index for fuzzy clustering. Pattern Recogn. Lett. **26**(9), 1275–1291 (2005)
24. N.R. Pal, J.C. Bezdek, On cluster validity for the fuzzy c-means model. IEEE Trans. Fuzzy Syst. **3**(3), 370–379 (1995)
25. M.K. Pakhira, S. Bandyopadhyay, U. Maulik, Validity index for crisp and fuzzy clusters. Pattern Recogn. **37**(3), 487–501 (2004)
26. D.W. Kim, K.H. Lee, D. Lee, Fuzzy cluster validation index based on inter-cluster proximity. Pattern Recog. Lett. **24**(15), 2561–2574 (2003)
27. C.H. Wu, C.S. Ouyang, L.W. Chen, L.W. Lu, A new fuzzy clustering validity index with a median factor for centroid-based clustering. IEEE Trans. Fuzzy Syst. **23**(3), 701–718 (2015)

28. K.S. Chuang, H.L. Tzeng, S. Chen, J. Wu, T.J. Chen, Fuzzy c-means clustering with spatial information for image segmentation. Comput. Med. Imaging Graph. **30**(1), 9–15 (2006)
29. N.N. Gopal, M. Karnan, Diagnose brain tumor through MRI using image processing clustering algorithms such as fuzzy C means along with intelligent optimization techniques, in *IEEE International Conference on Computational Intelligence and Computing Research* (India, 2010), pp. 1–4), 28 Dec
30. S. Rajput, M. Parashar, H.M. Dubey, M. Pandit, Optimization of benchmark functions and practical problems using crow search algorithm, in *IEEE Fifth International Conference on Eco-friendly Computing and Communication Systems* (India, 2016), pp. 73–78, 8 Dec
31. M.G. Omran, A. Salman, A.P. Engelbrecht, Dynamic clustering using particle swarm optimization with application in image segmentation. Pattern Anal. Appl. **8**(4), 332 (2006)
32. J.C. Bezdek, R. Ehrlich, W. Full, FCM: the fuzzy c-means clustering algorithm. Comput. Geosci. **10**(2–3), 191–203 (1998)
33. S. Parvathavarthini, N. Karthikeyani Visalakshi, S. Shanthi, Breast cancer detection using crow search optimization based Intuitionistic fuzzy clustering with neighborhood attraction. Asian Pac. J. Cancer Prev. APJCP **20**(1), 157 (2019)
34. D. Oliva, S. Hinojosa, E. Cuevas, G. Pajares, O. Avalos, J. Gálvez, Cross entropy based thresholding for magnetic resonance brain images using crow search algorithm. Expert Syst. Appl. **15**(79), 164–180 (2017)
35. D. Gupta, S. Sundaram, A. Khanna, A.E. Hassanien, V.H. De Albuquerque, Improved diagnosis of Parkinson's disease using optimized crow search algorithm. Comput. Electr. Eng. **1**(68), 412–424 (2018)

Intelligent System for the Visual Support of Caloric Intake of Food in Inhabitants of a Smart City Using a Deep Learning Model

José Mejía, Alberto Ochoa-Zezzatti, Roberto Contreras-Masse and Gilberto Rivera

Abstract In our days, a recurrent problem in the Smart Cities of all Latin America will be the degenerative illnesses linked to food, one of the main reasons for the lack of organization of each citizen, is not being able to adequately determine the caloric intake, a proposal for a solution , is the development of an application that has the capacity to be able by means of the recognition of patterns and a deep learning model, to be able to specify what percentage of the nutritional value of each meal is covered by the associated quantities, one of the advantages of using a food repository is that solutions based on deep learning are not ready-made solutions. A development process is necessary to acquire an adequate set of instances and to customize the intelligent system. The latter includes the customization of the user interface, as well as the way in which the system retrieves and processes the feeding scenarios later. The resulting scenarios can be shown to the user in different ways, and/or retrieved cases can be adapted to be reused later. This research is about an intelligent model for decision making based on deep learning to solve the existing problem in the planning of food distribution in the population of a Smart City, for this first, we mentioned the need for intelligent systems in the processes of decision-making, where they are necessary due to the limitations associated with conventional human decision-making processes, among them: human experience is very scarce with respect to being able to calculate in a correct way the caloric value of food intake and we must to consider that citizens in a smart city are tired of the burden of physical or mental work, in addition to human beings forget the crucial details of a problem, and many times are inconsistent in their daily decisions. Complexity and investment of the time necessary to make food decisions tend to be complex for health as well as the high frequency of decision making found in the distribution to supermarkets, which mostly supply the food of a population with tendency to increase of individuals of the groutier type when they eat their food late at night. We use an image repository from DataWorld to our research (https://data.world/).

Keywords Making-decision · Deep learning · Caloric intake of food

J. Mejía · A. Ochoa-Zezzatti (✉) · R. Contreras-Masse · G. Rivera
Universidad Autónoma de Ciudad Juáirez, Juáirez, Mexico
e-mail: alberto.ochoa@uacj.mx

D. Oliva and S. Hinojosa (eds.), *Applications of Hybrid Metaheuristic Algorithms for Image Processing*, Studies in Computational Intelligence 890,
https://doi.org/10.1007/978-3-030-40977-7_19

1 Introduction

Nowadays, Childhood morbid obesity in Mexico is a health problem characterized by excess body fat in children. This level of adiposity is a determining factor for their general health and well-being, making them prone to suffer other secondary pathologies and associated with social aspects such as discrimination in schools due to their physical condition. According to the data published by the World Health Organization (WHO), the prevalence of this disease internationally is increasing. In 2027 there were about 87 million overweight children, most of them in developing countries, especially in Latin America, leading Mexico.

As in adults, obesity requires several complementary tests for its diagnosis. One of the most commonly used scales is known as body mass index or BMI.2 also known as the Quetelet index, it takes the weight and height of the person as a reference, assessing its proportionality based on the mathematical formula: mass divided by the height squared. However, the weighting of the resulting figure varies in the case of children. Unlike in adults, the body mass index has to be moved to a percentile table, taking into account two essential criteria: the age and sex of the child in question. According to the same, it is established that children with a body mass index between 85 and 95 on that scale are overweight, in the present investigation is to propose a means to deal with the health consequences, through the implementation of an intelligent application that allows to indicate in visual form, the caloric intake of each child in their day to day.

In recent times, Deep Learning has evolved computer vision algorithms [1] and specially the segmentation algorithms used to determine the of various objects of interest within an image [2–4]. In particular, the segmentation of food has been revised in several works of literature to develop systems to help users measure their calorie in-take. In [5] it is considered a system based on deep learning for recognize the food and determine portion sizes, also the system is implemented on a mobile device. In [6] it is proposed an assistant calorie measurement system using deep convolutional neural networks that can run on smart-phones, allowing the user to take a picture of the food and measure the amount of calorie intake automatically. The method shows high accuracy on single food portions. In [7] it is proposed a CNN-based food image recognition system addressing the problem of food recognition in images. In [8] was developed a deep learning-based visual food recognition and adapted it to the design of a food recognition system employing computing-based service to overcome some inherent problems of traditional mobile cloud computing paradigm.

2 Proposal Methodology

According to the tests, it is feasible to use a mobile application, to be able to determine the caloric consumption of each child, using real-time photographs of each meal made during the day and considering a nutritional pyramid, the most convenient time of the day is when it begins, since together to establish specific parameters of food intake should be considered caloric intake and its representation in the population pyramid. This research tries to specify an alternative route to food supplements, among other advantages. The thought of not knowing how many calories are associated with our food is of great importance, it is because of them that Artificial Intelligence has a series of tools at hand to achieve this problem, which in this case is technology, information, processing images and some mathematics. It should never be considered that the qualitative aspects are not important and having a healthy figure allows children to interact in a better way, that is why each of the components of the intelligent application, all of them linked to the objective of this project, consists of a mobile information system for geo-referenced location of food intake in children with morbid infantile obesity (SIGOMI) on levels of insecurity in the correct calculation of caloric intake at each moment of the day, which is presented in detail below , this research includes the analysis by means of deep learning to determine the danger of a continuous excess of caloric intakes higher than those established by the WHO for each population group in a border society and to avoid a contingency occurring during a specific time associated with bad eating habits. We carried out an exhaustive analysis of other similar investigations, the only similar context is explained in [9], where the insecurity of a group of children who perform higher than normal intakes and in fast-paced sites is calculated, by means of the delivery of products food in different places with random programming, but this research does not consider real statistics in time and the perspective to suggest a scenario of caloric intake associated with age, height and current weight and with a diet appropriate for their population group, a proposal model of this research is shown in Fig. 1.

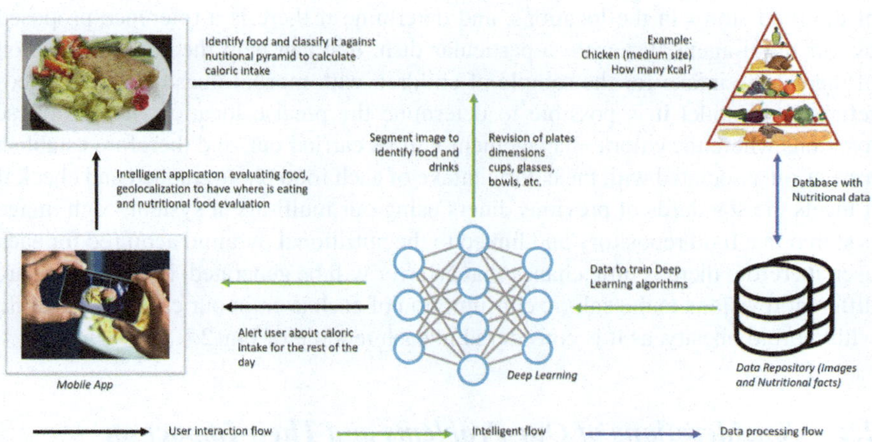

Fig. 1 Proposal model associated with our research

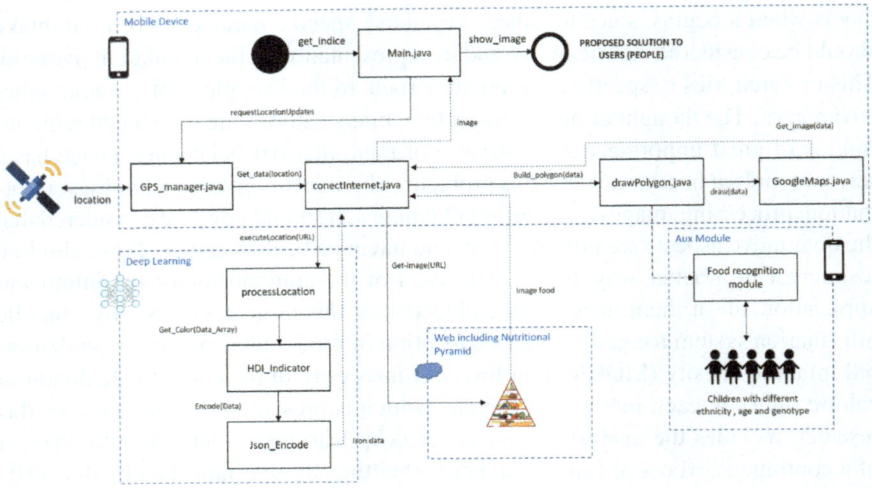

Fig. 2 Modular description of the components of our mobile computing application developed. *Source* Own design

2.1 Design of the Mobile Application

A decisive aspect of our research is the geo-reference of the location where food is consumed, this to determine the amount of associated caloric consumption, using our mobile application associated with Ubiquitous Computation it is possible to consider a decisive aspect that is the calories that are associated with the menus of the food stores in the locations, and determine if there is a reference proposed by our multi-agent system to a particular dish, that can help the correct selection of the caloric intake for the sample of children with morbid obesity. Using a geo-referencing model it is possible to determine the precise location of the diverse locations where the caloric management will be carried out, and therefore establish a criterion associated with the caloric intake of each food photographed and check if it meets the standards of previous diners using our multi-agent system, each image is stored in a food repository and linked to the nutritional pyramid acquired for each user, therefore there is little chance that an error will be generated, because there are different revisions to the caloric consumption of each user, in our case of each child with morbid obesity, as it is considered in each module of Fig. 2.

2.2 Considerations of Our Problems and Their Impact on Society

According to recent studies by the Pan American Health Organization, 17% of the infancy population in Ciudad Juarez—a frontier society—has characteristics of child morbidity, so that, in the wake of nutritional disorder, there is a problem of social

Fig. 3 A morbid obesity in children is characterized by anxiety problems, especially in single children. *Source* Cambridge News

psychology, characterized by the anxiety of having to eat, whenever there is a situation of stress, as can be seen in Fig. 3, this problem of children psychology generates a parallel problem that is the social stigma of not being left in the children stage, something that even leads to suicide in children, boys and girls.

In our days, Mexico now the tenth largest population with 128.7 million has the second place in the ranking of overweight population this affects the health system, as can be seen in Fig. 4, as the population grows in age, the Food has a deteriorating effect on their health and generates that a greater percentage of children and young people are affected by this paradigmatic change of diet, the high caloric intake of foods with saturated fats and types of salt harmful to health begin to affect the population as it is characterized a pattern of food that includes soft drinks with a refill system and ice cream and sweets with high concentrations of sugar.

The Organization for Economic Co-operation and Development (OECD), composed of 47 societies of high and medium economy, has counts of statistics of overweight in the target population of this study, as can be seen in Fig. 5, and is decisive for the consideration of this study that the young population has the highest morbid obesity growth in children, having food problems associated with caloric intake and food from commercial chains, which in many cases do not take care of caloric intake, nor present in detail such consumption.

If a determinant public policy associated with the decrease in the population of children with morbid obesity is not considered, it will not be possible to detail which nutritional aspects to consider could be decisive for the objective population, that is why determining a date established by the same OECD for the 2027 the critical route to follow in order to achieve the reduction of the caloric intake in overweight

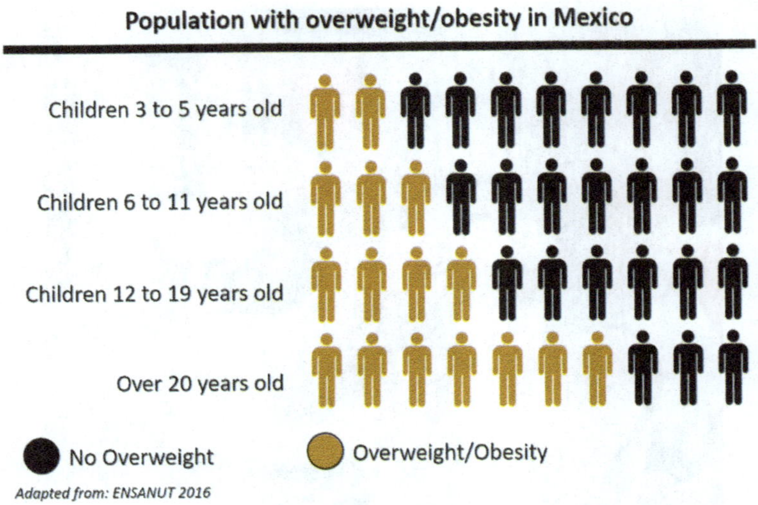

Fig. 4 Proportion of children with morbid childhood obesity, in Mexico. As the population gets old, the proportion of overweight increases

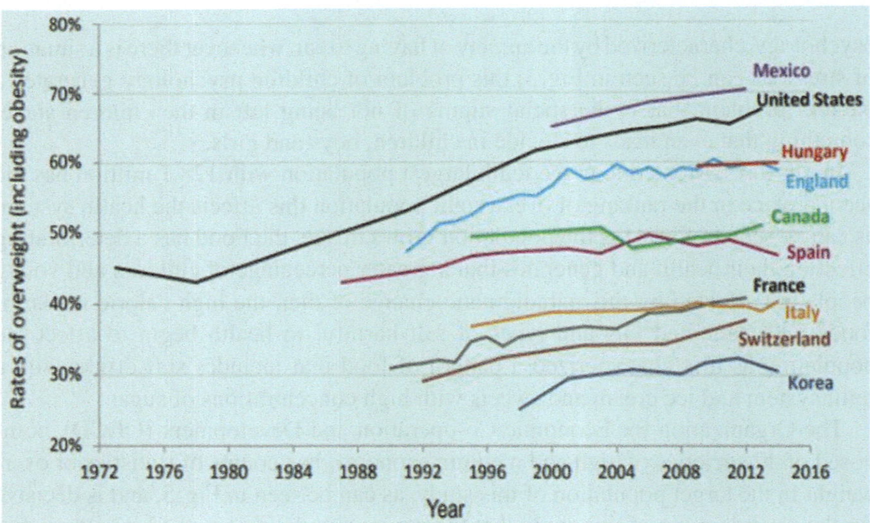

Fig. 5 Comparison of OECD countries with percentages of childhood obesity for at least one year prior to 13 years of age. *Source* OECD Statistics

children should be expensive, but not before establishing a particular nutritional management system associated with the size and weight indicated for each child and a child. specific exercise system to be carried out by the children of each of the 32 societies in Mexico, which considers socio-cultural and socio-economic aspects of each region.

2.3 Food Recognition Module

In this section more details of the food recognition stage are given, since it is a critical component of the proposed system (Fig. 2). The model has an interchange with recognition of food in the input image and the rest of data of our system. The proposed food recognition stage consists of a deep learning architecture. An input image is feed to the system and is subsequently segmented in several groups which are then classified as type of food or background. The architecture is based on Mask Regional Convolutional Neural Network (R-CNN) [10] which extends R-CNN [11] by adding an object mask to the existing branch for bounding box recognition.

Mask R-CNN is a deep learning architecture to solve instance segmentation, it uses the module for classification of fast-RCNN but with an extra module to create masks. Mask R-CNN adopts an identical first stage of Region Proposal Network, however, it adds a binary mask for each stage to produce binary mask from the input image. In this architecture, each module of the network has its own assigned loss, allowing the network to generate masks for every class without competition among classes. Figure 6, shows the network architecture. In this paper it is proposed a new architecture for the mask module in mask R-CNN, since the proposed network is focused on food recognition only, the network is simplifies to make it more easy to train and to have less computational burden as compared with for example ResNeXt [12] type networks of depth 50 or 101 layers which is a common choice used in for the

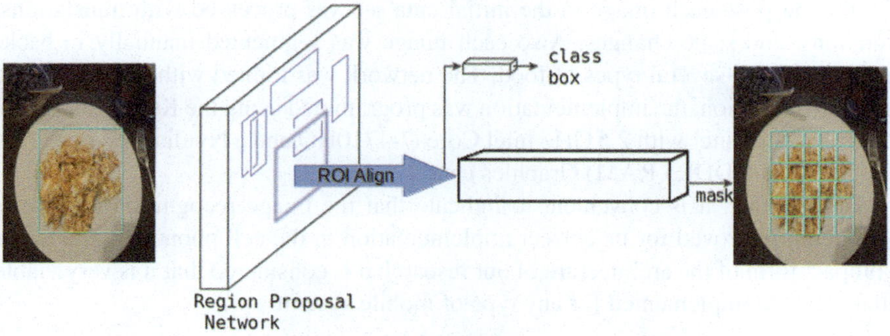

Fig. 6 Simplified diagram of the mask R-CNN network used for classification/segmentation of images. The network generates three types of output, the box where the food is, a binary mask delimiting the food, and the types (classes) of food found

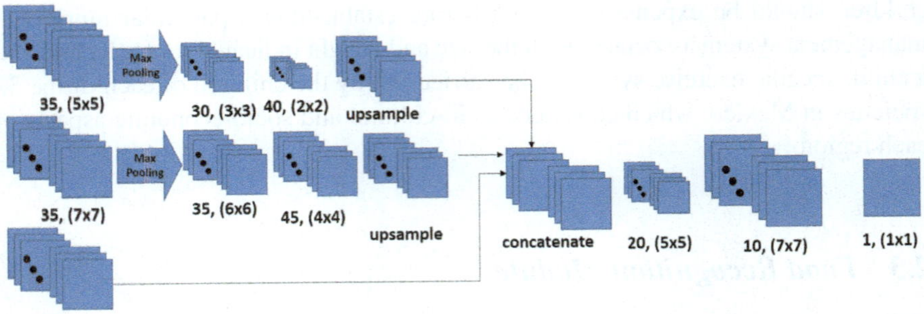

Fig. 7 The proposed architecture for the mask module. It consists of three channel of convolutional layers which are concatenated to have a rich feature map with information at different scales

module implementation in mask R-CNN implementations [10]. Thus the proposed architecture consists of three channels of convolutional networks, each channel is aimed to select different sizes of features: a large size channel with kernels of size 11×11, a medium size channel with output filter of size 4×4, and finally a small size channel with output filters of 2×2. The outputs of the different channels are concatenated in a feature map which contains information at different scales of the input image, note that for the concatenation the output of the medium and small channels are up sampled so that the output of the three channels have equal size. Subsequently the output is feed to a pair of deconvolution layers to recover the size of the original image and finally a convolution layer of 1×1 filter size to have a gray level image with the masks of food regions. Each mask of the output image has a gray level that indicates the class of the region that the mask encloses. The details of the described architecture are shown in Fig. 7.

In Fig. 8 shows several images used to train the network. To increment the transmission velocity to the server and reduce the computational cost, the input images were reduced to a fixed size of 128×128 pixels. This initial dataset was processed with data augmentation techniques [13, 14] to have a final data set of 4000 images, for this purpose each image in the initial data set was processed with translations, rotations, and scale changes. Also each image was segmented manually in background and the several types of food. The network was trained with a mean square error loss function, the implementation was programmed using the Keras framework [15] on a computer with 2.5 GHz Intel Core i7-4710HQ and a Nvidia GeForce GTX 980M (4 GB GDDR5 RAM) Graphics card.

Besides this, it is convenient to indicate that the image recognition module is still being improved for its correct implementation in the cell phone but due to the compact form of the architecture of our research it is considered that it is very viable that it can be implemented for any type of mobile system.

Fig. 8 Sample images from the input data. As can be seen, the dishes can consist of the same type of food, such as meat, chicken; or a combination like salads

3 Design of Experiments

To know the operation of the proposed model, we used the estimation of the sum of the food separately and finally the caloric combinations of a food that involves combinations of them. These proposals of food menus competed with each other in the category of proposal for the improvement of nutrition and therefore of the reduction of child morbid obesity and were evaluated by the intelligent application developed through their representatives in the deep learning model. For the estimation of the matrix of the selection of a menu according to the caloric intake, 30 runs of the experiment were conducted which were carried out under the same conditions, and for the group of finalist proposals, a design of experiments according to the attributes of each proposal to obtain a better estimate of the final classification. In each run, 350,000 function evaluations were performed. The average over the 30 runs was calculated for each proposal. Then, the average ranking was obtained to determine the 24 best proposals that will compete in the final ranking. Three measures are calculated from the 30 runs: mean, median and interquartile range. The interquartile range has an amplitude of 50% of the whole value of the median (second quartile Q2), which is calculated through the Q1 lower quartile (first quartile) and upper quartile (Q3 quartile third quartile). In descriptive statistics, a quartile is any of the three values that divide the ordered data into four equal parts, so that each part represents 1/4th of the population of the sample. The difference between the upper and lower quartiles is called the interquartile range. In the results section, the estimation of our approach to the analysis of food menu proposals for children with a new eating disorder is presented.

3.1 Experimentation

In order to be able similar, the most efficient arrangement of individuals associated with the food in a social network, we developed an atmosphere able to store the data of each one of the representing individuals of each society—including their food—, this with the purpose of distributing of an optimal form to each one of the evaluated

Table 1 Orthogonal array

A	B	C	D	E	F	G	H	Color
H	H	H	H	H	H	H	L	1
H	H	H	H	H	H	L	H	2
H	H	H	H	H	L	H	H	3
H	H	H	H	L	H	H	H	3
...	...	...	...	...	...	...	...	...

societies. One of the most interesting characteristics observed in this experiment was the diversity of the cultural patterns established by each community related with the food. The scenes structured associated with the agents cannot be reproduced in general, since they only represent a little while moment in the space and time of the different societies. These represent a unique form and innovating of adaptive behavior which solves a computational problem that it does not try to clustering the societies only with a factor associated with his external appearance (attributes of each society), trying to solve a computational problem that involves a complex change between the existing relations. The generated configurations can be metaphorically related to the knowledge of the behavior of the community with respect to an optimization problem (to select culturally 47 similar societies and their food, without being of the same quadrant associated with Nutritional pyramid). The main experiment consisted of detailing each one of the 568 diverse kind of food or their combinations, with 500 agents, and one condition of unemployment of 50 epochs, this allowed us to generate the best selection of each Quadrant and their possible location in a Diorama associated with food, which was obtained after comparing the different cultural and social similarities from each community, and to evaluate with Multiple Matching Model each one of them. The developed tool classified each one of the societies pertaining to each quadrant, with different wardrobe for societies that included linguistic identity and for societies only with cultural identity, this permit identifies changes in the time respect at other societies. The design of the experiment consists in an orthogonal array test, with the interactions between the variables: emotional control, ability to fight, intelligence, agility, force, resistance, social leadership, and speed. These variables are studied in a range of emotions associated with the food, another variable is the color of food represented by numbers (0–256). The orthogonal array is $L - N(2^8)$, in other words, 8 factors in N executions, N is defined by the combination of possible values of the 8 variables an the possible range of color (To see Table 1).

For measure the performance of the network we use accuracy for the recognition module in R-CNN and for the segmentation mask stage we used the Intersection over Union (IoU) which is a common metric [16, 17] used to evaluated segmentation accuracy. This metric computes the total of pixels in the intersection between the set of predicted pixels (A_{pred}) and ground truth pixels (A_{GT}) for each class, and divides by the number of pixels in their union, as is shown in Eq. 1.

$$\text{IoU} = \frac{A_{pred} \cap A_{GT}}{A_{pred} \cup A_{GT}} \quad (1)$$

4 Results

The so-called Somatotype is a crucial element in our research, because the caloric intake continues and the existing genetics determines bodies of a certain form, which require different treatments in the use of reduction of felling, and therefore the existing determination of the real possibility to continue in a process of caloric reduction, which will be effective in the near future, as it can be the youth, the seven different types of somatotypes can be visualized in Fig. 9.

One of the most relevant aspects of our research, is that if you adjust the caloric intake prior to adolescence, each of the individuals will not be the target of bullying in their youth, considering that they could reduce at least one or two sizes to the utilizer our proposal of application of mobile computing.

4.1 Results of the Segmentation Network

In Fig. 11 is shown the output from the segmentation on several images and Fig. 10 shows the output of both the segmentation combined with the classification output. The averages of IoU of the segmented regions for each type of food are shown in Table 2. Most classes of food have an average above of 0.9 of accuracy, except for liquids, water, coffee; this could be because of to the recipient used to contain the liquid, which could have a large variation in size, color and shape.

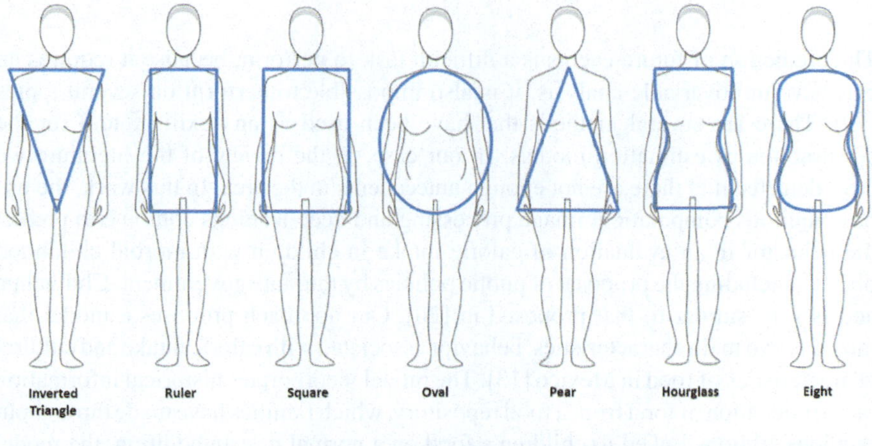

Fig. 9 A subsequent aspect of our research is the establishment of somatotype models in girls and therefore an aspect of the type of clothing, which can be designed for everyday use, as proposed in [18]

Fig. 10 Segmentation results of the proposed network

5 Multivariable Analysis

The prediction of future events is a difficult task to perform, because it requires an extensive multivariable analysis, it is also impossible to perform on several topics [19]. There are several methods that have been used as an auxiliary tool for the construction of estimation models. In our case, in the review of the literature we have detected that there are not enough antecedents in the area. In this work, the use of ubiquitous computation, image processing and deep leanings combine to predict the behavior in an evaluation of caloric intake in children with morbid childhood obesity, including the proposal of public policies by the state government. Chihuahua and is very similar to that proposed in [20]. Our approach proposes a model that includes two main characteristics: behavior associated with caloric intake and cultural characteristics of food in Mexico [13]. The model incorporates historical information on the allocation of food from a food repository, which families have made throughout previous editions linked to children's food on a normal day. In addition, the model includes information on the intrinsic characteristics of the nutritional menu that represents each proposal of caloric intake.

Fig. 11 Output images of the mask stage combined with the classification stage

Table 2 Measures of intersection over union (IoU) for each class

Class	IoU	Class	IoU	Class	IoU
Background	0.98484454	Egg	0.94114587	Vegetable	0.96853764
Juice	0.87009111	Coffee	0.81870388	Beans	0.90149943
Beef	0.8459461	Cereal	0.93650956	Fruit	0.97946966
Yogurt	0.97838868	Chicken	0.9701942	Cheese	0.98694365
Cream	0.99432157	Bread	0.95197437	Potato	0.9634425
Tortilla	0.9580226	Rice	0.94042164	Water	0.84217442

6 Conclusions and Future Research

The contribution of greater value of our research is related to a ubiquitous computer system capable of determining in real time the caloric intake of a child from the related food and immediately compare it with the population pyramid that should be covered as a function of his size, age, sex and physical complexion, as is proposed in [14]. Also, our proposed recognition module is compact and has an average IoU accuracy above of 90% for most types of food. A future research would be to analyze the proposals of nutritional menus from small companies such as the Faroe Islands, Guernsey, Jersey, Liechtenstein, Kosovo or Gibraltar, who face problems of various food types including the obtaining of tempura and fresh fruits very different from countries with more than 1 million inhabitants and with extensive cultivation extensions. The OECD currently uses this type of innovative methods based on Artificial Intelligence to correctly characterize and evaluate the different opinions of different societies in the context of being able to listen to all existing voices, including those of minorities, something similar could be used in Ciudad Juáirez in where 37.14% of the population was not born in the state of Chihuahua - and because of this, the consumption of dishes from Zacatecas, Durango and Coahuila have a total and complete caloric intake, reaching 46.89% of the people who do not In the City, an approach based on these two major minorities could improve citizen participation.

References

1. I. Goodfellow, Y. Bengio, A. Courville, *Deep Learning* (MIT Press, 2016)
2. L.C. Chen, G. Papandreou, I. Kokkinos, K. Murphy, A.L. Yuille, Deeplab: semantic image segmentation with deep convolutional nets, atrous convolution, and fully connected CRFS. IEEE Trans. Pattern Anal. Mach. Intell. **40**(4), 834–848 (2018)
3. G. Lin, A. Milan, C. Shen, I.D. Reid, RefineNet: multi-path refinement networks for high-resolution semantic segmentation, in *IEEE Conference on Computer Vision and Pattern Recognition (CVPR)*, vol. 1, no. 2 (2017), p. 5
4. H. Zhao, J. Shi, X. Qi, X. Wang, J. Jia, Pyramid scene parsing network, in *IEEE Conference on Computer Vision and Pattern Recognition (CVPR)* (2017), pp. 2881–2890
5. M. De Bonis, G. Amato, F. Falchi, C. Gennaro, P. Manghi, *Deep Learning Techniques for Visual Food Recognition on a Mobile App. MISSI* (2018), pp. 303–312
6. L. Herranz, W. Min, S. Jiang, Food recognition and recipe analysis: integrating visual content, context and external knowledge (2018). arXiv:1801.07239
7. W. Hui, M. Merler, R. Uceda-Sosa, J.R. Smith, Learning to make better mistakes: semantics-aware visual food recognition. ACM Multimed. 172–176 (2016)
8. K. Yanai, T. Maruyama, Y. Kawano, A cooking recipe recommendation system with visual recognition of food ingredients. IJIM **8**(2), 28–34 (2014)
9. C. Liu, Y. Cao, Y. Luo, G. Chen, V. Vokkarane, Y. Ma, Deepfood: Deep learning-based food image recognition for computer-aided dietary assessment, in *International Conference on Smart Homes and Health Telematics* (Springer, Cham, 2016), pp. 37–48
10. P. Kuhad, A. Yassine, S. Shimohammadi, Using distance estimation and deep learning to simplify calibration in food calorie measurement, in *2015 IEEE International Conference on Computational Intelligence and Virtual Environments for Measurement Systems and Applications (CIVEMSA)* (IEEE, 2015)
11. P. Pouladzadeh, P. Kuhad, S.V.B. Peddi, A. Yassine, S. Shirmohammadi, Food calorie measurement using deep learning neural network, in *IEEE International Instrumentation and Measurement Technology Conference Proceedings (I2MTC)*(2016), (pp. 1–6)
12. S. Xie, R. Girshick, P. Dollár, Z. Tu, K. He, Aggregated residual transformations for deep neural networks, in *IEEE Conference on Computer Vision and Pattern Recognition* (2017)
13. O. Ronneberger, P. Fischer, T. Brox, U-net: convolutional networks for biomedical image segmentation, in *International Conference on Medical Image Computing and Computer-Assisted Intervention* (Springer, Cham, 2015), pp. 234–241
14. X. Cui, V. Goel, B. Kingsbury, Data augmentation for deep neural network acoustic modeling. IEEE/ACM Trans. Audio Speech Lang. Process. (TASLP) **23**(9), 1469–1477 (2015)
15. F. Chollet, et al. Keras (2015)

16. G. Papandreou, L.C. Chen, K.P. Murphy, A.L. Yuille, Weakly-and semi-supervised learning of a deep convolutional network for semantic image segmentation, in *Proceedings of the IEEE International Conference on Computer Vision* (2015), pp. 1742–1750
17. G. Papandreou, L.C. Chen, K. Murphy, A.L. Yuille, Weakly-and semi-supervised learning of a DCNN for semantic image segmentation (2015). arXiv:1502.02734
18. C. Liu, Y. Cao, Y. Luo, G. Chen, V. Vokkarane, M. Yunsheng, P. Hou, A new deep learning-based food recognition system for dietary assessment on an edge computing service infrastructure. IEEE Trans. Serv. Comput. **11**(2), 249–261 (2018)
19. K. He, G. Gkioxari, P. Dollár, R. Girshick, Mask r-cnn, in *2017 IEEE International Conference on Computer Vision (ICCV)* (IEEE, 2017), pp. 2980–2988
20. S. Ren, K. He, R. Girshick, J. Sun, Faster r-cnn: towards real-time object detection with region proposal networks, in *Advances in Neural Information Processing Systems* (2015), pp. 91–99

Image Thresholding with Metaheuristic Algorithms for Cerebral Injuries

Ángel Chavarin, Jorge Gálvez and Omar Avalos

Abstract Cerebral injuries are a constant in high-risk sports like football, boxing or extreme sports; similarly traffic accidents with motorcycles. Many times an injury of this type goes unnoticed because it is too small, unfortunately, many people die due to complications. The detection of brain traumatisms can be done with digital image processing. The cerebral tissues have different densities, as well as clear delimitations among them; an abnormality can be seen in an MRI indicating what the problem is when segmenting an image of a damaged brain and compare with a healthy brain the present damage can be diagnosed. This chapter makes a revision of the cuckoo search (CS) which is an optimization algorithm that presents ideal features for the thresholding of images as previous experiments have shown their effectiveness. The objective function Kapur is used because it is a strong option for image segmentation. Besides, a comparison is made with some optimization algorithms.

Keywords Image thresholding · Image segmentation · Kapur's entropy · Metaheuristic algorithm · Cuckoo search

1 Introduction

The digital processing of images has benefited the medical sciences, and particularly, has been of great help in radiology and MRI diagnosis [1]. Thanks to the image segmentation techniques, a large amount of redundant information can be eliminated, in a way that allows a simpler and faster analysis with the relevant information. In segmentation, thresholding approaches are widely used because they are easy to implement and allow to discriminate the information selectively [2].

Á. Chavarin (✉) · J. Gálvez · O. Avalos
CUCEI, Universidad de Guadalajara, Guadalajara, Jalisco, Mexico
e-mail: angel.chavarin@alumnos.udg.mx

J. Gálvez
e-mail: jorge.galvez@academicos.udg.mx

O. Avalos
e-mail: omar.avalos@academicos.udg.mx

D. Oliva and S. Hinojosa (eds.), *Applications of Hybrid Metaheuristic Algorithms for Image Processing*, Studies in Computational Intelligence 890,
https://doi.org/10.1007/978-3-030-40977-7_20

Nowadays, there are many people practicing high-risk sports, where head accidents are frequent. After a head concussion it is almost certain that an injury of greater or lesser severity will develop [3, 4]. Other causes of cerebral injuries are cardiovascular, congenital or chronic diseases; for example heart-attacks, anoxia, infections, and even strokes. The abuse of drugs like crack, heroin, or methamphetamines also can produce a brain damage [5, 6].

Thresholding consists in the separation of regions or objects from an image to obtain information from it and make its analysis easier [7]. This separation can be given by the difference between the values of the pixels or, separated based on the intensity of a group of pixels in a certain area. Segmentation has many practical applications in science, as well as, histopathology [8], botany [9], geography [10] or even advanced electronics [11].

When segmenting, the simplest technique is to use image thresholding; the most elementary technique is to substitute a pixel from the image with a black pixel if its intensity is smaller than a previously established T value and all above that T is substitute with a white pixel. The threshold can be either manually set or with the help of an algorithm [12]. When more thresholds are needed, it is necessary to use multi-thresholding. From this it follows that there are 2 types of thresholding:

- bi-level thresholding
- multi-thresholding

The bi-level thresholding, as the name implies, separates the pixels present in an image into two different levels; while multi-thresholding allows more than two levels generating as many as needed. In the brand new algorithms is usual to use threshold based in histograms; a histogram is a plot that allows discovering the underlying frequency distribution of a set of continuous data. Image thresholding based on histograms suffer a major limitation [11]; some methods are inefficient due to their parametric nature since it is computationally demanding the identification of their parameters.

The most representative thresholding method is the Otsu's method, which was originally developed as a bi-level method and later it was adapted to multi-level obtaining good results [12]. Furthermore, the more complex the image is, the more difficult this method becomes. So, the research community has been working on developing better methods at a lower computational cost. However, to be able to segment images, any strategy cannot be used because not all are optimal. Already exist some criteria called entropies [13]; these are criteria of the information's theory and that they are used to analyzing the information of a data set [14]. Some of the most used are the Shannon entropy, Tsallis entropy, minimum crossing entropy, among others.

All of the approaches to image segmentation need an efficient optimization method in order to correctly identify the best set of thresholds of an image to maximize or minimize the entropy evaluated. Several authors state that an optimization algorithm is an algorithm that allows obtaining the best solution based on a set of data of a given problem and such a problem can have several valid solutions. Different optimization algorithms such as particle swarm optimization (PSO), differential evolution (DE) or artificial bee colony (ABC) are good options but not the best.

2 Kapur's Entropy for Multilevel Thresholding

Supposing that $P = (p_1, p_2, p_3, \ldots, p_n) \in \Delta_n$ Where $\Delta_n = \left\{(p_1, p_2, \ldots, p_n) \vee \geq 0, i = 1, 2, \ldots, n, n \geq 2, \sum_{i=1}^{n}, p_i = 1\right\}$ Is a set of discrete finite n-ary probability distributions [8] then the entropy of the total image can be defined as:

$$H(P) = \sum_{i=1}^{n} p_i \log_2 p_i \tag{1}$$

where i denotes an 8-bit gray level digital image of dimension M × N. P is the normalized histogram for an image with L = 256 Gy levels. Now, if there are n–1 thresholds (t), partitioning the normalized histogram into classes, then the entropy for each class may be computed as

$$H_1(t) = -\sum_{i=0}^{t_1} \frac{p_i}{P_1} \ln\left(\frac{p_i}{P_1}\right) \tag{2}$$

$$H_2(t) = -\sum_{i=t_1+1}^{t_2} \frac{p_i}{P_2} \ln\left(\frac{p_i}{P_2}\right),$$

$$H_n(t) = -\sum_{i=t_{n-1}+1}^{L-1} \frac{p_i}{P_1} \ln\left(\frac{p_i}{P_1}\right)$$

where:

$$P_1(t) = \sum_{i=0}^{t_1} p_i, p_2(t) = \sum_{i=t_1+1}^{t_2} p_i, \ldots, p_n = \sum_{i=t_{n-1}+1}^{L-1} p_i \tag{3}$$

To ease the computation, two dummy thresholds $t_0 = 0$, $t_n = L - 1$ are introduced with $t_0 < t_1 < \ldots < t_{n-1} < t_n$. The dummy thresholds are the origin and the end of the histogram and they help to form valid segmented classes. The first threshold t_1 Will from the first class C_1 in the interval formed by $[t_0, t_1]$ While t_n Aid to the formation of the last class. Then the optimal threshold value can be found by:

$$\varphi(t_1, t_2, \ldots, t_n) = Arg\max([H_1(t) + H_2(t) + \cdots + H_n(t)]) \tag{4}$$

3 Lévy Flights in Nature

The animals look for a source of food in a random or quasi-random way in nature; actually, the feeding of the animal is a random march because the movement is based on the current location and the probability of transition of the next place. The probability of the chosen directions is modeled mathematically [15].

Several authors have shown that the flight behavior of various species of animals and insects demonstrates the characteristics of Lévy flights. A Lévy flight is a random walk in which the length of the steps is calculated according to a heavy-tailed probability of distribution. The distance from the origin of the random walk tends to a stable distribution after a large number of steps.

4 Cuckoo Search Optimization Algorithm

In nature, there is a bird called cuckoo (Fig. 1). This bird has a harmful attitude toward other birds, which is brood parasitism. It consists of that the female of the cuckoo arrives at a nest of another species that already has eggs and the absent parents are to place there a single egg. In this way, you do not have to worry about raising your offspring; they do not take any nest, but the nests of species that have similar eggs in a way that mimic each other [15]. This reduces the probability if the eggs being abandoned.

Fig. 1 Common Cuckoo (Cuculus canorus) captured at Borit, Gojal, and Gilgit-Baltistan, Pakistan by Imran Shah (2018)

Each egg in the nest represents a solution. And a cuckoo egg symbolizes a new solution; the goal is to use the best solutions (cuckoos) to replace the worst solutions. In its simplest form, each nest has an egg. The algorithm can be extended to a group of solutions (several eggs). In this way, it can be used for more complex problems. The algorithm is based on the following 3 ideas:

- Each cuckoo lays one egg at a time in a randomly selected nest;
- The best nests with eggs of better quality will be transferred to the next generation;
- The number of available host nests is arranged, when the laying of a cuckoo egg is discovered by the host bird with a probability $p_a \in (0, 1)$. In this case, the host bird can throw away the egg or leave the nest to build a completely new nest in a new position.

For a maximization problem, the quality or fitness of a solution can simply be proportional to the objective function. Similarly, other forms of fitness can be defined by the fitness function in genetic algorithms [15].

When new solutions are generated x (t + 1) for, say, a cuckoo that performs a Lévy flight.

$$X_i^{t+1} = X_i^{t\vee + a \oplus Lévy(\lambda)} \quad (5)$$

where $a > 0$ is the size of the step that should be related to the scales of the interest problem and in most cases, we can use a = 1. The above equation is essentially the stochastic equation for the random walk. In general, a random walk is a Markov chain whose next state or location only depends on its current location (the first term of the previous equation) and the transition probability (the second term) and the product means sentry wise multiplications. This is very similar to those used by the PSO, but the random gait via Lévy flights essentially provides a random walk while the length of the random step is drawn from a Lévy distribution.

$$Lévyu = t - \lambda, \ (1 < \lambda \leq 3) \quad (6)$$

That has an infinite variety with an infinite average. Here the steps essentially form a power law walk step-length distribution with a heavy-tailed. Some of the new solutions should be generated by the Lévy walk method around the best solution obtained so far, this will speed up the local search. However, another important part of the new solution should be generated by remote field randomization and whose locations should be as far away as possible from the best current solution, this will prevent the system from being trapped in some local optimum.

The CS is a population-based algorithm similar to the GA or the PSO that allows a more efficient randomization since the step length is heavy-tailed, in which any large step is possible; An advantage over the other algorithms is that the number of parameters to be adjusted is lower and therefore this allows a better adaptation to a greater class of optimization problems. In addition, each nest can represent a set of solutions so the CS can be extended to the type of meta-population algorithms.

4.1 Explanation of CS Algorithm

To solve an optimization problem, it is necessary that the values of the problem variables be formed as an array. In GA and PSO, they are called "chromosomes" and "position of the particle" respectively; in the CS algorithm, it is called habitat. In an N-dimensional optimization problem, a habitat is a 1 × N arrangement, representing the current position of the cuckoo. This arrangement is defined as $habitat = [x_1, x_2, \ldots, x_N]$.

Each of the values of the variable $(x_1, x_2, \ldots, x_N)$ is a floating point number. The benefit of habitat is obtained by evaluating the gain function *fpat* a habitat of. Then, $profit = cost(habitat) = fp(x_1, x_2, \ldots, x_N)$. as can be seen, the CS algorithm is an algorithm that maximizes a beneficial function.

In this phase, an initial population is randomly produced in a solution space for the problem to solve [16]:

$$x_{i,j}^{0} = a_j + rand \times \left(b_j - a_j\right)(i?1, 2, \ldots, N; j = 1, 2, \ldots, D) \tag{7}$$

where $x_{i,j}^{0}$ is the initialization solution; a_j and b_j Represent the upper and lower limits of the solution space respectively. N is the population size, D is the size of the solution space and Rand is a uniformly distributed random number in the range [0, 1].

To generate new candidate solutions, Lévy flights can be expressed as follows:

$$x_i^{t+1} = x_i^t + step \times l\acute{e}vy(\lambda) \tag{8}$$

$$step = \alpha \times \left(x_i^t - x_{best}\right) \tag{9}$$

where x_i^t is the current solution, x_i^{t+1} represents the new solution, λ denotes the power coefficient, α is the step size factor, depending on the scale of the problem to be solved, and step represents the random search factor, which uses the information about the best current solution x_{best}. $L\acute{e}vy(\lambda)$ represents a random number extracted from Lévy's distribution and is given as:

$$levy(\lambda) = \frac{m}{|n|^{\frac{1}{\beta}}} \tag{10}$$

where *m* and *n* represent 2 random numbers following the normal distribution, *β* represents a distribution parameter.

In biased random walk, a fraction of the new solutions is generated by the following equation:

$$x_i^{t+1} = x_i^t + r \times \left(x_{r1}^t - x_{r2}^t\right) \tag{11}$$

where r is the scale factor bounded in the range [0, 1], x_{r1} And x_{r2} Are two solutions chosen randomly in the current population.

5 Results

In this chapter, the Kapur function was used as an objective function, with different optimization algorithms, so as to obtain varied results in the experimentation.

A set of public images of cerebral pathologies of randomly selected images (Fig. 2) was used; in order to obtain images with high detail density for later analysis and to verify the functioning of the CS, a comparison was made between this and the PSO, SFS and FA. So that the superiority of the former could be demonstrated over the latter. The set of images used was not modified in any of the cases so that they were in a situation of equal conditions.

In all the algorithms the following specifications were used in the code:

- 50 iterations
- Set of 10 black and white images in JPG format
- Same number of thresholds k = {2, 3, 4, 5}.

Table 1 shows the pseudo-code of the cuckoo search algorithm. It can be seen that it is much simpler than others who, although they do the same, are more extensive in their application, such as the SFS or the PSO. It is notable that the algorithm uses less computational time, and although the PSO has a similar computational time, it has inferior results. Then there is the SFS, which, although it has slightly similar results to the CS, takes some hours to reach the results; this is because it explores the search space a lot more than the CS.

When experimenting, the quality of the results and the time used by each algorithm put to the test were taken into account; because the quality of a result cannot be seen only with human observation, was made use of the maximum signal to noise ratio (PSNR) (Table 2) as well as the structural similarity index (SSIM) (Table 3) to measure the quality of the image.

The software used in this chapter was MATLAB 2018a; therefore, there could be features not present in previous versions; it is very important to emphasize that the best result gives thanks to efficiency coding of the algorithm, however, a good processor will allow to obtain results in a faster way. The features of the computer used in the experimentation will also determine the efficiency of the algorithm; the hardware components used have the following characteristics:

- Processor: RYZEN 5 1600 @ 3.2 GHz
- RAM: 20 GB DDR4
- Graphic card: NVIDIA GeForce GT 1030.

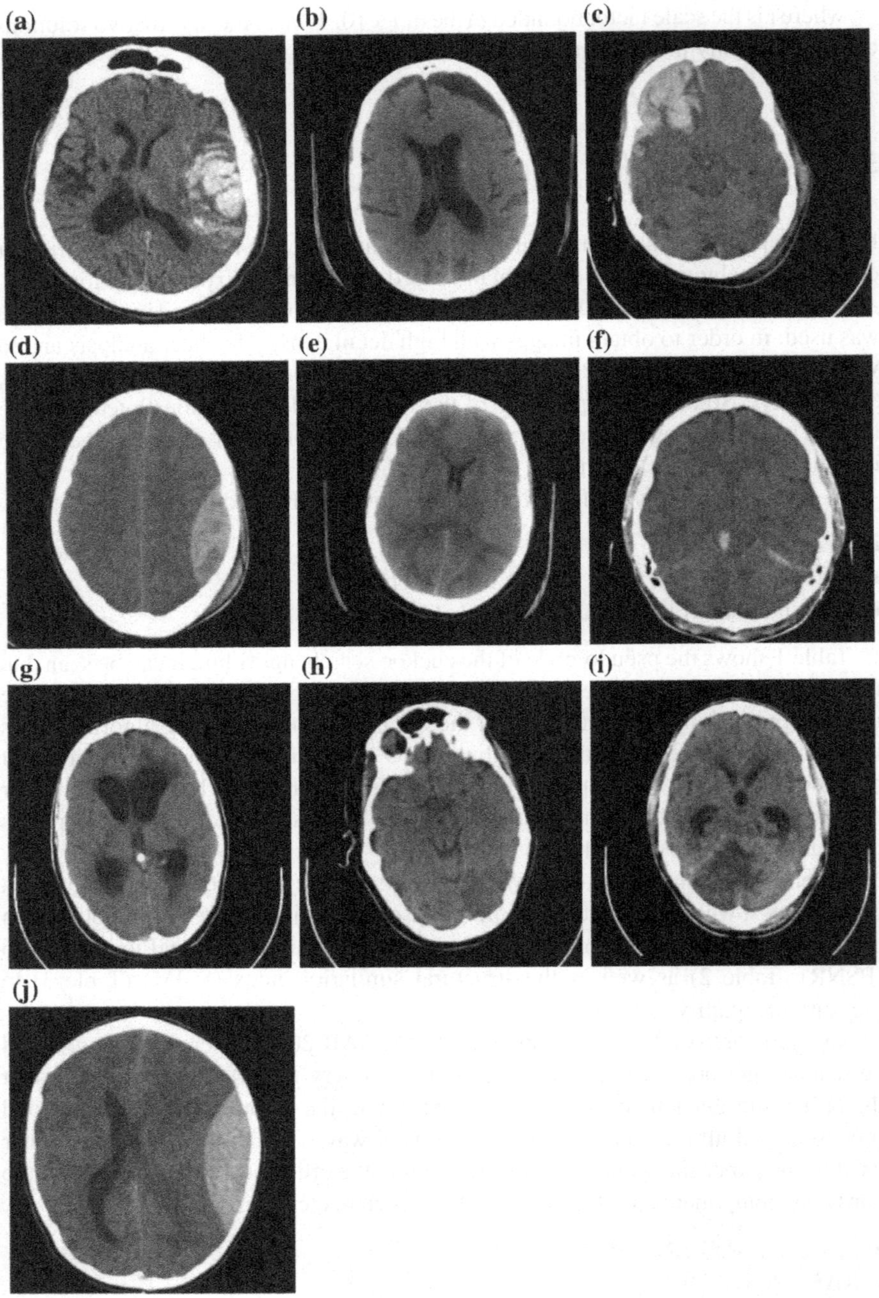

Fig. 2 selected images of brain injuries: **a**, **c** concussion, **b** chronic SDH, **d** EDH, **e** embolism, **f** hemorrhage, **g** hydrocephalus, **h** infarct, **i** stroke, **j** traumatic brain injury

Table 1 Cuckoo search pseudo algorithm [17]

```
Begin
    Objective function f(x)
    Generate an initial population of n host nest
    Evaluate fitness and rank eggs
        While (t > max_generation) or stop criterion
            t = t + 1
            Get a cuckoo randomly/generate next solution by Lévy flights
            Evaluate quality/fitness
            Choose a random nest j
            If (F_i > F_j)
                Replace j by the new solution
            End if
            Worst nest is abandoned with probability P_a And new nest is built
            Evaluate fitness, rank the solutions and find current best
        End while
    Post process results and visualization
End
```

As can be seen in Fig. (3), as the number of the thresholds increases, more and more objects can be seen in the image, as well as the severity of the damage present. In Fig. (3a) You see only an abnormal white spot that represents the blow to the right with separate regions. Despite looking small, it is a very serious injury. As the level of thresholds increases, you can see how the surrounding tissues were damaged and, therefore, the actual severity of the brain contusion, taking the same shape as a diffuse spot inside the skull. It is noteworthy that this process, although it takes a bit to find the optimal result, only 1 image can be used and the lesion can be found in a couple of minutes or less.

5.1 Peak Signal-To-Noise Ratio (PSNR) and Structural Similarity Index (SSIM) Metrics

Table 2 shows the results obtained from the PSNR test applied to each algorithm. The PSNR test (Peak Signal-to-Noise Ratio) is a metric that consists of calculating the relationship between the maximum possible energy of a signal and the noise that affects its reliable representation [18]. It is assumed that an image $f(i, j)$ containing $N \times M$ pixels and a reconstructed image $F(i, j)$ where F is reconstructed by decoding the encoded version of $f(i, j)$. Error metrics are calculated only with the luminance signal, so the values of $f(i, j)$ range from black (0) to white (255). Now, this is calculated by a series of steps:

Table 2 Comparison PSNR between different algorithms (dB)

Image	Levels	CS	FA	SFS	PSO
		PSNR	PSNR	PSNR	PSNR
20110106-Ct	2	**16.6627**	14.3353	11.4611	13.2451
	3	**18.5936**	15.9539	15.5317	13.4419
	4	**20.0226**	19.1797	18.8178	13.8525
	5	**20.4244**	20.4058	20.2546	13.9395
Chronic-SDH	2	**17.6104**	16.9630	12.4793	16.5644
	3	20.0337	**20.1639**	14.1013	15.7147
	4	**22.1984**	21.4391	19.9936	15.6805
	5	**23.5305**	23.0891	20.5005	15.2603
Contusion	2	**17.9268**	17.6132	12.3819	14.9604
	3	**20.9159**	19.3421	16.3183	14.8807
	4	20.8785	**21.0450**	19.6249	16.8234
	5	**24.0510**	23.5026	20.9266	158,658
Edh-1	2	17.9245	**19.6379**	12.3054	14.6461
	3	**22.4655**	21.3579	12.6472	15.8280
	4	**21.6769**	21.4665	19.2610	14.9133
	5	**23.8829**	22.9581	20.1961	15.1170
Embolism	2	18.7617	**19.1282**	13.5387	14.5645
	3	**21.4313**	20.4319	17.6394	15.4144
	4	**21.2767**	20.0757	19.9549	16.8021
	5	**23.5919**	23.2137	21.6532	16.6891
Hemorrhage	2	18.2593	**18.6964**	11.6806	14.5909
	3	20.3489	**20.6273**	12.0059	15.1448
	4	20.5696	**20.6350**	18.9737	15.1080
	5	22.8030	**22.9324**	20.0883	16.2235
Hydrocephalus	2	**19.5402**	19.4819	12.6060	14.8016
	3	**21.4459**	20.1327	17.8573	16.4002
	4	**22.2851**	21.7129	20.7974	17.6394
	5	**24.5962**	21.8320	21.3369	16.9634
Infarct	2	**19.7080**	16.2953	12.6043	15.0028
	3	**21.7341**	19.5203	16.4519	14.9918
	4	22.9637	**23.1625**	19.0658	15.5938
	5	**23.7235**	22.9008	20.4509	15.7397
Stroke	2	**18.1808**	16.1507	12.5089	14.8104
	3	**20.2839**	19.9772	16.8495	16.4092
	4	**22.0775**	20.9237	19.0784	16.4699

(continued)

Table 2 (continued)

Image	Levels	CS	FA	SFS	PSO
		PSNR	PSNR	PSNR	PSNR
	5	**23.2205**	22.7707	21.5340	17.0010
VAQ_20101-8	2	**16.5266**	15.3588	11.1469	14.3827
	3	18.0488	**19.1279**	11.2200	14.6704
	4	19.9860	**21.9658**	13.9818	15.4242
	5	21.2030	**23.2335**	19.8684	14.8223

1. The mean square error (MSE) of the reconstructed image is calculated as follows:

$$MSE = \sum \frac{[f(i,j) - F(i,j)]^2}{N \times M} \tag{12}$$

2. The RMSE is the square root of the MSE, so its square root is calculated:

$$RMSE = \sqrt{MSE} \tag{13}$$

3. PSNR is calculated:

$$PSNR = 20\log_{10}\left(\frac{255}{RMSE}\right) \tag{14}$$

The PSNR is expressed in decibels, so that the higher the value, the better the quality of the reconstructed image.

Table 3 shows the results obtained from the SSIM test applied to each algorithm. The SSIM test (structural similarity index) is a well-known quality metric used to measure the similarity between two images. Is considered to be correlated with the quality perception of the human visual system (HVS) [18]. The SSIM is designed by modeling any image distortion as a luminance distortion and contrast distortion. SSIM is defined as:

$$SSIM(f,g) = l(f,g)c(f,g)s(f,g) \tag{15}$$

Table 3 Comparison SSIM between different algorithms

Image	Levels	CS	FA	SFS	PS
		SSIM	SSIM	SSIM	SSIM
20110106-Ct	2	**0.7177**	0.7062	0.5056	0.7072
	3	0.7936	**0.8114**	0.7812	0.7003
	4	**0.8684**	0.8519	0.8311	0.7312
	5	**0.9052**	0.8505	0.8850	0.7378
Chronic-SDH	2	**0.8450**	0.8409	0.6202	0.8120
	3	**0.9052**	0.8873	0.6779	0.8306
	4	**0.9118**	0.9019	0.9051	0.8369
	5	0.9141	**0.9310**	0.9159	0.8122
Contusion	2	**0.8730**	0.8677	0.6680	0.8433
	3	**0.8977**	0.8799	0.8700	0.8470
	4	**0.9269**	0.9283	0.9098	0.8819
	5	**0.9505**	0.9432	0.9317	0.8793
Edh-1	2	0.8980	**0.9015**	0.6680	0.8497
	3	**0.9377**	0.9309	0.6803	0.8880
	4	**0.9373**	0.9247	0.9267	0.8795
	5	0.9486	**0.9535**	0.9344	0.8765
Embolism	2	**0.8930**	0.8744	0.7005	0.8620
	3	**0.9274**	0.9133	0.8954	0.8499
	4	**0.9320**	0.9187	0.9150	0.8951
	5	**0.9567**	0.9420	0.9261	0.8860
Hemorrhage	2	0.8795	**0.8803**	0.5992	0.8400
	3	**0.9258**	0.9145	0.6131	0.8516
	4	**0.9346**	0.9246	0.8982	0.8737
	5	**0.9539**	0.9407	0.9155	0.8554
Hydrocephalus	2	**0.8692**	0.8659	0.6723	0.8170
	3	0.9155	**0.9659**	0.8912	0.8272
	4	0.9218	**0.9390**	0.9163	0.8396
	5	**0.9347**	0.9302	0.9277	0.8463
Infarct	2	**0.8776**	0.8663	0.6842	0.8437
	3	**0.9341**	0.9234	0.8790	0.8476
	4	0.9378	**0.9399**	0.9063	0.8571
	5	**0.9615**	0.9462	0.9381	0.8785
Stroke	2	**0.8644**	0.8488	0.6905	0.8302
	3	**0.9128**	0.9095	0.8666	0.8433
	4	**0.9298**	0.9174	0.9062	0.8611

(continued)

Table 3 (continued)

Image	Levels	CS	FA	SFS	PS
		SSIM	SSIM	SSIM	SSIM
	5	**0.9500**	0.9485	0.9394	0.8444
VAQ_20101-8	2	**0.8114**	0.8100	0.4821	0.7901
	3	**0.8581**	0.8782	0.4840	0.7960
	4	**0.9016**	0.8828	0.6064	0.8478
	5	0.9054	**0.9144**	0.9042	0.8476

(a) (b) (c) (d)

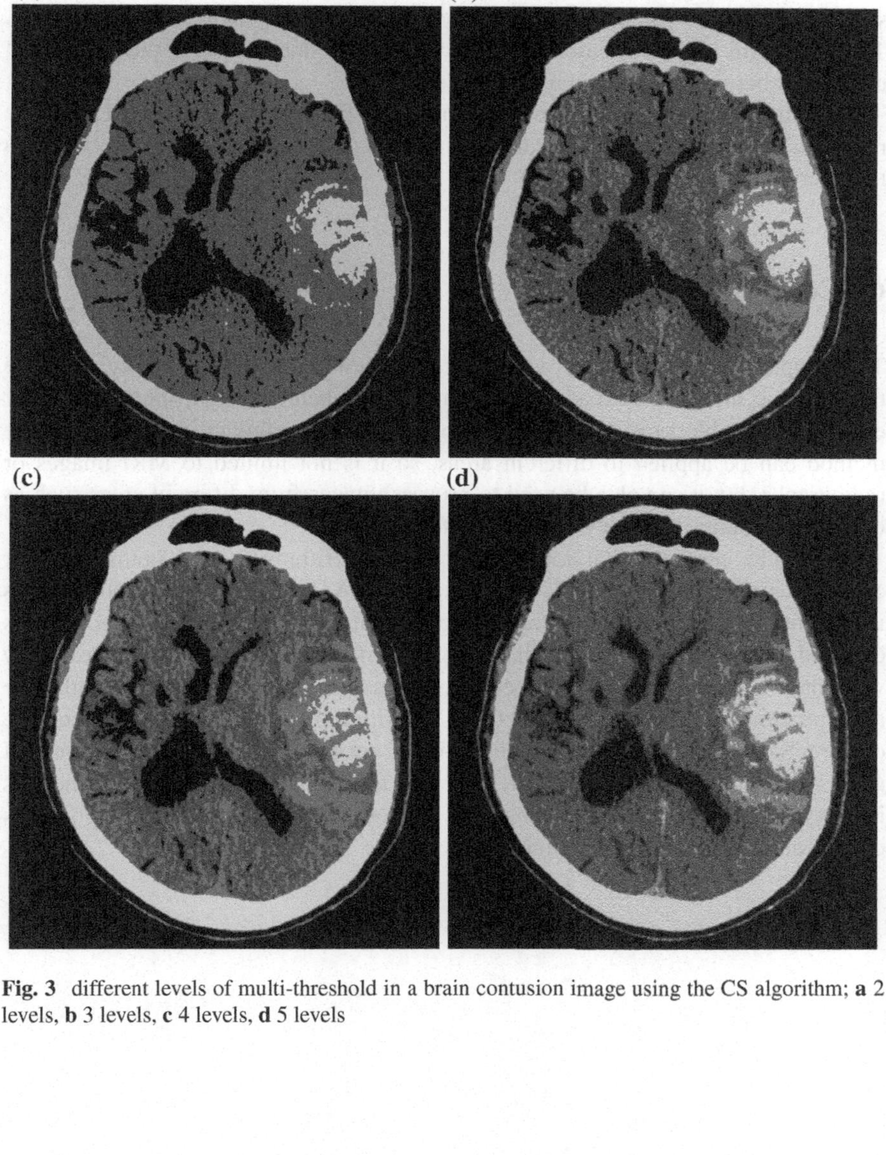

Fig. 3 different levels of multi-threshold in a brain contusion image using the CS algorithm; **a** 2 levels, **b** 3 levels, **c** 4 levels, **d** 5 levels

where:

$$\begin{cases} l(f,g) = \frac{2\mu_f\mu_g + C_1}{\mu_F^2 + \mu_G^2 + C1} \\ c(f,g) = \frac{2\sigma_f\sigma_g + C_2}{\sigma_f^2 + \sigma_g^2 + C_2} \\ s(f,g) = \frac{(\sigma_{fg} + C_3)}{(\sigma_f\sigma_g + C_3)} \end{cases} \quad (16)$$

The first term of Eq. 16 is the luminance comparison function that measures the closeness of the luminance average between the two images $(\mu_f \wedge \mu_g)$. This factor is equal to 1 only if $\mu_f = \mu_g$. The second term is the contrast comparison function. Measure the closeness of the contrast between the two images. It is measured with the standard deviation σ_f and σ_g. This term is maximum and equal to 1 only if $\sigma_f = \sigma_g$. The third term is the structure comparison function that measures the correlation coefficient between the two images f and g.

Note that σ_{fg} is the co-variance between f and g. Positive values of SSIM are in the range [0, 1] and a value of 0 indicates that there is no correlation, while a 1 indicates that there is a complete correlation.

6 Conclusions

This chapter makes a brief but very concise review regarding the theory of image thresholding, allowing the reader to implement a metaheuristic algorithm using Kapur entropy to be able to process clinical images. Beyond clinical images, this same method can be applied to different areas, so it is not limited to MRI images or radiographs, but it can also be used to process images from different areas such as orography, botany, histology, and many others.

The reader can intuit that not only can these algorithms be used, but many other existing ones such as the ABC, the CSA, the GSA, the HSA, among others. It is important to note again that the results depend directly on the way an algorithm is encoded, so 2 programmers may have slightly different results due to the way they write their code. It is evident that better results can be obtained over time, either by an improvement in the hardware or by an algorithm that improves the performance of the CS. In any case, the possibility should not be closed to an alternative that gives better quality results, after all, there will always be room for improvement as in everything; it will be left to the reader's discretion whether to try other algorithms or with another objective functions.

References

1. A. Patel, B. van Ginneken, F.J.A. Meijer, E.J. van Dijk, M. Prokop, R. Manniesing, Robust cranial cavity segmentation in CT and CT perfusion images of trauma and suspected stroke patients. Med. Image Anal. **36**, 216–228 (2017)
2. M.A. Elaziz, S. Lu, Many-objectives multilevel thresholding image segmentation using knee evolutionary algorithm. Expert Syst. Appl. **125**, 305–316 (2019)
3. S.L. Zuckerman et al., Structural brain injury in sports-related concussion. Neurosurg. Focus (2012)
4. A.E. Valasek, T.M. McCambridge, Sports-related concussion, in *Pediatric and Adolescent Concussion: Diagnosis, Management and Outcomes* (2012)
5. J.L. Cadet, I.N. Krasnova, S. Jayanthi, J. Lyles, Neurotoxicity of substituted amphetamines: molecular and cellular mechanisms. Neurotox. Res. (2007)
6. M.S. Gold et al., Methamphetamine- and trauma-Induced brain injuries: comparative cellular and molecular neurobiological substrates. Biol Psychiatry (2009)
7. N.R. Pal, S.K. Pal, A review on image segmentation techniques. Pattern Recognit. **26**(9), 1277–1294 (1993)
8. S. Hinojosa, K.G. Dhal, M.A. Elaziz, D. Oliva, E. Cuevas, Entropy-based imagery segmentation for breast histology using the stochastic fractal search. Neurocomputing **321**, 201–215 (2018)
9. T. Wang et al., SegRoot: a high throughput segmentation method for root image analysis. Comput. Electron. Agric. **162**, 845–854 (2019)
10. R. Xu, S. Niu, Q. Chen, Z. Ji, D. Rubin, Y. Chen, Automated geographic atrophy segmentation for SD-OCT images based on two-stage learning model. Comput. Biol. Med. **105**, 102–111 (2019)
11. S. Patra, R. Gautam, A. Singla, A novel context sensitive multilevel thresholding for image segmentation. Appl. Soft Comput. **23**, 122–127 (2014)
12. L. Shapiro, G. Stockman, Binary image analysis, in *Computer Vision* (2001)
13. P. Upadhyay, J.K. Chhabra, Kapur's entropy based optimal multilevel image segmentation using crow search algorithm. Appl. Soft Comput., 105522 (2019)
14. R.M. Gray, *Entropy and Information Theory* (2011)
15. A.S. Joshi, O. Kulkarni, G.M. Kakandikar, V.M. Nandedkar, Cuckoo search optimization—a review. Mater. Today Proc. **4**(8), 7262–7269 (2017)
16. J. Cheng, L. Wang, Y. Xiong, Ensemble of cuckoo search variants. Comput. Ind. Eng. **135**, 299–313 (2019)
17. W.C.E. Lim, K. Ganesan, S.G. Ponnambalam, PCB drill path optimization by combinatorial cuckoo search algorithm, **2014** (2014)
18. A. Horé, D. Ziou, Image quality metrics: PSNR vs. SSIM, in *Proceedings—International Conference on Pattern Recognition* (2010)

Generative Adversarial Network and Retinal Image Segmentation

Talha Iqbal and Hazrat Ali

Abstract With the advancements in medical equipment technologies, our capabilities to perform diagnosis from retinal images are increasing. However, the need for having more retinal image data is also rising. To investigate and detect possible abnormalities, the accessibility of the retinal images database is vital. While there exist a number of retinal images datasets mainly available from research institutions, the number is scarce and additional data is desired. As a matter of fact, it is not always feasible to collect more data from patients. This issue can be addressed by producing synthetic retinal images from already existing limited data. Hence, it is desirable to develop and validate machine learning algorithms which may generate additional synthetic retinal image data. Generative models provide for generating retinal images. Generative models can learn essential concealed features of data which can be utilized for generating new retinal images having close resemblance to the original images. This chapter provides the reader a detailed discussion on the theoretical and mathematical model of generative adversarial networks (GANs), their use cases and their capability to generate synthetic retinal images.

Keywords Deep learning · Generative adversarial networks · Medical imaging

1 Human Eye

1.1 The Anatomy of the Eye

The 80% of information fed to the brain is received from eyes [18] and retina tissue is the most important section of the eye. Retina is a lively metabolic muscle having

---

T. Iqbal
Lambe Institute of Translational Research, National University of Ireland, Galway, Ireland
e-mail: t.iqbal1@nuigalway.ie

H. Ali (✉)
Department of Electrical and Computer Engineering, COMSATS University Islamabad, Abbottabad Campus, Islamabad, Pakistan
e-mail: hazratali@cuiatd.edu.pk

D. Oliva and S. Hinojosa (eds.), *Applications of Hybrid Metaheuristic Algorithms for Image Processing*, Studies in Computational Intelligence 890,
https://doi.org/10.1007/978-3-030-40977-7_21

blood supply. One may see and interpret its movement directly. The position and functionality of retina carry vital role, which is why physicians and doctors look at it to detect and identify most of the eye related diseases. We can also detect diseases which are visible in the retinal images, such as diabetic retinopathy. To understand the information about the diseases, it is useful to familiarize the reader with the structure of the eye [18].

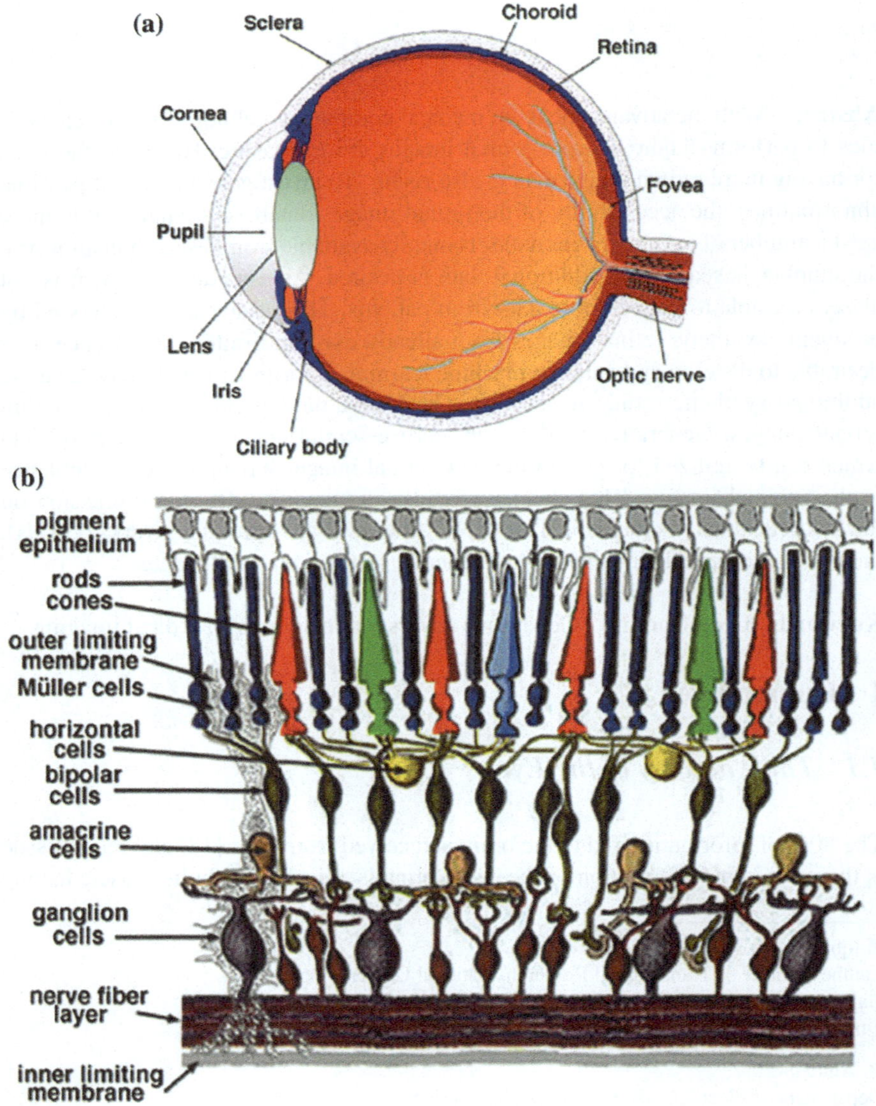

Fig. 1 Structure of eye

Eye is round shape in structure and has three major layers: first one is the outer layer which is made of sclera and cornea. The second one is center layer that includes ciliary body, iris, and choroid. The last layer is inner layer which is composed of retina. An incoming light ray enters the eye and travels to the retina after passing the cornea, pupil and lens. The retina collects incoming light rays and converts them into specific neural signal. The neural signal is then read by the brain via optic nerve.

Figure 1 shows the structure of one eye and all the layers in retina. For more comprehensive information about the eye and structure of eye, readers are referred to [18].

2 Overall Framework for Generation of Synthetic Images

Typically, the generation of synthetic retinal data (images) has three stages, as presented in Fig. 2. The first stage is the pre-processing of the dataset. Here we deal with the rescaling and cropping of the image. By variation in the resolution of retinal images, having a preprocessing step has become essential. A key point to ponder here is that training process has high computational cost due to fairly large scale retinal images. So, the retinal images are downscaled. Moreover, the black regions in all retinal images are cropped. The second step is operation of generative adversarial models (GANs) to produce synthetic retinal dataset (images). Both the methods are explained in detail in the next sections.

2.1 Introduction to the Generative Models

For better understanding of fundamentals of the generative models, it is useful to differentiate the generative models.

A computer can be trained to recognize the data and make interpretations based on that data using different machine learning approaches. There are two different types of learning methodologies, first is supervised and second is unsupervised. In supervised learning, we have a label for each data occurrence. Here, the objective of the model is to recognize the data conditioned to a particular target and assign a new target value

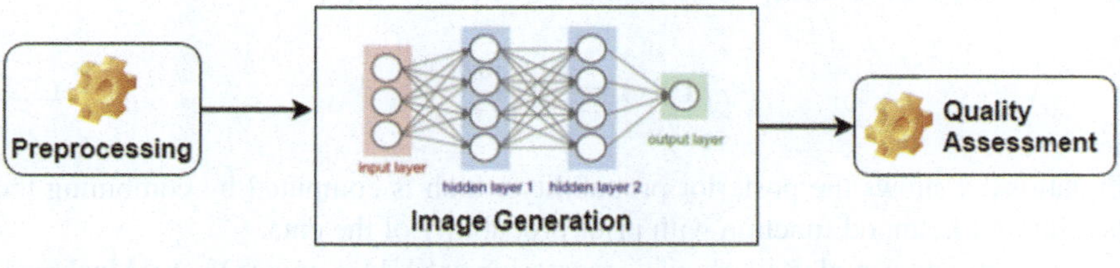

Fig. 2 Framework of generating new retinal images

to a newly input data. If these target values are different types/classes/categories, then such problems are called classification problems. For example, we have a spam filtering problem and we want to assign a label i.e., spam or non-spam to the new incoming emails, then this is a classification problem and we can use the classification algorithm like logistic regression algorithm to solve this problem. On the other hand, if we have continuous value targets, then it is a regression problem. For example, the prediction of rates of exchange for different currencies. The aim is to calculate the rate of exchange depending upon certain factors such as a major political change in a country, a natural disaster or a specific boom in exports by a country. As the exchange rate is a continuous value, the problem of prediction can be addressed by using regression techniques such as linear regression.

In unsupervised learning, the data does not have target labels. There are no defined classes to be allocated to an input. The idea is to cluster the input data into different groups. Examples of unsupervised learning include clustering the customers based on their patterns of online purchasing or clustering group of people as supporters of *Manchester United* based on their tweets pattern. Both the previously mentioned tasks falls in the category of unsupervised learning. Both generator and discriminator models are extensively applied in supervised learning problems [2, 22]. Each one of the two approaches a given problem with a different perspective. Generator model tries to learn the distribution of input dataset and thus learns how the data are generated for classification. Generative models have the capability to generate new samples from given data. Discriminative models on the other hand, discriminate the data using prior probabilities. Figure 3 summarizes a general hierarchy of different types of machine learning models.

Suppose, we have an input as x and our aim is to label this data into related class denoted by y. Say, this is a classification problem. In order to perform the required classification, we use a deep generator model that acquires joint probability of x and y i.e. $p(x, y)$. The discriminator, on the other hand learns class probability $p(y|x)$. This prior probability is actually the decision boundary between two or more output classes. The joint probability calculation is done with Eq. 2.1:

$$p(x, y) = p(x|y)p(y) \tag{2.1}$$

Here conditional probability is denoted as $p(x|y)$ and $p(y)$ is the marginal distribution of the class.

Generative method learns $p(x|y)$ and $p(y)$ to assign a given data to a suitable class. Then $p(y|x)$ is computed by Bayes Rule [2]:

$$p(y|x) = \frac{p(x|y)p(y)}{p(x)} \tag{2.2}$$

Equation 2.2 shows the posterior probability which is computed by combining the maximum likelihood function with prior probability of the data.

Discriminative models like logistic regression or SVM (Support Vector Machines) [2] perform classification using training dataset as input, by following these steps:

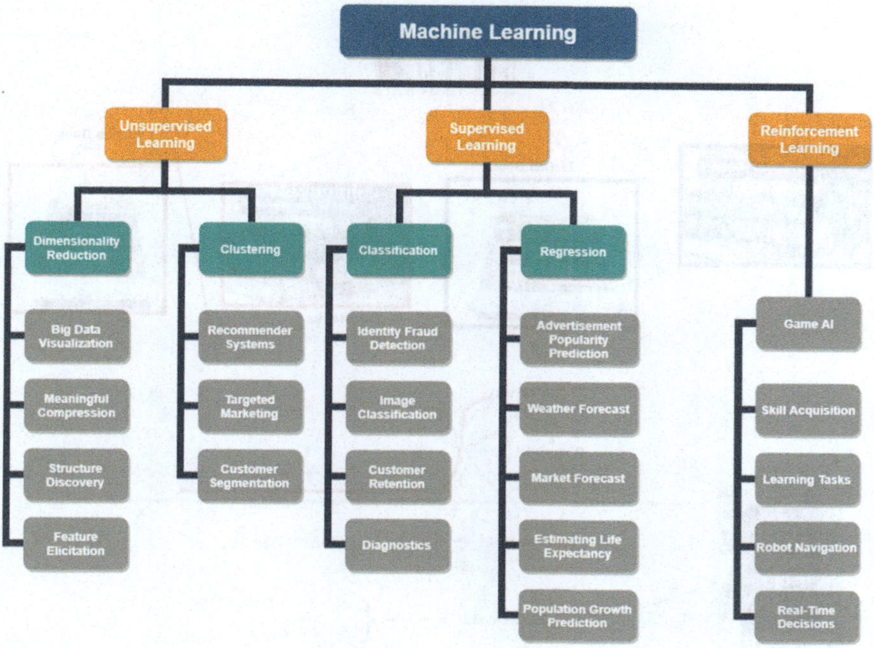

Fig. 3 Machine learning types and applications

- Find a decision boundary by training the model.
- After the training, the model can be utilized to classify new data samples. For example as shown in Fig. 4, the entry (book) is classified as class A (say *history* books) or class B (say *fiction* books).

The technique mentioned above mainly describes how a discriminator works. A Bayesian classifier solves the classification task as if we are using generative modeling [2]:

- First of all we select all the entries available under the label of class A. The model learns the pattern of class A data i.e., it learns $p(x|y = 0)$.
- Next, we select all the entries available under the label of class B and build a model which learns the pattern of class B data i.e., it learns $p(x|y = 1)$.
- The new data is assigned a class label by looking at the new entry and predicting the class to which the new data is closer.

2.2 *Parameter Estimation*

After given the insight of generative and discriminative models, next we discuss how generative model produces novel synthetic/fake data from the existing/real data.

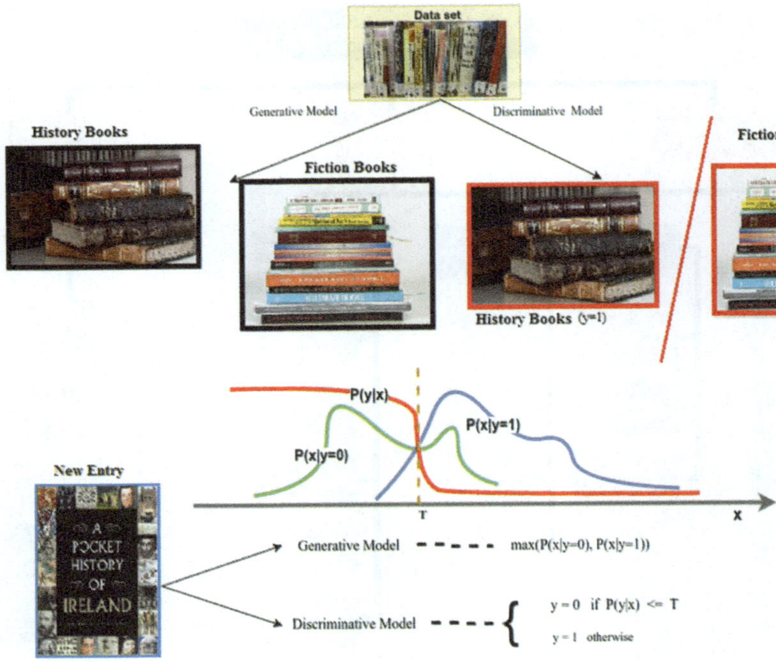

Fig. 4 Difference between the generator and discriminator models

Suppose we have two categories of books. Let class A be the *history* books and class B be *fiction* books. Consider that, we have seen all the class entries and have learnt some tricks to replicate all the entries to look like class A or class B entry. Now, as an expert, we are able to differentiate any input of class A from class B. After the model has learnt about both the entries, we will make our new data conditioned on the entry sample we have in our hand. Our goal is to produce/generate the new entry looking as close as possible to the entries in class A. In generative models, the parameter estimation is a crucial step to create new samples.

Figure 4 explains generator and discriminator models in machine learning classification problem. Generative model acquires the features of both classes and discriminative model learns and draws decision boundary between both classes.

There are two stages to generate new data using generative model; the first step is estimation of the parameters for the model and second step is to generate new samples from these estimated parameters. The estimator uses maximum likelihood estimator (MLE) i.e., to find the parameters which best describe the given data. For example, suppose our data is x and the parameter assigned to it is θ. This parameter value can be mean (denoted by μ) or variance (denoted by σ^2) of the given data. This means the distribution of given input must be Gaussian and its density function be denoted as $p(x|\theta)$. Maximum Likelihood Estimator finds the parameters, which can maximize this density function, as shown in Eq. 2.3:

$$\hat{\theta} = \arg_{\theta} \max p(x|\theta) \tag{2.3}$$

For the production of the new data we do sampling as $\hat{x} \rightarrow p(x|\theta)$ after the parameters are estimated.

3 Generative Models

With enhanced computational resources, innovations have been made on constructing complex neural networks with deep architectures [26]. With the advent of new and complex models, we need large power for manipulating huge data-sets as well as running complex models. This development has made the researcher community to take steps and make some substantial improvements in different deep learning algorithms to solve challenging machine learning and computer vision problems [8, 26].

The latest development in deep learning domain is the work on deep generative models. Generative Adversarial Networks (GANs) [14], a recently proposed model, attracted the research community to investigate its use on artificial data generation task. In the medical imaging domain, GANs can be used to generate realistic looking retinal images.

3.1 Generative Adversarial Networks (GANs)

GANs provide an implementation of generative models for artificial or fake image generation. In this section, we discuss GAN model in detail and also summarize previous studies on use of GAN model.

Goodfellow et al. [14], in 2014, presented a generative adversarial network structure as a substitute to generative model estimators, and named it GANs. GANs are made using two types of deep neural networks i.e. Generator network, denoted as (G) and Discriminator network, denoted as (D). Both networks compete against each other to maximize their own gains. Generator (G) generates output image from some random noise z while discriminator (D) discriminates the data into two classes i.e., the data generated by (G) and the data coming from the training set. So as described in [14], the training of GANs can be explained in analogy with the competition between a criminal and the police. Generator G is the criminal and tries to generate forged money and uses it without been discovered while discriminator D is acting as the police who tries to identify the fake money. For both the models, the basic objective is to improve their skills. GANs structure is demonstrated in Fig. 5.

In GANs, generator G acquires the distribution P_g and the input data x as [14]:

1. $P_z(z)$ is the prior input noise variables and is defined before the training.
2. We construct a representation of the data space like $G(z; \theta_g)$.

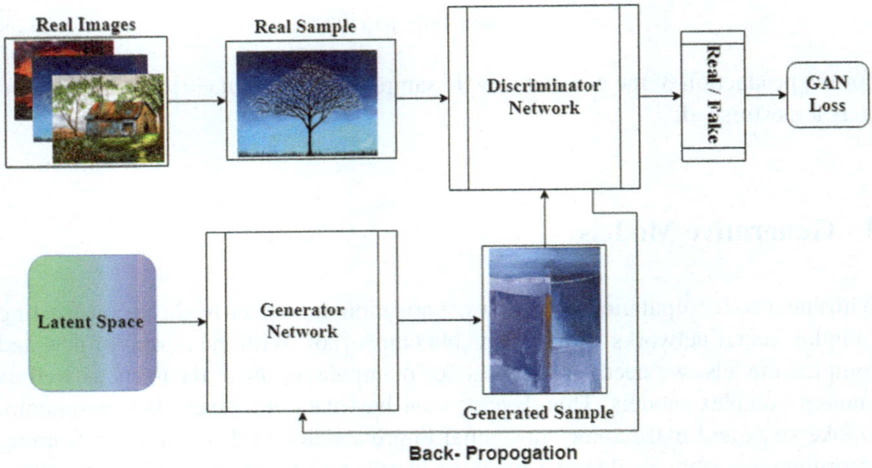

Fig. 5 General structure of generative adversarial networks

3. We design one more network of multi-layer perceptron $D(x; \theta_d)$. Here discriminator D provides only a single scalar value at output. This output is greater (close to 1) if the sample is derived from real data and is small (close to zero) for the synthetic/fake image.
4. Finally, we train both the D (discriminator) and the G (generator) simultaneously. The discriminator D maximizes probability of correctly assigned label to new input while generator G aims to minimize the term $log(1 - D(G(z)))$.

Discriminator and generator networks play a min-max game that can be expressed using $V(D, G)$ function and is defined as [14]:

$$\min_G \max_D V(D, G) = E_{x \to P_{data}(x)}[\log D(x)] + E_{z \to P_z(z)}[\log(1 - D(G(z)))] \tag{3.1}$$

Here, discriminator D tries to maximize $D(x)$ and minimize $D(G(z))$. So, the objective function of D becomes as:

$$\max_D V(D, G) = E_{x \to P_{data}(x)}[\log D(x)] + E_{z \to P_z(z)}[\log(1 - D(G(z)))] \tag{3.2}$$

Generator G minimizes the objective function shown in Eq. 3.1 and is represented as:

$$\min_G V(D, G) = E_{z \to P_z(z)}[\log(1 - D(G(z)))] \simeq \max_G E_{z \to P_z(z)}[\log(D(G(z)))] \tag{3.3}$$

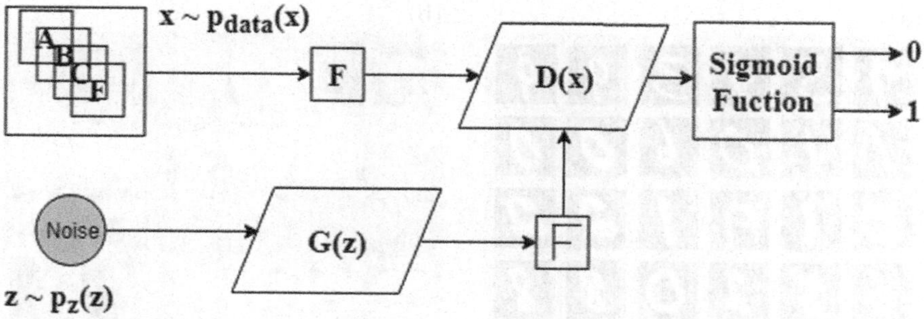

Fig. 6 A visualization for mathematical representation of GAN

Both generator and discriminator plays min-max game and adjusts their parameters based on each other decisions and actions [13]. Note that neither of the models directly controls the other model parameters. This min-max game stops at Nash Equilibrium point [13]. The mathematical justification of GANs is briefly shown in Fig. 6. Algorithm 1 shows the training procedure implemented by Ian Goodfellow in [14].

GANs were initially tested on MNIST and CIFAR-10 datasets [14]. We can see that GANs are capable of generating synthetic images which are a lot similar to real images (see Fig. 7).

3.2 *Steps for Designing Generative Adversarial Network*

Before designing architecture of a GAN, we must consider the point; it is proven that there are certain blocks of CNN, which help in accelerating computation and avoiding mode collapses of the whole network.

The GAN reported in [14] is based on simple neural networks. Shortcomings of GANs such as unstable behavior during training and collapse of model have motivated the research community to explore and design new methods of GANs training. As the Convolutional Neural Network based deep neural networks are powerful approach for classification problem [8, 26], most variants of GANs are based on CNN designs [9, 27, 35, 38]. Thus, better training mechanism for CNN based models has been revealed. Predominantly, Radford et al. [35] performed a wonderful study to explore best possible approaches to accelerate the training of GANs while ensuring stability. Some changes introduced in [13, 35] are:

1. Use ReLU as an activation function.
2. Use Batch Normalization.

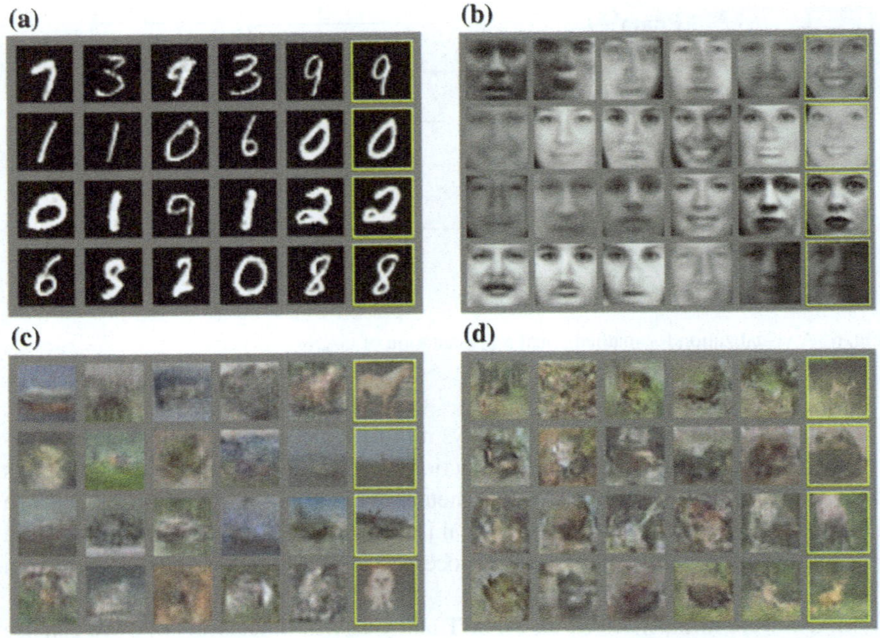

Fig. 7 Generated samples: **a** MNIST dataset; **b** TFD dataset; **c**, **d** CIFAR-10 dataset [14]

3. Use Leaky ReLU.
4. For upsampling, use Deconvolution/UpSampling Layer.

Due to the nonlinearity while ensuring stability activation functions, there are critical design patterns to build a CNN architectures. We have four types of activation functions that are used in the deep neural networks [37]:

- **Sigmoid**: The input to a sigmoid function is real number. The sigmoid converts the input into the range between 0 and 1. It is shown in the previous GANs that sigmoid generates blurry and saturated images [9, 38]. Furthermore, it is not a preferred choice as it generates a non-centered output.
- **Tanh**: This function transforms real values to numbers in the range between -1 and 1. We have zero centered output and the produced images are sharp.
- **ReLU**: It speeds up the convergence of deep neural network [3]. It was used in most of the early GAN designs [9, 13, 27, 35, 38]. ReLU is computationally less expensive. ReLU is also required to be use with care as it may lead to the *knockout problem* i.e., it causes a dying neuron problem [3].

Algorithm 1: Minibatch based training procedure of GAN as defined in [7]. The discriminator is updated k times, which is a hyperparameter and Goodfellow used $k = 1$ by considering computation time.

foreach *number iterations* **do**
  foreach k *steps* **do**
   - Sample minibatch of m noise samples $\{z^{(1)}, ..., z^{(m)}\}$ from noise prior $p_g(z)$.
   - Sample minibatch of m noise examples $\{x^{(1)}, ..., x^{(m)}\}$ from data generating distribution $p_{data}(x)$.
   - Update the discriminator by ascending stochastic gradient descent: $\nabla_{\theta_d} \frac{1}{m} \sum_{i=1}^{m} [\log D(x^{(i)}) + \log(1 - D(G(z^{(i)})))]$

  end
  - Sample minibatch of m noise samples $\{z^{(1)}, ..., z^{(m)}\}$ from noise prior $p_g(z)$.
  - Update the generator by descending stochastic gradient descent: $\nabla_{\theta_g} \frac{1}{m} \sum_{i=1}^{m} [\log(1 - D(G(z^{(i)})))]$

end

Another vital block in the architecture of GAN is *batch normalization* [19]. It decreases the network's training time. Generally, normalization is considered as a pre-processing step. A common issue in most of the neural networks is the initialization of weights, which may lead to over-fitting [26]. Compared to dropout layer used previously batch normalization allows fast learning and gives better accuracy along with prevention of over-fitting problem [13]. Batch normalization can also be thought of as regularization term which permits the flow of data in-between different layers to be whitened.

One more key ingredient for constructing a GAN is the Up-Sampling layer [35]. Up-sampling layers take an input image and up-scales the images to greater dimensions via calculating each pixel value using bilinear interpolation.

3.3 Generative Adversarial Networks Application

The basic GAN model uses neural network as the building blocks. Denton et al. [9] extended the GANs to CNN and named it LAPGAN. Here, images are produced with the use of Laplacian pyramid structure [34]. The architecture in LAPGAN uses series of GANs. Each of the GAN works at a certain scale value of Laplacian pyramid and captures some unique structures of the images.

The critical problem incurred while dealing with GANs is that discriminator takes fairly long time for minimization of the loss function. Moreover, it does not provide any facility of controlling the learning process, which helps to speed up the overall learning process and also helps us to generate more realistic data examples. To overcome this problem, Gauthier [38] designed a GAN for face generation task on the basis of previous knowledge. The work in [38] added a conditioning aptitude to GANs and thus referred as cGAN,, which is illustrated in Fig. 8.

A noticeable factor is conditional data vector y which have massive impact on the training of generator G and discriminator D models. Here facial expression and some additional face related structures such as age and race are selected as y. Function in Eq. 3.1 is re-defined when we include y with generator G and discriminator D:

$$\min_{G}\max_{D} v(D, G) = E_{x,y\rightarrow P_{data}(x,y)}[\log D(x, y)] + E_{y\rightarrow P_y, Z\rightarrow P_Z(Z)}[\log(1 - D(G(Z, y), y))] \quad (3.4)$$

cGAN and GANs have used the LFW (Labeled Faces in the Wild) dataset [38]. It is presented that cGAN generated better realistic looking images than GANs. The structure shown in Fig. 8, has both generator G and discriminator D in the last layer.

The capability of GANs to learn the details of the input images and produce realistic looking images at the output has motivated researchers and scientists to inspect its potential use in different fields. For example, InfoGAN model [5] is proposed that learns the conditional variable all by itself. The model in [5] suggests using mutual information for enhanced capacity. This conditional variable could be categorical class information or continuous variables like rotated angles and edge structure. This depends on the kind of problem we are dealing with. Weidong et al. [42] was motivated by cGAN and presented Semi-Latent GAN. The model in [42] can generate face images which can express emotions such as sad face, male face or female face, face color, etc. It modifies facial images conditioned on above facial attributes. The study in [27] has utilized GANs for production of high resolution images from a poor quality images. The reported GAN can produce super resolution output images from its input pair.

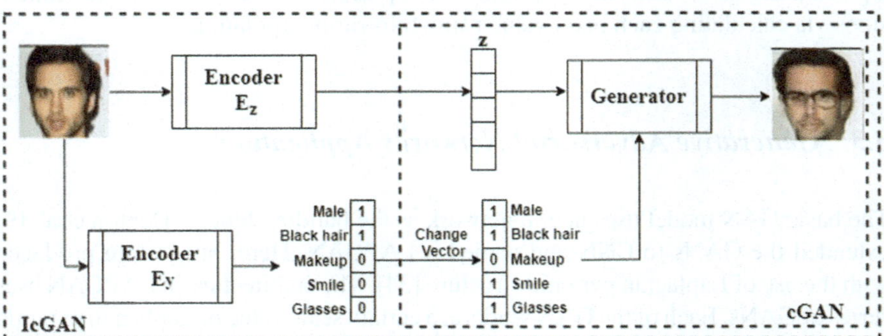

Fig. 8 Basic structure of cGAN

In the field of computer vision, automatic captioning of images is an interesting task. In this problem, the image context is defined textually conditioned on some features from the input data sample [26]. To use GANs for such problems, Zhang et al. proposed to invert the defined problem and established a variant known as StackGAN [43] that produces output images just from the text descriptions. StackGAN is divided into two stages of GANs which are stacked on each other. The first GAN takes a text description and produces some basic shapes such as edges, colors and corners. Then, the second GAN takes the output from the first GAN and processes a given description to generate an output image with better resolution. GANs have also been used on astro-physical images dataset [36] for the purpose of feature extraction problem.

With this background on applications of GANs for diverse tasks, we now present the work done done on generation of synthetic retinal images using GAN.

4 Synthetic Retinal Image Generation

In recent times, strong interest has emerged in the use of computer-aided imaging tools [1, 6, 7, 10, 16, 17, 24, 25, 29, 41]. Computer aided diagnosis relies on advanced machine learning and computer vision techniques [4]. Today, majority of the medical professionals use computer-aided medical images for diagnosis purposes. Retinal vessel network analysis gives us information about the status of general system and conditions of the eyes. Ophthalmologists can diagnose early sign of vascular burden due to hypertension and diabetes as well as vision threatening retinal diseases like Retinal Artery Occlusion (RAO) and Retinal Vein Occlusion (RVO) from abnormality in vascular structure [32]. To augment such analysis, automatic vessels' segmentation methods have been extensively studied. Recently, deep learning methods have shown potential to produce promising results with higher accuracy, occasionally better performance than medical specialist in the field of medical imaging [11]. Deep learning also improves efficiency of analyzing data due to its computational and automated nature. However, most of the medical images are often 3 dimensional (e.g. MRI and CT). It is difficult as well as inefficient to produce manually annotated images. In general, medical images are inadequate, expensive and offer restricted use due to legal issues (patient privacy). Moreover, the datasets of medical images available publicly often lack consistency in size and annotation. This makes them less useful for training of neural networks, which are data-hungry. This directly limits the development of medical diagnosis systems. So, generation of synthetic images along with their segmented images is helpful in medical image analysis and diagnosis systems.

For generation of synthesized images, different variants of GANs have been reported in [9, 15, 31, 35]. DCGAN [35] introduced set of constraints which stabilized the training of the model. CGAN [31] trained the model and generated output conditioned to some auxiliary information. LAPGAN [9] uses cascade formation of convolutional neural networks within framework of Laplacian pyramid for the generation of the new images. InfoGAN [15] learns disentangled representations in unsupervised manner. GANs have performed well on small medical image datasets as discussed in [33]. The authors in [33] have used GANs for unsupervised adaptation of the multi-modal medical images. GANs are useful for many applications like unsupervised representation learning [39] or image-to-image translation [12]. Typically, vessel segmentation task is considered as image translation problem where segmented vessel map is produced at output using fundoscopic image as an input to the model. We can have clearer and sharper vessel segmented masks, if we constrain our output to resemble the annotation done by human experts.

Recent work in the domain of medical imaging has shown possibility of improved performance even on small datasets. This has become possible through provision of some prior knowledge in a deep neural network [30]. U-net [11] architecture is popular for segmentation of bio-medical images, which shows how strongly an augmented data can be utilized to cope with low amount of training data available to train deep networks. Data augmentation is easy to implement and gives good results but it is only able to give fixed variations for any given dataset and requires the augmentation to fit in the given dataset. Impressive results are achieved by Gatys et al. [12] by application of deep learning algorithm. Similar approach with modifications has been used by [21, 40], reducing the computational complexity. More traditional approaches for segmentation of filamentary structured images have been reported in [23, 28].

A recent development on retinal images generation using GANs is reported in [20]. The authors in [20] have introduced a variant of GAN called as the MI-GAN that focuses on retinal vessels image segmentation and generation. These synthesized images are convincingly realistic looking (see Figs. 9 and 10). The authors claim to enhance the image segmentation performance by including the generated retinal

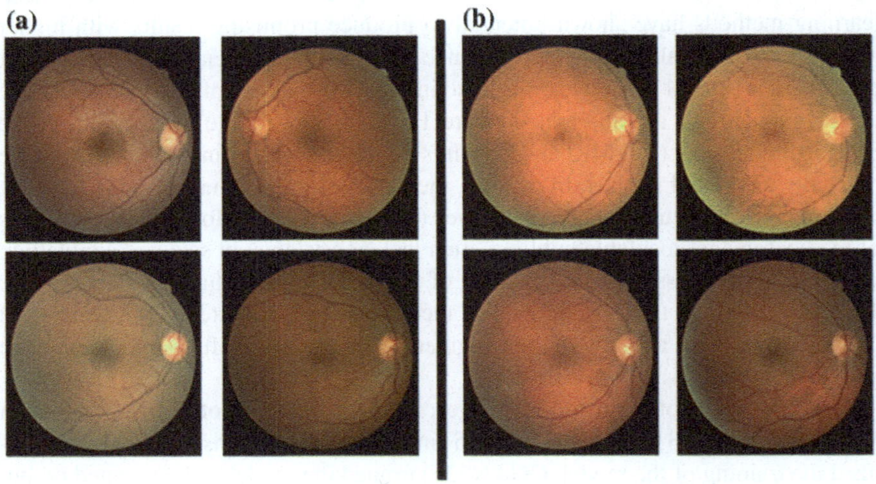

Fig. 9 DRIVE Dataset: **a** Original (Input) Images, **b** Generated (Output) Images, as reported in [20]

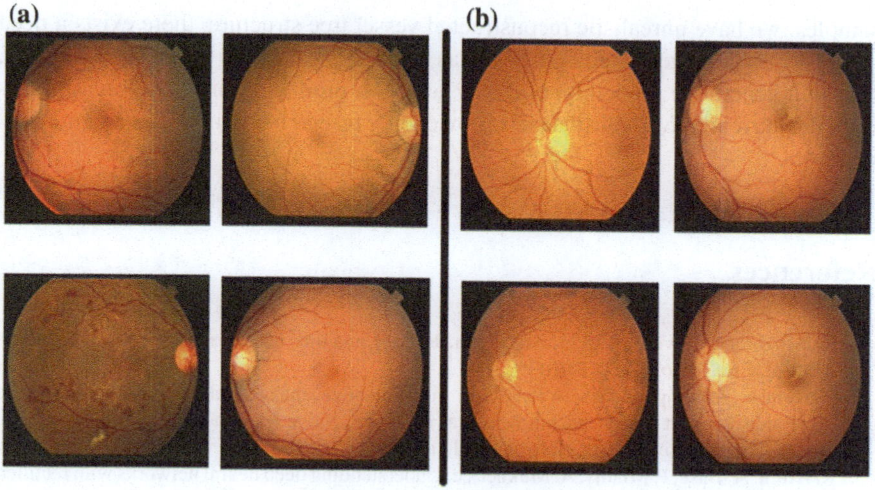

Fig. 10 STARE Dataset: **a** Original (Input) Images, **b** Generated (Output) Images, as reported in [20]

images into training set of the algorithm. The MI-GAN is capable of learning useful features from a small training set. The subset used to train MI-GAN consisted of only 10 examples from each dataset namely DRIVE and STARE. They have compared their model with few state of the art algorithms like Kernel Boost, N^4-Fields, DRIU and Wavelets methods and their algorithm outperformed all other existing models in terms of AUC ROC, AUC PR and Dice co-efficient. The MI-GAN achieved less false positive rate at fine vessels and have drawn more clearer segmentation lines, as compared to other methods. Their proposed future work involves investigation into datasets of different bio-medical images for interplay of synthesized images, domain adaptation tasks and segmentation of the medical images.

5 Conclusion

The generation and reconstruction of retinal images from existing retinal images dataset is a valuable research direction as it provides large dataset to the scientists for experimentation, validation and auxiliary advancement in the retina's vascular tree structure examination algorithms. Specifically, researchers focus on reconstructing of the phantom retinal images using chromatic colors like from RBG pairs. The available chromatic retinal data can also be utilized to generate chromatic retinal images along with reconstruction of spectral retinal data. The synthesis of retinal images can be accomplished by the use of GANs. For example, we have seen that the fovea and optical disc are synthesized quite well. However, the synthesis of vessel tree structure is challenging and is an open research direction. As in the majority of samples, we have unrealistic reconstructed vessel tree structure, there exists a need of large retinal images database (as well as other medical image datasets). Therefore, by using generative modeling, one may focus on producing diverse retinal images from the currently available datasets provided by research communities and medical institutes.

References

1. M.D. Abràmoff, M.K. Garvin, M. Sonka, Retinal imaging and image analysis. IEEE Rev. Biomed. Eng. **3**, 169–208 (2010)
2. P. Alonso, A comparison between some discriminative and generative classifiers (logistic regression, support vector machines, neural networks, Naive Bayes and Bayesian networks). Master Thesis, University of Helsinki (2015)
3. R. Arora, A. Basu, P. Mianjy, A. Mukherjee, Understanding deep neural networks with rectified linear units. arXiv preprint arXiv:1611.01491 (2016)
4. C. Becker, R. Rigamonti, V. Lepetit, P. Fua, Supervised feature learning for curvilinear structure segmentation, in *International Conference on Medical Image Computing and Computer-Assisted Intervention* (Springer, Berlin, 2013), pp. 526–533
5. X. Chen, Y. Duan, R. Houthooft, J. Schulman, I. Sutskever, P. Abbeel, InfoGAN: interpretable representation learning by information maximizing generative adversarial nets, in *Advances in Neural Information Processing Systems*, pp. 2172–2180 (2016)
6. N. Chhabra, R. Dutta, Low quality iris detection in smart phone: a survey. Int. J. Comput. Sci. Mobile Comput. **5**(4), 271–276 (2016)
7. T.E. de Carlo, E. Moult, W. Choi, N. KWaheed, J.S. Duker, J.G. Fujimoto, Non-neovascular age-related macular degeneration, in *Clinical OCT Angiography Atlas*, 1st edn. (Jaypee Brothers Medical Publishers, New Delhi, India, 2015), pp. 75–88
8. L. Deng, Y. Dong et al., Deep learning: methods and applications. Found. Trends Signal Process. **7**(3–4), 197–387 (2014)
9. E.L Denton, S. Chintala, R. Fergus, et al., Deep generative image models using a Laplacian pyramid of adversarial networks, in *Advances in Neural Information Processing Systems*, pp. 1486–1494 (2015)
10. H.L. Feng, S. Sharma, S. Stinnett, S. Asrani, P. Mruthyunjaya, *Identification of Posterior Segment Pathology With en Face Retinal Imaging Using Multicolor Confocal Scanning Laser Ophthalmoscopy* (Retina, Philadelphia, PA, 2018)
11. Y. Ganin, V. Lempitsky, N^4-fields: neural network nearest neighbor fields for image transforms, in *Asian Conference on Computer Vision* (Springer, Berlin, 2014), pp. 536–551
12. L.A. Gatys, A.S. Ecker, M. Bethge, A neural algorithm of artistic style. arXiv preprint arXiv:1508.06576 (2015)
13. I. Goodfellow, NIPS 2016 tutorial: generative adversarial networks. arXiv preprint arXiv:1701.00160 (2016)
14. I. Goodfellow, J. Pouget-Abadie, M. Mirza, B. Xu, D. Warde-Farley, S. Ozair, A. Courville, Y. Bengio, Generative adversarial nets, in *Advances in Neural Information Processing Systems*, pp. 2672–2680 (2014)
15. I. Gulrajani, F. Ahmed, M. Arjovsky, V. Dumoulin, A.C. Courville, Improved training of Wasserstein GANs, in *Advances in Neural Information Processing Systems*, pp. 5769–5779 (2017)
16. T.-C. Ho, M.-S. Chen, J.-S. Huang, Y.-F. Shih, H. Ho, Y.-H. Huang, Foveola nonpeeling technique in internal limiting membrane peeling of myopic foveoschisis surgery. Retina **32**(3), 631–634 (2012)

17. F.G. Holz, R.F. Spaide, *Medical Retina: Focus on Retinal Imaging* (Springer Science & Business Media, Berlin, 2010)
18. M.I. Iacono, E. Neufeld, E. Akinnagbe, K. Bower, J. Wolf, I.V. Oikonomidis, D. Sharma, B. Lloyd, B.J. Wilm, M. Wyss et al., MIDA: a multimodal imaging-based detailed anatomical model of the human head and neck. PloS one **10**(4), 1–35 (2015)
19. S. Ioffe, C. Szegedy, Batch normalization: accelerating deep network training by reducing internal covariate shift, in *Proceedings of the 32nd International Conference on International Conference on Machine Learning*, vol. 37, ICML'15, pp. 448–456 (2015)
20. T. Iqbal, H. Ali, Generative adversarial network for medical images (MI-GAN). J. Med. Syst. **42**(11), 231 (2018)
21. J. Johnson, A. Alahi, L. Fei-Fei, Perceptual losses for real-time style transfer and super-resolution, in *European Conference on Computer Vision* (Springer, Berlin, 2016), pp. 694–711
22. M.A. Kiasari, G.-J. Jang, M. Lee, Novel iterative approach using generative and discriminative models for classification with missing features. Neurocomputing **225**, 23–30 (2017)
23. C. Kirbas, F. Quek, A review of vessel extraction techniques and algorithms. ACM Comput. Surveys (CSUR) **36**(2), 81–121 (2004)
24. L. Laaksonen et al., Spectral retinal image processing and analysis for ophthalmology. Acta Universitatis Lappeenrantaensis (2016)
25. D. Le Bihan, Looking into the functional architecture of the brain with diffusion MRI, in *International Congress Series*, vol. 1290 (Elsevier, Amsterdam, 2006), pp. 1–24
26. Y. LeCun, Y. Bengio, G. Hinton, Deep learning. Nature **521**(7553), 436–444 (2015)
27. C. Ledig, L. Theis, F. Huszár, J. Caballero, A. Cunningham, A. Acosta, A. Aitken, A. Tejani, J. Totz, Z. Wang, et al., Photo-realistic single image super-resolution using a generative adversarial network, in *2017 IEEE Conference on Computer Vision and Pattern Recognition (CVPR)*, pp. 105–114 (2017)
28. D. Lesage, E.D. Angelini, I. Bloch, G. Funka-Lea, A review of 3D vessel lumen segmentation techniques: models, features and extraction schemes. Med. Image Anal. **13**(6), 819–845 (2009)
29. Y.B. Liang, Q. Zhou, D.S. Friedman, L.X. Guo, L.P. Sun, Q.F. Zong, X.D. Yang, N.L. Wang, A population-based assessment of 24-hour ocular perfusion pressure among patients with primary open angle glaucoma: the Handan eye study. Asia-Pacif. J. Ophthalmol. **5**(2), 127–132 (2016)
30. G. Litjens, T. Kooi, B.E. Bejnordi, A. A. Adiyoso Setio, F. Ciompi, M. Ghafoorian, J.A.W.M. van der Laak, B. van Ginneken, C.I. Sánchez, A survey on deep learning in medical image analysis. Med. Image Anal. 42:60–88 (2017)
31. M. Mirza, S. Osindero, Conditional generative adversarial nets. arXiv preprint arXiv:1411.1784 (2014)
32. U. Nguyen, L. Laaksonen, H. Uusitalo, L. Lensu, Reconstruction of retinal spectra from RGB data using a RBF network, in *2016 6th International Conference on Image Processing Theory Tools and Applications (IPTA)* (IEEE, New York, 2016), pp. 1–6
33. H. Peng, M. Hawrylycz, J. Roskams, S. Hill, N. Spruston, E. Meijering, G.A. Ascoli, BigNeuron: large-scale 3D neuron reconstruction from optical microscopy images. Neuron **87**(2), 252–256 (2015)
34. T. Rabie, M. Baziyad, I. Kamel, Enhanced high capacity image steganography using discrete wavelet transform and the Laplacian pyramid. Multimedia Tools Appl. pp. 1–26 (2018)
35. A. Radford, L. Metz, S. Chintala, Unsupervised representation learning with deep convolutional generative adversarial networks. arXiv preprint arXiv:1511.06434 (2015)
36. K. Schawinski, C. Zhang, H. Zhang, L.Fowler, G.K. Santhanam, Generative adversarial networks recover features in astrophysical images of galaxies beyond the deconvolution limit. Monthly Notices of the R. Astron. Soc.: Lett. 467(1):L110–L114 (2017)
37. J. Schmidhuber, Deep learning in neural networks: an overview. Neural Networks **61**, 85–117 (2015)
38. Y. Song, J. Zhu, X. Wang, H. Qi, Talking face generation by conditional recurrent adversarial network. arXiv preprint arXiv:1804.04786 (2018)
39. C. Szegedy, S. Ioffe, V. Vanhoucke, A.A. Alemi, Inception-v4, inception-resnet and the impact of residual connections on learning, in *AAAI*, vol. 4, p. 12 (2017)

40. D. Ulyanov, V. Lebedev, A. Vedaldi, V.S. Lempitsky, Texture networks: feed-forward synthesis of textures and stylized images, in *ICML*, pp. 1349–1357 (2016)
41. V. Velayutham, P.Z. Benitez-Aguirre, M.E. Craig, G. Liew, T.Y. Wong, A.J. Jenkins, K.C. Donaghue, Innovative technology shows impact of glycaemic control on peripheral retinal vessels in adolescents with type 1 diabetes. Diabetologia **60**(10), 2103–2110 (2017)
42. W. Yin, Y. Fu, L. Sigal, X. Xue, Semi-latent GAN: learning to generate and modify facial images from attributes. arXiv preprint arXiv:1704.02166 (2017)
43. H. Zhang, T. Xu, H. Li, S. Zhang, X. Huang, X. Wang, D. Metaxas, StackGAN: text to photo-realistic image synthesis with stacked generative adversarial networks, in *International Conference on Computer Vision (ICCV)*, pp. 5907–5915 (2017)

Retraction Note to: Ethnic Characterization in Amalgamated People for Airport Security Using a Repository of Images and Pigeon-Inspired Optimization (PIO) Algorithm for the Improvement of Their Results

Alberto Ochoa-Zezzatti, José Mejía, Roberto Contreras-Masse, Erwin Martínez and Andrés Hernández

Retraction Note to: Chapter "Ethnic Characterization in Amalgamated People for Airport Security Using a Repository of Images and Pigeon-Inspired Optimization (PIO) Algorithm for the Improvement of Their Results" in: D. Oliva and S. Hinojosa (eds.), *Applications of Hybrid Metaheuristic Algorithms for Image Processing*, Studies in Computational Intelligence 890, https://doi.org/10.1007/978-3-030-40977-7_5

The editors have retracted this chapter because it significantly overlaps with a previously published article by Duan and Qiao [1] and a previously published conference paper by Cruz-Álvarez et al. [2]. Additionally, a small section of the chapter appears to be generated by computer software, potentially Mathgen. The editors therefore no longer have confidence in the validity of this chapter.

The authors have not responded to correspondence regarding this retraction notice.

[1] Duan, H. and Qiao, P. (2014), "Pigeon-inspired optimization: a new swarm intelligence optimizer for air robot path planning", International Journal of Intelligent Computing and Cybernetics, Vol. 7 No. 1, pp. 24-37. https://doi.org/10.1108/IJICC-02-2014-0005

[2] Cruz-Álvarez V.R., Montes-Gonzalez F., Ochoa A., Palacios-Leyva R.E. (2012) Distribution and Selection of Colors on a Diorama to Represent Social Issues

The retracted version of this chapter can be found at https://doi.org/10.1007/978-3-030-40977-7_5

D. Oliva and S. Hinojosa (eds.), *Applications of Hybrid Metaheuristic Algorithms for Image Processing*, Studies in Computational Intelligence 890,
https://doi.org/10.1007/978-3-030-40977-7_22

Using Cultural Algorithms and Graph Coloring. In: Omatu S., De Paz Santana J., González S., Molina J., Bernardos A., Rodríguez J. (eds) Distributed Computing and Artificial Intelligence. Advances in Intelligent and Soft Computing, vol 151. Springer, Berlin, Heidelberg. https://doi.org/10.1007/978-3-642-28765-7_8

The retracted chapter and book have been updated with the changes.

The manufacturer's authorised representative in the EU is Springer Nature Customer Service Centre GmbH, Europaplatz 3, 69115 Heidelberg, Germany. If you have any concerns regarding our products, please contact ProductSafety@springernature.com

Printed and bound by CPI Group (UK) Ltd, Croydon, CR0 4YY
22/09/2026
02221048-0011